STUDY GUIDE
AND
SELECTED SOLUTIONS MANUAL

Karen Timberlake
Los Angeles Valley College

Mark Quirie
Algonquin College

BASIC
CHEMISTRY

FIFTH EDITION

Editor in Chief: Jeanne Zalesky
Executive Editor: Terry Haugen
Senior Acquisitions Editor: Scott Dustan
Project Manager: Laura Perry
Program Manager: Lisa Pierce
Editorial Assistant: Lindsey Pruett
Project Management Team Lead: David Zielonka
Program Management Team Lead: Kristen Flathman
Manufacturing Buyer: Maura Zaldivar-Garcia
Cover Photo Credit: Sergio Stakhnyk/Shutterstock

www.pearsonhighered.com

ISBN 10: 0-134-16726-0
ISBN 13: 978-0-134-16726-8

Contents

Contents

Preface

This Study Guide and Selected Solutions Manual is intended to accompany *Basic Chemistry, Fifth Edition*. Our purpose is to provide you with additional learning resources to increase your understanding of the key concepts in the text. Each chapter correlates with a chapter in the text. For every Section in each chapter, there are Learning Exercises and Answers that focus on problem solving, which in turn promote an understanding of the chemical principles. A Checklist of chapter concepts and a multiple-choice Practice Test provide a review of the chapter content. In the Selected Solutions, we have included the answers and solutions to all the odd-numbered problems in the text.

I hope that this Study Guide and Selected Solutions Manual will help in the learning of chemistry. If you wish to make comments or corrections, or ask questions, you can send us an e-mail message at *khemist@aol.com.*

Karen and William Timberlake
Mark Quirie

To the Student

One must learn by doing the thing;
though you think you know it, you
have no certainty until you try.
—*Sophocles*

Here you are in a chemistry class with your textbook in front of you. Perhaps you have already been assigned some reading or some problems to do in the book. Looking through the chapter, you may see words, terms, and pictures that are new to you. This may very well be your first experience with a science class like chemistry. At this point you may have some questions about what you can do to learn chemistry. This *Study Guide and Selected Solutions Manual* is written with those considerations in mind.

Learning chemistry is similar to learning something new such as tennis or skiing or driving. If I asked you how you learn to play tennis or ski or drive a car, you would probably tell me that you would need to practice often. It is the same with learning chemistry; understanding the chemical ideas and successfully solving the problems depends on the time and effort you invest in it. If you practice every day, you will find that learning chemistry is an exciting experience and a way to understand the current issues of the environment, health, and medicine.

Manage Your Study Time

I often recommend a study system to students in which you read one section of the text and immediately practice the questions and problems that are at the end of that section. In this way, you concentrate on a small amount of information and use what you learned to answer questions. This helps you to organize and review the information without being overwhelmed by the entire chapter. It is important to understand each section, because they build like steps. Information presented in each chapter proceeds from the basic to the more complex. Perhaps you can only study three or four sections of the chapter. As long as you also practice doing some problems at the same time, the information will stay with you.

Form a Study Group

I highly recommend that you form a study group in the first week of your chemistry class. Working with your peers will help you use the language of chemistry. Scheduling a time to meet each week will help you study and prepare to discuss problems. You will be able to teach some things to other students in the group, and sometimes they will help you understand a topic that puzzles you. You won't always understand a concept right away. Your group will help you see your way through it. Most of all, a study group creates a strong support system when students help each other successfully complete the class.

Go to Office Hours

Try to go to your tutor's and/or professor's office hours. Your professor wants you to understand and enjoy learning this material. Often a tutor is assigned to a class or there are tutors available at your college. Don't be intimidated. Going to see a tutor or your professor is one of the best ways to clarify what you need to learn in chemistry.

Using This Study Guide and Selected Solutions Manual

Now you are ready to sit down and study chemistry. Let's go over some methods that can help you learn chemistry. This *Study Guide* is written specifically to help you understand and practice the chemical concepts that are presented in your class and in your text. The following features are part of this *Study Guide and Selected Solutions Manual:*

1. Learning Goals
The Learning Goals give you an overview of what each section in the chapter is about and what you can expect to accomplish when you complete your study and learning of that section.

2. Key Terms
Key Terms appear throughout each chapter in the Study Guide. As you review the description of the Key Terms, you will have an overview of the topics you have studied in that chapter. Because many of the Key Terms may be new to you, this is an opportunity to review their meaning.

3. Chapter Sections
Each chapter section begins with a list of the important ideas to guide you through each of the learning activities. When you are ready to begin your study, read the matching section in the textbook and review the *Guides to Problem Solving* and the *Sample Problems* in the text. Included are the *Key Math Skills* and *Core Chemistry Skills.*

4. Learning Exercises
The Learning Exercises give you an opportunity to practice problem solving related to the chemical principles in the chapter. Each set of Learning Exercises reviews one chemical principle. The answers are found immediately following each Learning Exercise. Check your answers right away. If they don't match the answer in the Study Guide, review that section of the text again. It is important to make corrections before you go on. Chemistry involves a layering of skills such that each one must be understood before the next one can be learned.

5. Checklist
Use the Checklist to check your understanding of the Learning Goals. This Checklist gives you an overview of the major topics in each section. If something does not sound familiar, go back and review it. One aspect of being a strong problem-solver is the ability to evaluate your knowledge and understanding as you study.

6. Practice Test
A Practice Test is found at the end of each chapter. When you feel you have learned the material in a chapter, you can check your understanding by taking the Practice Test. The Practice Test questions are keyed to the Learning Goals, which allows you to identify the sections you may need to review. Answers are found at the end of the Practice Test. If the results of the Practice Test indicate that you know the material, you are ready to proceed to the next chapter.

7. Selected Answers and Solutions
The Selected Answers and Solutions to the odd-numbered problems for each chapter in the text follow each Practice Test. For certain chapters, Selected Answers to the Combining Ideas for groups of chapters are included.

Resources

Basic Chemistry, Fifth edition, provides an integrated teaching and learning package of support material for both students and professors.

Name of Supplement	Available in Print	Available Online	Instructor or Student Supplement	Description
Study Guide and Selected Solutions Manual (ISBN 0134167260)	✓		Resource for Students	The Study Guide and Selected Solutions Manual, by Karen Timberlake and Mark Quirie, promotes active learning through a variety of exercises with answers as well as practice tests that are connected directly to the learning goals of the textbook. Complete solutions to odd-numbered problems are included.
MasteringChemistry® (www.masteringchemistry.com) (ISBN 0134177150)		✓	Resource for Students and Instructors	MasteringChemistry® from Pearson is the leading online teaching and learning system designed to improve results by engaging students before, during, and after class with powerful content. Ensure that students arrive ready to learn by assigning educationally effective content before class, and encourage critical thinking and retention with in-class resources such as Learning Catalytics. Students can further master concepts after class through traditional homework assignments that provide hints and answer-specific feedback. The Mastering gradebook records scores for all automatically graded assignments while diagnostic tools give instructors access to rich data to assess student understanding and misconceptions.
MasteringChemistry with Pearson eText (ISBN 0133899306)		✓	Resource for Students	The fifth edition of *Basic Chemistry* features a Pearson eText enhanced with media within Mastering. In conjunction with Mastering assessment capabilities, Interactive Videos, and 3D animations will improve student engagement and knowledge retention. Each chapter will contain a balance of interactive animations, videos, sample calculations, and self-assessments/quizzes embedded directly in the eText. Additionally, the Pearson eText offers students the power to create notes, highlight text in different colors, create bookmarks, zoom, and view single or multiple pages.
Instructor's Solutions Manual–Download Only (ISBN 0134167279)		✓	Resource for Instructors	Prepared by Mark Quirie, the solutions manual highlights chapter topics, and includes answers and solutions for all questions and problems in the text.
Instructor Resource Materials–Download Only (ISBN 0134167252)		✓	Resource for Instructors	Includes all the art, photos, and tables from the book in JPEG format for use in classroom projection or when creating study materials and tests. In addition, the instructors can access modifiable PowerPoint™ lecture outlines. Also available are downloadable files of the Instructor's Solutions Manual and a set of "clicker questions" designed for use with classroom-response systems. Also visit the Pearson Education catalog page for Timberlake's *Basic Chemistry, Fifth edition*, at www.pearsonhighered.com to download available instructor supplements.
TestGen Test Bank–Download Only (ISBN 0133891895)		✓	Resource for Instructors	Prepared by William Timberlake, this resource includes more than 2000 questions in multiple-choice, matching, true/false, and short-answer format.

Name of Supplement	Available in Print	Available Online	Instructor or Student Supplement	Description
Laboratory Manual by Karen Timberlake (ISBN 0321811852)	✓		Resource for Students	This best-selling lab manual coordinates 35 experiments with the topics in *Basic Chemistry*, *Fifth edition*, uses laboratory investigations to explore chemical concepts, develop skills of manipulating equipment, reporting data, solving problems, making calculations, and drawing conclusions.
Online Instructor Manual for Laboratory Manual (ISBN 0321812859)		✓	Resource for Students	This manual contains answers to report sheet pages for the Laboratory Manual and a list of the materials needed for each experiment with amounts given for 20 students working in pairs, available for download at www.pearsonhighered.com.

1
Chemistry in Our Lives

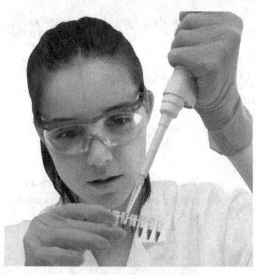

Credit: nyaivanova / Shutterstock

When a female victim was found dead in her home, samples of her blood and stomach contents were sent to Sarah, who works as a forensic scientist. After using a variety of chemical tests, Sarah concluded that the victim was poisoned when she ingested ethylene glycol. Since the initial symptoms of ethylene glycol poisoning are similar to those of alcohol intoxication, the victim was unaware of the poisoning. When ethylene glycol is oxidized, the products can cause kidney failure and may be toxic to the body. Two weeks later, police arrested the victim's husband after finding a bottle of ethylene glycol in the laundry room of the couple's home. How does this result show the use of the scientific method?

LOOKING AHEAD

1.1 Chemistry and Chemicals

Learning Goal: Define the term chemistry and identify substances as chemicals.

- Chemistry is the study of the composition, structure, properties, and reactions of matter.
- A chemical is a substance that has the same composition and properties wherever it is found.

♦ **Learning Exercise 1.1**

Is each of the following a chemical?

a. _____ ascorbic acid (vitamin C)

b. _____ time for a radioisotope to lose half its activity

c. _____ sodium fluoride in toothpaste

d. _____ amoxicillin, an antibiotic

e. _____ distance of 2.5 mi walked on a treadmill during exercise

f. _____ carbon nanotubes used to transport medication to cancer cells

Answers   **a.** yes   **b.** no   **c.** yes   **d.** yes   **e.** no   **f.** yes

1.2 Scientific Method: Thinking Like a Scientist

Learning Goal: Describe the activities that are part of the scientific method.

- The scientific method is a process of making observations, forming a hypothesis, and testing the hypothesis with experiments.
- When the results of the experiments are analyzed, a conclusion or theory is made as to whether the hypothesis is *true* or *false*.

Scientific Method

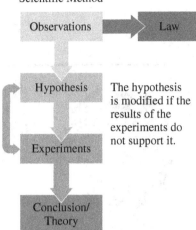

The scientific method develops a conclusion or theory using observations, hypotheses, and experiments.

◆ **Learning Exercise 1.2A**

Identify each of the following as an observation (O), a hypothesis (H), or an experiment (E):

a. _____ Sunlight is necessary for the growth of plants.

b. _____ Plants in the shade are shorter than plants in the sun.

c. _____ Plant leaves are covered with aluminum foil and their growth is measured.

d. _____ Fertilizer is added to plants.

e. _____ Ozone slows plant growth by interfering with photosynthesis.

f. _____ Ozone causes brown spots on plant leaves.

Answers   **a.** H   **b.** O   **c.** E   **d.** E   **e.** H   **f.** O

Students make observations in the chemistry laboratory.
Credit: Argonne National Laboratory

◆ **Learning Exercise 1.2B**

Identify each of the following as an observation (O), a hypothesis (H), an experiment (E), or a conclusion (C):

a. _____ One half-hour after eating a wheat roll, Cynthia experiences stomach cramps.

b. _____ The next morning, Cynthia eats a piece of toast and one half-hour later has an upset stomach.

c. _____ Cynthia thinks she may be gluten intolerant.

d. _____ Cynthia tries a gluten-free roll and does not have any stomach cramps.

e. _____ Because Cynthia does not have an upset stomach after eating a gluten-free roll, she believes that she is gluten intolerant.

Answers   **a.** O   **b.** O   **c.** H   **d.** E   **e.** C

1.3 Learning Chemistry: A Study Plan

Learning Goal: Develop a study plan for learning chemistry.

- Components of the text that promote learning include: *Looking Ahead, Learning Goals, Chapter Readiness, Key Math Skills, Core Chemistry Skills, Try It First, Connect, Engage, Analyze the Problems, Sample Problems, Study Checks with Answers, Guides to Problem Solving (GPS), Chemistry Link to Health, Chemistry Link to the Environment, Questions and Problems, Applications, Follow Ups, Concept Maps, Chapter Reviews, Key Terms, Review of Key Math Skills, Review of Core Chemistry Skills, Understanding the Concepts, Additional Questions and Problems, Challenge Problems, Answers to Selected Questions and Problems, Combining Ideas,* and *Glossary/Index.*
- An active learner continually interacts with chemical concepts while reading the text and attending class.
- Working with a study group clarifies ideas and facilitates problem solving.

♦ **Learning Exercise 1.3**

Which of the following activities would be included in a successful study plan for learning chemistry?

a. _____ attending class once in a while

b. _____ working problems with friends from class

c. _____ attending review sessions

d. _____ planning a regular study time

e. _____ not doing the assigned problems

f. _____ visiting the instructor during office hours

g. _____ working the Sample Problem before checking the solution

Studying in a group can be beneficial to learning.

Answers   **a.** no   **b.** yes   **c.** yes   **d.** yes   **e.** no   **f.** yes   **g.** yes

1.4 Key Math Skills for Chemistry

Learning Goal: Review math concepts used in chemistry: place values, positive and negative numbers, percentages, solving equations, and interpreting graphs.

- Basic math skills are important in learning chemistry.
- In a number, we identify the place value of each digit.
- A positive number is greater than zero and has a positive sign ($+$); a negative number is less than zero and has a negative sign ($-$).
- A percentage is calculated as the parts divided by the whole, then multiplied by 100%.
- A linear equation is solved by rearranging it to place the unknown value on one side.
- A graph represents the relationship between two variables, which are plotted on perpendicular axes.

Study Note

On your calculator, there are four keys that are used for basic mathematical operations. The change sign $\boxed{+/-}$ key is used to change the sign of a number.

To practice these basic calculations on the calculator, work through the problem going from left to right doing the operations in the order they occur. If your calculator has a change sign $\boxed{+/-}$ key, a negative number is entered by entering the number and then pressing the change sign $\boxed{+/-}$ key. At the end, press the equals $\boxed{=}$ key or ANS or ENTER.

Addition and Subtraction

Example 1: $15 - 8 + 2 =$

Solution: $15 \boxed{-} 8 \boxed{+} 2 \boxed{=} 9$

Example 2: $4 + (-10) - 5 =$

Solution: $4 \boxed{+} 10 \boxed{+/-} \boxed{-} 5 \boxed{=} -11$

Multiplication and Division

Example 3: $2 \times (-3) =$

Solution: $2 \boxed{\times} 3 \boxed{+/-} \boxed{=} -6$

Example 4: $\dfrac{8 \times 3}{4} =$

Solution: $8 \boxed{\times} 3 \boxed{\div} 4 \boxed{=} 6$

♦ **Learning Exercise 1.4A**

⚙ **KEY MATH SKILL**

a. Identify the place value for each of the digits in the number 825.10.   Identifying Place Values

Digit	Place Value
8	
2	
5	
1	
0	

Answer

Digit	Place Value
8	hundreds
2	tens
5	ones
1	tenths
0	hundredths

b. A tablet contains 0.325 g of aspirin. Identify the place value for each of the digits in the number 0.325 g.

Digit	Place Value
3	
2	
5	

Answer

Digit	Place Value
3	tenths
2	hundredths
5	thousandths

♦ **Learning Exercise 1.4B**

⚙ **KEY MATH SKILL**

Solve the following:

Using Positive and Negative Numbers in Calculations

$$\frac{-14 + 22}{-4} = ?$$

Answer

$$\frac{-14 + 22}{-4} = -2$$

♦ **Learning Exercise 1.4C**

✪ **KEY MATH SKILL**

Calculating Percentages

There are 20 patients on the 3rd floor of a hospital. If 4 patients were discharged today, what percentage of patients were discharged?

Answer

$$\frac{4 \text{ patients discharged}}{20 \text{ patients}} \times 100\% = 20\% \text{ patients discharged}$$

Study Note

1. Solve an equation for a particular variable by performing the same mathematical operation on *both* sides of the equation.
2. If you eliminate a symbol or number by subtracting, you need to subtract that same symbol or number on the opposite side.
3. If you eliminate a symbol or number by adding, you need to add that same symbol or number on the opposite side.
4. If you cancel a symbol or number by dividing, you need to divide both sides by that same symbol or number.
5. If you cancel a symbol or number by multiplying, you need to multiply both sides by that same symbol or number.

♦ **Learning Exercise 1.4D**

✪ **KEY MATH SKILL**

Solving Equations

Solve the following equation for y:

$$3y - 8 = 22$$

Answer

$$3y - 8 = 22$$
$$3y - 8 + 8 = 22 + 8$$
$$3y = 30$$
$$\frac{3y}{3} = \frac{30}{3}$$
$$y = 10$$

♦ **Learning Exercise 1.4E**

✪ **KEY MATH SKILL**

Interpreting Graphs

Use the following graph to answer questions **a** and **b**:

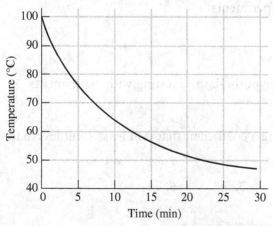

Temperature versus Time for Cooling of Coffee

a. How many minutes elapse when coffee cools from 100 °C to 65 °C?

b. What is the temperature after the coffee has cooled for 15 min?

Temperature versus Time for Cooling of Coffee

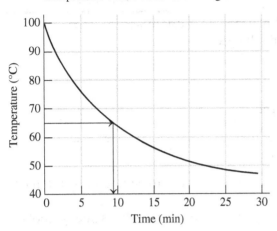

Temperature versus Time for Cooling of Coffee

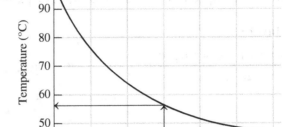

Answers

a. Draw a horizontal line from 65 °C on the temperature axis to intersect the curved line. Then drop a vertical line to the time axis and read the time where it crosses the *x* axis, which is estimated to be 9 min.

b. Draw a vertical line from 15 min on the time axis to intersect the curved line. Draw a horizontal line from the intersect to the temperature axis and read the temperature where it crosses the *y* axis, which is estimated to be 57 °C.

1.5 Writing Numbers in Scientific Notation

Learning Goal: Write a standard number in scientific notation and vice versa.

• Large and small numbers can be written using scientific notation in which the decimal point is moved to give a coefficient of at least 1 but less than 10 and the number of spaces moved is shown as a power of 10.

• A value written in scientific notation has two parts: a number that is at least 1 but less than 10, called a coefficient, and a power of 10.

• A large number will have a positive power of 10, while a small number will have a negative power of 10.

• For numbers greater than 10, the decimal point is moved to the left to give a positive power of 10.

• For numbers less than 1, the decimal point is moved to the right to give a negative power of 10.

Key Terms for Sections 1.1 to 1.5

Match each of the following key terms with the correct description:

a. scientific method	**b.** experiment	**c.** scientific notation
d. conclusion/theory	**e.** chemistry	**f.** hypothesis

1. _____ an explanation of nature validated by many experiments

2. _____ the study of substances and how they interact

3. _____ a possible explanation of a natural phenomenon

4. _____ the process of making observations, writing a hypothesis, and testing with experiments

5. _____ a procedure used to test a hypothesis

6. _____ a form of writing large and small numbers using a coefficient that is at least 1 but less than 10, followed by a power of 10

Answers **1.** d **2.** e **3.** f **4.** a **5.** b **6.** c

Study Note

1. For a number greater than 10, the decimal point is moved to the left to give a number that is at least 1 but less than 10 and a positive power of 10. For a number less than 1, the decimal point is moved to the right to give a number that is at least 1 but less than 10 and a negative power of 10.
2. To express 2500 in scientific notation, we can write it as 2.5×1000. Because 1000 is the same as 10^3, we can express 2500 in scientific notation as 2.5×10^3.
3. To express 0.082 in scientific notation, we move the decimal point two places to the right and write 8.2×0.01. Because 0.01 is the same as 10^{-2}, we can express 0.082 in scientific notation as 8.2×10^{-2}.

Guide to Writing a Number in Scientific Notation	
STEP 1	Move the decimal point to obtain a coefficient that is at least 1 but less than 10.
STEP 2	Express the number of places moved as a power of 10.
STEP 3	Write the product of the coefficient multiplied by the power of 10.

♦ Learning Exercise 1.5A

Write each of the following numbers in scientific notation:

⚛ KEY MATH SKILL

Converting between Standard Numbers and Scientific Notation

a. 24 100 _____

b. 825 _____

c. 230 000 _____

d. 53 000 _____

e. 0.002 _____

f. 0.000 001 5 _____

g. 0.08 _____

h. 0.000 24 _____

Answers
a. 2.41×10^4　　**b.** 8.25×10^2　　**c.** 2.3×10^5　　**d.** 5.3×10^4
e. 2×10^{-3}　　**f.** 1.5×10^{-6}　　**g.** 8×10^{-2}　　**h.** 2.4×10^{-4}

♦ Learning Exercise 1.5B

Which number in each of the following pairs is larger?

a. 2500 or 2.5×10^2 _____

b. 0.04 or 4×10^{-3} _____

c. 65 000 or 6.5×10^5 _____

d. 0.000 35 or 3.5×10^{-3} _____

e. 300 000 or 3×10^6 _____

f. 0.002 or 2×10^{-4} _____

Answers
a. 2500　　**b.** 0.04　　**c.** 6.5×10^5
d. 3.5×10^{-3}　　**e.** 3×10^6　　**f.** 0.002

♦ Learning Exercise 1.5C

Write each of the following as a standard number:

Examples: $2 \times 10^2 = 200$ and $3.2 \times 10^{-4} = 0.000\ 32$

a. 4×10^3 _____

b. 5.2×10^4 _____

c. 1.8×10^5 _____

d. 8×10^{-3} _____

e. 6×10^{-2} _____

f. 3.1×10^{-5} _____

Answers
a. 4000　　**b.** 52 000　　**c.** 180 000
d. 0.008　　**e.** 0.06　　**f.** 0.000 031

Checklist for Chapter 1

You are ready to take the Practice Test for Chapter 1. Be sure you have accomplished the following learning goals for this chapter. If not, review the section listed at the end of the goal. Then apply your new skills and understanding to the Practice Test.

After studying Chapter 1, I can successfully:

_____ Describe a substance as a chemical. (1.1)

_____ Identify the steps of the scientific method. (1.2)

_____ Make a study plan for successfully learning chemistry. (1.3)

_____ Use math skills needed for chemistry: place values, positive and negative numbers, percentages, solving equations, and interpreting graphs. (1.4)

_____ Write numbers in scientific notation using a coefficient and a power of 10. (1.5)

Practice Test for Chapter 1

The chapter sections to review are shown in parentheses at the end of each question.

1. Which of the following would be described as a chemical? (1.1)
   A. sleeping
   B. salt
   C. singing
   D. listening to a concert
   E. energy

2. Which of the following is not a chemical? (1.1)
   A. aspirin
   B. sugar
   C. feeling cold
   D. glucose
   E. vanilla

For questions 3 through 7, identify each statement as an observation (O), a hypothesis (H), or an experiment (E): (1.2)

3. _____ More sugar dissolves in 50 mL of hot water than in 50 mL of cold water.

4. _____ Samples containing 20 g of sugar each are placed separately in a glass of cold water and a glass of hot water.

5. _____ Sugar consists of white crystals.

6. _____ Water flows downhill.

7. _____ Drinking 10 glasses of water a day will help me lose weight.

For questions 8 through 12, answer yes or no. (1.3)

To learn chemistry, I will:

8. _____ work the problems in the chapter and check answers

9. _____ attend some classes but not all

10. _____ form a study group

11. _____ set up a regular study time

12. _____ wait until the night before the exam to start studying

13. Solve for the value of a in the equation $5a + 10 = 30$. (1.4)
    A. 2
    B. 4
    C. 8
    D. 20
    E. 40

14. Give the answer for the problem $\dfrac{22 + 23}{-4 + 1}$. (1.4)

    A. 5   **B.** −5   **C.** −7   **D.** −15   **E.** 15

15. At the zoo, there are 8 lions, 6 tigers, and 2 panthers. What percentage of the animals at the zoo are lions? (1.4)

    A. 50%   **B.** 30%   **C.** 20%   **D.** 10%   **E.** 5%

16. Use the following graph to determine the solubility of carbon dioxide in water, in g CO_2/100 g water, at 40 °C: (1.4)

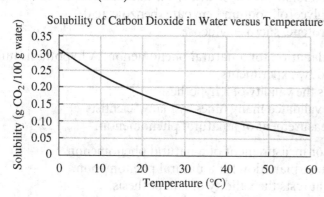

    A. 0.10   **B.** 0.15   **C.** 0.20   **D.** 0.25   **E.** 0.30

17. The contents found in a container at a victim's home were identified as 65.0 g of ethylene glycol in 182 g of solution. What is the percentage of ethylene glycol in the solution? (1.4)

    A. 15.3%   **B.** 26.3%   **C.** 35.7%   **D.** 48.0%   **E.** 65.0%

18. For breakfast, a person eats 55 g of eggs, 65 g of banana, and 85 g of milk. What percentage of the breakfast was milk? (1.4)

    A. 85%   **B.** 65%   **C.** 41%   **D.** 32%   **E.** 27%

19. The number 24 000 written in scientific notation is (1.5)

    A. 24   **B.** 24×10^3 **C.** 2.4×10^3 **D.** 2.4×10^{-3} **E.** 2.4×10^4

20. The number 0.005 written in scientific notation is (1.5)

    A. 5   **B.** 5×10^{-3} **C.** 5×10^{-2} **D.** 0.5×10^{-4} **E.** 5×10^3

21. The number 2.6×10^4 written as a standard number is (1.5)

    A. 26   **B.** 2.6   **C.** 260   **D.** 26 000   **E.** 0.000 26

22. The number 8.7×10^{-3} written as a standard number is (1.5)

    A. 0.0087   **B.** 8700   **C.** 0.87   **D.** 0.000 87   **E.** 870

Answers to the Practice Test

1. B	**2.** C	**3.** O	**4.** E	**5.** O
6. O	**7.** H	**8.** yes	**9.** no	**10.** yes
11. yes	**12.** no	**13.** B	**14.** D	**15.** A
16. A	**17.** C	**18.** C	**19.** E	**20.** B
21. D	**22.** A			

Selected Answers and Solutions to Text Problems

1.1 **a.** Chemistry is the study of the composition, structure, properties, and reactions of matter.

b. A chemical is a substance that has the same composition and properties wherever it is found.

1.3 Many chemicals are listed on a vitamin bottle such as vitamin A, vitamin B_3, vitamin B_{12}, vitamin C, and folic acid.

1.5 Typical items found in a medicine cabinet and some of the chemicals they contain are as follows:

Antacid tablets: calcium carbonate, cellulose, starch, stearic acid, silicon dioxide

Mouthwash: water, alcohol, thymol, glycerol, sodium benzoate, benzoic acid

Cough suppressant: menthol, beta-carotene, sucrose, glucose

1.7 **a.** A hypothesis proposes a possible explanation for a natural phenomenon. A hypothesis must be stated in a way that it can be tested by experiments.

b. An experiment is a procedure that tests the validity of a hypothesis.

c. A theory is a hypothesis that has been validated many times by many scientists.

d. An observation is a description or measurement of a natural phenomenon.

1.9 **a.** An observation (O) is a description or measurement of a natural phenomenon.

b. A hypothesis (H) proposes a possible explanation for a natural phenomenon.

c. An experiment (E) is a procedure that tests the validity of a hypothesis.

d. An observation (O) is a description or measurement of a natural phenomenon.

e. An observation (O) is a description or measurement of a natural phenomenon.

f. A conclusion (C) is an explanation of an observation that has been validated by repeated experiments that support a hypothesis.

1.11 **a.** An observation (O) is a description or measurement of a natural phenomenon.

b. A hypothesis (H) proposes a possible explanation for a natural phenomenon.

c. An experiment (E) is a procedure that tests the validity of a hypothesis.

d. An experiment (E) is a procedure that tests the validity of a hypothesis.

1.13 There are several things you can do that will help you successfully learn chemistry, including forming a study group, going to lecture, working *Sample Problems* and *Study Checks*, working *Questions and Problems* and checking answers, reading the assignment ahead of class, going to the instructor's office hours, and keeping a problem notebook.

1.15 Ways you can enhance your learning of chemistry include:

a. forming a study group.

c. visiting the professor during office hours.

e. answering the Engage question.

1.17 **a.** The bolded 8 is in the thousandths place.

b. The bolded 6 is in the ones place.

c. The bolded 6 is in the hundreds place.

1.19 **a.** $15 - (-8) = 15 + 8 = 23$

b. $-8 + (-22) = -30$

c. $4 \times (-2) + 6 = -8 + 6 = -2$

1.21 **a.** $\dfrac{21 \text{ flu shots}}{25 \text{ patients}} \times 100\% = 84\%$ received flu shots

b. total grams of alloy $= 56$ g silver $+ 22$ g copper $= 78$ g of alloy

$\dfrac{56 \text{ g silver}}{78 \text{ g alloy}} \times 100\% = 72\%$ silver

   c. total number of coins = 11 nickels + 5 quarters + 7 dimes = 23 coins

$$\frac{7 \text{ dimes}}{23 \text{ coins}} \times 100\% = 30\% \text{ dimes}$$

1.23 **a.**

$$4a + 4 = 40$$
$$4a + \cancel{4} - \cancel{4} = 40 - 4$$
$$4a = 36$$
$$\frac{\cancel{4}a}{\cancel{4}} = \frac{36}{4}$$
$$a = 9$$

    b.

$$\frac{a}{6} = 7$$
$$\cancel{6}\left(\frac{a}{\cancel{6}}\right) = 6(7)$$
$$a = 42$$

1.25 **a.** The graph shows the relationship between the temperature of a cup of tea and time.
    b. The vertical axis measures temperature, in °C.
    c. The values on the vertical axis range from 20 °C to 80 °C.
    d. As time increases, the temperature decreases.

1.27 **a.** Move the decimal point four places to the left to give 5.5×10^4.
    b. Move the decimal point two places to the left to give 4.8×10^2.
    c. Move the decimal point six places to the right to give 5×10^{-6}.
    d. Move the decimal point four places to the right to give 1.4×10^{-4}.
    e. Move the decimal point three places to the right to give 7.2×10^{-3}.
    f. Move the decimal point five places to the left to give 6.7×10^5.

1.29 **a.** The standard number is 1.2 times the power of 10^4 or 10 000, which gives 12 000.
    b. The standard number is 8.25 times the power of 10^{-2} or 0.01, which gives 0.0825.
    c. The standard number is 4 times the power of 10^6 or 1 000 000, which gives 4 000 000.
    d. The standard number is 5.8 times the power of 10^{-3} or 0.001, which gives 0.0058.

1.31 **a.** 7.2×10^3, which is also 72×10^2, is larger than 8.2×10^2.
    b. 3.2×10^{-2}, which is also 320×10^{-4}, is larger than 4.5×10^{-4}.
    c. 1×10^4 or 10 000 is larger than 1×10^{-4} or 0.0001.
    d. 6.8×10^{-2} or 0.068 is larger than 0.000 52.

1.33 $\dfrac{120 \text{ g ethylene glycol}}{450 \text{ g liquid}} \times 100\% = 27\% \text{ ethylene glycol}$

1.35 No. All of these ingredients are chemicals.

1.37 Yes. Sherlock's investigation includes making observations (gathering data), formulating a hypothesis, testing the hypothesis, and modifying it until one of the hypotheses is validated.

1.39 **a.** Describing the appearance of a patient is an observation (O).
    b. Formulating a reason for the extinction of dinosaurs is a hypothesis (H).
    c. Measuring the completion time of a race is an observation (O).

1.41 **a.** When two negative numbers are added, the answer has a negative sign.
    b. When a positive and negative number are multiplied, the answer has a negative sign.

1.43 **a.** An observation (O) is a description or measurement of a natural phenomenon.
    b. A hypothesis (H) proposes a possible explanation for a natural phenomenon.
    c. A conclusion (C) is an explanation of an observation that has been validated by repeated experiments that support a hypothesis.

1.45 If experimental results do not support your hypothesis, you should:
   b. modify your hypothesis.

1.47 A successful study plan would include:
   b. working the *Sample Problems* as you go through a chapter.
   c. going to your professor's office hours.

1.49 **a.** $4 \times (-8) = -32$
   b. $-12 - 48 = -12 + (-48) = -60$
   c. $\dfrac{-168}{-4} = 42$

1.51 total number of gumdrops $=$ 16 orange $+$ 8 yellow $+$ 16 black $=$ 40 gumdrops
   a. $\dfrac{8 \text{ yellow gumdrops}}{40 \text{ total gumdrops}} \times 100\% = 20\%$ yellow gumdrops
   b. $\dfrac{16 \text{ black gumdrops}}{40 \text{ total gumdrops}} \times 100\% = 40\%$ black gumdrops

1.53 **a.** Move the decimal point five places to the left to give 1.2×10^5.
   b. Move the decimal point seven places to the right to give 3.4×10^{-7}.
   c. Move the decimal point two places to the right to give 6.6×10^{-2}.
   d. Move the decimal point three places to the left to give 2.7×10^3.

1.55 **a.** Move the decimal point five places to the left to give 0.000 026.
   b. Move the decimal point two places to the right to give 650.
   c. Move the decimal point one place to the left to give 0.37.
   d. Move the decimal point five places to the right to give 530 000.

1.57 **a.** An observation (O) is a description or measurement of a natural phenomenon.
   b. A hypothesis (H) proposes a possible explanation for a natural phenomenon.
   c. An experiment (E) is a procedure that tests the validity of a hypothesis.
   d. A hypothesis (H) proposes a possible explanation for a natural phenomenon.

1.59 **a.**
$$2x + 5 = 41$$
$$2x + \cancel{5} - \cancel{5} = 41 - 5$$
$$2x = 36$$
$$\frac{\cancel{2}x}{\cancel{2}} = \frac{36}{2}$$
$$x = 18$$

   b.
$$\frac{5x}{3} = 40$$
$$\cancel{3}\left(\frac{5x}{\cancel{3}}\right) = 3(40)$$
$$5x = 120$$
$$\frac{\cancel{5}x}{\cancel{5}} = \frac{120}{5}$$
$$x = 24$$

1.61 **a.** The graph shows the relationship between the solubility of carbon dioxide in water and temperature.
   b. The vertical axis measures the solubility of carbon dioxide in water ($g\ CO_2/100\ g$ water).
   c. The values on the vertical axis range from 0 to $0.35\ g\ CO_2/100\ g$ water.
   d. As temperature increases, the solubility of carbon dioxide in water decreases.

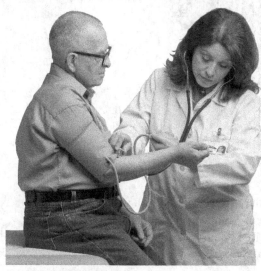

2

Chemistry and Measurements

Greg has been taking Inderal for high blood pressure. Every two months, he has his blood pressure taken by a registered nurse. During his last visit, his blood pressure was lower, 130/85. Thus, the dosage of Inderal was reduced to 45 mg to be taken twice daily. If one tablet contains 15 mg of Inderal, how many tablets does Greg need to take in one day?

Credit: AVAVA / Shutterstock

CHAPTER READINESS*

✢ Key Math Skills

- Identifying Place Values (1.4)
- Using Positive and Negative Numbers in Calculations (1.4)
- Calculating Percentages (1.4)
- Converting between Standard Numbers and Scientific Notation (1.5)

*These Key Math Skills from the previous chapter are listed here for your review as you proceed to the new material in this chapter.

LOOKING AHEAD

2.1 Units of Measurement
2.2 Measured Numbers and Significant Figures
2.3 Significant Figures in Calculations

2.4 Prefixes and Equalities
2.5 Writing Conversion Factors

2.6 Problem Solving Using Unit Conversion
2.7 Density

2.1 Units of Measurement

Learning Goal: Write the names and abbreviations for the metric or SI units used in measurements of length, volume, mass, temperature, and time.

- In science, physical quantities are described in units of the metric or International System of Units (SI).
- Some important units of measurement are meter (m) for length, liter (L) for volume, gram (g) and kilogram (kg) for mass, degree Celsius (°C) and kelvin (K) for temperature, and second (s) for time.

Key Terms for Section 2.1

Match each of the following key terms with the correct description:

a. International System
   of Units
b. volume
c. gram
d. second
e. meter
f. liter
g. mass

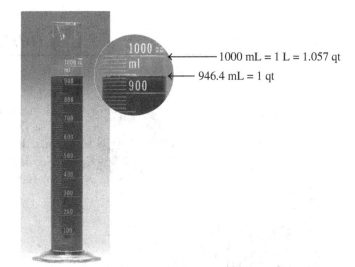

1. _____ the metric unit for volume that is slightly larger than a quart

2. _____ the metric unit used in measurements of mass

3. _____ the amount of space occupied by a substance

4. _____ the metric and SI unit for time

5. _____ a measure of the quantity of material in an object

Volume is the space occupied by a substance. In the metric system, volume is based on the liter (L), which is equal to 1.057 quarts (qt).

Credit: Pearson Education

6. _____ the official system of measurement used by scientists and used in most countries of the world

7. _____ the metric unit for length that is slightly longer than a yard

Answers   **1.** f   **2.** c   **3.** b   **4.** d   **5.** g   **6.** a   **7.** e

♦ **Learning Exercise 2.1**

Match the measurement with the quantity in each of the following:

a. length   b. volume   c. mass   d. temperature   e. time

1. _____ 45 g   2. _____ 8.2 m   3. _____ 215 °C   4. _____ 45 L

5. _____ 825 K   6. _____ 8.8 cm   7. _____ 140 kg   8. _____ 50 s

Answers   **1.** c   **2.** a   **3.** d   **4.** b
              5. d   **6.** a   **7.** c   **8.** e

2.2   Measured Numbers and Significant Figures

Learning Goal: Identify a number as measured or exact; determine the number of significant figures in a measured number.

On an electronic balance, the digital readout gives the mass of a nickel, which is 5.01 g.

Credit: Richard Megna / Fundamental Photographs, NYC

- A measured number is obtained when you use a measuring device.
- An exact number is obtained by counting items or from a definition within the same measuring system.
- There is uncertainty in every measured number but not in exact numbers.
- Significant figures in a measured number are all the digits, including the estimated digit.
- A number is a significant figure if it is not a zero.
- A zero is significant when it occurs between nonzero digits, at the end of a decimal number, or in the coefficient of a number written in scientific notation.

- A zero is not significant if it is at the beginning of a decimal number or used as a placeholder in a large number without a decimal point.

A volume of 10. mL of cough syrup is measured for a patient.

♦ Learning Exercise 2.2A

Label the number(s) in each of the following statements as measured (M) or exact (E) and give a reason for your answer:

a. _____ There are 7 days in 1 week.

b. _____ A pulmonary treatment lasts for 25 min.

c. _____ There are 1000 g in 1 kg.

d. _____ The chef used 12 kg of potatoes.

e. _____ There are 4 books on the shelf.

f. _____ The height of a patient is 1.8 m.

Answers
a. E (definition)
b. M (use a watch)
c. E (metric definition)
d. M (use a scale)
e. E (counted)
f. M (use a metric ruler)

Study Note

Significant figures (abbreviated SFs) are all the numbers reported in a measurement, including the estimated digit. Zeros are significant unless they are placeholders appearing at the beginning of a decimal number or in a large number without a decimal point.

4.255 g (four SFs)   0.0040 m (two SFs)   46 500 L (three SFs)   150. mL (three SFs)

♦ Learning Exercise 2.2B

State the number of significant figures in each of the following measured numbers:

a. 35.24 g _____   b. 8.0×10^{-5} m _____

c. 55 000 m _____   d. 600. mL _____

e. 5.025 L _____   f. 0.006 kg _____

g. 2.680×10^5 mm _____   h. 25.0 °C _____

▸ **CORE CHEMISTRY SKILL**

Counting Significant Figures

Answers   a. 4   b. 2   c. 2   d. 3   e. 4   f. 1   g. 4   h. 3

2.3   Significant Figures in Calculations

Learning Goal: Adjust calculated answers to give the correct number of significant figures.

- In multiplication or division, the final answer is written so that it has the same number of significant figures as the measurement with the fewest significant figures.

- In addition or subtraction, the final answer is written so that it has the same number of decimal places as the measurement with the fewest decimal places.

- To evaluate a calculator answer, count the significant figures in the measurements and round off the calculator answer properly.

- Answers for chemical calculations rarely use all the numbers that appear in the calculator display. Exact numbers are not included in the determination of the number of significant figures (SFs).

Study Note

1. To round off a number when the first digit to be dropped is *4 or less*, keep the digits you need and drop all the digits that follow.

    Round off 42.8254 to three SFs → 42.8 (drop 254)

2. To round off a number when the first digit to be dropped is *5 or greater*, keep the proper number of digits and increase the last retained digit by 1.

    Round off 8.4882 to two SFs → 8.5 (drop 882; increase the last retained digit by 1)

3. When rounding off large numbers without decimal points, maintain the value of the answer by adding nonsignificant zeros as placeholders.

    Round off 356 835 to three SFs → 357 000 (drop 835; increase the last retained digit by 1; add placeholder zeros)

♦ **Learning Exercise 2.3A**

KEY MATH SKILL

Round off each of the following to give two significant figures:

Rounding Off

a. 88.75 m _____
b. 0.002 923 g _____
c. 50.525 s _____
d. 1.672 L _____
e. 0.001 055 8 kg _____
f. 82 080 mL _____

Answers **a.** 89 m **b.** 0.0029 g **c.** 51 s **d.** 1.7 L **e.** 0.0011 kg **f.** 82 000 mL

Study Note

1. An answer obtained from multiplying and dividing has the same number of significant figures as the measurement with the fewest significant figures.

    1.5 × 32.546 = 48.819 → 49 *Answer rounded off to two SFs*
    Two SFs   Five SFs

2. An answer obtained from adding or subtracting has the same number of decimal places as the measurement with the fewest decimal places.

    82.223 + 4.1 = 86.323 → 86.3 *Answer rounded off to the tenths place*
    Thousandths place   Tenths place

♦ **Learning Exercise 2.3B**

CORE CHEMISTRY SKILL

Solve each problem and give the answer with the correct number of significant figures or decimal places:

Using Significant Figures in Calculations

a. $1.3 \times 71.5 =$

b. $\dfrac{8.00}{4.00} =$

c. $\dfrac{(0.082)(25.4)}{(0.116)(3.4)} =$

d. $\dfrac{3.05 \times 1.86}{118.5} =$

e. $\dfrac{376}{0.0073} =$

f. $38.520 - 11.4 =$

g. $4.2 + 8.15 =$

h. $102.56 + 8.325 - 0.8825 =$

Answers   **a.** 93   **b.** 2.00   **c.** 5.3   **d.** 0.0479
            e. 52 000 (5.2 × 10⁴)   **f.** 27.1   **g.** 12.4   **h.** 110.00

2.4   Prefixes and Equalities

Learning Goal: Use the numerical values of prefixes to write a metric equality.

- In the metric system, larger and smaller units use prefixes to change the size of the unit by factors of 10. For example, a prefix such as *centi* or *milli* preceding the unit *meter* gives a smaller length than a meter. A prefix such as *kilo* added to *gram* gives a unit that measures a mass that is 1000 times *greater* than a gram.
- An equality contains two units that measure the *same quantity* such as length, volume, mass, or time.
- Some common metric equalities are: 1 m = 100 cm; 1 L = 1000 mL; 1 kg = 1000 g.

Metric and SI Prefixes

Prefix	Symbol	Numerical Value	Scientific Notation	Equality
Prefixes That Increase the Size of the Unit				
peta	P	1 000 000 000 000 000	10^{15}	$1\ Pg = 1 \times 10^{15}\ g$ $1\ g = 1 \times 10^{-15}\ Pg$
tera	T	1 000 000 000 000	10^{12}	$1\ Ts = 1 \times 10^{12}\ s$ $1\ s = 1 \times 10^{-12}\ Ts$
giga	G	1 000 000 000	10^{9}	$1\ Gm = 1 \times 10^{9}\ m$ $1\ m = 1 \times 10^{-9}\ Gm$
mega	M	1 000 000	10^{6}	$1\ Mg = 1 \times 10^{6}\ g$ $1\ g = 1 \times 10^{-6}\ Mg$
kilo	k	1 000	10^{3}	$1\ km = 1 \times 10^{3}\ m$ $1\ m = 1 \times 10^{-3}\ km$
Prefixes That Decrease the Size of the Unit				
deci	d	0.1	10^{-1}	$1\ dL = 1 \times 10^{-1}\ L$ $1\ L = 10\ dL$
centi	c	0.01	10^{-2}	$1\ cm = 1 \times 10^{-2}\ m$ $1\ m = 100\ cm$
milli	m	0.001	10^{-3}	$1\ ms = 1 \times 10^{-3}\ s$ $1\ s = 1 \times 10^{3}\ ms$
micro	μ*	0.000 001	10^{-6}	$1\ \mu g = 1 \times 10^{-6}\ g$ $1\ g = 1 \times 10^{6}\ \mu g$
nano	n	0.000 000 001	10^{-9}	$1\ nm = 1 \times 10^{-9}\ m$ $1\ m = 1 \times 10^{9}\ nm$
pico	p	0.000 000 000 001	10^{-12}	$1\ ps = 1 \times 10^{-12}\ s$ $1\ s = 1 \times 10^{12}\ ps$
femto	f	0.000 000 000 000 001	10^{-15}	$1\ fs = 1 \times 10^{-15}\ s$ $1\ s = 1 \times 10^{15}\ fs$

*In medicine, the abbreviation mc for the prefix micro is used because the symbol μ may be misread, which could result in a medication error. Thus, 1 μg would be written as 1 mcg.

♦ **Learning Exercise 2.4A**

Match the items in column A with those from column B.

A	B
1. _____ megameter	**a.** nanometer
2. _____ 0.1 m	**b.** decimeter
3. _____ millimeter	**c.** 10^{-6} m
4. _____ centimeter	**d.** 0.01 m
5. _____ 10^{-9} m	**e.** 1000 m
6. _____ micrometer	**f.** 10^{-3} m
7. _____ kilometer	**g.** 10^{6} m

Some Typical Laboratory Test Values

Substance in Blood	Typical Range
Albumin	3.5–5.0 g/dL
Ammonia	20–70 mcg/dL
Calcium	8.5–10.5 mg/dL
Cholesterol	105–250 mg/dL
Iron (male)	80–160 mcg/dL
Protein (total)	6.0–8.0 g/dL

Answers **1.** g   **2.** b   **3.** f   **4.** d
5. a   **6.** c   **7.** e

Laboratory results for blood work are often reported in mass per deciliter (dL).

♦ **Learning Exercise 2.4B**

Complete each of the following metric relationships:

a. 1 L = _____ mL   **b.** 1 L = _____ dL

c. 1 m = _____ cm   **d.** 1 s = _____ ms

e. 1 kg = _____ g   **f.** 1 m = _____ mm

g. 1 mg = _____ mcg   **h.** 1 dL = _____ L

Answers **a.** 1000   **b.** 10   **c.** 100   **d.** 1000
e. 1000   **f.** 1000   **g.** 1000   **h.** 0.1

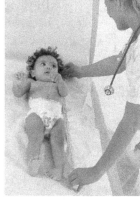

The metric length of 1 m is the same length as 10 dm, 100 cm, or 1000 mm.

CHASSENET / BSIP SA / Alamy

♦ **Learning Exercise 2.4C**

For each of the following pairs, which is the larger unit?

a. milliliter or microliter _____   **b.** g or kg _____

c. nm or m _____   **d.** centimeter or decimeter _____

e. kilosecond or second _____   **f.** PL or pL _____

Answers **a.** milliliter   **b.** kg   **c.** m   **d.** decimeter   **e.** kilosecond   **f.** PL

2.5   Writing Conversion Factors

Learning Goal: Write a conversion factor for two units that describe the same quantity.

•   A conversion factor represents an equality expressed in the form of a fraction.

•   Two conversion factors can be written for any relationship between equal quantities.

For the metric–U.S. equality 2.54 cm = 1 in., the corresponding conversion factors are:

$$\frac{2.54 \text{ cm}}{1 \text{ in.}} \quad \text{and} \quad \frac{1 \text{ in.}}{2.54 \text{ cm}}$$

- A percentage (%) is written as a conversion factor by expressing matching units in the relationship as the parts of a specific substance in 100 parts of the whole.
- An equality may be stated that applies only to a specific problem.
- The ratio of parts per million (ppm) is the same as the milligrams of a substance per kilogram (mg/kg).
- The ratio of parts per billion (ppb) equals the micrograms per kilogram (μg/kg, mcg/kg).

Some Common Equalities

Quantity	Metric (SI)	U.S.	Metric–U.S.
Length	1 km = 1000 m 1 m = 1000 mm 1 cm = 10 mm	1 ft = 12 in. 1 yd = 3 ft 1 mi = 5280 ft	2.54 cm = 1 in. (exact) 1 m = 39.37 in. 1 km = 0.6214 mi
Volume	1 L = 1000 mL 1 dL = 100 mL 1 mL = 1 cm³ 1 mL = 1 cc*	1 qt = 4 cups 1 qt = 2 pt 1 gal = 4 qt	946.4 mL = 1 qt 1 L = 1.057 qt 473.2 mL = 1 pt 3.785 L = 1 gal 5 mL = 1 tsp* 15 mL = 1 T (tbsp)*
Mass	1 kg = 1000 g 1 g = 1000 mg 1 mg = 1000 mcg*	1 lb = 16 oz	1 kg = 2.205 lb 453.6 g = 1 lb 28.35 g = 1 oz
Time	1 h = 60 min 1 min = 60 s	1 h = 60 min 1 min = 60 s	

*Used in medicine.

Study Note

Metric conversion factors are obtained from metric prefixes. For example, the metric equality 1 m = 100 cm is written as two factors:

$$\frac{1 \text{ m}}{100 \text{ cm}} \quad \text{and} \quad \frac{100 \text{ cm}}{1 \text{ m}}$$

For equality $(1 \text{ in.})^2 = (2.54 \text{ cm})^2$, the corresponding conversion factors are:

$$\frac{(1 \text{ in.})^2}{(2.54 \text{ cm})^2} \quad \text{and} \quad \frac{(2.54 \text{ cm})^2}{(1 \text{ in.})^2}$$

◆ **Learning Exercise 2.5A**

Write the equality and two conversion factors for each of the following pairs of units:

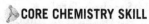
CORE CHEMISTRY SKILL

Writing Conversion Factors from Equalities

    a. millimeters and meters   **b.** kilograms and grams

    c. kilograms and pounds   **d.** seconds and minutes

e. kilometers and meters   **f.** milliliters and quarts

g. deciliters and liters   **h.** square millimeters and square centimeters

Answers

a.   1 m $=$ 1000 mm

$$\frac{1000 \text{ mm}}{1 \text{ m}} \text{ and } \frac{1 \text{ m}}{1000 \text{ mm}}$$

b.   1 kg $=$ 1000 g

$$\frac{1000 \text{ g}}{1 \text{ kg}} \text{ and } \frac{1 \text{ kg}}{1000 \text{ g}}$$

c.   1 kg $=$ 2.205 lb

$$\frac{2.205 \text{ lb}}{1 \text{ kg}} \text{ and } \frac{1 \text{ kg}}{2.205 \text{ lb}}$$

d.   1 min $=$ 60 s

$$\frac{60 \text{ s}}{1 \text{ min}} \text{ and } \frac{1 \text{ min}}{60 \text{ s}}$$

e.   1 km $=$ 1000 m

$$\frac{1000 \text{ m}}{1 \text{ km}} \text{ and } \frac{1 \text{ km}}{1000 \text{ m}}$$

f.   1 qt $=$ 946.4 mL

$$\frac{946.4 \text{ mL}}{1 \text{ qt}} \text{ and } \frac{1 \text{ qt}}{946.4 \text{ mL}}$$

g.   1 L $=$ 10 dL

$$\frac{10 \text{ dL}}{1 \text{ L}} \text{ and } \frac{1 \text{ L}}{10 \text{ dL}}$$

h. $(1 \text{ cm})^2 = (10 \text{ mm})^2$

$$\frac{(10 \text{ mm})^2}{(1 \text{ cm})^2} \text{ and } \frac{(1 \text{ cm})^2}{(10 \text{ mm})^2}$$

Study Note

1. Sometimes, a statement within a problem gives an equality that is true only for that problem. Then conversion factors can be written that are true only for that problem. For example, a problem states that there are 50 mg of vitamin B in a tablet. The equality and its conversion factors are

   1 tablet $=$ 50 mg of vitamin B

   $$\frac{50 \text{ mg vitamin B}}{1 \text{ tablet}} \text{ and } \frac{1 \text{ tablet}}{50 \text{ mg vitamin B}}$$

2. If a problem gives a percentage ($\%$), it can be stated as parts per 100 parts. For example, a candy bar contains 45% by mass chocolate. This percentage ($\%$) equality and its conversion factors are written using the same unit.

   45 g of chocolate $=$ 100 g candy bar

   $$\frac{45 \text{ g chocolate}}{100 \text{ g candy bar}} \text{ and } \frac{100 \text{ g candy bar}}{45 \text{ g chocolate}}$$

3. When a problem gives ppm or ppb, it can be stated as parts per million (ppm), which is mg/kg, or parts per billion (ppb), which is μg/kg. For example, the level of nitrate in the Los Angeles water supply is 2.5 ppm. The equality and its conversion factors are 2.5 mg of nitrate $=$ 1 kg of water.

   $$\frac{2.5 \text{ mg nitrate}}{1 \text{ kg water}} \text{ and } \frac{1 \text{ kg water}}{2.5 \text{ mg nitrate}}$$

♦ **Learning Exercise 2.5B**

Write the equality and two conversion factors for each of the following statements:

a. A cheese contains 55% fat by mass.

b. The Daily Value (DV) for zinc is 15 mg.

c. A 125-g steak contains 45 g of protein.

d. One iron tablet contains 65 mg of iron.

e. A car travels 14 miles on 1 gal of gasoline.

f. A water sample contains 5.4 ppb of arsenic.

Answers

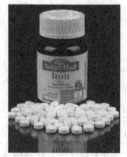

One tablet contains 65 mg of iron.

Credit: Editorial Image, LLC / Alamy

a. 100 g of cheese = 55 g of fat

$$\frac{55 \text{ g fat}}{100 \text{ g cheese}} \quad \text{and} \quad \frac{100 \text{ g cheese}}{55 \text{ g fat}}$$

b. 15 mg of zinc = 1 day

$$\frac{15 \text{ mg zinc}}{1 \text{ day}} \quad \text{and} \quad \frac{1 \text{ day}}{15 \text{ mg zinc}}$$

c. 125 g of steak = 45 g of protein

$$\frac{45 \text{ g protein}}{125 \text{ g steak}} \quad \text{and} \quad \frac{125 \text{ g steak}}{45 \text{ g protein}}$$

d. 1 tablet = 65 mg of iron

$$\frac{65 \text{ mg iron}}{1 \text{ tablet}} \quad \text{and} \quad \frac{1 \text{ tablet}}{65 \text{ mg iron}}$$

e. 1 gal = 14 mi

$$\frac{14 \text{ mi}}{1 \text{ gal}} \quad \text{and} \quad \frac{1 \text{ gal}}{14 \text{ mi}}$$

f. 1 kg of water = 5.4 μg of arsenic

$$\frac{5.4 \text{ }\mu\text{g arsenic}}{1 \text{ kg water}} \quad \text{and} \quad \frac{1 \text{ kg water}}{5.4 \text{ }\mu\text{g arsenic}}$$

♦ **Learning Exercise 2.5C**

Write the equality and two conversion factors, and identify the numbers as exact or give the number of significant figures for each of the following:

a. One tablet of calcium contains 315 mg of calcium.

b. The EPA level for fluoride in drinking water is 4 ppm.

c. One tablespoon of cough syrup contains 15 mL of cough syrup.

Answers

a. 1 tablet = 315 mg of calcium

$$\frac{315 \text{ mg calcium}}{1 \text{ tablet}} \quad \text{and} \quad \frac{1 \text{ tablet}}{315 \text{ mg calcium}}$$

The 315 mg is measured: It has three SFs. The 1 tablet is exact.

b. 1 kg of water = 4 mg of fluoride

$$\frac{4 \text{ mg fluoride}}{1 \text{ kg water}} \quad \text{and} \quad \frac{1 \text{ kg water}}{4 \text{ mg fluoride}}$$

The 4 mg is measured: It has one SF. The 1 kg is exact.

c. 1 tablespoon of syrup = 15 mL of syrup

$$\frac{15 \text{ mL syrup}}{1 \text{ tablespoon syrup}} \quad \text{and} \quad \frac{1 \text{ tablespoon syrup}}{15 \text{ mL syrup}}$$

The 15 mL is measured: It has two SFs. The 1 tablespoon is exact.

2.6   Problem Solving Using Unit Conversion

Learning Goal: Use conversion factors to change from one unit to another.

- Conversion factors can be used to change a quantity expressed in one unit to a quantity expressed in another unit.
- In the process of problem solving, a given unit is multiplied by one or more conversion factors until the needed unit is obtained.

Guide to Problem Solving Using Conversion Factors	
STEP 1	State the given and needed quantities.
STEP 2	Write a plan to convert the given unit to the needed unit.
STEP 3	State the equalities and conversion factors.
STEP 4	Set up the problem to cancel units and calculate the answer.

Example: A patient receives 2850 mL of saline solution. How many liters is 2850 mL?

Solution: **STEP 1 State the given and needed quantities.**

Analyze the Problem	Given	Need	Connect
	2850 mL	liters	metric factor (L/mL)

STEP 2 Write a plan to convert the given unit to the needed unit.

$$\text{milliliters} \xrightarrow{\text{Metric factor}} \text{liters}$$

STEP 3 State the equalities and conversion factors.

$$1 \text{ L} = 1000 \text{ mL}$$

$$\frac{1000 \text{ mL}}{1 \text{ L}} \quad \text{and} \quad \frac{1 \text{ L}}{1000 \text{ mL}}$$

STEP 4 Set up the problem to cancel units and calculate the answer.

$$2850 \text{ mL} \times \frac{1 \text{ L}}{1000 \text{ mL}} = 2.85 \text{ L}$$

Three SFs   Exact   Three SFs

♦ **Learning Exercise 2.6A**

Use metric conversion factors to solve each of the following problems.
For **e** to **h**, complete boxes for Given, Need, Connect, and Answer.

a. 189 mL = _____ L

b. 2.7 cm = _____ mm

⮞ **CORE CHEMISTRY SKILL**

Using Conversion Factors

c. 0.274 m = _____ cm

d. 0.076 kg = _____ g

e. How many meters tall is a person whose height is 175 cm?

Analyze the Problem	Given	Need	Connect	Answer

f. There is 0.646 L of water in a teapot. How many milliliters is that?

Analyze the Problem	Given	Need	Connect	Answer

g. If a ring contains 13 500 mg of gold, how many grams of gold are in the ring?

Analyze the Problem	Given	Need	Connect	Answer

h. You walked a distance of 1.5 km on the treadmill at the gym. How many meters did you walk?

Analyze the Problem	Given	Need	Connect	Answer

Answers   **a.** 0.189 L   **b.** 27 mm   **c.** 27.4 cm   **d.** 76 g

e.

Analyze the Problem	Given	Need	Connect	Answer
	175 cm	meters	metric factor (m/cm)	1.75 m

f.

Analyze the Problem	Given	Need	Connect	Answer
	0.646 L	milliliters	metric factor (mL/L)	646 mL

g.

Analyze the Problem	Given	Need	Connect	Answer
	13 500 mg	grams	metric factor (g/mg)	13.5 g

h.

Analyze the Problem	Given	Need	Connect	Answer
	1.5 km	meters	metric factor (m/km)	1500 m

♦ Learning Exercise 2.6B

Use metric–U.S. conversion factors to solve each of the following problems:

a. 880 g = _____ lb

b. 4.6 qt = _____ mL

c. 1.50 ft = _____ cm

d. 8.10 in. = _____ mm

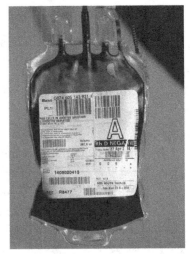

1 pt of blood contains 473.2 mL.

Credit: Syner-Comm / Alamy

Answers   **a.** 1.9 lb   **b.** 4400 mL   **c.** 45.7 cm
      d. 206 mm

♦ Learning Exercise 2.6C

Use conversion factors to solve each of the following problems.
For **c** to **f**, complete boxes for Given, Need, Connect, and Answer.

a. 120 mm = _____ in.

b. 245 lb = _____ kg

c. Your friend has a height of 6 ft 3 in. What is your friend's height in meters?

Analyze the Problem	Given	Need	Connect	Answer

d. In a triple-bypass surgery, a patient requires 3.00 pt of whole blood. How many milliliters of blood were given?

Analyze the Problem	Given	Need	Connect	Answer

e. A doctor orders 0.450 g of a sulfa drug. On hand are 150-mg tablets. How many tablets of the sulfa drug are needed?

Analyze the Problem	Given	Need	Connect	Answer

f. A mouthwash contains 27% alcohol by volume. How many milliliters of alcohol are in a 1.18-L bottle of mouthwash?

Analyze the Problem	Given	Need	Connect	Answer

Answers   **a.** 4.7 in.   **b.** 111 kg

c.

	Given	Need	Connect	Answer
Analyze the Problem	6 ft 3 in	meters	factors (in./ft, m/in.)	1.9 m

d.

	Given	Need	Connect	Answer
Analyze the Problem	3.00 pt	milliliters	factor (mL/pt)	1420 mL

e.

	Given	Need	Connect	Answer
Analyze the Problem	0.450 g, 150-mg tablet	tablets	factors (mg/g, tablet/mg)	3 tablets

f.

	Given	Need	Connect	Answer
Analyze the Problem	1.18 L, 27% (v/v)	milliliters of alcohol	factors (mL/L, % alcohol)	320 mL

2.7 Density

Learning Goal: Calculate the density of a substance; use the density to calculate the mass or volume of a substance.

- The density of a substance is the ratio of its mass to its volume, usually in units of g/cm^3 or g/mL.

$$\text{Density} = \frac{\text{mass of substance}}{\text{volume of substance}}$$

- The volume of 1 mL is equal to the volume of 1 cm^3.
- Specific gravity is a relationship between the density of a substance and the density of water.

Key Terms for Sections 2.2 to 2.7

Match each of the following key terms with the correct description:

- **a.** equality
- **b.** conversion factor
- **c.** measured number
- **d.** exact number
- **e.** density
- **f.** prefix
- **g.** significant figures
- **h.** specific gravity

1. _____ number obtained by counting

2. _____ part of the name of a metric unit that precedes the base unit and specifies the size of the measurement

3. _____ ratio in which the numerator and denominator are quantities from an equality

4. _____ relationship between two units that measure the same quantity

5. _____ digits recorded in a measurement

6. _____ number obtained when a quantity is determined by using a measuring device

7. _____ relationship of the mass of an object to its volume expressed as grams per cubic centimeter (g/cm^3) or grams per milliliter (g/mL)

8. _____ the relationship between the density of a substance and the density of water

Answers   **1.** d   **2.** f   **3.** b   **4.** a   **5.** g   **6.** c   **7.** e   **8.** h

♦ **Learning Exercise 2.7A**

Calculate the density for each of the following:

    a. What is the density, in grams per milliliter, of glycerol if a 200.0-mL sample has a mass of 252 g?

    b. A person with diabetes may produce 5 to 12 L of urine per day. Calculate the density, in grams per milliliter, of a 100.0 mL-urine sample that has a mass of 100.2 g.

    c. A solid has a mass of 5.5 oz. When placed in a graduated cylinder with a water level of 55.2 mL, the object causes the water level to rise to 73.8 mL. What is the density of the object, in grams per milliliter?

Answers   **a.** 1.26 g/mL   **b.** 1.002 g/mL   **c.** 8.4 g/mL

Guide to Using Density	
STEP 1	State the given and needed quantities.
STEP 2	Write a plan to calculate the needed quantity.
STEP 3	Write the equalities and their conversion factors including density.
STEP 4	Set up the problem to calculate the needed quantity.

Example: Density can be used as a factor to convert between the mass (g) and volume (mL) of a substance. The density of silver is 10.5 g/mL. What is the mass, in grams, of 6.0 mL of silver?

Solution: **STEP 1 State the given and needed quantities.**

Analyze the Problem	Given	Need	Connect
	6.0 mL of silver; density 10.5 g/mL	grams of silver	density factor

      STEP 2 Write a plan to calculate the needed quantity.

$$\text{milliliters} \xrightarrow{\text{Density factor}} \text{grams}$$

STEP 3 Write the equalities and their conversion factors including density.

1 mL of silver = 10.5 g of silver

$$\frac{10.5 \text{ g}}{1 \text{ mL}} \text{ and } \frac{1 \text{ mL}}{10.5 \text{ g}}$$

STEP 4 Set up the problem to calculate the needed quantity.

$$6.0 \text{ mL silver} \times \frac{10.5 \text{ g silver}}{1 \text{ mL silver}} = 63 \text{ g of silver}$$

Density factor

♦ Learning Exercise 2.7B

Use density or specific gravity as a conversion factor to solve each of the following:

a. A sugar solution has a density of 1.20 g/mL. What is the mass, in grams, of 0.225 L of the solution?

CORE CHEMISTRY SKILL

Using Density as a Conversion Factor

b. A piece of pure gold weighs 0.26 lb. If gold has a density of 19.3 g/mL, what is the volume, in milliliters, of the piece of gold?

c. A salt solution has a specific gravity of 1.15 and a volume of 425 mL. What is the mass, in grams, of the solution?

d. A 600.-g sample of a glucose solution has a density of 1.28 g/mL. What is the volume, in liters, of the sample?

Answers   **a.** 270. g   **b.** 6.1 mL   **c.** 489 g   **d.** 0.469 L

Checklist for Chapter 2

You are ready to take the Practice Test for Chapter 2. Be sure you have accomplished the following learning goals for this chapter. If not, review the section listed at the end of the goal. Then apply your new skills and understanding to the Practice Test.

After studying Chapter 2, I can successfully:

_____ Write the names and abbreviations for the metric (SI) units of measurement. (2.1)

_____ Identify a number as a measured number or an exact number. (2.2)

_____ Determine the number of significant figures in measured numbers. (2.2)

_____ Report a calculated answer with the correct number of significant figures. (2.3)

_____ Write an equality from the relationship between two equal quantities. (2.4)

_____ Write two conversion factors for an equality. (2.5)

_____ Use conversion factors to change from one unit to another unit. (2.6)

_____ Calculate the density or specific gravity of a substance or use density or specific gravity to calculate the mass or volume. (2.7)

Practice Test for Chapter 2

The chapter sections to review are shown in parentheses at the end of each question.

1. Which of the following is a metric measurement of volume? (2.1)
   A. kilogram   **B.** kilowatt   **C.** kiloliter   **D.** kilometer   **E.** kiloquart

2. Which of the following is a metric measurement of length? (2.1)
   A. centiliter   **B.** centigram   **C.** centiyard   **D.** centifoot   **E.** centimeter

3. The exact number in the following is (2.2)
   A. 4.5 L   **B.** 4 syringes   **C.** 6 qt   **D.** 15 g   **E.** 2.0 m

4. The measured number in the following is (2.2)
   A. 1 book   **B.** 2 cars   **C.** 4 flowers   **D.** 5 rings   **E.** 4 g

5. The number of significant figures in 105.4 m is (2.2)
   A. 1   **B.** 2   **C.** 3   **D.** 4   **E.** 5

6. The number of significant figures in 0.000 82 g is (2.2)
   A. 1   **B.** 2   **C.** 3   **D.** 4   **E.** 5

7. The calculator answer 5.7805 rounded to two significant figures is (2.2)
   A. 5   **B.** 5.7   **C.** 5.8   **D.** 5.78   **E.** 6.0

8. The calculator answer 3486.512 rounded to three significant figures is (2.2)
   A. 4000   **B.** 3500   **C.** 349   **D.** 3487   **E.** 3490

9. The answer for $16.0 \div 8.0$ with the correct number of significant figures is (2.3)
   A. 2   **B.** 2.0   **C.** 2.00   **D.** 0.2   **E.** 5.0

10. The answer for $58.5 + 9.158$ with the correct number of decimal places is (2.3)
    A. 67   **B.** 67.6   **C.** 67.7   **D.** 67.66   **E.** 67.658

11. The answer for $\dfrac{2.5 \times 3.12}{4.6}$ with the correct number of significant figures is (2.3)
    A. 0.54   **B.** 7.8   **C.** 0.85   **D.** 1.7   **E.** 1.69

12. Which of these prefixes has the largest value? (2.4)
    A. centi   **B.** deci   **C.** milli   **D.** kilo   **E.** micro

13. What is the decimal equivalent of the prefix *centi*? (2.4)
    A. 0.001   **B.** 0.01   **C.** 0.1   **D.** 10   **E.** 100

14. Which of the following is the smallest unit of mass measurement? (2.4)
    A. gram   **B.** milligram   **C.** kilogram   **D.** decigram   **E.** centigram

15. Which volume is the largest? (2.4)
    A. mL   **B.** dL   **C.** cm^3 **D.** L   **E.** kL

16. Which of the following is a conversion factor? (2.5)

    A. 12 in.   **B.** 3 ft   **C.** 20 m   **D.** $\dfrac{1000\ g}{1\ kg}$ **E.** 2 cm^3

17. Which is the correct conversion factor that relates milliliters to liters? (2.5)

    A. $\dfrac{1000\ mL}{1\ L}$ **B.** $\dfrac{100\ mL}{1\ L}$ **C.** $\dfrac{10\ mL}{1\ L}$ **D.** $\dfrac{0.01\ mL}{1\ L}$ **E.** $\dfrac{0.001\ mL}{1\ L}$

18. Which is the conversion factor for millimeters and centimeters? (2.5)

    A. $\dfrac{1\ mm}{1\ cm}$ **B.** $\dfrac{10\ mm}{1\ cm}$ **C.** $\dfrac{100\ cm}{1\ mm}$ **D.** $\dfrac{100\ mm}{1\ cm}$ **E.** $\dfrac{10\ cm}{1\ mm}$

19. A doctor orders 20 mg of prednisone. If a tablet contains 5 mg of prednisone, how many tablets should be given? (2.6)

    A. 1   **B.** 2   **C.** 3   **D.** 4   **E.** 5

20. 294 mm is equal to (2.6)

    A. 2940 m   **B.** 29.4 m   **C.** 2.94 m   **D.** 0.294 m   **E.** 0.0294 m

21. A doctor orders 200 mg of penicillin to be injected. If it is available as 1000 mg of penicillin per 5 mL, how many milliliters should be given? (2.6)

    A. 1 mL   **B.** 2 mL   **C.** 3 mL   **D.** 4 mL   **E.** 5 mL

22. What is the volume, in liters, of 800. mL of blood plasma? (2.6)

    A. 0.8 L   **B.** 0.80 L   **C.** 0.800 L   **D.** 8.0 L   **E.** 80.0 L

23. What is the mass, in kilograms, of a 22-lb toddler? (2.6)

    A. 10. kg   **B.** 48 kg   **C.** 10 000 kg   **D.** 0.048 kg   **E.** 22 000 kg

24. The number of milliliters in 2 dL of an antiseptic liquid is (2.6)

    A. 20 mL   **B.** 200 mL   **C.** 2000 mL   **D.** 20 000 mL   **E.** 500 000 mL

25. What is the height, in meters, of a person who is 5 ft 4 in. tall? (2.6)

    A. 64 m   **B.** 25 m   **C.** 14 m   **D.** 1.6 m   **E.** 1.3 m

26. How many ounces are in 1500 grams of carrots? (2.6)

    A. 94 oz   **B.** 53 oz   **C.** 24 000 oz   **D.** 33 oz   **E.** 3.3 oz

27. How many quarts of orange juice are in 255 mL of juice? (2.6)

    A. 0.255 qt   **B.** 269 qt   **C.** 236 qt   **D.** 0.269 qt   **E.** 3.71 qt

28. Your doctor places you on a 2200-kcal diet, with 19% of the kilocalories from fat. How many kilocalories are you allowed from fat? (2.6)

    A. 19 kcal   **B.** 2200 kcal   **C.** 1900 kcal   **D.** 420 kcal   **E.** 4200 kcal

29. 1.5 ft is the same length, in centimeters, as (2.6)

    A. 46 cm   **B.** 7.1 cm   **C.** 18 cm   **D.** 3.8 cm   **E.** 0.59 cm

30. The quantity of lead in water was 4.6 ppb. What mass, in micrograms, of lead is in 2.0 kg of water? (2.6)

    A. 9.2 μg   **B.** 4.6 μg   **C.** 2.3 μg   **D.** 0.92 μg   **E.** 0.46 μg

31. How many milliliters of a salt solution with a density of 1.8 g/mL are needed to provide 450 g of salt solution? (2.7)

    A. 250 mL   **B.** 25 mL   **C.** 810 mL   **D.** 450 mL   **E.** 45 mL

32. Three liquids have densities of 1.15 g/mL, 0.79 g/mL, and 0.95 g/mL. When the liquids, which do not mix, are poured into a graduated cylinder, the liquid at the top has a density of (2.7)

    A. 1.15 g/mL   **B.** 1.00 g/mL   **C.** 0.95 g/mL   **D.** 0.79 g/mL   **E.** 0.16 g/mL

33. A sample of oil has a mass of 65 g and a volume of 80.0 mL. What is the specific gravity of the oil? (2.7)

  A. 1.5   **B.** 1.4   **C.** 1.2   **D.** 0.90   **E.** 0.81

34. What is the mass, in grams, of a 15.0-mL sample of an antihistamine solution with a density of 1.04 g/mL? (2.7)

  A. 104 g   **B.** 10.4 g   **C.** 15.6 g   **D.** 1.56 g   **E.** 14.4 g

35. Ethanol has a density of 0.785 g/mL. What is the mass, in grams, of 0.250 L of ethanol? (2.7)

  A. 196 g   **B.** 158 g   **C.** 318 g   **D.** 0.253 g   **E.** 0.160 g

36. A patient has a blood volume of 6.8 qt. If the density of blood is 1.06 g/mL, what is the mass, in grams, of the patient's blood? (2.7)

  A. 1060 g   **B.** 6400 g   **C.** 6800 g   **D.** 1500 g   **E.** 990 g

Answers to the Practice Test

1. C	**2.** E	**3.** B	**4.** E	**5.** D
6. B	**7.** C	**8.** E	**9.** B	**10.** C
11. D	**12.** D	**13.** B	**14.** B	**15.** E
16. D	**17.** A	**18.** B	**19.** D	**20.** D
21. A	**22.** C	**23.** A	**24.** B	**25.** D
26. B	**27.** D	**28.** D	**29.** A	**30.** A
31. A	**32.** D	**33.** E	**34.** C	**35.** A
36. C				

Selected Answers and Solutions to Text Problems

2.1 **a.** The abbreviation for the unit gram is g.
    b. The abbreviation for the unit liter is L.
    c. The abbreviation for the unit degree Celsius is °C.
    d. The abbreviation for the unit pound is lb.
    e. The abbreviation for the unit second is s.

2.3 **a.** A liter is a unit of volume.
    b. A centimeter is a unit of length.
    c. A kilometer is a unit of length.
    d. A second is a unit of time.

2.5 **a.** The unit is a meter, which is a unit of length.
    b. The unit is a gram, which is a unit of mass.
    c. The unit is a milliliter, which is a unit of volume.
    d. The unit is a second, which is a unit of time.
    e. The unit is a degree Celsius, which is a unit of temperature.

2.7 **a.** The unit is a second, which is a unit of time.
    b. The unit is a kilogram, which is a unit of mass.
    c. The unit is a gram, which is a unit of mass.
    d. The unit is a degree Celsius, which is a unit of temperature.

2.9 The estimated digit is the last digit reported in a measurement.
    a. In 8.6 m, the estimated digit is the 6 in the tenths place (0.6).
    b. In 45.25 g, the estimated digit is the 5 in the hundredths place (0.05).
    c. In 29 °C, the estimated digit is the 9 in the ones place.

2.11 Measured numbers are obtained using some type of measuring device. Exact numbers are numbers obtained by counting items or using a definition that compares two units in the same measuring system.
    a. The value 67.5 kg is a measured number; measurement of mass requires a measuring device.
    b. The value 2 tablets is obtained by counting, making it an exact number.
    c. The values in the metric definition 1 L = 1000 mL are exact numbers.
    d. The value 1720 km is a measured number; measurement of distance requires a measuring device.

2.13 Measured numbers are obtained using some type of measuring device. Exact numbers are numbers obtained by counting items or using a definition that compares two units in the same measuring system.
    a. 6 oz of hamburger meat is a measured number (3 hamburgers is a counted/exact number).
    b. Neither are measured numbers (both 1 table and 4 chairs are counted/exact numbers).
    c. Both 0.75 lb of grapes and 350 g of butter are measured numbers.
    d. Neither are measured numbers (the values in a definition are exact numbers).

2.15 **a.** Zeros at the beginning of a decimal number are not significant.
    b. Zeros between nonzero digits are significant.
    c. Zeros at the end of a decimal number are significant.
    d. Zeros in the coefficient of a number written in scientific notation are significant.
    e. Zeros used as placeholders in a large number without a decimal point are not significant.

2.17 **a.** All five numbers are significant figures (5 SFs).
  b. Only the two nonzero numbers are significant (2 SFs); the preceding zeros are placeholders.
  c. Only the two nonzero numbers are significant (2 SFs); the zeros that follow are placeholders.
  d. All three numbers in the coefficient of a number written in scientific notation are significant (3 SFs).
  e. All four numbers to the right of the decimal point, including the last zero in the decimal number, are significant (4 SFs).
  f. All three numbers, including the zeros at the end of a decimal number, are significant (3 SFs).

2.19 Both measurements in part **b** have three significant figures, and both measurements in part **c** have two significant figures.

2.21 **a.** 5000 L is the same as 5 × 1000 L, which is written in scientific notation as 5.0×10^3 L with two significant figures.
  b. 30 000 g is the same as 3 × 10 000 g, which is written in scientific notation as 3.0×10^4 g with two significant figures.
  c. 100 000 m is the same as 1 × 100 000 m, which is written in scientific notation as 1.0×10^5 m with two significant figures.
  d. 0.000 25 cm is the same as $2.5 \times \dfrac{1}{10\ 000}$ cm, which is written in scientific notation as 2.5×10^{-4} cm.

2.23 **a.** 1.607 kg has four significant figures.
  b. 130 mcg has two significant figures.
  c. 4.02×10^6 red blood cells has three significant figures.

2.25 A calculator often gives more digits than the number of significant figures allowed in the answer.

2.27 **a.** 1.85 kg; the last digit is dropped since it is 4 or less.
  b. 88.2 L; since the fourth digit is 4 or less, the last three digits are dropped.
  c. 0.004 74 cm; since the fourth significant digit (the first digit to be dropped) is 5 or greater, the last retained digit is increased by 1 when the last four digits are dropped.
  d. 8810 m; since the fourth significant digit (the first digit to be dropped) is 5 or greater, the last retained digit is increased by 1 when the last digit is dropped (a nonsignificant zero is added at the end as a placeholder).
  e. 1.83×10^5 s; since the fourth digit is 4 or less, the last digit is dropped. The $\times 10^5$ is retained so that the magnitude of the answer is not changed.

2.29 **a.** To round off 56.855 m to three significant figures, drop the final digits 55 and increase the last retained digit by 1 to give 56.9 m.
  b. To round off 0.002 282 g to three significant figures, drop the final digit 2 to give 0.002 28 g.
  c. To round off 11 527 s to three significant figures, drop the final digits 27 and add two zeros as placeholders to give 11 500 s (1.15×10^4 s).
  d. To express 8.1 L to three significant figures, add a significant zero to give 8.10 L.

2.31 **a.** $45.7 \times 0.034 = 1.6$ Two significant figures are allowed since 0.034 has 2 SFs.
  b. $0.002\ 78 \times 5 = 0.01$ One significant figure is allowed since 5 has 1 SF.
  c. $\dfrac{34.56}{1.25} = 27.6$ Three significant figures are allowed since 1.25 has 3 SFs.
  d. $\dfrac{(0.2465)(25)}{1.78} = 3.5$ Two significant figures are allowed since 25 has 2 SFs.
  e. $(2.8 \times 10^4)(5.05 \times 10^{-6}) = 0.14$ or 1.4×10^{-1} Two significant figures are allowed since 2.8×10^4 has 2 SFs.
  f. $\dfrac{(3.45 \times 10^{-2})(1.8 \times 10^5)}{(8 \times 10^3)} = 0.8$ or 8×10^{-1} One significant figure is allowed since 8×10^3 has 1 SF.

2.33 **a.** 45.48 cm + 8.057 cm = 53.54 cm Two decimal places are allowed since 45.48 cm has two decimal places.

b. 23.45 g + 104.1 g + 0.025 g = 127.6 g One decimal place is allowed since 104.1 g has one decimal place.

c. 145.675 mL − 24.2 mL = 121.5 mL One decimal place is allowed since 24.2 mL has one decimal place.

d. 1.08 L − 0.585 L = 0.50 L Two decimal places are allowed since 1.08 L has two decimal places.

2.35 The km/h markings indicate how many kilometers (how much distance) will be traversed in 1 hour's time if the speed is held constant. The mph (mi/h) markings indicate the same distance traversed *but measured in miles* during the 1 hour of travel.

2.37 **a.** mg **b.** dL **c.** km **d.** fg

2.39 **a.** centiliter **b.** kilogram **c.** millisecond **d.** gigameter

2.41 **a.** 0.01 **b.** 1 000 000 000 000 (or 1×10^{12})
c. 0.001 (or 1×10^{-3}) **d.** 0.1

2.43 **a.** decigram **b.** microgram
c. kilogram **d.** centigram

2.45 **a.** 1 m = 100 cm **b.** 1 m = 1×10^9 nm
c. 1 mm = 0.001 m **d.** 1 L = 1000 mL

2.47 **a.** kilogram, since 10^3 g is greater than 10^{-3} g
b. milliliter, since 10^{-3} L is greater than 10^{-6} L
c. km, since 10^3 m is greater than 10^0 m
d. kL, since 10^3 L is greater than 10^{-1} L
e. nanometer, since 10^{-9} m is greater than 10^{-12} m

2.49 A conversion factor can be inverted to give a second conversion factor: $\dfrac{1\ m}{100\ cm}$ and $\dfrac{100\ cm}{1\ m}$

2.51 **a.** 1 m = 100 cm; $\dfrac{100\ cm}{1\ m}$ and $\dfrac{1\ m}{100\ cm}$

b. 1 g = 1×10^9 ng; $\dfrac{1 \times 10^9\ ng}{1\ g}$ and $\dfrac{1\ g}{1 \times 10^9\ ng}$

c. 1 kL = 1000 L; $\dfrac{1000\ L}{1\ kL}$ and $\dfrac{1\ kL}{1000\ L}$

d. 1 s = 1000 ms; $\dfrac{1000\ ms}{1\ s}$ and $\dfrac{1\ s}{1000\ ms}$

e. $(1\ m)^3 = (100\ cm)^3$; $\dfrac{(100\ cm)^3}{(1\ m)^3}$ and $\dfrac{(1\ m)^3}{(100\ cm)^3}$

2.53 a. 1 yd = 3 ft; $\dfrac{3\text{ ft}}{1\text{ yd}}$ and $\dfrac{1\text{ yd}}{3\text{ ft}}$; the 1 yd and 3 ft are both exact (U.S. definition).

b. 1 kg = 2.205 lb; $\dfrac{2.205\text{ lb}}{1\text{ kg}}$ and $\dfrac{1\text{ kg}}{2.205\text{ lb}}$; the 2.205 lb is measured: it has 4 SFs; the 1 kg is exact.

c. 1 min = 60 s; $\dfrac{60\text{ s}}{1\text{ min}}$ and $\dfrac{1\text{ min}}{60\text{ s}}$; the 1 min and 60 s are both exact.

d. 1 gal of gasoline = 27 mi; $\dfrac{27\text{ mi}}{1\text{ gal gasoline}}$ and $\dfrac{1\text{ gal gasoline}}{27\text{ mi}}$; the 27 mi is measured: it has 2 SFs; the 1 gal is exact.

e. 100 g of sterling = 93 g of silver; $\dfrac{93\text{ g silver}}{100\text{ g sterling}}$ and $\dfrac{100\text{ g sterling}}{93\text{ g silver}}$; the 93 g is measured: it has 2 SFs; the 100 g is exact.

2.55 a. 1 s = 3.5 m; $\dfrac{3.5\text{ m}}{1\text{ s}}$ and $\dfrac{1\text{ s}}{3.5\text{ m}}$; the 3.5 m is measured: it has 2 SFs; the 1 s is exact.

b. 4700 mg of potassium = 1 day; $\dfrac{4700\text{ mg potassium}}{1\text{ day}}$ and $\dfrac{1\text{ day}}{4700\text{ mg potassium}}$; the 4700 mg is measured: it has 2 SFs; the 1 day is exact.

c. 1 gal of gasoline = 46.0 km; $\dfrac{46.0\text{ km}}{1\text{ gal gasoline}}$ and $\dfrac{1\text{ gal gasoline}}{46.0\text{ km}}$; the 46.0 km is measured: it has 3 SFs; the 1 gal is exact.

d. 1 kg of plums = 29 mcg of pesticide; $\dfrac{29\text{ mcg pesticide}}{1\text{ kg plums}}$ and $\dfrac{1\text{ kg plums}}{29\text{ mcg pesticide}}$; the 29 mcg is measured: it has 2 SFs; the 1 kg is exact.

e. 100 g of crust = 28.2 g of silicon; $\dfrac{28.2\text{ g silicon}}{100\text{ g crust}}$ and $\dfrac{100\text{ g crust}}{28.2\text{ g silicon}}$; the 28.2 g is measured: it has 3 SFs; the 100 g is exact.

2.57 a. 1 tablet = 630 mg of calcium; $\dfrac{630\text{ mg calcium}}{1\text{ tablet}}$ and $\dfrac{1\text{ tablet}}{630\text{ mg calcium}}$; the 630 mg is measured: it has 2 SFs; the 1 tablet is exact.

b. 60 mg of vitamin C = 1 day; $\dfrac{60\text{ mg vitamin C}}{1\text{ day}}$ and $\dfrac{1\text{ day}}{60\text{ mg vitamin C}}$; the 60 mg is measured: it has 1 SF; the 1 day is exact.

c. 1 tablet = 50 mg of atenolol; $\dfrac{50\text{ mg atenolol}}{1\text{ tablet}}$ and $\dfrac{1\text{ tablet}}{50\text{ mg atenolol}}$; the 50 mg is measured: it has 1 SF; the 1 tablet is exact.

d. 1 tablet = 81 mg of aspirin; $\dfrac{81\text{ mg aspirin}}{1\text{ tablet}}$ and $\dfrac{1\text{ tablet}}{81\text{ mg aspirin}}$; the 81 mg is measured: it has 2 SFs; the 1 tablet is exact.

2.59 a. 5 mL of syrup = 10 mg of Atarax; $\dfrac{10\text{ mg Atarax}}{5\text{ mL syrup}}$ and $\dfrac{5\text{ mL syrup}}{10\text{ mg Atarax}}$

b. 1 tablet = 0.25 g of Lanoxin; $\dfrac{0.25\text{ g Lanoxin}}{1\text{ tablet}}$ and $\dfrac{1\text{ tablet}}{0.25\text{ g Lanoxin}}$

c. 1 tablet = 300 mg of Motrin; $\dfrac{300\text{ mg Motrin}}{1\text{ tablet}}$ and $\dfrac{1\text{ tablet}}{300\text{ mg Motrin}}$

2.61 When using a conversion factor, you are trying to cancel existing units and arrive at a new (needed) unit. The conversion factor must be properly oriented so that the unit in the denominator cancels the preceding unit in the numerator.

2.63 **a.** **Given**   44.2 mL   **Need**   liters

   Plan   mL → L $\dfrac{1\ L}{1000\ mL}$

   Set-up $44.2\ \cancel{mL} \times \dfrac{1\ L}{1000\ \cancel{mL}} = 0.0442\ L$ (3 SFs)

   b. **Given**   8.65 m   **Need**   nanometers

   Plan   m → nm $\dfrac{1 \times 10^9\ nm}{1\ m}$

   Set-up $8.65\ \cancel{m} \times \dfrac{1 \times 10^9\ nm}{1\ \cancel{m}} = 8.65 \times 10^9\ nm$ (3 SFs)

   c. **Given** 5.2×10^8 g   **Need**   megagrams

   Plan   g → Mg $\dfrac{1\ Mg}{1 \times 10^6\ g}$

   Set-up $5.2 \times 10^8\ \cancel{g} \times \dfrac{1\ Mg}{1 \times 10^6\ \cancel{g}} = 5.2 \times 10^2\ Mg$ (2 SFs)

   d. **Given**   0.72 ks   **Need**   milliseconds

   Plan   ks → s → ms $\dfrac{1000\ s}{1\ ks}$ $\dfrac{1000\ ms}{1\ s}$

   Set-up $0.72\ \cancel{ks} \times \dfrac{1000\ \cancel{s}}{1\ \cancel{ks}} \times \dfrac{1000\ ms}{1\ \cancel{s}} = 7.2 \times 10^5\ ms$ (2 SFs)

2.65 **a.** **Given**   3.428 lb   **Need**   kilograms

   Plan   lb → kg $\dfrac{1\ kg}{2.205\ lb}$

   Set-up $3.428\ \cancel{lb} \times \dfrac{1\ kg}{2.205\ \cancel{lb}} = 1.555\ kg$ (4 SFs)

   b. **Given**   1.6 m   **Need**   inches

   Plan   m → in. $\dfrac{39.37\ in.}{1\ m}$

   Set-up $1.6\ \cancel{m} \times \dfrac{39.37\ in.}{1\ \cancel{m}} = 63\ in.$ (2 SFs)

   c. **Given**   4.2 L   **Need**   quarts

   Plan   L → qt $\dfrac{1.057\ qt}{1\ L}$

   Set-up $4.2\ \cancel{L} \times \dfrac{1.057\ qt}{1\ \cancel{L}} = 4.4\ qt$ (2 SFs)

   d. **Given**   0.672 ft   **Need**   millimeters

   Plan   ft → in. → cm → mm $\dfrac{12\ in.}{1\ ft}$ $\dfrac{2.54\ cm}{1\ in.}$ $\dfrac{10\ mm}{1\ cm}$

   Set-up $0.672\ \cancel{ft} \times \dfrac{12\ \cancel{in.}}{1\ \cancel{ft}} \times \dfrac{2.54\ \cancel{cm}}{1\ \cancel{in.}} \times \dfrac{10\ mm}{1\ \cancel{cm}} = 205\ mm$ (3 SFs)

2.67 **a.** **Given** 175 cm   **Need** meters

Plan cm → m $\dfrac{1 \text{ m}}{100 \text{ cm}}$

Set-up 175 c̶m̶ × $\dfrac{1 \text{ m}}{100 \text{ c̶m̶}}$ = 1.75 m (3 SFs)

b. **Given** 5000 mL   **Need** liters

Plan mL → L $\dfrac{1 \text{ L}}{1000 \text{ mL}}$

Set-up 5000 m̶L̶ × $\dfrac{1 \text{ L}}{1000 \text{ m̶L̶}}$ = 5 L (1 SF)

c. **Given** 0.0055 kg   **Need** grams

Plan kg → g $\dfrac{1000 \text{ g}}{1 \text{ kg}}$

Set-up 0.0055 k̶g̶ × $\dfrac{1000 \text{ g}}{1 \text{ k̶g̶}}$ = 5.5 g (2 SFs)

d. **Given** 3500 cm³   **Need** cubic meters

Plan cm³ → m³ $\dfrac{(1 \text{ m})^3}{(100 \text{ cm})^3}$

Set-up 3500 c̶m̶³ × $\dfrac{(1 \text{ m})^3}{(100 \text{ c̶m̶})^3}$ = 3.5 × 10⁻³ m³ (2 SFs)

2.69 **a.** **Given** 0.500 qt   **Need** milliliters

Plan qt → mL $\dfrac{946.4 \text{ mL}}{1 \text{ qt}}$

Set-up 0.500 q̶t̶ × $\dfrac{946.4 \text{ mL}}{1 \text{ q̶t̶}}$ = 473 mL (3 SFs)

b. **Given** 175 lb   **Need** kilograms

Plan lb → kg $\dfrac{1 \text{ kg}}{2.205 \text{ lb}}$

Set-up 175 l̶b̶ × $\dfrac{1 \text{ kg}}{2.205 \text{ l̶b̶}}$ = 79.4 kg (3 SFs)

c. **Given** 74 kg body mass, 15% body fat   **Need** pounds of body fat

Plan kg of body mass → kg of body fat → lb of body fat
(percent equality: 100 kg of body mass = 15 g of body fat)

$\dfrac{15 \text{ kg body fat}}{100 \text{ kg body mass}}$ $\dfrac{2.205 \text{ lb body fat}}{1 \text{ kg body fat}}$

Set-up 74 k̶g̶ ̶b̶o̶d̶y̶ ̶m̶a̶s̶s̶ × $\dfrac{15 \text{ k̶g̶ ̶b̶o̶d̶y̶ ̶f̶a̶t̶}}{100 \text{ k̶g̶ ̶b̶o̶d̶y̶ ̶m̶a̶s̶s̶}}$ × $\dfrac{2.205 \text{ lb body fat}}{1 \text{ k̶g̶ ̶b̶o̶d̶y̶ ̶f̶a̶t̶}}$

= 24 lb of body fat (2 SFs)

d. **Given** 10.0 oz of fertilizer, 15% nitrogen   **Need** grams of nitrogen

Plan oz of fertilizer → g of fertilizer → g of nitrogen
(percent equality: 100 g of fertilizer = 15 g of nitrogen)

$\dfrac{28.35 \text{ g}}{1 \text{ oz}}$ $\dfrac{15 \text{ g nitrogen}}{100 \text{ g fertilizer}}$

Set-up 10.0 o̶z̶ ̶f̶e̶r̶t̶i̶l̶i̶z̶e̶r̶ × $\dfrac{28.35 \text{ g̶ ̶f̶e̶r̶t̶i̶l̶i̶z̶e̶r̶}}{1 \text{ o̶z̶ ̶f̶e̶r̶t̶i̶l̶i̶z̶e̶r̶}}$ × $\dfrac{15 \text{ g nitrogen}}{100 \text{ g̶ ̶f̶e̶r̶t̶i̶l̶i̶z̶e̶r̶}}$ = 43 g of nitrogen (2 SFs)

2.71 **a. Given** 250 L of water   **Need**   gallons of water

  Plan L → gal $\dfrac{1 \text{ gal}}{3.785 \text{ L}}$

  Set-up $250 \cancel{\text{ L}} \times \dfrac{1 \text{ gal}}{3.785 \cancel{\text{ L}}} = 66 \text{ gal (2 SFs)}$

b. Given 0.024 g of sulfa drug, 8-mg tablets   **Need**   number of tablets

  Plan g of sulfa drug → mg of sulfa drug → number of tablets

  $\dfrac{1000 \text{ mg}}{1 \text{ g}}$ $\dfrac{1 \text{ tablet}}{8 \text{ mg sulfa drug}}$

  Set-up $0.024 \cancel{\text{ g sulfa drug}} \times \dfrac{1000 \cancel{\text{ mg}}}{1 \cancel{\text{ g}}} \times \dfrac{1 \text{ tablet}}{8 \cancel{\text{ mg sulfa drug}}} = 3 \text{ tablets (1 SF)}$

c. Given 34-lb child, 115 mg of ampicillin/kg of body mass   **Need**   milligrams of ampicillin

  Plan lb of body mass → kg of body mass → mg of ampicillin

  $\dfrac{1 \text{ kg}}{2.205 \text{ lb}}$ $\dfrac{115 \text{ mg ampicillin}}{1 \text{ kg body mass}}$

  Set-up $34 \cancel{\text{ lb body mass}} \times \dfrac{1 \cancel{\text{ kg body mass}}}{2.205 \cancel{\text{ lb body mass}}} \times \dfrac{115 \text{ mg ampicillin}}{1 \cancel{\text{ kg body mass}}}$

  $= 1800 \text{ mg of ampicillin (2 SFs)}$

d. Given 4.0 oz of ointment   **Need**   grams of ointment

  Plan oz → g $\dfrac{28.35 \text{ g}}{1 \text{ oz}}$

  Set-up $4.0 \cancel{\text{ oz}} \times \dfrac{28.35 \text{ g}}{1 \cancel{\text{ oz}}} = 110 \text{ g of ointment (2 SFs)}$

2.73 **a. Given** 500. mL of IV saline solution, 80. mL/h   **Need**   infusion time in hours

  Plan mL of IV saline solution → hours $\dfrac{1 \text{ h}}{80. \text{ mL saline solution}}$

  Set-up $500. \cancel{\text{ mL saline solution}} \times \dfrac{1 \text{ h}}{80. \cancel{\text{ mL saline solution}}} = 6.3 \text{ h (2 SFs)}$

b. Given 72.6-lb child, 1.5. mg of Medrol/kg of body mass, 20. mg of Medrol/mL of solution   **Need**   milliliters of Medrol solution

  Plan lb of body mass → kg of body mass → mg of Medrol → mL of solution

  $\dfrac{1 \text{ kg}}{2.205 \text{ lb}}$ $\dfrac{1.5 \text{ mg Medrol}}{1 \text{ kg body mass}}$ $\dfrac{1 \text{ mL solution}}{20. \text{ mg Medrol}}$

  Set-up $72.6 \cancel{\text{ lb body mass}} \times \dfrac{1 \cancel{\text{ kg body mass}}}{2.205 \cancel{\text{ lb body mass}}} \times \dfrac{1.5 \cancel{\text{ mg Medrol}}}{1 \cancel{\text{ kg body mass}}} \times \dfrac{1 \text{ mL solution}}{20. \cancel{\text{ mg Medrol}}}$

  $= 2.5 \text{ mL of Medrol solution (2 SFs)}$

2.75 **a. Given** 175 lb of body weight, $LD_{50} = 3300$ mg/kg   **Need**   grams of table salt

  Plan lb of body weight → kg of body mass → mg of table salt → g of table salt

  $\dfrac{1 \text{ kg body mass}}{2.205 \text{ lb body weight}}$ $\dfrac{3300 \text{ mg table salt}}{1 \text{ kg body mass}}$ $\dfrac{1 \text{ g table salt}}{1000 \text{ mg table salt}}$

  Set-up

  $175 \cancel{\text{ lb body weight}} \times \dfrac{1 \cancel{\text{ kg body mass}}}{2.205 \cancel{\text{ lb body weight}}} \times \dfrac{3300 \cancel{\text{ mg table salt}}}{1 \cancel{\text{ kg body mass}}} \times \dfrac{1 \text{ g table salt}}{1000 \cancel{\text{ mg table salt}}}$

  $= 260 \text{ g of table salt (2 SFs)}$

b. Given 175 lb of body weight, $LD_{50} = 6$ mg/kg   **Need**   grams of sodium cyanide

  Plan lb of body weight → kg of body mass → mg of sodium cyanide → g of sodium cyanide

  $\dfrac{1 \text{ kg body mass}}{2.205 \text{ lb body weight}}$ $\dfrac{6 \text{ mg sodium cyanide}}{1 \text{ kg body mass}}$ $\dfrac{1 \text{ g sodium cyanide}}{1000 \text{ mg sodium cyanide}}$

Set-up

$$175 \text{ lb body weight} \times \frac{1 \text{ kg body mass}}{2.205 \text{ lb body weight}} \times \frac{6 \text{ mg sodium cyanide}}{1 \text{ kg body mass}} \times \frac{1 \text{ g sodium cyanide}}{1000 \text{ mg sodium cyanide}}$$

$$= 0.5 \text{ g of sodium cyanide } (1 \text{ SF})$$

c. Given 175 lb of body weight, $LD_{50} = 1100$ mg/kg   **Need** grams of aspirin

    Plan lb of body weight → kg of body mass → mg of aspirin → g of aspirin

$$\frac{1 \text{ kg body mass}}{2.205 \text{ lb body weight}} \quad \frac{1100 \text{ mg aspirin}}{1 \text{ kg body mass}} \quad \frac{1 \text{ g aspirin}}{1000 \text{ mg aspirin}}$$

Set-up

$$175 \text{ lb body weight} \times \frac{1 \text{ kg body mass}}{2.205 \text{ lb body weight}} \times \frac{1100 \text{ mg aspirin}}{1 \text{ kg body mass}} \times \frac{1 \text{ g aspirin}}{1000 \text{ mg aspirin}}$$

$$= 88 \text{ g of aspirin } (2 \text{ SFs})$$

2.77 Density is the mass of a substance divided by its volume. $\text{Density} = \dfrac{\text{mass (grams)}}{\text{volume (mL)}}$

The densities of solids and liquids are usually stated in g/mL or g/cm³, so in some problems the units will need to be converted.

a. $\text{Density} = \dfrac{\text{mass (grams)}}{\text{volume (mL)}} = \dfrac{24.0 \text{ g}}{20.0 \text{ mL}} = 1.20 \text{ g/mL } (3 \text{ SFs})$

b. Given 0.250 lb of butter, 130.3 mL   **Need** density (g/mL)

    Plan lb → g, then calculate density $\dfrac{453.6 \text{ g}}{1 \text{ lb}}$

    Set-up $\quad 0.250 \text{ lb} \times \dfrac{453.6 \text{ g}}{1 \text{ lb}} = 113.4 \text{ g } (3 \text{ SFs allowed})$

$$\therefore \text{ Density} = \frac{\text{mass}}{\text{volume}} = \frac{113.4 \text{ g}}{130.3 \text{ mL}} = 0.870 \text{ g/mL } (3 \text{ SFs})$$

c. Given 12.00 mL initial volume, 13.45 mL final volume, 4.50 g   **Need** density (g/mL)

    Plan calculate volume by difference, then calculate density

    Set-up volume of gem: 13.45 mL total − 12.00 mL water = 1.45 mL

$$\therefore \text{ Density} = \frac{\text{mass}}{\text{volume}} = \frac{4.50 \text{ g}}{1.45 \text{ mL}} = 3.10 \text{ g/mL } (3 \text{ SFs})$$

d. $\text{Density} = \dfrac{\text{mass}}{\text{volume}} = \dfrac{3.85 \text{ g}}{3.00 \text{ mL}} = 1.28 \text{ g/mL } (3 \text{ SFs})$

2.79 a. Given 514.1 g, 114 cm³   **Need** density (g/mL)

    Plan convert volume cm³ → mL, then calculate density

    Set-up $\quad 114 \text{ cm}^3 \times \dfrac{1 \text{ mL}}{1 \text{ cm}^3} = 114 \text{ mL}$

$$\therefore \text{ Density} = \frac{\text{mass}}{\text{volume}} = \frac{514.1 \text{ g}}{114 \text{ mL}} = 4.51 \text{ g/mL } (3 \text{ SFs})$$

b. Given 0.100 pt, 115.25 g initial, 182.48 g final   **Need** density (g/mL)

    Plan pt → mL, then calculate mass by difference, then calculate density

$$\frac{473.2 \text{ mL}}{1 \text{ pt}}$$

    Set-up $\quad 0.100 \text{ pt} \times \dfrac{473.2 \text{ mL}}{1 \text{ pt}} = 47.3 \text{ mL } (3 \text{ SFs})$

    mass of syrup = 182.48 g − 115.25 g = 67.23 g

$$\therefore \text{ Density} = \frac{\text{mass}}{\text{volume}} = \frac{67.23 \text{ g}}{47.3 \text{ mL}} = 1.42 \text{ g/mL } (3 \text{ SFs})$$

c. **Given**   8.51 kg, 3.15 L   **Need**   density (g/mL)

  Plan   kg → g   and   L → mL, then calculate density $\dfrac{1000\ g}{1\ kg}$ $\dfrac{1000\ mL}{1\ L}$

  Set-up $8.51\ \cancel{kg} \times \dfrac{1000\ g}{1\ \cancel{kg}} = 8510\ g\ (3\ SFs)$ and

  $3.15\ \cancel{L} \times \dfrac{1000\ mL}{1\ \cancel{L}} = 3150\ mL\ (3\ SFs)$

  $\therefore$ Density $= \dfrac{mass}{volume} = \dfrac{8510\ g}{3150\ mL} = 2.70\ g/mL\ (3\ SFs)$

2.81   In these problems, the density is used as a conversion factor.

  a. **Given**   1.50 kg of ethanol   **Need**   liters of ethanol

  Plan   kg → g → mL → L $\dfrac{1000\ g}{1\ kg}$ $\dfrac{1\ mL}{0.79\ g}$ $\dfrac{1\ L}{1000\ mL}$

  Set-up $1.50\ \cancel{kg\ alcohol} \times \dfrac{1000\ \cancel{g}}{1\ \cancel{kg\ alcohol}} \times \dfrac{1\ \cancel{mL}}{0.79\ \cancel{g}} \times \dfrac{1\ L}{1000\ \cancel{mL}}$

  $= 1.9\ L$ of ethanol (2 SFs)

  b. **Given**   6.5 mL of mercury   **Need**   grams of mercury

  Plan   mL → g $\dfrac{13.6\ g}{1\ mL}$

  Set-up $6.5\ \cancel{mL} \times \dfrac{13.6\ g}{1\ \cancel{mL}} = 88\ g$ of mercury (2 SFs)

  c. **Given**   225 cm^3 of silver   **Need**   ounces of silver

  Plan   cm^3 → mL → g → oz $\dfrac{1\ mL}{1\ cm^3}$ $\dfrac{10.5\ g}{1\ cm^3}$ $\dfrac{1\ oz}{28.35\ g}$

  Set-up $225\ \cancel{cm^3} \times \dfrac{1\ \cancel{mL}}{1\ \cancel{cm^3}} \times \dfrac{10.5\ \cancel{g}}{1\ \cancel{mL}} \times \dfrac{1\ oz}{28.35\ \cancel{g}} = 83.3\ oz$ of silver (3 SFs)

2.83   a. **Given**   74.1 cm^3 of copper   **Need**   grams of copper

  Plan   cm^3 → mL → g $\dfrac{1\ mL}{1\ cm^3}$ $\dfrac{8.92\ g}{1\ mL}$

  Set-up $74.1\ \cancel{cm^3} \times \dfrac{1\ \cancel{mL}}{1\ \cancel{cm^3}} \times \dfrac{8.92\ g}{1\ \cancel{mL}} = 661\ g$ of copper (3 SFs)

  b. **Given**   12.0 gal of gasoline   **Need**   kilograms of gasoline

  Plan   gal → qt → mL → g → kg $\dfrac{4\ qt}{1\ gal}$ $\dfrac{946.4\ mL}{1\ qt}$ $\dfrac{0.74\ g}{1\ mL}$ $\dfrac{1\ kg}{1000\ g}$

  Set-up $12.0\ \cancel{gal} \times \dfrac{4\ \cancel{qt}}{1\ \cancel{gal}} \times \dfrac{946.4\ \cancel{mL}}{1\ \cancel{qt}} \times \dfrac{0.74\ \cancel{g}}{1\ \cancel{mL}} \times \dfrac{1\ kg}{1000\ \cancel{g}} = 34\ kg$ of gasoline (2 SFs)

  c. **Given**   27 g of ice   **Need**   cubic centimeters of ice

  Plan   g → mL → cm^3 $\dfrac{1\ mL}{0.92\ g}$ $\dfrac{1\ cm^3}{1\ mL}$

  Set-up $27\ \cancel{g} \times \dfrac{1\ \cancel{mL}}{0.92\ \cancel{g}} \times \dfrac{1\ cm^3}{1\ \cancel{mL}} = 29\ cm^3$ (2 SFs)

2.85   Because the density of aluminum is 2.70 g/cm^3, silver is 10.5 g/cm^3, and lead is 11.3 g/cm^3, we can identify the unknown metal by calculating its density as follows:

  Density $= \dfrac{mass\ of\ metal}{volume\ of\ metal} = \dfrac{217\ g}{19.2\ cm^3} = 11.3\ g/cm^3$ (3 SFs)

  $\therefore$ the metal is lead.

2.87 a. Specific gravity $= \dfrac{\text{density of substance}}{\text{density of water}} = \dfrac{1.030 \text{ g/mL}}{1.00 \text{ g/mL}} = 1.03$ (3 SFs)

  b. Density $= \dfrac{\text{mass of glucose solution}}{\text{volume of glucose solution}} = \dfrac{20.6 \text{ g}}{20.0 \text{ mL}} = 1.03$ g/mL (3 SFs)

  c. Specific gravity $= \dfrac{\text{density of substance}}{\text{density of water}}$

  ∴ Density of substance = specific gravity × density of water
  = 0.92 × 1.00 g/mL = 0.92 g/mL
  Mass of substance = volume of substance × density of substance
  = 750 mL × 0.92 g/mL = 690 g (2 SFs)

  d. Specific gravity $= \dfrac{\text{density of substance}}{\text{density of water}}$

  ∴ Density of substance = specific gravity × density of water
  = 0.850 × 1.00 g/mL = 0.850 g/mL

  ∴ Volume of solution $= \dfrac{\text{mass of solution}}{\text{density of solution}} = \dfrac{325 \text{ g}}{0.850 \text{ g/mL}} = 382$ mL (3 SFs)

2.89 a. 42 mcg of iron = 1 dL of blood; $\dfrac{42 \text{ mcg iron}}{1 \text{ dL blood}}$ and $\dfrac{1 \text{ dL blood}}{42 \text{ mcg iron}}$

  b. Given   8.0-mL blood sample, 42 mcg of iron/dL   **Need**   micrograms of iron

  Plan   mL of blood → dL of blood → mcg of iron $\dfrac{1 \text{ dL blood}}{100 \text{ mL blood}}$ $\dfrac{42 \text{ mcg iron}}{1 \text{ dL blood}}$

  Set-up   8.0 mL blood $\times \dfrac{1 \text{ dL blood}}{100 \text{ mL blood}} \times \dfrac{42 \text{ mcg iron}}{1 \text{ dL blood}} = 3.4$ mcg of iron (2 SFs)

2.91 Both measurements in part **c** have two significant figures, and both measurements in part **d** have four significant figures.

2.93 a. The number of legs is a counted number; it is exact.
  b. The height is measured with a ruler or tape measure; it is a measured number.
  c. The number of chairs is a counted number; it is exact.
  d. The area is measured with a ruler or tape measure; it is a measured number.

2.95 61.5 °C

2.97 a. length $= 38.4$ in. $\times \dfrac{2.54 \text{ cm}}{1 \text{ in.}} = 97.5$ cm (3 SFs)

  b. length $= 24.2$ in. $\times \dfrac{2.54 \text{ cm}}{1 \text{ in.}} = 61.5$ cm (3 SFs)

  c. There are three significant figures in the length measurement.
  d. Area = length × width = 97.5 cm × 61.5 cm = 6.00×10^3 cm² (3 SFs)

2.99 a. Diagram 3; a cube that has a greater density than the water will sink to the bottom.
  b. Diagram 4; a cube with a density of 0.80 g/mL will be about two-thirds submerged in the water.
  c. Diagram 1; a cube with a density that is one-half the density of water will be one-half submerged in the water.
  d. Diagram 2; a cube with the same density as water will float just at the surface of the water.

2.101 Since all three solids have a mass of 10.0 g, the one with the smallest volume must have the highest density; the one with the largest volume will have the lowest density.

  A would be gold; it has the highest density (19.3 g/mL) and the smallest volume.
  B would be silver; its density is intermediate (10.5 g/mL) and the volume is intermediate.
  C would be aluminum; it has the lowest density (2.70 g/mL) and the largest volume.

2.103 The green cube has the same volume as the gray cube. However, the green cube has a larger mass on the scale, which means that its mass/volume ratio is larger. Thus, the density of the green cube is higher than the density of the gray cube.

2.105 a. To round off 0.000 012 58 L to three significant figures, drop the final digit 8 and increase the last retained digit by 1 to give 0.000 012 6 L or 1.26×10^{-5} L.
   b. To round off 3.528×10^2 kg to three significant figures, drop the final digit 8 and increase the last retained digit by 1 to give 353 kg (3.53×10^2 kg).
   c. To express 125 111 m to three significant figures, drop the final digits 111 and add three zeros as placeholders to give 125 000 m (or 1.25×10^5 m).
   d. To express 34.9673 s to three significant figures, drop the final digits 673 and increase the last retained digit by 1 to give 35.0 s.

2.107 a. The total mass is the sum of the individual components of the dessert.
   137.25 g + 84 g + 43.7 g = 265 g. No places to the right of the decimal point are allowed since the mass of the fudge sauce (84 g) has no digits to the right of the decimal point.
   b. Given   grams of dessert from part **a**   **Need**   pounds of dessert

   Plan   g → lb $\dfrac{1 \text{ lb}}{453.6 \text{ g}}$

   Set-up   265 g dessert (total) $\times \dfrac{1 \text{ lb}}{453.6 \text{ g}} = 0.584$ lb of dessert (3 SFs)

2.109 Given   1.95 euros/kg of grapes, \$1.14/euro   **Need**   cost in dollars per pound

   Plan   euros/kg → euros/lb → \$/lb $\dfrac{1.95 \text{ euros}}{1 \text{ kg grapes}} \quad \dfrac{1 \text{ kg}}{2.205 \text{ lb}} \quad \dfrac{\$1.14}{1 \text{ euro}}$

   Set-up $\dfrac{1.95 \text{ euros}}{1 \text{ kg grapes}} \times \dfrac{1 \text{ kg}}{2.205 \text{ lb}} \times \dfrac{\$1.14}{1 \text{ euro}} = \$1.01/\text{lb}$ of grapes (3 SFs)

2.111 Given   4.0 lb of onions   **Need**   number of onions

   Plan   lb → g → number of onions $\dfrac{453.6 \text{ g}}{1 \text{ lb}} \quad \dfrac{1 \text{ onion}}{115 \text{ g}}$

   Set-up   4.0 lb onions $\times \dfrac{453.6 \text{ g}}{1 \text{ lb}} \times \dfrac{1 \text{ onion}}{115 \text{ g}} = 16$ onions (2 SFs)

2.113 This problem requires several conversion factors. Let's take a look first at a possible unit plan. When you write out the unit plan, be sure you know a conversion factor you can use for each step.
   Given   7500 ft   **Need**   minutes

   Plan   ft → in. → cm → m → min $\dfrac{12 \text{ in.}}{1 \text{ ft}} \quad \dfrac{2.54 \text{ cm}}{1 \text{ in.}} \quad \dfrac{1 \text{ m}}{100 \text{ cm}} \quad \dfrac{1 \text{ min}}{55.0 \text{ m}}$

   Set-up   7500 ft $\times \dfrac{12 \text{ in.}}{1 \text{ ft}} \times \dfrac{2.54 \text{ cm}}{1 \text{ in.}} \times \dfrac{1 \text{ m}}{100 \text{ cm}} \times \dfrac{1 \text{ min}}{55.0 \text{ m}} = 42$ min (2 SFs)

2.115 Given   215 mL initial, 285 mL final volume, density of lead 11.3 g/mL   **Need**   grams of lead

   Plan   calculate the volume by difference and mL → g $\dfrac{11.3 \text{ g}}{1 \text{ mL}}$

   Set-up   The difference between the initial volume of the water and its volume with the lead object will give us the volume of the lead object: 285 mL total − 215 mL water = 70. mL of lead, then

   70. mL lead $\times \dfrac{11.3 \text{ g lead}}{1 \text{ mL lead}} = 790$ g of lead (2 SFs)

2.117 **Given** 1.2 kg of gasoline   **Need** cubic centimeters of gasoline

Plan kg → g → mL → cm³ $\dfrac{1000\text{ g}}{1\text{ mL}}$ $\dfrac{1\text{ mL}}{0.74\text{ g}}$ $\dfrac{1\text{ cm}^3}{1\text{ mL}}$

Set-up $1.2\ \cancel{\text{kg}} \times \dfrac{1000\ \cancel{\text{g}}}{1\ \cancel{\text{kg}}} \times \dfrac{1\ \cancel{\text{mL}}}{0.74\ \cancel{\text{g}}} \times \dfrac{1\text{ cm}^3}{1\ \cancel{\text{mL}}} = 1600\text{ cm}^3\ (1.6 \times 10^3\text{ cm}^3)$ of gasoline (2 SFs)

2.119 a. **Given** 8.0 oz   **Need** number of crackers

Plan oz → number of crackers $\dfrac{6\text{ crackers}}{0.50\text{ oz}}$

Set-up $8.0\ \cancel{\text{oz}} \times \dfrac{6\text{ crackers}}{0.50\ \cancel{\text{oz}}} = 96$ crackers (2 SFs)

b. **Given** 10 crackers, 4 g of fat/serving   **Need** ounces of fat
Plan number of crackers → servings → g of fat → oz of fat

$\dfrac{1\text{ serving}}{6\text{ crackers}}$ $\dfrac{4\text{ g fat}}{1\text{ serving}}$ $\dfrac{1\text{ oz}}{28.35\text{ g}}$

Set-up $10\ \cancel{\text{crackers}} \times \dfrac{1\ \cancel{\text{serving}}}{6\ \cancel{\text{crackers}}} \times \dfrac{4\ \cancel{\text{g}}\text{ fat}}{1\ \cancel{\text{serving}}} \times \dfrac{1\text{ oz}}{28.35\ \cancel{\text{g}}} = 0.2$ oz of fat (1 SF)

c. **Given** 2400 mg of sodium, 140 mg of sodium/serving   **Need** number of servings

Plan mg of sodium → servings $\dfrac{1\text{ serving}}{140\text{ mg sodium}}$

Set-up $2400\ \cancel{\text{mg sodium}} \times \dfrac{1\text{ serving}}{140\ \cancel{\text{mg sodium}}} = 17$ servings (2 SFs)

2.121 **Given** 10 days, 4 tablets/day, 250-mg tablets   **Need** ounces of amoxicillin
Plan days → tablets → mg of amoxicillin → g → oz of amoxicillin

$\dfrac{4\text{ tablets}}{1\text{ day}}$ $\dfrac{250\text{ mg amoxicillin}}{1\text{ tablet}}$ $\dfrac{1\text{ g}}{1000\text{ mg}}$ $\dfrac{1\text{ oz}}{28.35\text{ g}}$

Set-up $10\ \cancel{\text{days}} \times \dfrac{4\ \cancel{\text{tablets}}}{1\ \cancel{\text{day}}} \times \dfrac{250\ \cancel{\text{mg}}\ \cancel{\text{amoxicillin}}}{1\ \cancel{\text{tablet}}} \times \dfrac{1\ \cancel{\text{g}}}{1000\ \cancel{\text{mg}}} \times \dfrac{1\text{ oz}}{28.35\ \cancel{\text{g}}}$

$= 0.35$ oz of amoxicillin (2 SFs)

2.123 **Given** 5.0 mL of elixir, 30. mg of phenobarbital/7.5 mL
Need milligrams of phenobarbital

Plan mL of elixir → mg of phenobarbital $\dfrac{30.\text{ mg phenobarbital}}{7.5\text{ mL elixir}}$

Set-up $5.0\ \cancel{\text{mL elixir}} \times \dfrac{30.\text{ mg phenobarbital}}{7.5\ \cancel{\text{mL elixir}}} = 20.$ mg of phenobarbital (2 SFs)

2.125 Because the balance can measure mass to 0.001 g, the mass should be reported to 0.001 g. You should record the mass of the object as 31.075 g.

2.127 **Given** 3.0-h trip   **Need** gallons of gasoline

Plan h → mi → km → L → gal $\dfrac{55\text{ mi}}{1\text{ h}}$ $\dfrac{1\text{ km}}{0.6214\text{ mi}}$ $\dfrac{1\text{ L}}{11\text{ km}}$ $\dfrac{1\text{ gal}}{3.785\text{ L}}$

Set-up $3.0\ \cancel{\text{h}} \times \dfrac{55\ \cancel{\text{mi}}}{1\ \cancel{\text{h}}} \times \dfrac{1\ \cancel{\text{km}}}{0.6214\ \cancel{\text{mi}}} \times \dfrac{1\ \cancel{\text{L}}}{11\ \cancel{\text{km}}} \times \dfrac{1\text{ gal}}{3.785\ \cancel{\text{L}}}$

$= 6.4$ gal of gasoline (2 SFs)

2.129 Given   1.50 L of gasoline   **Need**   cm³ of olive oil

Plan   L of gasoline → mL of gasoline → g of gasoline → g of olive oil → mL of olive oil → cm³ of olive oil

(equality from question: 1 g of olive oil = 1 g of gasoline)

$$\frac{1000 \text{ mL gasoline}}{1 \text{ L gasoline}} \quad \frac{0.74 \text{ g gasoline}}{1 \text{ mL gasoline}} \quad \frac{1 \text{ mL olive oil}}{0.92 \text{ g olive oil}} \quad \frac{1 \text{ cm}^3 \text{ olive oil}}{1 \text{ mL olive oil}}$$

Set-up $1.50 \text{ L gasoline} \times \dfrac{1000 \text{ mL}}{1 \text{ L}} \times \dfrac{0.74 \text{ g gasoline}}{1 \text{ mL gasoline}} = 1110 \text{ g of gasoline}$

$1110 \text{ g olive oil} \times \dfrac{1 \text{ mL olive oil}}{0.92 \text{ g olive oil}} \times \dfrac{1 \text{ cm}^3}{1 \text{ mL}} = 1200 \text{ cm}^3 \, (1.2 \times 10^3 \text{ cm}^3) \text{ of olive oil (2 SFs)}$

2.131 Given   1.50 g of silicon, 3.00 in. diameter, density 2.33 g/cm³   **Need**   thickness (mm)

Plan $g \to cm^3$ $\dfrac{1 \text{ cm}^3}{2.33 \text{ g silicon}}$ and $d(\text{in.}) \to r(\text{in.}) \to r(\text{cm})$ $r = \dfrac{d}{2}$ $\dfrac{2.54 \text{ cm}}{1 \text{ in.}}$

then rearrange the volume equation for thickness (cm) → thickness (mm)

$$V = \pi r^2 h \to h = \frac{V}{\pi r^2} \qquad \frac{10 \text{ mm}}{1 \text{ cm}}$$

Set-up   Volume of wafer: $1.50 \text{ g} \times \dfrac{1 \text{ cm}^3}{2.33 \text{ g}} = 0.644 \text{ cm}^3$

Radius of wafer: $\dfrac{3.00 \text{ in.}}{2} \times \dfrac{2.54 \text{ cm}}{1 \text{ in.}} = 3.81 \text{ cm}$

$\therefore h = \dfrac{V}{\pi r^2} = \dfrac{0.644 \text{ cm}^3}{\pi (3.81 \text{ cm})^2} = 0.0141 \text{ cm} \times \dfrac{10 \text{ mm}}{1 \text{ cm}} = 0.141 \text{ mm (3 SFs)}$

2.133 a. Given   180 lb of body weight   **Need**   cups of coffee

Plan

lb of body weight → kg of body mass → mg of caffeine → fl oz of coffee → cups of coffee

$$\frac{1 \text{ kg body mass}}{2.205 \text{ lb body weight}} \quad \frac{192 \text{ mg caffeine}}{1 \text{ kg body mass}} \quad \frac{6 \text{ fl oz coffee}}{100. \text{ mg caffeine}} \quad \frac{1 \text{ cup coffee}}{12 \text{ fl oz coffee}}$$

Set-up $180 \text{ lb body weight} \times \dfrac{1 \text{ kg body mass}}{2.205 \text{ lb body weight}} \times \dfrac{192 \text{ mg caffeine}}{1 \text{ kg body mass}}$

$= 1.567 \times 10^4 \text{ mg of caffeine (2 SFs allowed)}$

$1.567 \times 10^4 \text{ mg caffeine} \times \dfrac{6 \text{ fl oz coffee}}{100. \text{ mg caffeine}} \times \dfrac{1 \text{ cup coffee}}{12 \text{ fl oz coffee}}$

$= 78 \text{ cups of coffee (2 SFs)}$

b. Given   milligrams of caffeine from part **a**   **Need**   cans of cola

Plan   mg of caffeine → cans of cola $\dfrac{1 \text{ can cola}}{50. \text{ mg caffeine}}$

Set-up $1.567 \times 10^4 \text{ mg caffeine} \times \dfrac{1 \text{ can cola}}{50. \text{ mg caffeine}} = 310 \text{ cans of cola (2 SFs)}$

c. Given   milligrams of caffeine from part **a**   **Need**   number of No-Doz tablets

Plan   mg of caffeine → No-Doz tablets $\dfrac{1 \text{ No-Doz tablet}}{200. \text{ mg caffeine}}$

Set-up $1.567 \times 10^4 \text{ mg caffeine} \times \dfrac{1 \text{ No-Doz tablet}}{200. \text{ mg caffeine}} = 78 \text{ No-Doz tablets (2 SFs)}$

2.135 a. Given 65 kg of body mass, 3.0% fat   **Need** pounds of fat

    Plan kg of body mass → kg of fat → lb of fat

        (percent equality: 100 kg of body mass = 3.0 kg of fat)

$$\frac{3.0 \text{ kg fat}}{100 \text{ kg body mass}} \qquad \frac{2.205 \text{ lb fat}}{1 \text{ kg fat}}$$

    Set-up $65 \text{ kg body mass} \times \dfrac{3.0 \text{ kg fat}}{100 \text{ kg body mass}} \times \dfrac{2.205 \text{ lb fat}}{1 \text{ kg fat}} = 4.3$ lb of fat (2 SFs)

b. Given 3.0 L of fat   **Need** pounds of fat

    Plan L → mL → g → lb $\qquad \dfrac{1000 \text{ mL}}{1 \text{ L}} \quad \dfrac{0.94 \text{ g}}{1 \text{ mL}} \quad \dfrac{1 \text{ lb}}{453.6 \text{ g}}$

    Set-up $3.0 \text{ L} \times \dfrac{1000 \text{ mL}}{1 \text{ L}} \times \dfrac{0.94 \text{ g}}{1 \text{ mL}} \times \dfrac{1 \text{ lb}}{453.6 \text{ g}} = 6.2$ lb of fat (2 SFs)

Matter and Energy

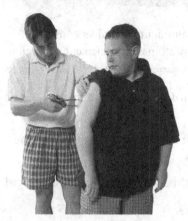

Since Charles had a visit with Daniel, a dietitian, Charles and his mother have been preparing healthier foods with smaller serving sizes. Charles has lost 6 lb. Yesterday he had a blood glucose test that showed his blood glucose level was normal. One of the meals that Charles likes consists of potatoes, carrots, and chicken. If Charles consumes a meal which contains 34 g of carbohydrate, 3 g of fat, and 25 g of protein, what is the energy content of the meal, in kilocalories and kilojoules?

Credit: Network Photographer / Alamy

CHAPTER READINESS*

⚛ Key Math Skills

- Using Positive and Negative Numbers in Calculations (1.4)
- Solving Equations (1.4)
- Interpreting Graphs (1.4)
- Converting between Standard Numbers and Scientific Notation (1.5)
- Rounding Off (2.3)

⚗ Core Chemistry Skills

- Counting Significant Figures (2.2)
- Using Significant Figures in Calculations (2.3)
- Writing Conversion Factors from Equalities (2.5)
- Using Conversion Factors (2.6)

*These Key Math Skills and Core Chemistry Skills from previous chapters are listed here for your review as you proceed to the new material in this chapter.

LOOKING AHEAD

3.1 Classification of Matter

Learning Goal: Classify examples of matter as pure substances or mixtures.

- Matter is anything that has mass and occupies space.
- A pure substance, whether it is an element or compound, has a definite composition.
- Elements are the simplest type of matter.
- A compound consists of atoms of two or more elements always chemically combined in the same proportion.
- Mixtures contain two or more substances that are physically, not chemically, combined.
- Mixtures are classified as homogeneous or heterogeneous.

Key Terms for Section 3.1

Match each of the following key terms with the descriptions:

    a. element   **b.** mixture   **c.** pure substance
    d. matter   **e.** compound

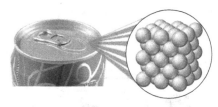

An aluminum container consists of many atoms of the element aluminum.

Credit: Norman Chan / Fotolia

1. _____ anything that has mass and occupies space

2. _____ an element or a compound that has a definite composition

3. _____ the physical combination of two or more substances that does not change the identities of the substances

4. _____ a pure substance consisting of two or more elements with a definite composition that can be broken down into simpler substances only by chemical methods

5. _____ a pure substance containing only one type of matter that cannot be broken down by chemical methods

Answers   **1.** d   **2.** c   **3.** b   **4.** e   **5.** a

---

♦ **Learning Exercise 3.1A**

Identify each of the following as an element (E) or a compound (C):

☞ **CORE CHEMISTRY SKILL**

Classifying Matter

    a. _____ iron   **b.** _____ carbon dioxide

    c. _____ potassium iodide   **d.** _____ gold

    e. _____ aluminum   **f.** _____ table salt (sodium chloride)

Answers   **a.** E   **b.** C   **c.** C   **d.** E   **e.** E   **f.** C

♦ **Learning Exercise 3.1B**

Identify each of the following as a pure substance (P) or mixture (M):

    a. _____ bananas and milk   **b.** _____ sulfur

    c. _____ silver   **d.** _____ a bag of raisins and nuts

    e. _____ pure water   **f.** _____ sand and water

Water consists of many molecules of H_2O, each containing two H atoms and one O atom.

Credit: Bomshtein / Shutterstock

Answers   **a.** M   **b.** P   **c.** P   **d.** M   **e.** P   **f.** M

♦ **Learning Exercise 3.1C**

Identify each of the following mixtures as homogeneous (Ho) or heterogeneous (He):

    a. _____ chocolate milk   **b.** _____ sand and water

    c. _____ orange soda   **d.** _____ a bag of raisins and nuts

    e. _____ air   **f.** _____ vinegar

Answers   **a.** Ho   **b.** He   **c.** Ho   **d.** He   **e.** Ho   **f.** Ho

3.2 States and Properties of Matter

Learning Goal: Identify the states and the physical and chemical properties of matter.

- The states of matter are solid, liquid, and gas.
- Physical properties are those characteristics of a substance that can change without affecting the identity of the substance.
- A substance undergoes a physical change when its shape, size, or state changes, but its composition does not change.
- Chemical properties are those characteristics that describe the ability of a substance to change into a new substance.
- Melting, freezing, boiling, and condensing are typical changes of state.

♦ Learning Exercise 3.2A

State whether the following statements describe a gas (G), a liquid (L), or a solid (S):

a. _____ There are no attractions among the particles.

b. _____ The particles are held close together in a definite pattern.

c. _____ This substance has a definite volume but no definite shape.

d. _____ The particles are moving extremely fast.

e. _____ This substance has no definite shape and no definite volume.

f. _____ The particles in this substance are vibrating slowly in fixed positions.

g. _____ This substance has a definite volume and a definite shape.

Answers   **a.** G   **b.** S   **c.** L   **d.** G   **e.** G   **f.** S   **g.** S

♦ Learning Exercise 3.2B

Classify each of the following as a physical (P) or chemical (C) property:

a. _____ Silver is shiny.

b. _____ Water is a liquid at 25 °C.

c. _____ Wood burns.

d. _____ Mercury is a very dense liquid.

e. _____ Helium is not reactive.

f. _____ Ice cubes float in water.

Answers   **a.** P   **b.** P   **c.** C   **d.** P   **e.** C   **f.** P

♦ Learning Exercise 3.2C

⬧ CORE CHEMISTRY SKILL

Identifying Physical and Chemical Changes

Classify each of the following changes as physical (P) or chemical (C):

a. _____ Sodium melts at 98 °C.

b. _____ Iron forms rust in air and water.

c. _____ Water condenses on a cold window.

d. _____ Fireworks explode when ignited.

e. _____ Gasoline burns in a car engine.

f. _____ Paper is cut to make confetti.

Answers   **a.** P   **b.** C   **c.** P   **d.** C   **e.** C   **f.** P

3.3 Temperature

Learning Goal: Given a temperature, calculate a corresponding temperature on another scale.

- Temperature is measured in degrees Celsius (°C) or kelvins (K). In the United States, the Fahrenheit scale (°F) is still in use.
- The equation $T_F = 1.8(T_C) + 32$ is used to convert a Celsius temperature to a Fahrenheit temperature.
- When rearranged for T_C, this equation is used to convert a Fahrenheit temperature to a Celsius temperature.

$$T_C = \frac{T_F - 32}{1.8}$$

- The temperature on the Celsius scale is related to the temperature on the Kelvin scale: $T_K = T_C + 273$.

Guide to Calculating Temperature	
STEP 1	State the given and needed quantities.
STEP 2	Write a temperature equation.
STEP 3	Substitute in the known values and calculate the new temperature.

◆ **Learning Exercise 3.3**

Calculate the temperature in each of the following problems:

a. To prepare yogurt, milk is warmed to 185 °F. What Celsius temperature is needed to prepare the yogurt?

b. A heat-sensitive vaccine is stored at −12 °C. What is that temperature on a Fahrenheit thermometer?

c. A patient has a temperature of 39.5 °C. What is that temperature on the Fahrenheit scale?

d. A patient in the emergency room with hypothermia has a temperature of 94 °F. What is the temperature on the Celsius scale?

e. The temperature in an autoclave used to sterilize surgical equipment is 253 °F. What is that temperature on the Celsius scale?

f. Liquid nitrogen at −196 °C is used to freeze and remove precancerous skin growths. What temperature will this be on the Kelvin scale?

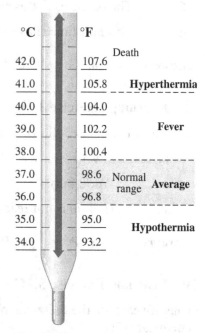

CORE CHEMISTRY SKILL

Converting between Temperature Scales

Very high temperatures (hyperthermia) can lead to convulsions and brain damage.

Credit: Digital Vision / Alamy

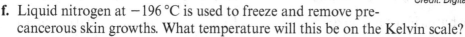

Answers **a.** 85.0 °C **b.** 10. °F **c.** 103.1 °F **d.** 34 °C **e.** 123 °C **f.** 77 K

3.4   Energy

Learning Goal: Identify energy as potential or kinetic; convert between units of energy.

- Energy is the ability to do work.
- Potential energy is stored energy, which is determined by position or chemical composition; kinetic energy is the energy of motion.
- The SI unit of energy is the joule (J), the metric unit is the calorie (cal). One calorie (cal) is equal to exactly 4.184 joules (J).

$$\frac{4.184 \text{ J}}{1 \text{ cal}} \text{ and } \frac{1 \text{ cal}}{4.184 \text{ J}}$$

Key Terms for Sections 3.2 to 3.4

Match each of the following key terms with the correct description:

  a. potential energy   **b.** joule   **c.** kinetic energy
  d. calorie   **e.** energy   **f.** physical change

1. _____ the SI unit of energy

2. _____ the energy of motion

3. _____ the amount of energy required to raise the temperature of 1 g of water by 1 °C

4. _____ a change in physical properties of a substance with no change in its composition

5. _____ the ability to do work

6. _____ a type of energy that is stored for future use

Answers   **1.** b   **2.** c   **3.** d   **4.** f   **5.** e   **6.** a

♦ Learning Exercise 3.4A

State whether each of the following statements describes potential (P) or kinetic (K) energy:

a. _____ a potted plant sitting on a ledge   **b.** _____ your breakfast cereal

c. _____ logs sitting in a fireplace   **d.** _____ a piece of candy

e. _____ an arrow shot from a bow   **f.** _____ a ski jumper standing at the top of the ski jump

g. _____ a jogger running   **h.** _____ a skydiver waiting to jump

i. _____ water flowing down a stream   **j.** _____ a bowling ball striking the pins

Answers   **a.** P   **b.** P   **c.** P   **d.** P   **e.** K
                 f. P   **g.** K   **h.** P   **i.** K   **j.** K

♦ Learning Exercise 3.4B

Match the words in column A with the descriptions in column B.

A	**B**
1. _____ calorie	**a.** the SI unit of work and energy
2. _____ kilocalorie	**b.** the heat needed to raise 1 g of water by 1 °C
3. _____ joule	**c.** 1000 cal

Answers   **1.** b   **2.** c   **3.** a

♦ **Learning Exercise 3.4C**

⚬ **CORE CHEMISTRY SKILL**

Using Energy Units

Convert each of the following energy units:

a. 58 000 cal to kcal

b. 3450 J to cal

c. 2.8 kJ to cal

d. 15 200 cal to kJ

Answers   **a.** 58 kcal   **b.** 825 cal   **c.** 670 cal   **d.** 63.6 kJ

3.5   Specific Heat

Learning Goal: Calculate the specific heat for a substance. Use specific heat to calculate heat loss or gain.

- Specific heat is the amount of energy required to change the temperature of 1 g of a substance by 1 °C.
- The specific heat for liquid water is 1.00 cal/g °C or 4.184 J/g °C.
- The heat, q, lost or gained can be calculated using the mass of the substance, the temperature difference, and its specific heat (SH): $q = \text{mass} \times \Delta T \times SH$.

Guide to Calculating Specific Heat	
STEP 1	State the given and needed quantities.
STEP 2	Write the relationship for specific heat.
STEP 3	Set up the problem to calculate the specific heat.

♦ **Learning Exercise 3.5A**

⚬ **CORE CHEMISTRY SKILL**

Calculating Specific Heat

Calculate the specific heat, in J/g °C, for each of the following:

a. a 15.2-g sample of a metal that absorbs 231 J when its temperature rises from 84.5 °C to 125.3 °C.

b. a 31.8-g sample of a metal that absorbs 816 J when its temperature rises from 23.7 °C to 56.2 °C.

c. a 38.2-g sample of a metal that absorbs 125 J when its temperature rises from 62.1 °C to 68.4 °C.

Answers   **a.** 0.372 J/g °C   **b.** 0.790 J/g °C   **c.** 0.52 J/g °C

♦ **Learning Exercise 3.5B**

a. Copper has a specific heat of 0.385 J/g °C. When 1250 J are added to a copper sample, its temperature rises from 24.6 °C to 61.3 °C. What is the mass, in grams, of the copper sample?

b. A sample of aluminum has a specific heat of 0.897 J/g °C. When 785 J are added to an aluminum sample, its temperature rises from 14 °C to 106 °C. What is the mass, in grams, of the aluminum sample?

c. Silver has a specific heat of 0.235 J/g °C. What is the temperature change, in degrees Celsius, for a 15.2-g piece of silver when 426 J is added?

Answers **a.** 88.5 g   **b.** 9.5 g   **c.** 119 °C

Guide to Using Specific Heat	
STEP 1	State the given and needed quantities.
STEP 2	Calculate the temperature change (ΔT).
STEP 3	Write the heat equation and needed conversion factors.
STEP 4	Substitute in the given values and calculate the heat, making sure units cancel.

◆ **Learning Exercise 3.5C**

The specific heat for water is 4.184 J/g °C. Calculate the kilojoules (kJ) gained or released during the following:

CORE CHEMISTRY SKILL

Using the Heat Equation

a. heating 20.0 g of water from 22 °C to 77 °C

b. heating 10.0 g of water from 12.4 °C to 67.5 °C

c. cooling 0.450 kg of water from 80.0 °C to 35.0 °C

d. cooling 125 g of water from 72.0 °C to 45.0 °C

Answers **a.** 4.6 kJ (gained)   **b.** 2.31 kJ (gained)
      c. 84.7 kJ (released)   **d.** 14.1 kJ (released)

3.6 Energy and Nutrition

Learning Goal: Use the energy values to calculate the kilocalories (kcal) or kilojoules (kJ) for a food.

- The nutritional unit Calorie (Cal) is the same amount of energy as 1 kcal, or 1000 calories.
- When a substance is burned in a calorimeter, the water that surrounds the reaction chamber absorbs the heat given off. The heat absorbed by the water is calculated, and the energy value for the substance (energy per gram) is determined.

- The energy values for three food types are: carbohydrate 4 kcal/g (17 kJ/g), fat 9 kcal/g (38 kJ/g), protein 4 kcal/g (17 kJ/g).
- The energy content of a food is the sum of kilocalories or kilojoules from carbohydrate, fat, and protein.

Key Terms for Sections 3.5 and 3.6

Match each of the following key terms with the correct description:

    a. energy value   **b.** specific heat   **c.** Calorie (Cal)   **d.** kilojoule

1. _____ a nutritional unit of energy equal to 1000 cal or 1 kcal

2. _____ a quantity of heat that changes the temperature of 1 g of a substance by 1 °C

3. _____ the number of kilocalories or kilojoules obtained per gram of carbohydrate, fat, or protein

4. _____ a unit of energy equal to 1000 J

Answers   **1.** c   **2.** b   **3.** a   **4.** d

Guide to Calculating the Energy from a Food	
STEP 1	State the given and needed quantities.
STEP 2	Use the energy value for each food type and calculate the kcal or kJ rounded off to the tens place.
STEP 3	Add the energy for each food type to give the total energy from the food.

♦ Learning Exercise 3.6A

Calculate the kilojoules (kJ) for each food using the following data (round off the kilojoules for each food type to the tens place):

Food	Carbohydrate (g)	Fat (g)	Protein (g)	kJ
a. green peas, cooked, 1 cup	19	1	9	_____
b. potato chips, 10 chips	10	8	1	_____
c. cream cheese, 8 oz	5	86	18	_____
d. lean hamburger, 3 oz	0	10	23	_____
e. banana, 1	26	0	1	_____

Answers   **a.** 510 kJ   **b.** 490 kJ   **c.** 3670 kJ
                    d. 770 kJ   **e.** 460 kJ

♦ **Learning Exercise 3.6B**

Use energy values to solve each of the following:

a. Complete the following table listing ingredients for a peanut butter sandwich (round off the kilocalories for each food type to the tens place):

	Carbohydrate (g)	Fat (g)	Protein (g)	kcal
bread, 2 slices	30	0	5	_____
peanut butter, 2 tbsp	10	13	8	_____
jelly, 2 tsp	10	0	0	_____
margarine, 1 tsp	0	7	0	_____

Total kcal in sandwich _____

b. How many kilocalories are in a single serving of pudding that contains 31 g of carbohydrate, 5 g of fat, and 5 g of protein (round off the kilocalories for each food type to the tens place)?

c. A ham and cheese sandwich provides 810 kcal. If the sandwich contains 42 g of carbohydrate and 46 g of protein, how many grams of fat are in the sandwich (round off the kilocalories for each food type to the tens place)?

d. A serving of breakfast cereal provides 220 kcal. In this serving, there are 6 g of fat and 8 g of protein. How many grams of carbohydrate are in the cereal (round off the kilocalories for each food type to the tens place)?

Answers

a. bread = 140 kcal; peanut butter = 190 kcal; jelly = 40 kcal; margarine = 60 kcal
   total kcal in sandwich = 430 kcal
b. carbohydrate = 120 kcal; fat = 50 kcal; protein = 20 kcal
   total = 190 kcal
c. carbohydrate = 170 kcal; protein = 180 kcal
   810 kcal − 170 kcal − 180 kcal = 460 kcal from fat

$$460 \; \text{kcal} \times \frac{1 \text{ g fat}}{9 \text{ kcal}} = 51 \text{ g of fat}$$

d. fat = 50 kcal; protein = 30 kcal
   50 kcal + 30 kcal = 80 kcal from fat and protein
   220 kcal − 50 kcal − 30 kcal = 140 kcal from carbohydrate

$$140 \; \text{kcal} \times \frac{1 \text{ g carbohydrate}}{4 \text{ kcal}} = 35 \text{ g of carbohydrate}$$

Checklist for Chapter 3

You are ready to take the Practice Test for Chapter 3. Be sure you have accomplished the following learning goals for this chapter. If not, review the section listed at the end of the goal. Then apply your new skills and understanding to the Practice Test.

After studying Chapter 3, I can successfully:

_____ Identify matter as a pure substance or mixture. (3.1)

_____ Identify a mixture as homogeneous or heterogeneous. (3.1)

_____ Identify the physical state of a substance as a solid, liquid, or gas. (3.2)

_____ Identify properties of a substance as physical or chemical. (3.2)

_____ Calculate a temperature in degrees Fahrenheit, degrees Celsius, or kelvins. (3.3)

_____ Describe some forms of energy. (3.4)

_____ Change a quantity in one energy unit to another energy unit. (3.4)

_____ Given the mass of a sample, specific heat, and the temperature change, calculate the heat lost or gained. (3.5)

_____ Using the energy values, calculate the energy, in kilocalories (kcal) or kilojoules (kJ), for a food sample. (3.6)

Practice Test for Chapter 3

The chapter sections to review are shown in parentheses at the end of each question.

For questions 1 through 4, classify each of the following as a pure substance (P) or a mixture (M): (3.1)

1. toothpaste _____

2. platinum _____

3. chromium _____

4. mouthwash _____

For questions 5 through 8, classify each of the following mixtures as homogeneous (Ho) or heterogeneous (He): (3.1)

5. noodle soup _____

6. salt water _____

7. chocolate chip cookie _____

8. mouthwash _____

9. Which of the following is a chemical property? (3.2)
   A. dynamite explodes
   B. a shiny metal
   C. a melting point of 110 °C
   D. rain on a cool day
   E. breaking up cement

For questions 10 through 12, answer with solid (S), liquid (L), or gas (G): (3.2)

10. _____ has a definite volume but takes the shape of a container

11. _____ does not have a definite shape or definite volume

12. _____ has a definite shape and a definite volume

13. Which of the following is a chemical property of silver? (3.2)
    A. density of 10.5 g/mL
    B. shiny
    C. melts at 961 °C
    D. good conductor of heat
    E. reacts to form tarnish

14. Which of the following is a physical property of silicon? (3.2)
    A. burns in chlorine
    B. has a black to gray color
    C. reacts with nitric acid
    D. reacts with oxygen to form sand
    E. used to make silicone

For questions 15 through 19, answer as physical change (P) or chemical change (C): (3.2)

15. _____ Butter melts in a hot pan.

16. _____ Iron forms rust with oxygen.

17. _____ Baking powder forms bubbles (CO_2) as a cake is baking.

18. _____ Water boils.

19. _____ Propane burns in a camp stove.

20. 105 °F = _____ °C (3.3)
    A. 73 °C   B. 41 °C   C. 58 °C   D. 90 °C   E. 221 °C

21. The melting point of gold is 1064 °C. The Fahrenheit temperature needed to melt gold would be (3.3)
    A. 129 °F   B. 623 °F   C. 1031 °F   D. 1913 °F   E. 1947 °F

22. The average daytime temperature on the planet Mercury is 683 K. What is this temperature on the Celsius scale? (3.3)
    A. 956 °C   B. 715 °C   C. 680. °C   D. 446. °C   E. 410. °C

23. Which of the following would be described as potential energy? (3.4)
    A. a car going around a racetrack
    B. a rabbit hopping
    C. oil in an oil well
    D. a moving merry-go-round
    E. a bouncing ball

24. Which of the following would be described as kinetic energy? (3.4)
    A. a car battery
    B. a can of tennis balls
    C. gasoline in a car fuel tank
    D. a box of matches
    E. a tennis ball crossing over the net

25. The number of joules needed to raise the temperature of 5.0 g of water from 25 °C to 55 °C is (3.5)
    A. 5.0 J   B. 36 J   C. 30.0 J   D. 335 J   E. 630 J

26. The number of kilojoules released when 15 g of water cools from 58 °C to 22 °C is (3.5)
    A. 0.13 kJ   B. 0.54 kJ   C. 2.3 kJ   D. 63 kJ   E. 150 kJ

For questions 27 through 29, consider a cup of milk that is 3.5% butterfat with an energy value of 160 kcal. In the cup of milk, there are 12 g of carbohydrate, 9.0 g of fat, and protein (round off the kilocalories for each food type to the tens place): (3.6)

27. The number of kilocalories provided by the carbohydrate is
    A. 10 kcal   B. 20 kcal   C. 40 kcal   D. 50 kcal   E. 80 kcal

28. The number of kilocalories provided by the fat is
    A. 9 kcal   B. 20 kcal   C. 50 kcal   D. 70 kcal   E. 80 kcal

29. Using your answers to questions 27 and 28, the number of grams of protein is
    A. 1 g   B. 5 g   C. 8 g   D. 10 g   E. 20 g

30. A person on a diet has a lunch of 3 oz of salmon (no carbohydrate, 8 g of fat, 16 g of protein), 1 cup of nonfat milk (12 g of carbohydrate, no fat, 9 g of protein), and 1 cup of raw carrots (11 g of carbohydrate, no fat, 2 g of protein). What is the total kilocalories for the lunch (round off the kilocalories for each food type to the tens place)? (3.6)
    A. 270 kcal   B. 200 kcal   C. 110 kcal   D. 90 kcal   E. 50 kcal

31. How many hours would the person in problem 30 have to swim to expend the calories from her lunch, if swimming uses 2100 kJ/h? (3.6)

    A. 5.7 h   **B.** 5.0 h   **C.** 2.8 h   **D.** 1.6 h   **E.** 0.54 h

Answers to the Practice Test

1. M	**2.** P	**3.** P	**4.** M	**5.** He
6. Ho	**7.** He	**8.** Ho	**9.** A	**10.** L
11. G	**12.** S	**13.** E	**14.** B	**15.** P
16. C	**17.** C	**18.** P	**19.** C	**20.** B
21. E	**22.** E	**23.** C	**24.** E	**25.** E
26. C	**27.** D	**28.** E	**29.** C	**30.** A
31. E				

Selected Answers and Solutions to Text Problems

3.1 A *pure substance* is matter that has a fixed or definite composition: either elements or compounds. In a *mixture*, two or more substances are physically mixed but not chemically combined.

   a. Baking soda is composed of one type of matter ($NaHCO_3$), which makes it a pure substance.

   b. A blueberry muffin is composed of several substances mixed together, which makes it a mixture.

   c. Ice is composed of one type of matter (H_2O molecules), which makes it a pure substance.

   d. Zinc is composed of one type of matter (Zn atoms), which makes it a pure substance.

   e. Trimix is a physical mixture of oxygen, nitrogen, and helium gases, which makes it a mixture.

3.3 *Elements* are the simplest type of pure substance, containing only one type of atom. *Compounds* contain two or more elements chemically combined in a specific proportion.

   a. A silicon (Si) chip is an element since it contains only one type of atom (Si).

   b. Hydrogen peroxide (H_2O_2) is a compound since it contains two elements (H, O) chemically combined.

   c. Oxygen (O_2) is an element since it contains only one type of atom (O).

   d. Rust (Fe_2O_3) is a compound since it contains two elements (Fe, O) chemically combined.

   e. Methane (CH_4) in natural gas is a compound since it contains two elements (C, H) chemically combined.

3.5 A *homogeneous mixture* has a uniform composition; a *heterogeneous mixture* does not have a uniform composition throughout the mixture.

   a. Vegetable soup is a heterogeneous mixture since it has chunks of vegetables.

   b. Seawater is a homogeneous mixture since it has a uniform composition.

   c. Tea is a homogeneous mixture since it has a uniform composition.

   d. Tea with ice and lemon slices is a heterogeneous mixture since it has chunks of ice and lemon.

   e. Fruit salad is a heterogeneous mixture since it has chunks of fruit.

3.7   **a.** A gas has no definite volume or shape.

   b. In a gas, the particles do not interact with each other.

   c. In a solid, the particles are held in a rigid structure.

3.9 A *physical property* is a characteristic of the substance such as color, shape, odor, luster, size, melting point, and density. A *chemical property* is a characteristic that indicates the ability of a substance to change into a new substance.

   a. Color and physical state are physical properties.

   b. The ability to react with oxygen is a chemical property.

   c. The temperature of a substance is a physical property.

   d. Milk souring describes chemical reactions and is thus a chemical property.

   e. Burning butane gas in oxygen forms new substances, which makes it a chemical property.

3.11 When matter undergoes a *physical change*, its state or appearance changes, but its composition remains the same. When a *chemical change* occurs, the original substance is converted into a new substance, which has different physical and chemical properties.

   a. Water vapor condensing is a physical change since the physical form of the water changes, but the composition of the substance does not.

   b. Cesium metal reacting is a chemical change since new substances form.

   c. Gold melting is a physical change since the physical state changes, but not the composition of the substance.

   d. Cutting a puzzle results in a physical change since the size and shape change, but not the composition of the substance.

   e. Grating cheese results in a physical change since the size and shape change, but not the composition of the substance.

3.13   a. The high reactivity of fluorine is a chemical property since it allows for the formation of new substances.

   b. The physical state of fluorine is a physical property.

   c. The color of fluorine is a physical property.

   d. The reactivity of fluorine with hydrogen is a chemical property since it allows for the formation of new substances.

   e. The melting point of fluorine is a physical property.

3.15   The Fahrenheit temperature scale is still used in the United States. A normal body temperature is 98.6 °F on this scale. To convert her 99.8 °F temperature to the equivalent reading on the Celsius scale, the following calculation must be performed:

$$T_C = \frac{(99.8 - 32)}{1.8} = \frac{67.8}{1.8} = 37.7\,°C \text{ (3 SFs) (1.8 and 32 are exact numbers)}$$

Because a normal body temperature is 37.0 on the Celsius scale, her temperature of 37.7 °C would indicate a mild fever.

3.17   To convert Celsius to Fahrenheit: $T_F = 1.8(T_C) + 32$

To convert Fahrenheit to Celsius: $T_C = \dfrac{(T_F - 32)}{1.8}$ (1.8 and 32 are exact numbers)

To convert Celsius to Kelvin: $T_K = T_C + 273$

To convert Kelvin to Celsius: $T_C = T_K - 273$

   a. $T_F = 1.8(T_C) + 32 = 1.8(37.0) + 32 = 66.6 + 32 = 98.6\,°F$

   b. $T_C = \dfrac{(T_F - 32)}{1.8} = \dfrac{(65.3 - 32)}{1.8} = \dfrac{33.3}{1.8} = 18.5\,°C$

   c. $T_K = T_C + 273 = -27 + 273 = 246\,K$

   d. $T_K = T_C + 273 = 62 + 273 = 335\,K$

   e. $T_C = \dfrac{(T_F - 32)}{1.8} = \dfrac{(114 - 32)}{1.8} = \dfrac{82}{1.8} = 46\,°C$

3.19   a. $T_C = \dfrac{(T_F - 32)}{1.8} = \dfrac{(106 - 32)}{1.8} = \dfrac{74}{1.8} = 41\,°C$

   b. $T_C = \dfrac{(T_F - 32)}{1.8} = \dfrac{(103 - 32)}{1.8} = \dfrac{71}{1.8} = 39\,°C$

   No, there is no need to phone the doctor. The child's temperature is less than 40.0 °C.

3.21   When the roller-coaster car is at the top of the ramp, it has its maximum potential energy. As it descends, potential energy is converted into kinetic energy. At the bottom, all of its energy is kinetic.

3.23   a. Potential energy is stored in the water at the top of the waterfall.

   b. Kinetic energy is displayed as the kicked ball moves.

   c. Potential energy is stored in the chemical bonds in the coal.

   d. Potential energy is stored when the skier is at the top of the hill.

3.25 a. Given 3500 cal   **Need** kilocalories

     Plan cal → kcal $\dfrac{1 \text{ kcal}}{1000 \text{ cal}}$

     Set-up 3500 ~~cal~~ $\times \dfrac{1 \text{ kcal}}{1000 \text{ ~~cal~~}} = 3.5 \text{ kcal} \ (2 \text{ SFs})$

b. Given 415 J   **Need** calories

     Plan J → cal $\dfrac{1 \text{ cal}}{4.184 \text{ J}}$

     Set-up 415 ~~J~~ $\times \dfrac{1 \text{ cal}}{4.184 \text{ ~~J~~}} = 99.2 \text{ cal} \ (3 \text{ SFs})$

c. Given 28 cal   **Need** joules

     Plan cal → J $\dfrac{4.184 \text{ J}}{1 \text{ cal}}$

     Set-up 28 ~~cal~~ $\times \dfrac{4.184 \text{ J}}{1 \text{ ~~cal~~}} = 120 \text{ J} \ (2 \text{ SFs})$

d. Given 4.5 kJ   **Need** calories

     Plan kJ → J → cal $\dfrac{1000 \text{ J}}{1 \text{ kJ}}$ $\dfrac{1 \text{ cal}}{4.184 \text{ J}}$

     Set-up 4.5 ~~kJ~~ $\times \dfrac{1000 \text{ ~~J~~}}{1 \text{ ~~kJ~~}} \times \dfrac{1 \text{ cal}}{4.184 \text{ ~~J~~}} = 1100 \text{ cal} \ (2 \text{ SFs})$

3.27 a. Given 3.0 h, 270 kJ/h   **Need** joules

     Plan h → kJ → J $\dfrac{1000 \text{ J}}{1 \text{ kJ}}$

     Set-up 3.0 ~~h~~ $\times \dfrac{270 \text{ ~~kJ~~}}{1.0 \text{ ~~h~~}} \times \dfrac{1000 \text{ J}}{1 \text{ ~~kJ~~}} = 8.1 \times 10^5 \text{ J} \ (2 \text{ SFs})$

b. Given 3.0 h, 270 kJ/h   **Need** kilocalories

     Plan h → kJ → J → cal → kcal $\dfrac{1000 \text{ J}}{1 \text{ kJ}}$ $\dfrac{1 \text{ cal}}{4.184 \text{ J}}$ $\dfrac{1 \text{ kcal}}{1000 \text{ cal}}$

     Set-up 3.0 ~~h~~ $\times \dfrac{270 \text{ ~~kJ~~}}{1.0 \text{ ~~h~~}} \times \dfrac{1000 \text{ ~~J~~}}{1 \text{ ~~kJ~~}} \times \dfrac{1 \text{ ~~cal~~}}{4.184 \text{ ~~J~~}} \times \dfrac{1 \text{ kcal}}{1000 \text{ ~~cal~~}} = 190 \text{ kcal} \ (2 \text{ SFs})$

3.29 Copper, which has the lowest specific heat of the samples, would reach the highest temperature.

3.31 a. Given heat = 312 J; mass = 13.5 g; $\Delta T = 83.6\,°\text{C} - 24.2\,°\text{C} = 59.4\,°\text{C}$

     Need specific heat (J/g °C)

     Plan $SH = \dfrac{\text{heat}}{\text{mass} \times \Delta T}$

     Set-up specific heat $(SH) = \dfrac{312 \text{ J}}{13.5 \text{ g} \times 59.4\,°\text{C}} = \dfrac{0.389 \text{ J}}{\text{g}\,°\text{C}} \ (3 \text{ SFs})$

b. Given heat = 345 J; mass = 48.2 g; $\Delta T = 57.9\,°\text{C} - 35.0\,°\text{C} = 22.9\,°\text{C}$

     Need specific heat (J/g °C)

     Plan $SH = \dfrac{\text{heat}}{\text{mass} \times \Delta T}$

     Set-up specific heat $(SH) = \dfrac{345 \text{ J}}{48.2 \text{ g} \times 22.9\,°\text{C}} = \dfrac{0.313 \text{ J}}{\text{g}\,°\text{C}} \ (3 \text{ SFs})$

3.33 a. Given $SH_{water} = 4.184 \text{ J/g °C}$; mass $= 25.0$ g; $\Delta T = 25.7\,°C - 12.5\,°C = 13.2\,°C$
Need heat (q) in joules
Plan heat $(q) = \text{mass} \times \Delta T \times SH$

Set-up heat $= 25.0\,\cancel{g} \times 13.2\,\cancel{°C} \times \dfrac{4.184 \text{ J}}{\cancel{g}\,\cancel{°C}} = 1380 \text{ J (3 SFs)}$

∴ 1380 J of energy are required.

b. Given $SH_{copper} = 0.385 \text{ J/g °C}$; mass $= 38.0$ g; $\Delta T = 246\,°C - 122\,°C = 124\,°C$
Need heat (q) in joules
Plan heat $(q) = \text{mass} \times \Delta T \times SH$

Set-up heat $= 38.0\,\cancel{g} \times 124\,\cancel{°C} \times \dfrac{0.385 \text{ J}}{\cancel{g}\,\cancel{°C}} = 1810 \text{ J (3 SFs)}$

∴ 1810 J of energy are required.

3.35 a. Given $SH_{gold} = 0.129 \text{ J/g °C}$; heat $= 225$ J; $\Delta T = 47.0\,°C - 15.0\,°C = 32.0\,°C$
Need mass in grams

Plan heat $(q) = \text{mass} \times \Delta T \times SH$ ∴ mass $= \dfrac{\text{heat}}{\Delta T \times SH}$

Set-up mass $= \dfrac{\text{heat}}{\Delta T \times SH} = \dfrac{225\,\cancel{J}}{32.0\,\cancel{°C} \times \dfrac{0.129\,\cancel{J}}{g\,\cancel{°C}}} = 54.5 \text{ g of gold (3 SFs)}$

b. Given $SH_{iron} = 0.452 \text{ J/g °C}$; heat $= -8.40$ kJ; $\Delta T = 82.0\,°C - 168.0\,°C = -86.0\,°C$
Need mass in grams

Plan heat $(q) = \text{mass} \times \Delta T \times SH$ ∴ mass $= \dfrac{\text{heat}}{\Delta T \times SH} \quad \dfrac{1000 \text{ J}}{1 \text{ kJ}}$

Set-up mass $= \dfrac{\text{heat}}{\Delta T \times SH} = \dfrac{(-8.40\,\cancel{kJ}) \times \dfrac{1000\,\cancel{J}}{1\,\cancel{kJ}}}{(-86.0\,\cancel{°C}) \times \dfrac{0.452\,\cancel{J}}{g\,\cancel{°C}}} = 216 \text{ g of iron (3 SFs)}$

3.37 a. Given $SH_{iron} = 0.452 \text{ J/g °C}$; heat $= 1580$ J; mass $= 20.0$ g
Need ΔT

Plan heat $(q) = \text{mass} \times \Delta T \times SH$ ∴ $\Delta T = \dfrac{\text{heat}}{\text{mass} \times SH}$

Set-up $\Delta T = \dfrac{\text{heat}}{\text{mass} \times SH} = \dfrac{1580\,\cancel{J}}{20.0\,\cancel{g} \times \dfrac{0.452\,\cancel{J}}{\cancel{g}\,°C}} = 175\,°C \text{ (3 SFs)}$

b. Given $SH_{water} = 4.184 \text{ J/g °C}$; heat $= 7.10$ kJ; mass $= 150.0$ g
Need ΔT

Plan heat $(q) = \text{mass} \times \Delta T \times SH$ ∴ $\Delta T = \dfrac{\text{heat}}{\text{mass} \times SH} \quad \dfrac{1000 \text{ J}}{1 \text{ kJ}}$

Set-up $\Delta T = \dfrac{\text{heat}}{\text{mass} \times SH} = \dfrac{7.10\,\cancel{kJ} \times \dfrac{1000\,\cancel{J}}{1\,\cancel{kJ}}}{150.0\,\cancel{g} \times \dfrac{4.184\,\cancel{J}}{\cancel{g}\,°C}} = 11.3\,°C \text{ (3 SFs)}$

3.39 a. Given 125 kJ **Need** kilocalories

Plan $\text{kJ} \rightarrow \text{J} \rightarrow \text{cal} \rightarrow \text{kcal} \quad \dfrac{1000 \text{ J}}{1 \text{ kJ}} \quad \dfrac{1 \text{ cal}}{4.184 \text{ J}} \quad \dfrac{1 \text{ kcal}}{1000 \text{ cal}}$

Set-up $125\,\cancel{kJ} \times \dfrac{1000\,\cancel{J}}{1\,\cancel{kJ}} \times \dfrac{1\,\cancel{cal}}{4.184\,\cancel{J}} \times \dfrac{1 \text{ kcal}}{1000\,\cancel{cal}} = 29.9 \text{ kcal (3 SFs)}$

b. Given 870. kJ **Need** kilocalories

Plan kJ → J → cal → kcal $\quad \dfrac{1000\ J}{1\ kJ} \quad \dfrac{1\ cal}{4.184\ J} \quad \dfrac{1\ kcal}{1000\ cal}$

Set-up $870.\ \cancel{kJ} \times \dfrac{1000\ \cancel{J}}{1\ \cancel{kJ}} \times \dfrac{1\ \cancel{cal}}{4.184\ \cancel{J}} \times \dfrac{1\ kcal}{1000\ \cancel{cal}} = 208\ kcal\ (3\ SFs)$

3.41 a. Given one cup of orange juice that contains 26 g of carbohydrate, 2 g of protein, and no fat
Need total energy in kilojoules

Food Type	Mass		Energy Value		Energy (rounded off to the tens place)
Carbohydrate	26 g	×	$\dfrac{17\ kJ}{1\ g}$	=	440 kJ
Protein	2 g	×	$\dfrac{17\ kJ}{1\ g}$	=	30 kJ
			Total energy content	=	470 kJ

b. Given one apple that provides 72 kcal of energy and contains no fat or protein
Need grams of carbohydrate

Plan kcal → g of carbohydrate $\quad \dfrac{1\ g\ carbohydrate}{4\ kcal}$

Set-up $72\ \cancel{kcal} \times \dfrac{1\ g\ carbohydrate}{4\ \cancel{kcal}} = 18\ g\ of\ carbohydrate\ (2\ SFs)$

c. Given one tablespoon of vegetable oil that contains 14 g of fat, and no carbohydrate or protein
Need total energy in kilocalories

Food Type	Mass		Energy Value		Energy (rounded off to the tens place)
Fat	14 g	×	$\dfrac{9\ kcal}{1\ g}$	=	130 kcal

d. Given one avocado that provides 405 kcal in total, and contains 13 g of carbohydrate and 5 g of protein
Need grams of fat

Food Type	Mass		Energy Value		Energy (rounded off to the tens place)
Carbohydrate	13 g	×	$\dfrac{4\ kcal}{1\ g}$	=	50 kcal
Fat	? g	×	$\dfrac{9\ kcal}{1\ g}$	=	? kcal
Protein	5 g	×	$\dfrac{4\ kcal}{1\ g}$	=	20 kcal
			Total energy content	=	405 kcal

∴ Energy from fat = 405 kcal − (50 kcal + 20 kcal) = 335 kcal

∴ $335\ \cancel{kcal} \times \dfrac{1\ g\ fat}{9\ \cancel{kcal}} = 37\ g\ of\ fat$

3.43 **Given** one cup of clam chowder that contains 16 g of carbohydrate, 12 g of fat, and 9 g of protein
Need total energy in kilocalories and kilojoules

Food Type	Mass		Energy Value		Energy (rounded off to the tens place)
Carbohydrate	16 g	×	$\dfrac{4 \text{ kcal (or } 17 \text{ kJ)}}{1 \text{ g}}$	=	60 kcal (or 270 kJ)
Fat	12 g	×	$\dfrac{9 \text{ kcal (or } 38 \text{ kJ)}}{1 \text{ g}}$	=	110 kcal (or 460 kJ)
Protein	9 g	×	$\dfrac{4 \text{ kcal (or } 17 \text{ kJ)}}{1 \text{ g}}$	=	40 kcal (or 150 kJ)
			Total energy content	=	210 kcal (or 880 kJ)

3.45 $3.2 \text{ L glucose solution} \times \dfrac{1000 \text{ mL solution}}{1 \text{ L solution}} \times \dfrac{5.0 \text{ g glucose}}{100. \text{ mL solution}} \times \dfrac{4 \text{ kcal}}{1 \text{ g glucose}}$

$= 640 \text{ kcal (rounded off to the tens place)}$

3.47 **a.** Using the energy totals from Table 3.9:

Breakfast	Energy
Banana, 1 medium	110 kcal
Milk, nonfat, 1 cup	90 kcal
Egg, 1 large	70 kcal
Total =	270 kcal

Lunch	Energy
Carrots, 1 cup	50 kcal
Beef, ground, 3 oz	220 kcal
Apple, 1 medium	60 kcal
Milk, nonfat, 1 cup	90 kcal
Total =	420 kcal

Dinner	Energy
Chicken, no skin, 6 oz	2 × 110 kcal
Potato, baked	100 kcal
Broccoli, 3 oz	30 kcal
Milk, nonfat, 1 cup	90 kcal
Total =	440 kcal

b. total kilocalories for one day = 270 kcal + 420 kcal + 440 kcal = 1130 kcal
c. Since Charles will maintain his weight if he consumes 1800 kcal per day, he will lose weight on this new diet (1130 kcal/day).
d. For Charles, his energy balance is 1130 kcal/day − 1800 kcal/day = −670 kcal/day.

$5.0 \text{ lb} \times \dfrac{3500 \text{ kcal}}{1.0 \text{ lb}} \times \dfrac{1 \text{ day}}{670 \text{ kcal}} = 26 \text{ days (2 SFs)}$

3.49 **a.** The diagram shows two different kinds of atoms chemically combined in a definite 2:1 ratio; it represents a compound.
   b. The diagram shows two different kinds of matter (atoms and molecules) physically mixed, not chemically combined; it represents a mixture.
   c. The diagram contains only one kind of atom; it represents an element.

3.51 A *homogeneous mixture* has a uniform composition; a *heterogeneous mixture* does not have a uniform composition throughout the mixture.
   a. Lemon-flavored water is a homogeneous mixture since it has a uniform composition (as long as there are no lemon pieces).
   b. Stuffed mushrooms are a heterogeneous mixture since there are mushrooms and chunks of filling.
   c. Eye drops are a homogeneous mixture since they have a uniform composition.

3.53 61.4 °C
$$T_F = 1.8(T_C) + 32 = 1.8(61.4) + 32 = 111 + 32 = 143\ °F\ (3\ SFs)$$

3.55 $T_C = \dfrac{(T_F - 32)}{1.8} = \dfrac{(155 - 32)}{1.8} = \dfrac{123}{1.8} = 68.3\ °C\ (3\ SFs)$
$$\text{(1.8 and 32 are exact numbers)}$$
$$T_K = T_C + 273 = 68.3 + 273 = 341\ K\ (3\ SFs)$$

3.57 **Given**   10.0 cm³ cubes of gold (SH_{gold} = 0.129 J/g °C), and
   aluminum ($SH_{aluminum}$ =0.897 J/g °C); ΔT = 25 °C − 15 °C = 10. °C
Need   energy in joules and calories

Plan   cm³ → g   Heat = mass × ΔT × SH $\dfrac{19.3\ \text{g gold}}{1\ cm^3}$ $\dfrac{2.70\ \text{g aluminum}}{1\ cm^3}$

Set-up   <u>for gold:</u> $10.0\ cm^3 \times \dfrac{19.3\ g}{1\ cm^3} \times 10.\ °C \times \dfrac{0.129\ J}{g\ °C} = 250\ J\ (2\ SFs)$

$10.0\ cm^3 \times \dfrac{19.3\ g}{1\ cm^3} \times 10.\ °C \times \dfrac{0.129\ J}{g\ °C} \times \dfrac{1\ cal}{4.184\ J} = 60.\ cal\ (2\ SFs)$

<u>for aluminum:</u> $10.0\ cm^3 \times \dfrac{2.70\ g}{1\ cm^3} \times 10.\ °C \times \dfrac{0.897\ J}{g\ °C} = 240\ J\ (2\ SFs)$

$10.0\ cm^3 \times \dfrac{2.70\ g}{1\ cm^3} \times 10.\ °C \times \dfrac{0.897\ J}{g\ °C} \times \dfrac{1\ cal}{4.184\ J} = 58\ cal\ (2\ SFs)$

Thus, the energy needed to heat each of the metals is almost the same.

3.59 **a.** **Given**   a meal consisting of: cheeseburger: 46 g of carbohydrate, 40. g of fat, 47 g of protein
   french fries: 47 g of carbohydrate, 16 g of fat, 4 g of protein
   chocolate shake: 76 g of carbohydrate, 10. g of fat, 10. g of protein
Need   energy from each food type in kilocalories
Plan   total grams of each food type, then g → kcal
Set-up   total carbohydrate = 46 g + 47 g + 76 g = 169 g
   total fat = 40. g + 16 g + 10. g = 66 g
   total protein = 47 g + 4 g + 10. g = 61 g

Food Type	Mass		Energy Value		Energy (rounded off to the tens place)
Carbohydrate	169 g	×	$\dfrac{4\ kcal}{1\ g}$	=	680 kcal
Fat	66 g	×	$\dfrac{9\ kcal}{1\ g}$	=	590 kcal
Protein	61 g	×	$\dfrac{4\ kcal}{1\ g}$	=	240 kcal

b. total energy for the meal $= 680 \text{ kcal} + 590 \text{ kcal} + 240 \text{ kcal} = 1510 \text{ kcal}$

c. **Given** 1510 kcal from meal   **Need** hours of sleeping to "burn off"

    Plan $\text{kcal} \rightarrow \text{h}$ $\dfrac{1 \text{ h sleeping}}{60 \text{ kcal}}$

    Set-up $1510 \cancel{\text{ kcal}} \times \dfrac{1 \text{ h sleeping}}{60 \cancel{\text{ kcal}}} = 25 \text{ h of sleeping (2 SFs)}$

d. **Given** 1510 kcal from meal   **Need** hours of running to "burn off"

    Plan $\text{kcal} \rightarrow \text{h}$ $\dfrac{1 \text{ h running}}{750 \text{ kcal}}$

    Set-up $1510 \cancel{\text{ kcal}} \times \dfrac{1 \text{ h running}}{750 \cancel{\text{ kcal}}} = 2.0 \text{ h of running (2 SFs)}$

3.61 *Elements* are the simplest type of pure substance, containing only one type of atom. *Compounds* contain two or more elements chemically combined in a definite proportion. In a *mixture*, two or more substances are physically mixed but not chemically combined.

    a. Carbon in pencils is an element since it contains only one type of atom (C).

    b. Carbon monoxide is a compound since it contains two elements (C, O) chemically combined.

    c. Orange juice is composed of several substances physically mixed together (e.g. water, sugar, citric acid), which makes it a mixture.

3.63 A *homogeneous mixture* has a uniform composition; a *heterogeneous mixture* does not have a uniform composition throughout the mixture.

    a. A hot fudge sundae is a heterogeneous mixture since it has ice cream, fudge sauce, and perhaps a cherry.

    b. Herbal tea is a homogeneous mixture since it has a uniform composition.

    c. Vegetable oil is a homogeneous mixture since it has a uniform composition.

3.65 **a.** A vitamin tablet is a solid.

    b. Helium in a balloon is a gas.

    c. Milk is a liquid.

    d. Air is a mixture of gases.

    e. Charcoal is a solid.

3.67 A *physical property* is a characteristic of the substance such as color, shape, odor, luster, size, melting point, and density. A *chemical property* is a characteristic that indicates the ability of a substance to change into a new substance.

    a. The luster of gold is a physical property.

    b. The melting point of gold is a physical property.

    c. The ability of gold to conduct electricity is a physical property.

    d. The ability of gold to form a new substance with sulfur is a chemical property.

3.69 When matter undergoes a *physical change*, its state or appearance changes, but its composition remains the same. When a *chemical change* occurs, the original substance is converted into a new substance, which has different physical and chemical properties.

    a. Plant growth produces new substances, so it is a chemical change.

    b. A change of state from solid to liquid is a physical change.

    c. Chopping wood into smaller pieces is a physical change.

    d. Burning wood, which forms new substances, results in a chemical change.

3.71 **a.** $T_C = \dfrac{(T_F - 32)}{1.8} = \dfrac{(134 - 32)}{1.8} = \dfrac{102}{1.8} = 56.7\,°C$

    $T_K = T_C + 273 = 56.7 + 273 = 330. \text{ K}$

b. $T_C = \dfrac{(T_F - 32)}{1.8} = \dfrac{(-69.7 - 32)}{1.8} = \dfrac{-101.7}{1.8} = -56.50\,°C$

$T_K = T_C + 273 = -56.50 + 273 = 217\,K$

3.73 $T_C = \dfrac{(T_F - 32)}{1.8} = \dfrac{(-15 - 32)}{1.8} = \dfrac{-47}{1.8} = -26\,°C$

$T_K = T_C + 273 = -26 + 273 = 247\,K$

3.75 Sand must have a lower specific heat than water since the same amount of heat causes a greater temperature change in the sand than in the water.

3.77 **Given**   mass = 0.50 g of oil; 18.9 kJ produced
Need   energy value (kcal/g)

Plan   kJ → kcal $\dfrac{1000\,J}{1\,kJ}$ $\dfrac{1\,cal}{4.184\,J}$ $\dfrac{1\,kcal}{1000\,cal}$

then energy value $= \dfrac{\text{heat in kcal}}{\text{mass in g}}$

Set-up $18.9\,\cancel{kJ} \times \dfrac{1000\,\cancel{J}}{1\,\cancel{kJ}} \times \dfrac{1\,\cancel{cal}}{4.184\,\cancel{J}} \times \dfrac{1\,kcal}{1000\,\cancel{cal}} = 4.52\,kcal\ (3\ \text{SFs})$

$\therefore$ energy value $= \dfrac{\text{heat produced}}{\text{mass of oil}} = \dfrac{4.52\,kcal}{0.50\,g\ oil} = 9.0\,kcal/g$ of vegetable oil (2 SFs)

3.79 **Given**   heat = 575 J; mass = 45.6 g; $\Delta T = 35\,°C - 21\,°C = 14\,°C$
Need   specific heat (J/g °C)

Plan $SH = \dfrac{\text{heat}}{\text{mass} \times \Delta T}$

Set-up   specific heat $(SH) = \dfrac{575\,J}{45.6\,g \times 14\,°C} = \dfrac{0.90\,J}{g\,°C}$ (2 SFs)

By comparing this value to those listed in Table 3.7, we identify the substance as aluminum.

3.81 **a.** **Given** $SH_{ethanol} = 2.46\,J/g\,°C$; mass = 15.0 g; $\Delta T = -42.0\,°C - 60.5\,°C = -102.5\,°C$
  Need   heat (q) in joules
  Plan   heat $(q) = \text{mass} \times \Delta T \times SH$

  Set-up   heat $= 15.0\,\cancel{g} \times (-102.5\,\cancel{°C}) \times \dfrac{2.46\,J}{\cancel{g}\,\cancel{°C}} = -3780\,J$ (3 SFs)

    $\therefore$ 3780 J of energy are lost.

b. **Given** $SH_{iron} = 0.452\,J/g\,°C$; mass = 125 g; $\Delta T = 55\,°C - 118\,°C = -63\,°C$
  Need   heat (q) in joules
  Plan   heat $(q) = \text{mass} \times \Delta T \times SH$

  Set-up   heat $= 125\,\cancel{g} \times (-63\,\cancel{°C}) \times \dfrac{0.452\,J}{\cancel{g}\,\cancel{°C}} = -3600\,J$ (2 SFs)

    $\therefore$ 3600 J of energy are lost.

3.83 **a.** **Given** $SH_{aluminum} = 0.897\,J/g\,°C$; heat = 8.80 kJ; $\Delta T = 26.8\,°C - 12.5\,°C = 14.3\,°C$
  Need   mass in grams

  Plan   heat $(q) = \text{mass} \times \Delta T \times SH$ $\therefore$ mass $= \dfrac{\text{heat}}{\Delta T \times SH}$ $\dfrac{1000\,J}{1\,kJ}$

  Set-up   mass $= \dfrac{\text{heat}}{\Delta T \times SH} = \dfrac{8.80\,\cancel{kJ} \times \dfrac{1000\,\cancel{J}}{1\,\cancel{kJ}}}{14.3\,\cancel{°C} \times \dfrac{0.897\,\cancel{J}}{g\,\cancel{°C}}} = 686\,g$ of aluminum (3 SFs)

b. Given $SH_{titanium} = 0.523 \text{ J/g °C}$; heat $= -14\,200 \text{ J}$; $\Delta T = 42\,°C - 185\,°C = -143\,°C$

Need mass in grams

Plan heat $(q) = \text{mass} \times \Delta T \times SH$ ∴ mass $= \dfrac{\text{heat}}{\Delta T \times SH}$

Set-up mass $= \dfrac{\text{heat}}{\Delta T \times SH} = \dfrac{(-14\,200\text{ J})}{(-143\,°C) \times \dfrac{0.523\text{ J}}{\text{g °C}}} = 190.\text{ g of titanium (3 SFs)}$

3.85 a. Given $SH_{gold} = 0.129 \text{ J/g °C}$; heat $= 7680 \text{ J}$; mass $= 85.0 \text{ g}$ **Need** ΔT

Plan heat $(q) = \text{mass} \times \Delta T \times SH$ ∴ $\Delta T = \dfrac{\text{heat}}{\text{mass} \times SH}$

Set-up $\Delta T = \dfrac{\text{heat}}{\text{mass} \times SH} = \dfrac{7680\text{ J}}{85.0\text{ g} \times \dfrac{0.129\text{ J}}{\text{g °C}}} = 700.\,°C \text{ (3 SFs)}$

b. Given $SH_{copper} = 0.385 \text{ J/g °C}$; heat $= 6.75 \text{ kJ}$; mass $= 50.0 \text{ g}$ **Need** ΔT $\dfrac{1000\text{ J}}{1\text{ kJ}}$

Plan heat $(q) = \text{mass} \times \Delta T \times SH$ ∴ $\Delta T = \dfrac{\text{heat}}{\text{mass} \times SH}$

Set-up $\Delta T = \dfrac{\text{heat}}{\text{mass} \times SH} = \dfrac{6.75\text{ kJ} \times \dfrac{1000\text{ J}}{1\text{ kJ}}}{50.0\text{ g} \times \dfrac{0.385\text{ J}}{\text{g °C}}} = 351\,°C \text{ (3 SFs)}$

3.87 Given 1 lb of body fat; 15% (m/m) water in body fat

Need kilocalories to "burn off"

Plan Because each gram of body fat contains 15% water, a person actually loses 85 grams of fat per hundred grams of body fat. (We considered 1 lb of fat as exactly 1 lb.)

lb of body fat → g of body fat → g of fat → kcal

$\dfrac{453.6\text{ g body fat}}{1\text{ lb body fat}}$ $\dfrac{85\text{ g fat}}{100\text{ g body fat}}$ $\dfrac{9\text{ kcal}}{1\text{ g fat}}$

Set-up $1\text{ lb body fat} \times \dfrac{453.6\text{ g body fat}}{1\text{ lb body fat}} \times \dfrac{85\text{ g fat}}{100\text{ g body fat}} \times \dfrac{9\text{ kcal}}{1\text{ g fat}} = 3500\text{ kcal (2 SFs)}$

3.89 Given mass $= 725 \text{ g of water}$; $SH_{water} = 4.184 \text{ J/g °C}$; $\Delta T = 37\,°C - 65\,°C = -28\,°C$

Need heat in kilojoules

Plan Heat $= \text{mass} \times \Delta T \times SH$, then J → kJ $\dfrac{1\text{ kJ}}{1000\text{ J}}$

Set-up Heat $= 725\text{ g} \times (-28\,°C) \times \dfrac{4.184\text{ J}}{\text{g °C}} \times \dfrac{1\text{ kJ}}{1000\text{ J}} = -85\text{ kJ (2 SFs)}$

∴ 85 kJ are lost from the water bottle and are transferred to the muscles.

3.91 Given 0.66 g of olive oil; 370 g of water; $SH_{water} = 1.00 \text{ cal/g °C}$; $\Delta T = 38.8\,°C - 22.7\,°C = 16.1\,°C$

Need energy value (in kcal/g)

Plan Heat $= \text{mass} \times \Delta T \times SH$ then cal → kcal $\dfrac{1\text{ kcal}}{1000\text{ cal}}$

Set-up Heat $= 370\text{ g} \times 16.1\,°C \times \dfrac{1\text{ cal}}{\text{g °C}} \times \dfrac{1\text{ kcal}}{1000\text{ cal}} = 6.0\text{ kcal}$

∴ energy value $= \dfrac{\text{heat produced}}{\text{mass of oil}} = \dfrac{6.0\text{ kcal}}{0.66\text{ g oil}} = 9.1\text{ kcal/g of olive oil (2 SFs)}$

3.93 **Given** 2.4×10^7 J/1.0 lb of oil; mass = 150 kg of water; SH_{water} = 4.184 J/g °C;
$\Delta T = 100\,°C - 22\,°C = 78\,°C$

Need kilograms of oil needed

Plan Heat = mass $\times \Delta T \times SH$ then J → lb of oil → kg of oil $\dfrac{1\ kg}{2.205\ lb}$

Set-up Heat = 150 kg water $\times \dfrac{1000\ g}{1\ kg} \times 78\,°C \times \dfrac{4.184\ J}{g\,°C} = 4.9 \times 10^7$ J (2 SFs)

$\therefore$ mass of oil = 4.9×10^7 J $\times \dfrac{1.0\ lb\ oil}{2.4 \times 10^7\ J} \times \dfrac{1.0\ kg\ oil}{2.205\ lb\ oil} = 0.93$ kg of oil (2 SFs)

3.95 **Given**

Copper	**Water**
mass = 70.0 g	mass = 50.0 g
initial temperature = 54.0 °C	initial temperature = 26.0 °C
final temperature = 29.2 °C	final temperature = 29.2 °C
$SH = ?$ J/g °C	SH = 4.184 J/g °C

Need specific heat (SH) of copper (J/g °C)

Plan heat lost by copper = −heat gained by water
For both, heat = mass $\times \Delta T \times SH$

Set-up <u>For water:</u>
heat gained = mass $\times \Delta T \times SH$ = 50.0 g $\times (29.2\,°C - 26.0\,°C) \times \dfrac{4.184\ J}{g\,°C} = 669$ J

<u>For copper:</u>
$\therefore$ heat lost by copper = −669 J

$SH = \dfrac{\text{heat lost}}{\text{mass} \times \Delta T} = \dfrac{-669\ J}{70.0\ g \times (29.2\,°C - 54.0\,°C)} = \dfrac{0.385\ J}{g\,°C}$ (3 SFs)

3.97 **a.** **Given** heat = 11 J; mass = 4.7 g; $\Delta T = 4.5\,°C$

Need specific heat (J/g °C)

Plan $SH = \dfrac{\text{heat}}{\text{mass} \times \Delta T}$

Set-up specific heat (SH) = $\dfrac{11\ J}{4.7\ g \times 4.5\,°C} = \dfrac{0.52\ J}{g\,°C}$ (2 SFs)

b. By comparing this calculated value to the specific heats given in Table 3.7, we would identify the unknown metal as titanium (SH = 0.523 J/g °C) rather than aluminum (SH = 0.897 J/g °C).

Selected Answers to Combining Ideas from Chapters 1 to 3

CI.1 **a.** There are 4 significant figures in the measurement 20.17 lb.

b. $20.17 \text{ lb} \times \dfrac{1 \text{ kg}}{2.205 \text{ lb}} = 9.147 \text{ kg} \ (4 \text{ SFs})$

c. $9.147 \text{ kg} \times \dfrac{1000 \text{ g}}{1 \text{ kg}} \times \dfrac{1 \text{ cm}^3}{19.3 \text{ g}} = 474 \text{ cm}^3 \ (3 \text{ SFs})$

d. $T_F = 1.8(T_C) + 32 = 1.8(1064) + 32 = 1947 \,^\circ\text{F} \ (4 \text{ SFs})$
$T_K = T_C + 273 = 1064 + 273 = 1337 \text{ K} \ (4 \text{ SFs})$

e. $\Delta T = T_{final} - T_{initial} = 1064 \,^\circ\text{C} - 500.\,^\circ\text{C} = 564 \,^\circ\text{C}$

$9.147 \text{ kg} \times \dfrac{1000 \text{ g}}{1 \text{ kg}} \times \dfrac{0.129 \text{ J}}{\text{g} \,^\circ\text{C}} \times \dfrac{1 \text{ cal}}{4.184 \text{ J}} \times 564 \,^\circ\text{C} \times \dfrac{1 \text{ kcal}}{1000 \text{ cal}} = 159 \text{ kcal} \ (3 \text{ SFs})$

f. $9.147 \text{ kg} \times \dfrac{1000 \text{ g}}{1 \text{ kg}} \times \dfrac{\$45.98}{1 \text{ g}} = \$421\,000 \ (3 \text{ SFs})$

CI.3 **a.** The water has its own shape in sample **B**.

b. Water sample **A** is represented by diagram **2**, which shows the particles in a random arrangement but close together.

c. Water sample **B** is represented by diagram **1**, which shows the particles fixed in a definite arrangement.

d. The state of matter indicated in diagram **1** is a <u>solid</u>; in diagram **2**, it is a <u>liquid</u>; and in diagram **3**, it is a <u>gas</u>.

e. The motion of the particles is slowest in diagram <u>**1**</u>.

f. The arrangement of particles is farthest apart in diagram <u>**3**</u>.

g. The particles fill the volume of the container in diagram <u>**3**</u>.

h. water $45\,^\circ\text{C} \longrightarrow 0\,^\circ\text{C}$: $\Delta T = 0\,^\circ\text{C} - 45\,^\circ\text{C} = -45\,^\circ\text{C}$
$q = m \times \Delta T \times SH$

$= 19 \text{ g} \times (-45\,^\circ\text{C}) \times \dfrac{4.184 \text{ J}}{\text{g} \,^\circ\text{C}} \times \dfrac{1 \text{ kJ}}{1000 \text{ J}} = -3.6 \text{ kJ} \ (2 \text{ SFs})$

$\therefore$ 3.6 kJ of heat is removed.

CI.5 **a.** $0.250 \text{ lb} \times \dfrac{453.6 \text{ g}}{1 \text{ lb}} \times \dfrac{1 \text{ cm}^3}{7.86 \text{ g}} = 14.4 \text{ cm}^3 \ (3 \text{ SFs})$

b. $30 \text{ nails} \times \dfrac{14.4 \text{ cm}^3}{75 \text{ nails}} = 5.76 \text{ cm}^3 = 5.76 \text{ mL} \ (3 \text{ SFs})$

New water level $= 17.6 \text{ mL water} + 5.76 \text{ mL} = 23.4 \text{ mL} \ (3 \text{ SFs})$

c. $\Delta T = T_f - T_i = 125\,^\circ\text{C} - 16\,^\circ\text{C} = 109\,^\circ\text{C}$
$q = m \times \Delta T \times SH$

$= 0.250 \text{ lb} \times \dfrac{453.6 \text{ g}}{1 \text{ lb}} \times 109\,^\circ\text{C} \times \dfrac{0.452 \text{ J}}{\text{g} \,^\circ\text{C}} = 5590 \text{ J} \ (3 \text{ SFs})$

d. iron $25\,^\circ\text{C} \longrightarrow 1535\,^\circ\text{C}$: $\Delta T = 1535\,^\circ\text{C} - 25\,^\circ\text{C} = 1510\,^\circ\text{C}$

$1 \text{ nail} \times \dfrac{0.250 \text{ lb Fe}}{75 \text{ nails}} \times \dfrac{453.6 \text{ g Fe}}{1 \text{ lb Fe}} \times \dfrac{0.452 \text{ J}}{\text{g} \,^\circ\text{C}} \times 1510.\,^\circ\text{C} = 1030 \text{ J} \ (3 \text{ SFs})$

Credit: Martin Harvey/Alamy

John is preparing to plant a new crop in his fields. Although the element nitrogen in the soil is important, he will also check the soil levels of the elements phosphorus and potassium, which are also important for a good yield. Tests show that the level of phosphorus in the soil is below 15 mg/kg, which is low. Tests also show that the potassium level in the soil is in the optimal range from 150 to 200 mg/kg. Thus, John will apply a fertilizer that contains phosphorus but no potassium. What are the symbols of the elements potassium, phosphorus, and nitrogen found in fertilizers?

CHAPTER READINESS*

❀ Key Math Skills

- Using Positive and Negative Numbers in Calculations (1.4)
- Calculating Percentages (1.4)
- Rounding Off (2.3)

⚗ Core Chemistry Skills

- Counting Significant Figures (2.2)
- Using Significant Figures in Calculations (2.3)

*These Key Math Skills and Core Chemistry Skills from previous chapters are listed here for your review as you proceed to the new material in this chapter.

LOOKING AHEAD

4.1 Elements and Symbols

Learning Goal: Given the name of an element, write its correct symbol; from the symbol, write the correct name.

- Chemical symbols are one- or two-letter abbreviations for the names of the elements; only the first letter of a chemical symbol is a capital letter.
- Element names are derived from planets, mythology, colors, minerals, and names of geographical locations and famous scientists.
- Of all the elements, only about 20 are essential for the well-being of the human body. Oxygen, carbon, hydrogen, and nitrogen make up 96% of human body mass.

♦ **Learning Exercise 4.1A**

Write the symbol for each of the following elements that are essential to health:

a. carbon _____   **b.** iron _____   **c.** sodium _____

d. phosphorus _____   **e.** oxygen _____   **f.** nitrogen _____

g. iodine _____   **h.** sulfur _____   **i.** potassium _____

j. cobalt _____   **k.** calcium _____   **l.** selenium _____

Answers   **a.** C   **b.** Fe   **c.** Na   **d.** P   **e.** O   **f.** N
              g. I   **h.** S   **i.** K   **j.** Co   **k.** Ca   **l.** Se

♦ **Learning Exercise 4.1B**

Write the name of the element represented by each of the following symbols:

a. Mg _____   **b.** Be _____

c. H _____   **d.** Si _____

e. Ag _____   **f.** Br _____

g. F _____   **h.** Zn _____

i. Cr _____   **j.** Al _____

The element silver is used for dental fillings.
Credit: Pearson Science / Pearson Education

Answers   **a.** magnesium   **b.** beryllium   **c.** hydrogen   **d.** silicon   **e.** silver
              f. bromine   **g.** fluorine   **h.** zinc   **i.** chromium   **j.** aluminum

4.2   The Periodic Table

Learning Goal: Use the periodic table to identify the group and the period of an element; identify the element as a metal, a nonmetal, or a metalloid.

- The periodic table is an arrangement of the elements into vertical columns and horizontal rows.
- Each vertical column is a group (or family) of elements that have similar properties.
- A horizontal row of elements is called a *period*.
- On the periodic table, the metals are located on the left of the heavy zigzag line, the nonmetals are to the right, and metalloids are found along the heavy zigzag line.
- The element hydrogen (H) located on the top left of the periodic table is a nonmetal.
- Main group, or representative, elements are found in Groups 1A (1), 2A (2), and 3A (13) to 8A (18). Transition elements are B-group elements (3 to 12).

Periodic Table of Elements

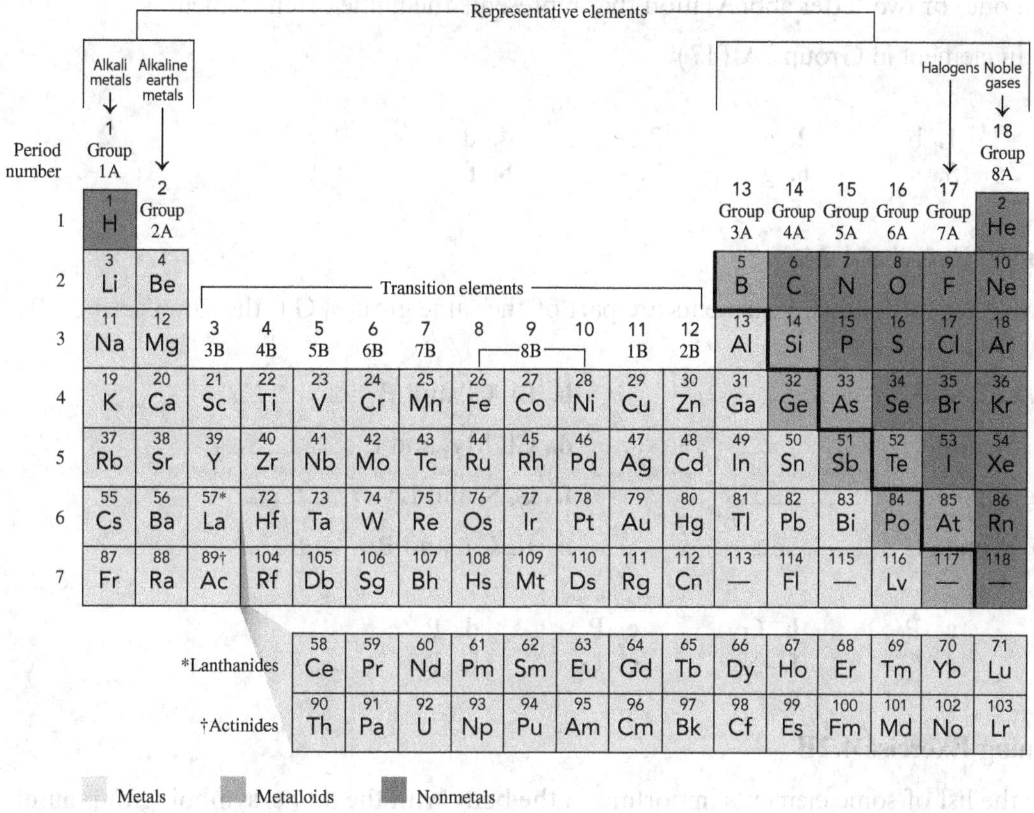

Study Note

1. The periodic table consists of horizontal rows called *periods* and vertical columns called *groups*, which are also called *families*.

2. Group 1A (1) contains the *alkali metals*, Group 2A (2) contains the *alkaline earth metals*, Group 7A (17) contains the *halogens*, and Group 8A (18) contains the *noble gases*.

Key Terms for Sections 4.1 and 4.2

Match each of the following key terms with the correct description:

a. nonmetal	**b.** period	**c.** chemical symbol	**d.** metal
e. group number	**f.** halogen	**g.** periodic table	**h.** transition element

1. _____ a horizontal row of elements on the periodic table

2. _____ an element with little or no luster that is a poor conductor of heat and electricity

3. _____ a number that appears at the top of each vertical column (group) in the periodic table

4. _____ an element that is shiny, malleable, ductile, and a good conductor of heat and electricity

5. _____ an element in the center of the periodic table that has a group number of 3 through 12

6. _____ an arrangement of elements by increasing atomic number

7. _____ a one- or two-letter abbreviation that represents the name of an element

8. _____ an element in Group 7A (17)

Answers	**1.** b	**2.** a	**3.** e	**4.** d
	5. h	**6.** g	**7.** c	**8.** f

♦ Learning Exercise 4.2A

Indicate whether the following elements are part of the same group (G), the same period (P), or neither (N):

a. Li, O, and F _____   **b.** Br, Cl, and F _____

c. Al, Si, and Cl _____   **d.** Cl, Mg, and P _____

e. Mg, Ca, and Ba _____   **f.** C, S, and Br _____

g. Sb, P, and As _____   **h.** K, Ca, and Br _____

Answers	**a.** P	**b.** G	**c.** P	**d.** P
	e. G	**f.** N	**g.** G	**h.** P

♦ Learning Exercise 4.2B

Complete the list of some elements important in the body with the name, symbol, group number, period number, and whether it is a metal, nonmetal, or metalloid in the following table:

Name and Symbol	Group Number	Period Number	Type of Element
	2A (2)	3	
Carbon, C			
	5A (15)	2	
Sulfur, S			
	5A (15)	5	
	1A (1)	1	

Answers

Name and Symbol	Group Number	Period Number	Type of Element
Magnesium, Mg	2A (2)	3	metal
Carbon, C	4A (14)	2	nonmetal
Nitrogen, N	5A (15)	2	nonmetal
Sulfur, S	6A (16)	3	nonmetal
Antimony, Sb	5A (15)	5	metalloid
Hydrogen, H	1A (1)	1	nonmetal

♦ **Learning Exercise 4.2C**

Identify each of the following elements as a metal (M), a nonmetal (NM), or a metalloid (ML):

a. Cl ____   b. P ____   c. Fe ____   d. Si ____   e. Al ____

f. O ____   g. Ca ____   h. Zn ____   i. Ge ____   j. Ag ____

Answers   a. NM   b. NM   c. M   d. ML   e. M
             f. NM   g. M   h. M   i. ML   j. M

♦ **Learning Exercise 4.2D**

Match the name of each of the following chemical groups with the elements K, Cl, He, Fe, Mg, Ne, Li, Cu, and Br:

a. halogen   _____

b. noble gas   _____

c. alkali metal   _____

d. alkaline earth metal _____

e. transition element _____

Answers   a. Cl, Br   b. He, Ne   c. K, Li   d. Mg   e. Fe, Cu

4.3   The Atom

Learning Goal: Describe the electrical charge and location in an atom for a proton, a neutron, and an electron.

• An atom is the smallest particle that retains the characteristics of an element.

• Atoms are composed of three subatomic particles. Protons have a positive charge (+), electrons carry a negative charge (−), and neutrons are electrically neutral.

• The protons and neutrons, with masses of about 1 amu, are found in the tiny, dense nucleus.

• The electrons are located outside the nucleus and have masses much smaller than the mass of a neutron or proton.

Dalton's Atomic Theory

1. All matter is made up of tiny particles called atoms.

2. All atoms of a given element are similar to one another and different from atoms of other elements.

3. Atoms of two or more different elements combine to form compounds. A particular compound is always made up of the same kinds of atoms and the same number of each kind of atom.

4. A chemical reaction involves the rearrangement, separation, or combination of atoms. Atoms are neither created nor destroyed during a chemical reaction.

◆ **Learning Exercise 4.3A**

True (T) or False (F): Each of the following statements is consistent with current atomic theory:

  a. All matter is composed of atoms.

  b. All atoms of an element are similar.

  c. Atoms can combine to form compounds.

  d. Most of the mass of the atom is outside of the nucleus.

Answers   **a.** T   **b.** T   **c.** T   **d.** F

◆ **Learning Exercise 4.3B**

Match the following terms with the correct statements:

  a. proton   **b.** neutron   **c.** electron   **d.** nucleus

  1. _____ is found in the nucleus of an atom

  2. _____ has a 1− charge

  3. _____ is found outside the nucleus

  4. _____ has a mass of about 1 amu

  5. _____ is the small, dense center of the atom

  6. _____ is neutral

  7. _____ is attracted to a proton

Aluminum foil consists of atoms of aluminum.
Credit: Russ Lappa / Pearson Education / Prentice Hall

Answers   **1.** a and b   **2.** c   **3.** c   **4.** a and b   **5.** d   **6.** b   **7.** c

4.4   Atomic Number and Mass Number

Learning Goal: Given the atomic number and the mass number of an atom, state the number of protons, neutrons, and electrons.

- The atomic number, which can be found on the periodic table, is the number of protons in a given atom of an element.
- In a neutral atom, the number of protons is equal to the number of electrons.
- The mass number is the total number of protons and neutrons in an atom.

Study Note

1. The *atomic number* is the number of protons in every atom of an element. In neutral atoms, the number of electrons equals the number of protons.
2. The *mass number* is the total number of neutrons and protons in the nucleus of an atom.
3. The number of neutrons is *mass number − atomic number*.

Example: How many protons and neutrons are in the nucleus of a strontium atom with a mass number of 89?

Solution: The atomic number of Sr is 38. Thus, an atom of Sr has 38 protons.
             Mass number − atomic number = number of neutrons: 89 − 38 = 51 neutrons

---

♦ Learning Exercise 4.4A

CORE CHEMISTRY SKILL
Counting Protons and Neutrons

Give the number of protons in each of the following neutral atoms:

a. an atom of carbon ____

b. an atom of the element with atomic number 15 ____

c. an atom with a mass number of 40 and atomic number 19 ____

d. an atom with 9 neutrons and a mass number of 19 ____

e. a neutral atom that has 18 electrons ____

Answers **a.** 6 **b.** 15 **c.** 19 **d.** 10 **e.** 18

♦ Learning Exercise 4.4B

Determine the number of neutrons in each of the following atoms:

a. a mass number of 42 and atomic number 20 ____

b. a mass number of 10 and 5 protons ____

c. a silicon atom with a mass number of 30 ____

d. a mass number of 9 and atomic number 4 ____

e. a mass number of 22 and 10 protons ____

f. a zinc atom with a mass number of 66 ____

Answers **a.** 22 **b.** 5 **c.** 16 **d.** 5 **e.** 12 **f.** 36

4.5 Isotopes and Atomic Mass

Learning Goal: Determine the number of protons, neutrons, and electrons in one or more of the isotopes of an element; calculate the atomic mass of an element using the percent abundance and mass of its naturally occurring isotopes.

- *Isotopes* are atoms of the same element that have the same number of protons but different numbers of neutrons.
- The atomic mass of an element is the weighted average mass of all the isotopes in a naturally occurring sample of that element.

Key Terms for Sections 4.3 to 4.5

Match each of the following key terms with the correct description:

a. electron **b.** atom **c.** atomic mass **d.** mass number **e.** neutron
f. isotope **g.** atomic number **h.** nucleus **i.** proton

1. ____ the total number of neutrons and protons in the nucleus of an atom

2. ____ the smallest particle of an element that retains the characteristics of the element

3. ____ a negatively charged subatomic particle having a minute mass

4. _____ a neutral subatomic particle found in the nucleus of an atom

5. _____ a number that is equal to the number of protons in an atom

6. _____ the compact, extremely dense center of an atom where the protons and neutrons are located

7. _____ a positively charged subatomic particle found in the nucleus of an atom

8. _____ the weighted average mass of all the naturally occurring isotopes of an element

9. _____ an atom that differs in mass number from another atom of the same element

Answers **1.** d   **2.** b   **3.** a   **4.** e   **5.** g
         6. h   **7.** i   **8.** c   **9.** f

Study Note

In the atomic symbol for a particular atom, the mass number appears in the upper left corner and the atomic number in the lower left corner of the atomic symbol. Mass number $\rightarrow$ $^{32}_{16}S$ Atomic number $\rightarrow$

This isotope is also written several other ways including sulfur-32, S-32, and ^{32}S.

♦ **Learning Exercise 4.5A**

Complete the following table for neutral atoms:

🔬 **CORE CHEMISTRY SKILL**

Writing Atomic Symbols for Isotopes

Atomic Symbol	Atomic Number	Mass Number	Number of Protons	Number of Neutrons	Number of Electrons
	12			12	
			20	22	
		55		29	
	35			45	
		35	17		
$^{120}_{50}Sn$					

Atoms of Mg

Isotopes of Mg

Answers

Atomic Symbol	Atomic Number	Mass Number	Number of Protons	Number of Neutrons	Number of Electrons
$^{24}_{12}Mg$	12	24	12	12	12
$^{42}_{20}Ca$	20	42	20	22	20
$^{55}_{26}Fe$	26	55	26	29	26
$^{80}_{35}Br$	35	80	35	45	35
$^{35}_{17}Cl$	17	35	17	18	17
$^{120}_{50}Sn$	50	120	50	70	50

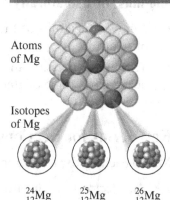

$^{24}_{12}Mg$ $^{25}_{12}Mg$ $^{26}_{12}Mg$

The nuclei of three naturally occurring magnesium isotopes have the same number of protons but different numbers of neutrons.
Credit: Pearson Science / Pearson Education

♦ **Learning Exercise 4.5B**

Identify the sets of atoms that are isotopes:

  a. $^{20}_{10}X$ **b.** $^{20}_{11}X$ **c.** $^{21}_{11}X$ **d.** $^{19}_{10}X$ **e.** $^{19}_{9}X$

Answer   Atoms **a** and **d** are isotopes (atomic number 10); atoms **b** and **c** are isotopes (atomic number 11).

Study Note

To calculate the atomic mass of an element, multiply the mass of each isotope by its percent abundance. Then add these values together. For example, a large sample of naturally occurring chlorine atoms consists of 75.76% of Cl-35 atoms (mass 34.97 amu) and 24.24% of Cl-37 atoms (mass 36.97 amu).

Isotope	Mass (amu)	×	Abundance (%)	=	Contribution to Atomic Mass
Cl-35	34.97	×	$\dfrac{75.76}{100}$	=	26.49 amu
Cl-37	36.97	×	$\dfrac{24.24}{100}$	=	8.962 amu
			Atomic mass of Cl	=	35.45 amu

♦ **Learning Exercise 4.5C**

Copper has two naturally occurring isotopes, $^{63}_{29}Cu$ and $^{65}_{29}Cu$. If that is the case, why is the atomic mass of copper listed as 63.55 on the periodic table?

Answer   Copper in nature consists of two isotopes that have different masses. The atomic mass is the weighted average of the individual masses of the two isotopes, which takes into consideration their percent abundance in the sample. The atomic mass does not represent the mass of any individual atom.

♦ **Learning Exercise 4.5D**

Rubidium is the element with atomic number 37.

  a. What is the symbol for rubidium? _____

  b. What is the group number for rubidium? _____

  c. What is the name of the family that includes rubidium? _____

  d. Rubidium has two naturally occurring isotopes, $^{85}_{37}Rb$ and $^{87}_{37}Rb$. How many protons, neutrons, and electrons are in each isotope?

e. In naturally occurring rubidium, 72.17% is $^{85}_{37}Rb$ with a mass of 84.91 amu; 27.83% is $^{87}_{37}Rb$ with a mass of 86.91 amu. Calculate the atomic mass for rubidium using the weighted average mass method.

Answers

a. Rb

b. Rubidium is in Group 1A (1).

c. Rubidium is in the family called the alkali metals.

d. The isotope $^{85}_{37}Rb$ has 37 protons, 48 neutrons, and 37 electrons. The isotope $^{87}_{37}Rb$ has 37 protons, 50 neutrons, and 37 electrons.

e. 84.91 amu (72.17/100) + 86.91 amu (27.83/100) = 61.28 amu + 24.19 amu = 85.47 amu for the atomic mass of Rb.

Checklist for Chapter 4

You are ready to take the Practice Test for Chapter 4. Be sure you have accomplished the following learning goals for this chapter. If not, review the section listed at the end of the goal. Then apply your new skills and understanding to the Practice Test.

After studying Chapter 4, I can successfully:

_____ Write the correct symbol or name for an element. (4.1)

_____ Use the periodic table to identify the group and period of an element. (4.2)

_____ Identify an element as a metal, nonmetal, or metalloid. (4.2)

_____ State the electrical charge, mass, and location in an atom for a proton, a neutron, and an electron. (4.3)

_____ Given the atomic number and mass number of an atom, state the number of protons, neutrons, and electrons. (4.4)

_____ State the number of protons, neutrons, and electrons in an isotope of an element. (4.5)

_____ Calculate the atomic mass of an element using the percent abundance and mass of its naturally occurring isotopes. (4.5)

Practice Test for Chapter 4

The chapter sections to review are shown in parentheses at the end of each question.

For questions 1 through 5, write the correct symbol for each of the elements listed: (4.1)

1. potassium _____

2. phosphorus _____

3. calcium _____

4. carbon _____

5. sodium _____

For questions 6 through 10, write the correct element name for each of the symbols listed: (4.1)

6. Fe _____

7. Cu _____

8. Cl _____

9. Pb _____

10. Ag _____

11. Some important elements in the body, C, N, and O, are part of a (4.2)
    A. period   B. group   C. neither

12. Some important elements in the body, Li, Na, and K, are part of a (4.2)
    A. period   B. group   C. neither

13. What is the classification of an atom with 15 protons and 17 neutrons? (4.2)
    A. metal   B. nonmetal   C. transition element   D. noble gas   E. halogen

14. What is the group number of the element with atomic number 3? (4.2)
    A. 1A (1)   B. 2A (2)   C. 3B (3)   D. 3A (13)   E. 5A (15)

For questions 15 through 17, indicate whether the statement is True or False: (4.3)

15. _____ The electrons of an atom are located in the nucleus.

16. _____ Most of the mass of an atom is due to the protons and neutrons.

17. _____ All matter is composed of atoms.

For questions 18 through 21, consider an atom with 12 protons and 13 neutrons: (4.4)

18. This atom has an atomic number of
    A. 12   B. 13   C. 23   D. 24   E. 25

19. This atom has a mass number of
    A. 12   B. 13   C. 23   D. 24   E. 25

20. This is an atom of
    A. carbon   B. sodium   C. magnesium   D. aluminum   E. manganese

21. The number of electrons in this atom is
    A. 12   B. 13   C. 23   D. 24   E. 25

For questions 22 through 25, consider an atom of calcium with a mass number of 42: (4.4)

22. This atom of calcium has an atomic number of
    A. 20   B. 22   C. 40   D. 41   E. 42

23. The number of protons in this atom of calcium is
    A. 20   B. 22   C. 40   D. 41   E. 42

24. The number of neutrons in this atom of calcium is
    A. 20   B. 22   C. 40   D. 41   E. 42

25. The number of electrons in this atom of calcium is
    A. 20   B. 22   C. 40   D. 41   E. 42

26. A platinum atom, $^{195}_{78}$Pt, has (4.4)
    A. $78\,p^+, 78\,n^0, 78\,e^-$ B. $195\,p^+, 195\,n^0, 195\,e^-$
    C. $78\,p^+, 195\,n^0, 78\,e^-$ D. $78\,p^+, 117\,n^0, 78\,e^-$
    E. $78\,p^+, 117\,n^0, 117\,e^-$

For questions 27 and 28, use the following list of atoms: (4.5)

$^{14}_{7}$G $^{16}_{8}$J $^{19}_{9}$M $^{16}_{7}$Q $^{18}_{8}$R

27. Which atom(s) is (are) isotopes of an atom with 8 protons and 9 neutrons?
    A. J   B. J, R   C. M, Q   D. M   E. Q

28. Which atom(s) is (are) isotopes of an atom with 7 protons and 8 neutrons?
    A. G   B. J   C. G, Q   D. J, R   E. none

Answers to the Practice Test

1. K	**2.** P	**3.** Ca	**4.** C	**5.** Na
6. iron	**7.** copper	**8.** chlorine	**9.** lead	**10.** silver
11. A	**12.** B	**13.** B	**14.** A	**15.** False
16. True	**17.** True	**18.** A	**19.** E	**20.** C
21. A	**22.** A	**23.** A	**24.** B	**25.** A
26. D	**27.** B	**28.** C		

Selected Answers and Solutions to Text Problems

4.1 **a.** Cu is the symbol for copper.
      b. Pt is the symbol for platinum.
      c. Ca is the symbol for calcium.
      d. Mn is the symbol for manganese.
      e. Fe is the symbol for iron.
      f. Ba is the symbol for barium.
      g. Pb is the symbol for lead.
      h. Sr is the symbol for strontium.

4.3 **a.** Carbon is the element with the symbol C.
      b. Chlorine is the element with the symbol Cl.
      c. Iodine is the element with the symbol I.
      d. Selenium is the element with the symbol Se.
      e. Nitrogen is the element with the symbol N.
      f. Sulfur is the element with the symbol S.
      g. Zinc is the element with the symbol Zn.
      h. Cobalt is the element with the symbol Co.

4.5 **a.** Sodium (Na) and chlorine (Cl) are in NaCl.
      b. Calcium (Ca), sulfur (S), and oxygen (O) are in $CaSO_4$.
      c. Carbon (C), hydrogen (H), chlorine (Cl), nitrogen (N), and oxygen (O) are in $C_{15}H_{22}ClNO_2$.
      d. Lithium (Li), carbon (C), and oxygen (O) are in Li_2CO_3.

4.7 **a.** C, N, and O are in Period 2.
      b. He is the element at the top of Group 8A (18).
      c. The alkali metals are the elements in Group 1A (1).
      d. Period 2 is the horizontal row of elements that ends with neon (Ne).

4.9 **a.** C is Group 4A (14), Period 2.
      b. He is the noble gas in Period 1.
      c. Na is the alkali metal in Period 3.
      d. Ca is Group 2A (2), Period 4.
      e. Al is Group 3A (13), Period 3.

4.11 On the periodic table, *metals* are located to the left of the heavy zigzag line, *nonmetals* are elements to the right, and *metalloids* (B, Si, Ge, As, Sb, Te, Po, and At) are located along the line.
      a. Ca is a metal.
      b. S is a nonmetal.
      c. Metals are shiny.
      d. An element that is a gas at room temperature is a nonmetal.
      e. Group 8A (18) elements are nonmetals.
      f. Br is a nonmetal.
      g. B is a metalloid.
      h. Ag is a metal.

4.13 **a.** Ca, an alkaline earth metal, is needed for bones and teeth, muscle contraction, and nerve impulses.
      b. Fe, a transition element, is a component of the oxygen carrier hemoglobin.
      c. K, an alkali metal, is the most prevalent ion (K^+) in cells, and is needed for muscle contraction and nerve impulses.
      d. Cl, a halogen, is the most prevalent negative ion (Cl^-) in fluids outside cells, and is a component of stomach acid (HCl).

4.15 **a.** A macromineral is an element essential to health that is present in the human body in amounts from 5 to 1000 g.
   b. Sulfur is a component of proteins, liver, vitamin B_1, and insulin.
   c. 86 g is a typical amount of sulfur in a 60.-kg adult.

4.17 **a.** The electron has the smallest mass.
   b. The proton has a 1+ charge.
   c. The electron is found outside the nucleus.
   d. The neutron is electrically neutral.

4.19 Rutherford determined that positively charged particles called protons are located in a very small, dense region of the atom called the nucleus.

4.21 **a.** True
   b. True
   c. True
   d. False; since a neutron has no charge, it is not attracted to a proton (a proton is attracted to an electron).

4.23 In the process of brushing your hair, strands of hair become charged with like charges that repel each other.

4.25 **a.** The atomic number is the same as the number of protons in an atom.
   b. Both are needed since the number of neutrons is the (mass number) − (atomic number).
   c. The mass number is the number of particles (protons + neutrons) in the nucleus.
   d. The atomic number is the same as the number of electrons in a neutral atom.

4.27 The atomic number defines the element and is found above the symbol of the element in the periodic table.
   a. Lithium, Li, has an atomic number of 3.
   b. Fluorine, F, has an atomic number of 9.
   c. Calcium, Ca, has an atomic number of 20.
   d. Zinc, Zn, has an atomic number of 30.
   e. Neon, Ne, has an atomic number of 10.
   f. Silicon, Si, has an atomic number of 14.
   g. Iodine, I, has an atomic number of 53.
   h. Oxygen, O, has an atomic number of 8.

4.29 The atomic number gives the number of protons in the nucleus of an atom. Since atoms are neutral, the atomic number also gives the number of electrons in the neutral atom.
   a. There are 18 protons and 18 electrons in a neutral argon atom.
   b. There are 25 protons and 25 electrons in a neutral manganese atom.
   c. There are 53 protons and 53 electrons in a neutral iodine atom.
   d. There are 48 protons and 48 electrons in a neutral cadmium atom.

4.31 The atomic number is the same as the number of protons in an atom and the number of electrons in a neutral atom; the atomic number defines the element. The number of neutrons is the (mass number) − (atomic number).

Name of the Element	Symbol	Atomic Number	Mass Number	Number of Protons	Number of Neutrons	Number of Electrons
Zinc	Zn	30	66	30	66 − 30 = 36	30
Magnesium	Mg	12	12 + 12 = 24	12	12	12
Potassium	K	19	19 + 20 = 39	19	20	19
Sulfur	S	16	16 + 15 = 31	16	15	16
Iron	Fe	26	56	26	56 − 26 = 30	26

4.33 **a.** Since the atomic number of strontium is 38, every Sr atom has 38 protons. An atom of strontium (mass number 89) has 51 neutrons ($89 - 38 = 51\,n$). Neutral atoms have the same number of protons and electrons. Therefore, 38 protons, 51 neutrons, 38 electrons.

b. Since the atomic number of chromium is 24, every Cr atom has 24 protons. An atom of chromium (mass number 52) has 28 neutrons ($52 - 24 = 28\,n$). Neutral atoms have the same number of protons and electrons. Therefore, 24 protons, 28 neutrons, 24 electrons.

c. Since the atomic number of sulfur is 16, every S atom has 16 protons. An atom of sulfur (mass number 34) has 18 neutrons ($34 - 16 = 18\,n$). Neutral atoms have the same number of protons and electrons. Therefore, 16 protons, 18 neutrons, 16 electrons.

d. Since the atomic number of bromine is 35, every Br atom has 35 protons. An atom of bromine (mass number 81) has 46 neutrons ($81 - 35 = 46\,n$). Neutral atoms have the same number of protons and electrons. Therefore, 35 protons, 46 neutrons, 35 electrons.

4.35 **a.** Since the number of protons is 15, the atomic number is 15 and the element symbol is P. The mass number is the sum of the number of protons and the number of neutrons, $15 + 16 = 31$. The atomic symbol for this isotope is $^{31}_{15}\text{P}$.

b. Since the number of protons is 35, the atomic number is 35 and the element symbol is Br. The mass number is the sum of the number of protons and the number of neutrons, $35 + 45 = 80$. The atomic symbol for this isotope is $^{80}_{35}\text{Br}$.

c. Since the number of electrons is 50, there must be 50 protons in a neutral atom. Since the number of protons is 50, the atomic number is 50, and the element symbol is Sn. The mass number is the sum of the number of protons and the number of neutrons, $50 + 72 = 122$. The atomic symbol for this isotope is $^{122}_{50}\text{Sn}$.

d. Since the element is chlorine, the element symbol is Cl, the atomic number is 17, and the number of protons is 17. The mass number is the sum of the number of protons and the number of neutrons, $17 + 18 = 35$. The atomic symbol for this isotope is $^{35}_{17}\text{Cl}$.

e. Since the element is mercury, the element symbol is Hg, the atomic number is 80, and the number of protons is 80. The mass number is the sum of the number of protons and the number of neutrons, $80 + 122 = 202$. The atomic symbol for this isotope is $^{202}_{80}\text{Hg}$.

4.37 **a.** Since the element is argon, the element symbol is Ar, the atomic number is 18, and the number of protons is 18. The atomic symbols for the isotopes with mass numbers of 36, 38, and 40 are $^{36}_{18}\text{Ar}$, $^{38}_{18}\text{Ar}$, and $^{40}_{18}\text{Ar}$, respectively.

b. They all have the same atomic number (the same number of protons and electrons).

c. They have different numbers of neutrons, which gives them different mass numbers.

d. The atomic mass of argon listed on the periodic table is the weighted average atomic mass of all the naturally occurring isotopes of argon.

e. The isotope Ar-40 ($^{40}_{18}\text{Ar}$) is the most prevalent in a sample of argon because its mass is closest to the average atomic mass of argon listed on the periodic table (39.95 amu).

4.39 The mass of an isotope is the mass of an individual atom. The atomic mass is the weighted average of all the naturally occurring isotopes of that element.

4.41 $^{69}_{31}\text{Ga}$ $68.93 \text{ amu} \times \dfrac{60.11}{100} = 41.43 \text{ amu}$

$^{71}_{31}\text{Ga}$ $70.92 \text{ amu} \times \dfrac{39.89}{100} = 28.29 \text{ amu}$

Atomic mass of Ga $= 69.72 \text{ amu (4 SFs)}$

4.43 Since the atomic mass of copper (63.55 amu) is closer to 63 amu, there are more atoms of $^{63}_{29}\text{Cu}$ in a sample of copper.

4.45 Since the atomic mass of thallium (204.4 amu) is closest to 205 amu, the more prevalent isotope of thallium is $^{205}_{81}\text{Tl}$.

4.47 **a.** Group 1A (1), alkali metals
   b. metal
   c. 19 protons
   d. The most prevalent isotope is K-39, $^{39}_{19}K$.
   e. $^{39}_{19}K$ $38.964 \text{ amu} \times \dfrac{93.26}{100} = 36.34 \text{ amu}$

   $^{40}_{19}K$ $39.964 \text{ amu} \times \dfrac{0.0117}{100} = 0.00468 \text{ amu}$

   $^{41}_{19}K$ $40.962 \text{ amu} \times \dfrac{6.73}{100} = 2.76 \text{ amu}$
   _____
   Atomic mass of K = 39.10 amu (4 SFs)

4.49 **a.** False; all atoms of a given element are identical to one another but different from atoms of other elements.
   b. True
   c. True
   d. False; atoms are never created or destroyed during a chemical reaction.

4.51 **a.** The atomic mass is the weighted average of the masses of all of the naturally occurring isotopes of the element. Isotope masses are based on the masses of all subatomic particles in the atom: protons (1), neutrons (2), and electrons (3), although almost all of that mass comes from the protons (1) and neutrons (2).
   b. The number of protons (1) is the atomic number.
   c. The protons (1) are positively charged.
   d. The electrons (3) are negatively charged.
   e. The number of neutrons (2) is the (mass number) − (atomic number).

4.53 **a.** $^{16}_{8}X$, $^{17}_{8}X$, and $^{18}_{8}X$ all have an atomic number of 8, so all have 8 protons.
   b. $^{16}_{8}X$, $^{17}_{8}X$, and $^{18}_{8}X$ all have an atomic number of 8, so all are isotopes of oxygen.
   c. $^{16}_{8}X$ and $^{16}_{9}X$ have mass numbers of 16, whereas $^{18}_{8}X$ and $^{18}_{10}X$ have mass numbers of 18.

4.55

	Atomic Symbol		
	$^{70}_{32}Ge$	$^{73}_{32}Ge$	$^{76}_{32}Ge$
Atomic Number	32	32	32
Mass Number	70	73	76
Number of Protons	32	32	32
Number of Neutrons	70 − 32 = 38	73 − 32 = 41	76 − 32 = 44
Number of Electrons	32	32	32

4.57 **a.** The diagram shows 4 protons and 5 neutrons in the nucleus of the element. Since the number of protons is 4, the atomic number is 4, and the element symbol is Be. The mass number is the sum of the number of protons and the number of neutrons, 4 + 5 = 9. The atomic symbol is $^{9}_{4}Be$.
   b. The diagram shows 5 protons and 6 neutrons in the nucleus of the element. Since the number of protons is 5, the atomic number is 5, and the element symbol is B. The mass number is the sum of the number of protons and the number of neutrons, 5 + 6 = 11. The atomic symbol is $^{11}_{5}B$.
   c. The diagram shows 6 protons and 7 neutrons in the nucleus of the element. Since the number of protons is 6, the atomic number is 6, and the element symbol is C. The mass number is the sum of the number of protons and the number of neutrons, 6 + 7 = 13. The atomic symbol is $^{13}_{6}C$.

   d. The diagram shows 5 protons and 5 neutrons in the nucleus of the element. Since the number of protons is 5, the atomic number is 5, and the element symbol is B. The mass number is the sum of the number of protons and the number of neutrons, $5 + 5 = 10$. The atomic symbol is $^{10}_{5}\text{B}$.

   e. The diagram shows 6 protons and 6 neutrons in the nucleus of the element. Since the number of protons is 6, the atomic number is 6, and the element symbol is C. The mass number is the sum of the number of protons and the number of neutrons, $6 + 6 = 12$. The atomic symbol is $^{12}_{6}\text{C}$.

   Representations **B** ($^{11}_{5}\text{B}$) and **D** ($^{10}_{5}\text{B}$) are isotopes of boron; **C** ($^{13}_{6}\text{C}$) and **E** ($^{12}_{6}\text{C}$) are isotopes of carbon.

4.59 The first letter of a symbol is a capital, but a second letter is lowercase. The symbol for cobalt is Co, but the symbols in CO are for the elements carbon and oxygen.

4.61 **a.** Br is in Group 7A (17), Period 4.
   b. Ar is in Group 8A (18), Period 3.
   c. Li is in Group 1A (1), Period 2.
   d. Ra is in Group 2A (2), Period 7.

4.63 On the periodic table, *metals* are located to the left of the heavy zigzag line, *nonmetals* are located to the right of the line, and *metalloids* (B, Si, Ge, As, Sb, Te, Po, and At) are located along the line.
   a. Zn is a metal.
   b. Co is a metal.
   c. Mn is a metal.
   d. I is a nonmetal.

4.65 **a.** False; the proton is a positively charged particle.
   b. False; the neutron has about the same mass as a proton.
   c. True
   d. False; the nucleus is the tiny, dense central core of an atom.
   e. True

4.67 **a.** Since the atomic number of cadmium is 48, every Cd atom has 48 protons. An atom of cadmium (mass number 114) has 66 neutrons ($114 - 48 = 66\,n$). Neutral atoms have the same number of protons and electrons. Therefore, 48 protons, 66 neutrons, 48 electrons.
   b. Since the atomic number of technetium is 43, every Tc atom has 43 protons. An atom of technetium (mass number 98) has 55 neutrons ($98 - 43 = 55\,n$). Neutral atoms have the same number of protons and electrons. Therefore, 43 protons, 55 neutrons, 43 electrons.
   c. Since the atomic number of gold is 79, every Au atom has 79 protons. An atom of gold (mass number 199) has 120 neutrons ($199 - 79 = 120\,n$). Neutral atoms have the same number of protons and electrons. Therefore, 79 protons, 120 neutrons, 79 electrons.
   d. Since the atomic number of radon is 86, every Rn atom has 86 protons. An atom of radon (mass number 222) has 136 neutrons ($222 - 86 = 136\,n$). Neutral atoms have the same number of protons and electrons. Therefore, 86 protons, 136 neutrons, 86 electrons.
   e. Since the atomic number of xenon is 54, every Xe atom has 54 protons. An atom of xenon (mass number 136) has 82 neutrons ($136 - 54 = 82\,n$). Neutral atoms have the same number of protons and electrons. Therefore, 54 protons, 82 neutrons, 54 electrons.

4.69

Name	Atomic Symbol	Number of Protons	Number of Neutrons	Number of Electrons
Selenium	$^{80}_{34}Se$	34	$80 - 34 = 46$	34
Nickel	$^{28+34}_{28}Ni$ or $^{62}_{28}Ni$	28	34	28
Magnesium	$^{12+14}_{12}Mg$ or $^{26}_{12}Mg$	12	14	12
Radium	$^{228}_{88}Ra$	88	$228 - 88 = 140$	88

4.71 **a.** The atomic number gives the number of <u>protons</u> in the nucleus.
 b. In an atom, the number of electrons is equal to the number of <u>protons</u>.
 c. Sodium and potassium are examples of elements called <u>alkali metals</u>.

4.73 **a.** The lightest alkali metal (Group 1A (1)) is lithium, which has an atomic number of 3 and the symbol Li.
 b. The heaviest noble gas (Group 8A (18)) is radon, which has an atomic number of 86 and the symbol Rn.
 c. The alkaline earth metal (Group 2A (2)) in Period 3 is magnesium, which has an atomic mass of 24.31 amu and the symbol Mg.
 d. The halogen (Group 7A (17)) with the fewest electrons is fluorine, which has an atomic mass of 19.00 amu and the symbol F.

4.75 The atomic number defines the element and is found above the symbol of the element in the periodic table.
 a. Nickel, Ni, has an atomic number of 28.
 b. Barium, Ba, has an atomic number of 56.
 c. Radium, Ra, has an atomic number of 88.
 d. Arsenic, As, has an atomic number of 33.
 e. Tin, Sn, has an atomic number of 50.
 f. Cesium, Cs, has an atomic number of 55.

4.77 The atomic number gives the number of protons in the nucleus of an atom. Since atoms are neutral, the atomic number also gives the number of electrons in the neutral atom.
 a. There are 25 protons and 25 electrons in a neutral manganese (Mn) atom.
 b. There are 15 protons and 15 electrons in a neutral phosphorus atom.
 c. There are 38 protons and 38 electrons in a neutral strontium (Sr) atom.
 d. There are 27 protons and 27 electrons in a neutral cobalt (Co) atom.
 e. There are 92 protons and 92 electrons in a neutral uranium atom.

4.79 **a.** Since the element is lead (Pb), the atomic number is 82, and the number of protons is 82. In a neutral atom, the number of electrons is equal to the number of protons, so there will be 82 electrons. The number of neutrons is the (mass number) $-$ (atomic number) $= 208 - 82 = 126\ n$. Therefore, 82 protons, 126 neutrons, 82 electrons.
 b. Since the element is lead (Pb), the atomic number is 82, and the number of protons is 82. The mass number is the sum of the number of protons and the number of neutrons, $82 + 132 = 214$. The atomic symbol for this isotope is $^{214}_{82}Pb$.
 c. Since the mass number is 214 (as in part **b**) and the number of neutrons is 131, the number of protons is the (mass number) $-$ (number of neutrons) $= 214 - 131 = 83\ p$. Since there are 83 protons, the atomic number is 83, and the element symbol is Bi (bismuth). The atomic symbol for this isotope is $^{214}_{83}Bi$.

4.81 **a.** Since the number of protons is 4, the atomic number is 4, and the element symbol is Be. The mass number is the sum of the number of protons and the number of neutrons, $4 + 5 = 9$. The atomic symbol for this isotope is $^{9}_{4}Be$.

b. Since the number of protons is 12, the atomic number is 12, and the element symbol is Mg. The mass number is the sum of the number of protons and the number of neutrons, $12 + 14 = 26$. The atomic symbol for this isotope is $^{26}_{12}Mg$.

c. Since the element is calcium, the element symbol is Ca, and the atomic number is 20. The mass number is given as 46. The atomic symbol for this isotope is $^{46}_{20}Ca$.

d. Since the number of electrons is 30, there must be 30 protons in a neutral atom. Since the number of protons is 30, the atomic number is 30, and the element symbol is Zn. The mass number is the sum of the number of protons and the number of neutrons, $30 + 40 = 70$. The atomic symbol for this isotope is $^{70}_{30}Zn$.

4.83 $^{28}_{14}Si$ $\quad$ 27.977 amu $\times \dfrac{92.23}{100} = 25.80$ amu

$\quad$ $^{29}_{14}Si$ $\quad$ 28.976 amu $\times \dfrac{4.68}{100} = 1.36$ amu

$\quad$ $^{30}_{14}Si$ $\quad$ 29.974 amu $\times \dfrac{3.09}{100} = 0.926$ amu

$\quad\quad\quad\quad$ Atomic mass of Si $= 28.09$ amu (4 SFs)

4.85 **a.** Since the element is gold (Au), the atomic number is 79, and the number of protons is 79. In a neutral atom, the number of electrons is equal to the number of protons, so there will be 79 electrons. The number of neutrons is the (mass number) − (atomic number) $= 197 − 79 = 118\ n$. In Au-197, there are 79 protons, 118 neutrons, 79 electrons.

b. Since the element is gold (Au), the atomic number is 79, and the number of protons is 79. The mass number is the sum of the number of protons and the number of neutrons, $79 + 116 = 195$. The atomic symbol for this isotope is $^{195}_{79}Au$.

c. Since the atomic number is 78, the number of protons is 78, and the element symbol is Pt. The mass number is the sum of the number of protons and the number of neutrons, $78 + 116 = 194$. The atomic symbol for this isotope is $^{194}_{78}Pt$.

4.87 $^{204}_{82}Pb$ $\quad$ 204.0 amu $\times \dfrac{1.40}{100} = 2.86$ amu

$\quad$ $^{206}_{82}Pb$ $\quad$ 206.0 amu $\times \dfrac{24.10}{100} = 49.65$ amu

$\quad$ $^{207}_{82}Pb$ $\quad$ 207.0 amu $\times \dfrac{22.10}{100} = 45.75$ amu

$\quad$ $^{208}_{82}Pb$ $\quad$ 208.0 amu $\times \dfrac{52.40}{100} = 109.0$ amu

$\quad\quad\quad\quad$ Atomic mass of Pb $= 207.3$ amu (4 SFs)

4.89 **Given** $\quad$ length $= 1$ in., sodium atoms (3.14×10^{-8} cm diameter)

$\quad\quad$ **Need** $\quad$ number of Na atoms

$\quad\quad$ **Plan** $\quad$ in. → cm → number of Na atoms

$$\dfrac{2.54\ cm}{1\ in.} \quad \dfrac{1\ atom\ Na}{3.14 \times 10^{-8}\ cm}$$

$\quad\quad$ **Set-up** $\quad$ $1\ \cancel{in.} \times \dfrac{2.54\ \cancel{cm}}{1\ \cancel{in.}} \times \dfrac{1\ atom\ Na}{3.14 \times 10^{-8}\ \cancel{cm}} = 8.09 \times 10^{7}$ atoms of Na (3 SFs)

5

Electronic Structure and Periodic Trends

Credit: Monty Rakusen/Getty images

Robert and Jennifer are materials engineers who develop and test various materials that are used in the manufacture of consumer goods. Materials engineers create new materials out of metals, ceramics, plastics, semiconductors, and combinations of materials called composites, which are used by the mechanical, chemical, and electrical industries. In their research, the two work with some of the elements in Groups 3A (13), 4A (14), and 5A (15) of the periodic table. These elements, such as silicon, have properties that make them good semiconductors.

A microchip requires growing a single crystal of a semiconductor, such as pure silicon. When small amounts of impurities are added to the crystalline structure, holes form through which electrons can travel with little obstruction. Microchips are manufactured for use in computers, cell phones, satellites, televisions, calculators, GPSs, and many other devices. What elements are found in Group 4A (14) on the periodic table?

CHAPTER READINESS*

⚛ Key Math Skill

- Converting between Standard Numbers and Scientific Notation (1.5)

🔬 Core Chemistry Skill

- Using Prefixes (2.4)

*The Key Math Skill and Core Chemistry Skill from previous chapters are listed here for your review as you proceed to the new material in this chapter.

LOOKING AHEAD

5.1 Electromagnetic Radiation

Learning Goal: Compare the wavelength, frequency, and energy of electromagnetic radiation.

- Electromagnetic radiation is energy that travels as waves in space at the speed of light.
- The speed of light (c) is 3.00×10^8 m/s.
- The wavelength (λ) is the distance between the crests of adjacent waves.
- The frequency (ν) is the number of waves that pass a point in 1 s.
- The electromagnetic spectrum is all forms of electromagnetic radiation, which can be arranged in order of decreasing wavelength.
- The wave equation, $c = \lambda\nu$, expresses the relationship of the speed of light (m/s) to wavelength (m) and frequency (s^{-1}).

♦ **Learning Exercise 5.1**

Using the electromagnetic spectrum, identify the type of radiation in each of the following pairs that would have the shorter wavelength:

a. microwave or infrared _____

b. radio waves or X-rays _____

c. gamma rays or radio waves _____

d. ultraviolet or microwave _____

Answers   **a.** infrared   **b.** X-rays   **c.** gamma rays   **d.** ultraviolet

5.2   Atomic Spectra and Energy Levels

Learning Goal: Explain how atomic spectra correlate with the energy levels in atoms.

- An atomic spectrum is a series of colored lines, each of which corresponds to a photon of a specific energy emitted by a heated element.
- A photon is a packet of energy with both particle and wave characteristics that travels at the speed of light.
- In an atom, the energy levels indicated by the principal quantum number, n, contain electrons of similar energies.
- The principal quantum number, n, increases as the energy of the electrons increases.
- When electrons change energy levels, photons of specific energies are absorbed or emitted.

♦ **Learning Exercise 5.2**

Identify the photon in each of the following pairs with the greater energy:

a. _____ infrared radiation or green light   **b.** _____ $\lambda = 10^9$ nm or $\lambda = 10^5$ nm

c. _____ violet light or red light   **d.** _____ $\lambda = 10^{-3}$ nm or $\lambda = 10^2$ nm

Answers   **a.** green light   **b.** $\lambda = 10^5$ nm   **c.** violet light   **d.** $\lambda = 10^{-3}$ nm

5.3   Sublevels and Orbitals

Learning Goal: Describe the sublevels and orbitals for the electrons in an atom.

- Within each energy level, electrons with identical energy are grouped in sublevels.
- The energies of the sublevels increase in order: $s < p < d < f$.
- The number of sublevels in each energy level is equal to its principal quantum number, n. For example, $n = 1$ has one sublevel (s), $n = 2$ has two sublevels (s, p), $n = 3$ has three sublevels (s, p, d), and $n = 4$ has four sublevels (s, p, d, f).
- An s sublevel can accommodate 2 electrons; a p sublevel can accommodate 6 electrons; a d sublevel can accommodate 10 electrons; and an f sublevel can accommodate 14 electrons.
- An orbital is a region in an atom where there is the greatest probability of finding an electron of certain energy.
- An orbital can contain a maximum of two electrons, which have opposite spins.

- An *s* orbital is spherical, and the three *p* orbitals have two lobes along three perpendicular axes. The *d* and *f* orbitals have more complex shapes.
- Each sublevel consists of a set of orbitals: an *s* sublevel consists of one orbital; a *p* sublevel consists of three orbitals; a *d* sublevel consists of five orbitals; and an *f* sublevel consists of seven orbitals.

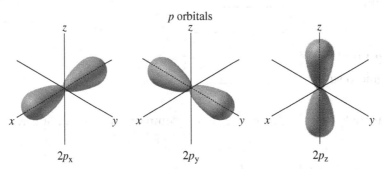

p orbitals

Each *p* orbital, which has two regions of high probability, is aligned along a different axis.

Key Terms for Sections 5.1 to 5.3

Match each of the following key terms with the correct description:

a. photon   **b.** wavelength   **c.** electromagnetic radiation   **d.** orbital

1. ____ the distance between the crests of two adjacent waves

2. ____ energy that travels as waves at the speed of light

3. ____ a packet of energy with both particle and wave characteristics that travels at the speed of light

4. ____ the region within an atom where an electron of a certain energy is most likely to be found

Answers   **1.** b   **2.** c   **3.** a   **4.** d

♦ Learning Exercise 5.3A

What is similar about the following orbitals?

    a. 4*s* and 4*d*

    b. 3*p* and 5*p*

Answers

a. The 4*s* and 4*d* orbitals are found in the same energy level with the same principal quantum number.
b. The 3*p* and 5*p* orbitals are the same type, which have the same dumbbell shape.

♦ Learning Exercise 5.3B

State the maximum number of electrons for each of the following:

    a. 3*p* sublevel ____   **b.** 3*d* sublevel ____

    c. 2*s* orbital ____   **d.** energy level *n* = 4 ____

    e. 1*s* sublevel ____   **f.** 4*p* orbital ____

    g. 5*p* sublevel ____   **h.** 4*f* sublevel ____

Answers   **a.** 6   **b.** 10   **c.** 2   **d.** 32   **e.** 2   **f.** 2   **g.** 6   **h.** 14

5.4 Orbital Diagrams and Electron Configurations

Learning Goal: Draw the orbital diagram and write the electron configuration for an element.

- An orbital diagram shows the orbitals in an atom as squares with arrows that represent the electrons. In the orbital diagram, two electrons in an orbital are shown with opposite spins; the first arrow is up and the second is down. In this example, the orbital diagram of helium consists of a square with two arrows.

$1s$

- The electron configuration shows the number of electrons in each sublevel in order of increasing energy. For example, the electron configuration for Mg (atomic number 12) is $1s^2 2s^2 2p^6 3s^2$.

- In the abbreviated electron configuration, the electron configuration of the preceding noble gas is replaced by its symbol in square brackets. For example, the abbreviated electron configuration for Mg is $[\text{Ne}]3s^2$.

Guide to Drawing Orbital Diagrams	
STEP 1	Draw boxes to represent the occupied orbitals.
STEP 2	Place a pair of electrons with opposite spins in each filled orbital.
STEP 3	Place the remaining electrons in the last occupied sublevel in separate orbitals.

♦ Learning Exercise 5.4A

Draw the orbital diagram for each of the following elements:

a. beryllium _____   **b.** carbon _____

c. sodium _____   **d.** nitrogen _____

e. fluorine _____   **f.** magnesium _____

Answers

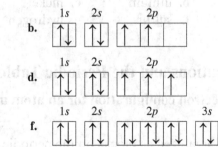

♦ Learning Exercise 5.4B

CORE CHEMISTRY SKILL
Writing Electron Configurations

Write the complete electron configuration for each of the following elements:

a. carbon ________________ **b.** calcium ________________

c. iron ________________ **d.** silicon ________________

e. chlorine ________________ **f.** phosphorus ________________

Answers **a.** $1s^22s^22p^2$ **b.** $1s^22s^22p^63s^23p^64s^2$ **c.** $1s^22s^22p^63s^23p^64s^23d^6$
d. $1s^22s^22p^63s^23p^2$ **e.** $1s^22s^22p^63s^23p^5$ **f.** $1s^22s^22p^63s^23p^3$

♦ Learning Exercise 5.4C

Write the abbreviated electron configuration for each of the elements in 5.4B:

a. carbon ____________ **b.** calcium ____________ **c.** iron ____________

d. silicon ____________ **e.** chlorine ____________ **f.** phosphorus ____________

Answers **a.** $[He]2s^22p^2$ **b.** $[Ar]3s^2$ **c.** $[Ar]4s^23d^6$
d. $[Ne]3s^23p^2$ **e.** $[Ne]3s^23p^5$ **f.** $[Ne]3s^23p^3$

♦ Learning Exercise 5.4D

Name the element with an electron configuration ending with each of the following notations:

a. $3p^6$ ____________ **b.** $2s^1$ ____________

c. $3d^8$ ____________ **d.** $4p^1$ ____________

e. $5p^5$ ____________ **f.** $3p^4$ ____________

g. $1s^1$ ____________ **h.** $6s^2$ ____________

Answers **a.** argon **b.** lithium **c.** nickel **d.** gallium
e. iodine **f.** sulfur **g.** hydrogen **h.** barium

5.5 Electron Configurations and the Periodic Table

Learning Goal: Write the electron configuration for an atom using the sublevel blocks on the periodic table.

- The electron sublevels are organized as blocks on the periodic table. The *s* block corresponds to Groups 1A (1) and 2A (2), the *p* block extends from Group 3A (13) to 8A (18), the *d* block contains the transition elements from 3B (3) to 2B (12), and the *f* block contains the inner transition elements.

Guide to Writing Electron Configurations Using Sublevel Blocks	
STEP 1	Locate the element on the periodic table.
STEP 2	Write the filled sublevels in order, going across each period.
STEP 3	Complete the configuration by counting the electrons in the last occupied sublevel block.

♦ **Learning Exercise 5.5A**

Use the sublevel blocks on the periodic table to write the complete electron configuration for each of the following:

a. S _____

b. Mn _____

c. Tc _____

d. Br _____

e. Sr _____

f. V _____

Answers   **a.** $1s^2 2s^2 2p^6 3s^2 3p^4$ **b.** $1s^2 2s^2 2p^6 3s^2 3p^6 4s^2 3d^5$
c. $1s^2 2s^2 2p^6 3s^2 3p^6 4s^2 3d^{10} 4p^6 5s^2 4d^5$ **d.** $1s^2 2s^2 2p^6 3s^2 3p^6 4s^2 3d^{10} 4p^5$
e. $1s^2 2s^2 2p^6 3s^2 3p^6 4s^2 3d^{10} 4p^6 5s^2$ **f.** $1s^2 2s^2 2p^6 3s^2 3p^6 4s^2 3d^3$

♦ **Learning Exercise 5.5B**

Use the sublevel blocks on the periodic table to write the abbreviated electron configuration for the each of the following elements:

> 🔬 **CORE CHEMISTRY SKILL**
>
> Using the Periodic Table to Write Electron Configurations

a. iodine   _____   **b.** barium   _____

c. zinc   _____   **d.** palladium   _____

e. rubidium   _____   **f.** selenium   _____

Answers   **a.** $[Kr]5s^2 4d^{10} 5p^5$ **b.** $[Xe]6s^2$ **c.** $[Ar]4s^2 3d^{10}$
d. $[Kr]5s^2 4d^8$ **e.** $[Kr]5s^1$ **f.** $[Ar]4s^2 3d^{10} 4p^4$

♦ **Learning Exercise 5.5C**

Give the symbol of the element that has each of the following:

a. six $3d$ electrons _____   **b.** first to fill four s orbitals completely _____

c. $[Kr]5s^2$ _____   **d.** $[Ar]4s^2 3d^{10} 4p^5$ _____

e. two $6p$ electrons _____   **f.** first to have nine completely filled p orbitals _____

Answers   **a.** Fe   **b.** Ca   **c.** Sr   **d.** Br   **e.** Pb   **f.** Kr

5.6   Trends in Periodic Properties

Learning Goal: Use the electron configurations of elements to explain the trends in periodic properties.

*   The physical and chemical properties of elements change going across each period and are repeated in each successive period.

*   Representative elements in a group have similar behavior.

*   The group number of a representative element gives the number of valence electrons in the atoms of that group.

*   A Lewis symbol is a convenient way to represent the valence electrons, which are shown as dots on the sides, top, or bottom of the symbol for the element. For beryllium, the Lewis symbol is:

    $\overset{\bullet}{\text{Be}} \cdot$

*   The atomic size of representative elements generally increases going down a group and decreases going from left to right across a period.

*   The ionization energy generally decreases going down a group and increases going from left to right across a period.

*   The metallic character increases going down a group and decreases going from left to right across a period.

Key Terms for Sections 5.4 to 5.6

CORE CHEMISTRY SKILL

Identifying Trends in Periodic Properties

Match each of the following key terms with the correct description:

a. *f* block   **b.** ionization energy   **c.** Lewis symbol   **d.** electron configuration
e. *d* block   **f.** *s* block   **g.** valence electrons

1. _____ the block of 14 elements in the rows at the bottom of the periodic table

2. _____ the elements in Groups 1A (1) and 2A (2) in which electrons fill the *s* orbitals

3. _____ the block of 10 elements from Groups 3B (3) to 2B (12)

4. _____ the representation of an atom that shows valence electrons around the symbol of the element

5. _____ the number of electrons in each sublevel in order of increasing energy

6. _____ a measure of how easily an element loses a valence electron

7. _____ the electrons in the highest energy level of an atom

Answers   **1.** a   **2.** f   **3.** e   **4.** c   **5.** d   **6.** b   **7.** g

♦ **Learning Exercise 5.6A**

⚛ **CORE CHEMISTRY SKILL**

Drawing Lewis Symbols

State the number of valence electrons, the group number of each element, and draw the Lewis symbol for each of the following:

Element	Valence Electrons	Group Number	Lewis Symbol
Sulfur			
Oxygen			
Magnesium			
Hydrogen			
Fluorine			
Aluminum			

Answers

Element	Valence Electrons	Group Number	Lewis Symbol
Sulfur	$6\,e^-$	Group 6A (16)	$\cdot\ddot{S}\colon$
Oxygen	$6\,e^-$	Group 6A (16)	$\cdot\ddot{O}\colon$
Magnesium	$2\,e^-$	Group 2A (2)	$Mg\cdot$
Hydrogen	$1\,e^-$	Group 1A (1)	$H\cdot$
Fluorine	$7\,e^-$	Group 7A (17)	$\cdot\ddot{F}\colon$
Aluminum	$3\,e^-$	Group 3A (13)	$\cdot\dot{Al}\cdot$

♦ **Learning Exercise 5.6B**

Indicate the element that has the larger atomic size.

a. _____ Mg or Ca   b. _____ Si or Cl

c. _____ Sr or Rb   d. _____ Br or Cl

e. _____ Li or Cs   f. _____ Li or N

g. _____ N or P   h. _____ As or Ca

Answers   **a.** Ca   **b.** Si   **c.** Rb   **d.** Br
              e. Cs   **f.** Li   **g.** P   **h.** Ca

♦ **Learning Exercise 5.6C**

Indicate the element that has the lower ionization energy.

a. _____ Mg or Na   b. _____ P or Cl

c. _____ K or Rb   d. _____ Br or F

e. _____ Li or O   f. _____ Sb or N

g. _____ K or Br   h. _____ S or Na

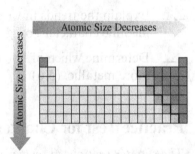

Atomic size increases going down a group and decreases going from left to right across a period.

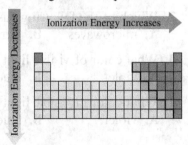

Ionization energy decreases going down a group and increases going from left to right across a period.

Answers   **a.** Na   **b.** P   **c.** Rb   **d.** Br
              e. Li   **f.** Sb   **g.** K   **h.** Na

♦ **Learning Exercise 5.6D**

Indicate the element that has more metallic character.

a. ____ K or Na   **b.** ____ O or Se

c. ____ Ca or Br   **d.** ____ I or F

e. ____ Li or N   **f.** ____ Pb or Ba

Answers   **a.** K   **b.** Se   **c.** Ca
              d. I   **e.** Li   **f.** Ba

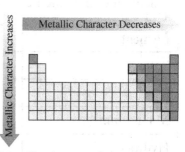

Metallic character increases going down a group and decreases going from left to right across a period.

Checklist for Chapter 5

You are ready to take the Practice Test for Chapter 5. Be sure you have accomplished the following learning goals for this chapter. If not, review the section listed at the end of the goal. Then apply your new skills and understanding to the Practice Test.

After studying Chapter 5, I can successfully:

____ Compare the wavelength, frequency, and energy of electromagnetic radiation. (5.1)

____ Explain how atomic spectra correlate with the energy levels in atoms. (5.2)

____ Describe the sublevels and orbitals for the electrons in an atom. (5.3)

____ Draw the orbital diagrams and write the electron configurations for hydrogen to argon. (5.4)

____ Use the sublevel blocks on the periodic table to write electron configurations. (5.5)

____ Explain the trends in valence electrons, group numbers, atomic size, ionization energy, and metallic character going down a group and across a period. (5.6)

____ Determine which of two elements has the greater atomic size, higher ionization energy, or is the more metallic. (5.6)

Practice Test for Chapter 5

The chapter sections to review are shown in parentheses at the end of each question.

1. Which of the following has the longest wavelength? (5.1)
   A. microwaves   **B.** ultraviolet   **C.** radio waves   **D.** infrared   **E.** visible

2. Which of the following has the highest frequency? (5.1)
   A. microwaves   **B.** ultraviolet   **C.** radio waves   **D.** infrared   **E.** visible

3. What color of visible light has the highest frequency? (5.1)
   A. violet   **B.** blue   **C.** green   **D.** yellow   **E.** red

4. What color of visible light has the longest wavelength? (5.1)
   A. violet   **B.** blue   **C.** green   **D.** yellow   **E.** red

5. Arrange the following in order of increasing energy: infrared, ultraviolet, radio waves, X-rays. (5.1)
   A. X-rays, infrared, ultraviolet, radio waves
   B. radio waves, ultraviolet, infrared, X-rays
   C. ultraviolet, infrared, radio waves, X-rays
   D. radio waves, infrared, ultraviolet, X-rays
   E. radio waves, X-rays, infrared, ultraviolet

6. The wavelength is (5.1)
   A. the height to the crest of a wave
   B. the length of a wave
   C. the distance between two adjacent crests of a wave
   D. the number of waves that pass a point in 1 s
   E. related directly to the frequency of a wave of electromagnetic radiation

7. The speed of light is 3.00×10^8 m/s. How many hours does it take for light to travel from Seattle to Mexico City, a distance of 3770 km? (5.1)
   A. 9.00×10^{12} h   B. 2.50×10^9 h   C. 3.60×10^{-1} h
   D. 3.49×10^{-6} h   E. 2.78×10^{-9} h

8. Atomic spectra (5.2)
   A. consist of lines of different colors
   B. are different for each of the elements
   C. occur when energy changes produce photons of certain wavelengths
   D. indicate that electrons in atoms have specific energy levels
   E. all of these

9. The maximum number of electrons that can be accommodated in energy level $n = 4$ is (5.3)
   A. 2   B. 4   C. 8   D. 18   E. 32

10. The sublevels that make up energy level $n = 3$ are (5.3)
    A. $1s2s3s$ B. $3s3p$ C. $3s3p3d$ D. $3s3p3d3f$ E. $3p3d3f$

11. The spherical volume within an atom in which an electron is likely to be found is called a(n) (5.3)
    A. d orbital   B. p orbital   C. s orbital   D. f orbital   E. g orbital

12. The number of orbitals in a $4p$ sublevel is (5.3)
    A. one   B. two   C. three   D. four   E. five

13. The energy level that consists of one s orbital and three p orbitals has principal quantum number (5.3)
    A. $n = 1$ B. $n = 2$ C. $n = 3$ D. $n = 4$ E. $n = 5$

14. The maximum number of electrons that the $5d$ sublevel can accommodate is (5.3)
    A. 2   B. 3   C. 6   D. 10   E. 14

15. In the orbital diagram of nitrogen, the $2p$ sublevel would be shown as (5.4)

    A.   B.   C.

    D.   E.

16. The complete electron configuration for silicon is (5.4)
    A. $1s^2 2s^2 2p^6 3s^4$ B. $1s^2 2s^2 2p^6 3s^2 3p^2$ C. $1s^2 2s^2 2p^6 3s^2 3d^2$
    D. $1s^2 2s^2 2p^6 3s^2 3p^4$ E. $1s^2 2s^2 2p^6 3p^4$

17. The complete electron configuration for oxygen is (5.4)
    A. $2s^22p^4$　　　　　B. $1s^22s^42p^4$　　　　　C. $1s^22s^6$
    D. $1s^22s^22p^23s^2$　　E. $1s^22s^22p^4$

18. The complete electron configuration for aluminum is (5.4)
    A. $1s^22s^22p^9$　　　　B. $1s^22s^22p^63p^5$　　　C. $1s^22s^22p^63s^23p^1$
    D. $1s^22s^22p^83p^1$　　E. $1s^22s^22p^63p^3$

For questions 19 through 23, match the final notation in the electron configuration with the following elements: (5.4)
    A. As　　　　　B. Rb　　　　　C. Na　　　　　D. Sn　　　　　E. Xe

19. $4p^3$ _____　　　　　　　　　　　20. $5s^1$ _____

21. $3s^1$ _____　　　　　　　　　　　22. $5p^6$ _____

23. $5p^2$ _____

24. The element that has an abbreviated electron configuration of $[Ar]4s^23d^6$ is (5.4)
    A. chromium　　B. iron　　　　C. krypton　　　D. calcium　　E. zinc

25. The element that begins filling the $6s$ sublevel is (5.5)
    A. lithium　　　B. rubidium　　C. cesium　　　D. scandium　　E. barium

26. The element that has 5 electrons in energy level $n = 4$ is (5.5)
    A. boron　　　　B. phosphorus　C. vanadium　　D. bromine　　E. arsenic

27. The number of electrons in the $5p$ sublevel of iodine is (5.5)
    A. 2　　　　　　B. 3　　　　　　C. 5　　　　　　D. 7　　　　　　E. 15

28. The number of valence electrons in gallium is (5.5)
    A. 1　　　　　　B. 2　　　　　　C. 3　　　　　　D. 13　　　　　E. 31

29. The atomic size of oxygen is larger than that of (5.6)
    A. lithium　　　B. sulfur　　　C. argon　　　　D. boron　　　E. fluorine

30. Of C, Si, Ge, Sn, and Pb, the element with the greatest ionization energy is (5.6)
    A. C　　　　　　B. Si　　　　　C. Ge　　　　　D. Sn　　　　　E. Pb

31. The Lewis symbol ·X· would be correct for an element in (5.6)
    A. Group 1A (1)　B. Group 2A (2)　C. Group 4A (14)　D. Group 6A (16)　E. Group 7A (17)

32. The Lewis symbol X· would be correct for (5.6)
    A. magnesium　　B. chlorine　　C. sulfur　　　D. cesium　　　E. nitrogen

33. Of Li, Na, K, Rb, and Cs, the element with the most metallic character is (5.6)
    A. Li　　　　　　B. Na　　　　　C. K　　　　　　D. Rb　　　　　E. Cs

Answers to the Practice Test

1. C	2. B	3. A	4. E	5. D
6. C	7. D	8. E	9. E	10. C
11. C	12. C	13. B	14. D	15. D
16. B	17. E	18. C	19. A	20. B
21. C	22. E	23. D	24. B	25. C
26. E	27. C	28. C	29. E	30. A
31. B	32. D	33. E		

Selected Answers and Solutions to Text Problems

5.1 The wavelength of UV light is the distance between crests of the wave of ultraviolet light.

5.3 "White" light has all the colors and wavelengths of the visible spectrum, including red and blue light. Red and blue light are visible light with specific wavelengths.

5.5 Ultraviolet radiation has a higher frequency and higher energy than infrared radiation.

5.7 **Given** $\lambda = 6.3 \times 10^{-5}$ cm **Need** meters and nanometers

Plan cm → m $\dfrac{1\text{ m}}{100\text{ cm}}$; cm → m → nm $\dfrac{1\text{ m}}{100\text{ cm}}\quad\dfrac{1 \times 10^9\text{ nm}}{1\text{ m}}$

Set-up $6.3 \times 10^{-5}\text{ cm} \times \dfrac{1\text{ m}}{100\text{ cm}} = 6.3 \times 10^{-7}\text{ m (2 SFs)}$

$6.3 \times 10^{-5}\text{ cm} \times \dfrac{1\text{ m}}{100\text{ cm}} \times \dfrac{1 \times 10^9\text{ nm}}{1\text{ m}} = 630\text{ nm (2 SFs)}$

5.9 AM radio has a longer wavelength than cell phones or infrared light.

5.11 In order of increasing wavelengths (shortest to longest): X-rays, blue light, microwaves

5.13 In order of increasing frequencies (lowest to highest): TV signal, microwaves, X-rays

5.15 Atomic spectra consist of a series of lines separated by dark sections, indicating that the energy emitted by the elements is not continuous.

5.17 Electrons can jump to higher energy levels when they absorb a photon.

5.19 **a.** A photon of green light has greater energy than a photon of yellow light.
b. A photon of blue light has greater energy than a photon of red light.

5.21 **a.** A 1*s* orbital is spherical.
b. A 2*p* orbital has two lobes.
c. A 5*s* orbital is spherical.

5.23 **a.** **1** and **2**; 1*s* and 2*s* orbitals have the same shape (spherical) and can contain a maximum of two electrons.
b. **3**; 3*s* and 3*p* sublevels are in the same energy level ($n = 3$).
c. **1** and **2**; 3*p* and 4*p* sublevels contain orbitals with the same two-lobed shape and can contain the same maximum number of electrons (six).
d. **1**, **2**, and **3**; all of the 3*p* orbitals have two lobes, can contain a maximum of two electrons, and are in energy level $n = 3$.

5.25 **a.** There are five orbitals in the 3*d* sublevel.
b. There is one sublevel in the $n = 1$ energy level.
c. There is one orbital in the 6*s* sublevel.
d. There are nine orbitals in the $n = 3$ energy level: one 3*s* orbital, three 3*p* orbitals, and five 3*d* orbitals.

5.27 **a.** Any orbital can hold a maximum of two electrons. Thus, a 2*p* orbital can have a maximum of two electrons.
b. The 3*p* sublevel contains three *p* orbitals, each of which can hold a maximum of two electrons, which gives a maximum of six electrons in the 3*p* sublevel.
c. Using $2n^2$, the calculation for the maximum number of electrons in the $n = 4$ energy level is $2(4)^2 = 2(16) = 32$ electrons.
d. The 5*d* sublevel contains five *d* orbitals, each of which can hold a maximum of two electrons, which gives a maximum of 10 electrons in the 5*d* sublevel.

5.29 The electron configuration shows the number of electrons in each sublevel of an atom. The abbreviated electron configuration uses the symbol of the preceding noble gas to show completed sublevels.

5.31 Determine the number of electrons and then fill orbitals in the following order: $1s^2 2s^2 2p^6 3s^2 3p^6$

  a. Boron is atomic number 5 and so it has 5 electrons. Fill orbitals as

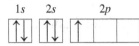

  b. Aluminum is atomic number 13 and so it has 13 electrons. Fill orbitals as

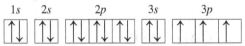

  c. Phosphorus is atomic number 15 and so it has 15 electrons. Fill orbitals as

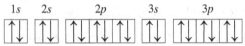

  d. Argon is atomic number 18 and so it has 18 electrons. Fill orbitals as

5.33 **a.** Ni $1s^2 2s^2 2p^6 3s^2 3p^6 4s^2 3d^8$
  b. Na $1s^2 2s^2 2p^6 3s^1$
  c. Li $1s^2 2s^1$
  d. Ti $1s^2 2s^2 2p^6 3s^2 3p^6 4s^2 3d^2$

5.35 **a.** Sn $[\text{Kr}] 5s^2 4d^{10} 5p^2$
  b. Cd $[\text{Kr}] 5s^2 4d^{10}$
  c. Se $[\text{Ar}] 4s^2 3d^{10} 4p^4$
  d. F $[\text{He}] 2s^2 2p^5$

5.37 **a.** Li is the element with one electron in the $2s$ sublevel.
  b. Ti is the element with two electrons in the $4s$ and two electrons in the $3d$ sublevels.
  c. Ge is the element with two electrons in the $4s$, ten electrons in the $3d$, and two electrons in the $4p$ sublevels.
  d. F is the element with two electrons in the $2s$ and five electrons in the $2p$ sublevels.

5.39 **a.** Al has three electrons in the energy level $n = 3$, $3s^2 3p^1$.
  b. C has two $2p$ electrons.
  c. Ar completes the $3p$ sublevel, $3p^6$.
  d. Be completes the $2s$ sublevel, $2s^2$.

5.41 On the periodic table, the s sublevel block is on the left, the p sublevel block is on the right, and the d sublevel block is in the center between the s and p blocks.
  a. As $1s^2 2s^2 2p^6 3s^2 3p^6 4s^2 3d^{10} 4p^3$
  b. Fe $1s^2 2s^2 2p^6 3s^2 3p^6 4s^2 3d^6$
  c. Co $1s^2 2s^2 2p^6 3s^2 3p^6 4s^2 3d^7$
  d. Kr $1s^2 2s^2 2p^6 3s^2 3p^6 4s^2 3d^{10} 4p^6$

5.43 On the periodic table, the *s* sublevel block is on the left, the *p* sublevel block is on the right, and the *d* sublevel block is in the center between the *s* and *p* blocks. The abbreviated electron configuration consists of the symbol of the preceding noble gas, followed by the electron configuration in the next period.
  a. Ti $[\text{Ar}]4s^2 3d^2$
  b. Br $[\text{Ar}]4s^2 3d^{10}4p^5$
  c. Ba $[\text{Xe}]6s^2$
  d. Pb $[\text{Xe}]6s^2 4f^{14}5d^{10}6p^2$

5.45 Use the final sublevel notation in the electron configuration to locate the element.
  a. P (ends in $3p^3$)   **b.** Co (ends in $3d^7$)
  c. Zn (ends in $3d^{10}$)   **d.** Bi (ends in $6p^3$)

5.47 **a.** Ga has three electrons in energy level $n = 4$; two are in the 4*s* block, and one is in the 4*p* block.
  b. N is the third element in the 2*p* block; it has three 2*p* electrons.
  c. Xe is the final element (6 electrons) in the 5*p* block.
  d. Zr is the second element in the 4*d* block; it has two 4*d* electrons.

5.49 **a.** Zn is the tenth element in the 3*d* block; it has 10 3*d* electrons.
  b. Na has one electron in the 3*s* block; the 2*p* block in Na is complete with six electrons.
  c. As is the third element in the 4*p* block; it has three 4*p* electrons.
  d. Rb is the first element in the 5*s* block; Rb has one 5*s* electron.

5.51 The group numbers 1A to 8A indicate the number of valence (outer) electrons for the elements in each vertical column.

5.53 **a.** An element with two valence electrons is in Group 2A (2).
  b. An element with five valence electrons is in Group 5A (15).
  c. An element with two valence electrons and five electrons in the *d* block is in Group 7B (7).
  d. An element with six valence electrons is in Group 6A (16).

5.55 **a.** Alkali metals, which are in Group 1A (1), have a valence electron configuration ns^1.
  b. Elements in Group 4A (14) have a valence electron configuration $ns^2 np^2$.
  c. Elements in Group 7A (17), have a valence electron configuration $ns^2 np^5$.
  d. Elements in Group 5A (15) have a valence electron configuration $ns^2 np^3$.

5.57 **a.** Aluminum in Group 3A (13) has three valence electrons.
  b. Any element in Group 5A (15) has five valence electrons.
  c. Barium in Group 2A (2) has two valence electrons.
  d. Each halogen in Group 7A (17) has seven valence electrons.

5.59 The number of dots is equal to the number of valence electrons, as indicated by the group number.
  a. Sulfur is in Group 6A (16); $\cdot\ddot{\underset{\cdot\cdot}{\text{S}}}\colon$

  b. Nitrogen is in Group 5A (15); $\cdot\ddot{\underset{\cdot}{\text{N}}}\cdot$

  c. Calcium is in Group 2A (2); $\overset{\cdot}{\text{Ca}}\cdot$

  d. Sodium is in Group 1A (1); $\text{Na}\cdot$

  e. Gallium is in Group 3A (13); $\cdot\overset{\cdot}{\text{Ga}}\cdot$

5.61 The atomic size of representative elements decreases going across a period from Group 1A to 8A and increases going down a group.
  a. In Period 3, Na, which is on the left, is larger than Cl.
  b. In Group 1A (1), Rb, which is farther down the group, is larger than Na.
  c. In Period 3, Na, which is on the left, is larger than Mg.
  d. In Period 5, Rb, which is on the left, is larger than I.

5.63 **a.** The atomic size of representative elements decreases from Group 1A to 8A: Mg, Al, Si.
  b. The atomic size of representative elements increases going down a group: I, Br, Cl.
  c. The atomic size of representative elements decreases from Group 1A to 8A: Sr, Sb, I.
  d. The atomic size of representative elements decreases from Group 1A to 8A: Na, Si, P.

5.65 **a.** In Br, the valence electrons are closer to the nucleus, so Br has a higher ionization energy than I.
  b. In Mg, the valence electrons are closer to the nucleus, so Mg has a higher ionization energy than Sr.
  c. Attraction for the valence electrons increases going from left to right across a period, giving P a higher ionization energy than Si.
  d. Attraction for the valence electrons increases going from left to right across a period, giving Xe a higher ionization energy than I.

5.67 **a.** Br, Cl, F; ionization energy decreases going down a group.
  b. Na, Al, Cl; going across a period from left to right, ionization energy increases.
  c. Cs, K, Na; ionization energy decreases going down a group.
  d. Ca, As, Br; going across a period from left to right, ionization energy increases.

5.69 Ca, Ga, Ge, Br; since metallic character decreases from left to right across a period, Ca, in Group 2A (2), will be the most metallic, followed by Ga, then Ge, and finally Br, in Group 7A (17).

5.71 Sr has a <u>lower</u> ionization energy and is <u>more</u> metallic than Sb.

5.73 Going down Group 6A (16),
  a. 1; the ionization energy <u>decreases</u>
  b. 2; the atomic size <u>increases</u>
  c. 2; the metallic character <u>increases</u>
  d. 3; the number of valence electrons <u>remains the same</u>

5.75 In Period 2, an atom of N compared to an atom of Li has a larger (greater)
  a. False; N has a smaller atomic size than Li.
  b. True
  c. True
  d. False; N has less metallic character than Li.
  e. True

5.77 **a.** 49 **b.** 49 **c.** $1s^2 2s^2 2p^6 3s^2 3p^6 4s^2 3d^{10} 4p^6 5s^2 4d^{10} 5p^1$; $[Kr] 5s^2 4d^{10} 5p^1$
  d. 3A (13), •İn•
  e. Indium has a larger atomic size than iodine.
  f. Iodine has a higher ionization energy than indium.

5.79 **a.** C has the longest wavelength.
  b. A has the shortest wavelength.
  c. A has the highest frequency.
  d. C has the lowest frequency.

5.81 **a.** An orbital with two lobes is a *p* orbital.
    b. A spherical orbital is an *s* orbital.
    c. An orbital with two lobes is a *p* orbital.

5.83 **a.** This orbital diagram is possible. The orbitals up to $3s^2$ are filled with electrons having opposite spins. The element represented would be magnesium.
    b. Not possible. The $2p$ sublevel would fill completely before the $3s$, and only 2 electrons are allowed in an orbital (not 3 as shown in the $3s$).

5.85 Atomic size increases going down a group. Li is **D** because it would be smallest. Na is **A**, K is **C**, and Rb is **B**.

5.87 A continuous spectrum from white light contains wavelengths of all energies. Atomic spectra are line spectra in which a series of lines corresponds to energy emitted when electrons drop from a higher energy level to a lower level.

5.89 The Pauli exclusion principle states that two electrons in the same orbital must have opposite spins.

5.91 A $4p$ orbital is possible because the $n = 4$ energy level has four sublevels, including a *p* sublevel. A $2d$ orbital is not possible because the $n = 2$ energy level has only *s* and *p* sublevels. There are no $3f$ orbitals because only *s*, *p*, and *d* sublevels are allowed for the $n = 3$ energy level. A $5f$ sublevel is possible in the $n = 5$ energy level because five sublevels are allowed, including $5f$.

5.93 **a.** The $3p$ sublevel starts to fill after completion of the $3s$ sublevel.
    b. The $5s$ sublevel starts to fill after completion of the $4p$ sublevel.
    c. The $4p$ sublevel starts to fill after completion of the $3d$ sublevel.
    d. The $4s$ sublevel starts to fill after completion of the $3p$ sublevel.

5.95 **a.** Iron is the sixth element in the $3d$ block; iron has six $3d$ electrons.
    b. Barium has a completely filled $5p$ sublevel, which is six $5p$ electrons.
    c. Iodine has a completely filled $4d$ sublevel, which is 10 $4d$ electrons.
    d. Radium has a filled $7s$ sublevel, which is two $7s$ electrons.

5.97 Ca, Sr, and Ba all have two valence electrons, ns^2, which places them in Group 2A (2), the alkaline earth metals.

5.99 **a.** X is a metal; Y and Z are nonmetals.
    b. X has the largest atomic size.
    c. Y has the highest ionization energy.
    d. Y has the smallest atomic size.

5.101 **a.** Phosphorus in Group 5A (15) has an electron configuration that ends with $3s^23p^3$.
    b. Lithium is the alkali metal that is highest in Group 1A (1) and has the smallest atomic size (H in Group 1A (1) is a nonmetal).
    c. Cadmium in Period 5 has a complete $4d$ sublevel with 10 electrons.
    d. Nitrogen at the top of Group 5A (15) has the highest ionization energy in that group.
    e. Sodium, the first element in Period 3, has the largest atomic size of that period.

5.103 Calcium has a greater number of protons than K; this greater nuclear charge makes it more difficult to remove an electron from Ca. The least tightly bound (valence) electron in Ca is farther from the nucleus than in Mg, and less energy is needed to remove it.

5.105 **a.** Since metallic character increases going down a group, Sb will be more metallic than As.
    b. Since metallic character decreases going across a period from left to right, Sn will be more metallic than Sb.
    c. Since metallic character decreases going across a period from left to right, P will be more metallic than Cl.
    d. Since metallic character increases going down a group and decreases going across a period from left to right, P will be more metallic than O.

5.107 a. Na on the far left of the heavy zigzag line on the periodic table is a metal.
   b. P is found in Group 5A (15).
   c. F at the top of Group 7A (17) and to the far right in Period 2 has the highest ionization energy.
   d. Na has the lowest ionization energy and loses an electron most easily.
   e. Cl is found in Period 3 in Group 7A.

5.109 a. Si $[Ne]3s^2 3p^2$; Group 4A (14)   **b.** Se $[Ar]4s^2 3d^{10} 4p^4$; Group 6A (16)
   c. Mn $[Ar]4s^2 3d^5$; Group 7B (7)   **d.** Sb $[Kr]5s^2 4d^{10} 5p^3$; Group 5A (15)

5.111 The number of dots is equal to the number of valence electrons, as indicated by the group number.
   a. Barium is in Group 2A (2); $\overset{\bullet}{Ba}\bullet$

   b. Fluorine is in Group 7A (17); $\cdot\overset{\bullet\bullet}{\underset{\bullet\bullet}{F}}\colon$

   c. Krypton is in Group 8A (18); $\colon\overset{\bullet\bullet}{\underset{\bullet\bullet}{Kr}}\colon$

   d. Arsenic is in Group 5A (15); $\cdot\overset{\bullet\bullet}{\underset{\bullet}{As}}\cdot$

5.113 a. Atomic size increases going down a group, which gives O the smallest atomic size in Group 6A (16).
   b. Atomic size decreases going from left to right across a period, which gives Ar the smallest atomic size in Period 3.
   c. Ionization energy decreases going down a group, which gives B the highest ionization energy in Group 3A (13).
   d. Ionization energy increases going from left to right across a period, which gives Na the lowest ionization energy in Period 3.
   e. Ru in Period 5 has six electrons in the $4d$ sublevel.

5.115 The series of lines separated by dark sections in atomic spectra indicate that the energy emitted by the elements is not continuous and that electrons are moving between discrete energy levels.

5.117 An energy level contains all the electrons with similar energy. A sublevel contains electrons with the same energy, while an orbital is the region around the nucleus where electrons of a certain energy are most likely to be found.

5.119 S has a larger atomic size than Cl; Cl is larger than F: S > Cl > F. F has a higher ionization energy than Cl; Cl has a higher ionization energy than S: F > Cl > S.

Ionic and Molecular Compounds

Sarah, a pharmacist, talked with her customer regarding questions about stomach upset from the low-dose aspirin, $C_9H_8O_4$, which he was taking to prevent a heart attack or stroke. Sarah explained that when he takes a low-dose aspirin, he needs to drink a glass of water. If his stomach is still upset, he can take the aspirin with food or milk each time. A few days later, Sarah talked with her customer again. He told her he followed her suggestions and that he no longer had any upset stomach from the low-dose aspirin. Is aspirin an ionic or a molecular compound?

Credit: Design Pics Inc./Alamy

CHAPTER READINESS*

✸✸ Key Math Skills

- Using Positive and Negative Numbers in Calculations (1.4)
- Solving Equations (1.4)

⚬ Core Chemistry Skills

- Writing Electron Configurations (5.4)
- Drawing Lewis Symbols (5.6)

*These Key Math Skills and Core Chemistry Skills from previous chapters are listed here for your review as you proceed to the new material in this chapter.

LOOKING AHEAD

6.1 Ions: Transfer of Electrons

Learning Goal: Write the symbols for the simple ions of the representative elements.

- The stability of the noble gases is associated with a stable electron configuration of 8 electrons, an octet, in their outermost energy level. Helium is stable with 2 electrons in the $n = 1$ energy level.
- Atoms of elements other than the noble gases achieve stability by losing, gaining, or sharing valence electrons with other atoms in the formation of compounds.
- Metals of the representative elements in Groups 1A (1), 2A (2), and 3A (13) achieve a noble gas electron configuration by losing their valence electron(s) to form positively charged cations with a charge of 1+, 2+, or 3+, respectively.
- Nonmetals in Groups 5A (15), 6A (16), and 7A (17) gain valence electrons to achieve an octet forming negatively charged anions with a charge of 3−, 2−, or 1−, respectively.

- *Ionic bonds* occur when the valence electrons of atoms of a metal are transferred to atoms of nonmetals. For example, sodium atoms lose electrons and chlorine atoms gain electrons to form the ionic compound NaCl.
- *Covalent bonds* form when atoms of nonmetals share valence electrons. In the molecular compounds H_2O and C_3H_8, atoms share electrons.

Study Note

When an atom loses or gains electrons, it acquires the electron configuration of the nearest noble gas. For example, sodium loses 1 electron, which gives the Na^+ ion the electron configuration of neon. Oxygen gains 2 electrons to give an oxide ion, O^{2-}, the electron configuration of neon.

♦ **Learning Exercise 6.1A**

The following metals, which are essential to health, lose electrons when they form ions. Indicate the group number, the number of electrons lost, and the ion (symbol and charge) for each of the following:

🦴 **CORE CHEMISTRY SKILL**

Writing Positive and Negative Ions

Metal	Group Number	Electrons Lost	Ion Formed
Magnesium			
Sodium			
Calcium			
Potassium			
Zinc			

Answers

Metal	Group Number	Electrons Lost	Ion Formed
Magnesium	2A (2)	2	Mg^{2+}
Sodium	1A (1)	1	Na^+
Calcium	2A (2)	2	Ca^{2+}
Potassium	1A (1)	1	K^+
Zinc	2B (12)	2	Zn^{2+}

♦ **Learning Exercise 6.1B**

The following nonmetals, which are essential to health, gain electrons when they form ions. Indicate the group number, the number of electrons gained, and the ion (symbol and charge) for each of the following:

Nonmetal	Group Number	Electrons Gained	Ion Formed
Chlorine			
Oxygen			
Nitrogen			
Iodine			
Sulfur			

Answers

Nonmetal	Group Number	Electrons Gained	Ion Formed
Chlorine	7A (17)	1	Cl^-
Oxygen	6A (16)	2	O^{2-}
Nitrogen	5A (15)	3	N^{3-}
Iodine	7A (17)	1	I^-
Sulfur	6A (16)	2	S^{2-}

6.2   Ionic Compounds

Learning Goal: Using charge balance, write the correct formula for an ionic compound.

- In the formulas of ionic compounds, the total positive charge is equal to the total negative charge. For example, the compound magnesium chloride, $MgCl_2$, contains Mg^{2+} and two Cl^-. The sum of the charges is zero: $1(2+) + 2(1-) = 0$.
- When two or more ions are needed for charge balance, that number is indicated by subscripts in the formula.
- A subscript of 1 is understood and not written.

For the following exercises, you may want to cut pieces of paper that represent typical positive and negative ions, as shown. To determine an ionic formula, place the smallest number of positive and negative pieces together to complete a geometric shape. Write the number of positive ions and negative ions as the subscripts for the formula.

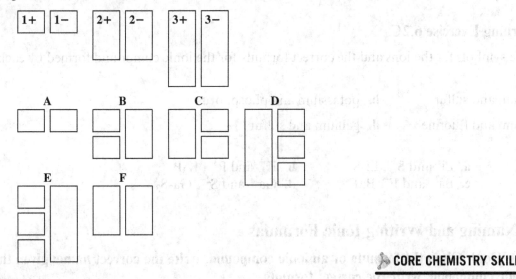

CORE CHEMISTRY SKILL

Writing Ionic Formulas

♦ **Learning Exercise 6.2A**

Write the letter (A to F) that matches the arrangement of ions in the following compounds:

Compound	Combination		Compound	Combination
a. $MgCl_2$	_____		**b.** Na_2S	_____
c. LiCl	_____		**d.** CaO	_____
e. K_3N	_____		**f.** $AlBr_3$	_____
g. MgS	_____		**h.** $BaCl_2$	_____

Answers   **a.** C   **b.** B   **c.** A   **d.** F   **e.** E   **f.** D   **g.** F   **h.** C

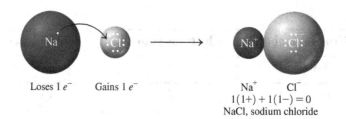

Loses 1 e^- Gains 1 e^-

Na^+ Cl^-
$1(1+) + 1(1-) = 0$
NaCl, sodium chloride

Study Note

You can check that the ionic formula you write is electrically neutral by multiplying each ionic charge by its subscript. When added together, their sum should equal zero. For example, the formula Na_2O gives $2(1+) + 1(2-) = (2+) + (2-) = 0$.

♦ **Learning Exercise 6.2B**

Write the correct formula for the ionic compound formed from each of the following pairs of ions:

a. Na^+ and Cl^- _____

b. K^+ and S^{2-} _____

c. Al^{3+} and O^{2-} _____

d. Mg^{2+} and Cl^- _____

e. Ca^{2+} and S^{2-} _____

f. Al^{3+} and Cl^- _____

g. Li^+ and N^{3-} _____

h. Ba^{2+} and P^{3-} _____

Answers   **a.** NaCl   **b.** K_2S **c.** Al_2O_3 **d.** $MgCl_2$
                e. CaS   **f.** $AlCl_3$ **g.** Li_3N **h.** Ba_3P_2

♦ **Learning Exercise 6.2C**

Write the symbols for the ions and the correct formula for the ionic compound formed by each of the following:

a. lithium and sulfur   **b.** potassium and phosphorus

c. barium and fluorine   **d.** gallium and sulfur

Answers   **a.** Li^+ and S^{2-}, Li_2S **b.** K^+ and P^{3-}, K_3P
                c. Ba^{2+} and F^-, BaF_2 **d.** Ga^{3+} and S^{2-}, Ga_2S_3

6.3   Naming and Writing Ionic Formulas

Learning Goal: Given the formula of an ionic compound, write the correct name; given the name of an ionic compound, write the correct formula.

- In naming ionic compounds, the positive ion is named first, followed by the name of the negative ion. The name of a representative metal ion in Group 1A (1), 2A (2), or 3A (13) is the same as its element name. The name of a nonmetal ion is obtained by replacing the end of its element name with *ide*.

- Most transition elements form cations with two or more ionic charges. Then the ionic charge must be written as a Roman numeral in parentheses after the name of the metal. For example, the cations of iron, Fe^{2+} and Fe^{3+}, are named iron(II) and iron(III). The ions of copper are Cu^+, copper(I), and Cu^{2+}, copper(II).

- The only transition elements with fixed charges are zinc, Zn^{2+}; silver, Ag^+; and cadmium, Cd^{2+}; no Roman numerals are used when naming their cations in ionic compounds.

Guide to Naming Ionic Compounds with Metals That Form a Single Ion	
STEP 1	Identify the cation and anion.
STEP 2	Name the cation by its element name.
STEP 3	Name the anion by using the first syllable of its element name followed by *ide*.
STEP 4	Write the name for the cation first and the name for the anion second.

♦ **Learning Exercise 6.3A**

Write the symbols for the ions and the correct name for each of the following ionic compounds:

Formula   **Ions**   **Name**

a. Cs_2O _____ _____   _____

b. $BaBr_2$ _____ _____   _____

c. Ca_3P_2 _____ _____   _____

d. Na_2S _____ _____   _____

Answers   **a.** Cs^+, O^{2-}, cesium oxide   **b.** Ba^{2+}, Br^-, barium bromide
             c. Ca^{2+}, P^{3-}, calcium phosphide   **d.** Na^+, S^{2-}, sodium sulfide

Guide to Naming Ionic Compounds with Variable Charge Metals	
STEP 1	Determine the charge of the cation from the anion.
STEP 2	Name the cation by its element name and use a Roman numeral in parentheses for the charge.
STEP 3	Name the anion by using the first syllable of its element name followed by *ide*.
STEP 4	Write the name for the cation first and the name for the anion second.

♦ **Learning Exercise 6.3B**

Write the name for each of the following ions:

a. Cl^- _____   **b.** Fe^{2+} _____   **c.** Cu^+ _____

d. Ag^+ _____   **e.** O^{2-} _____   **f.** Ca^{2+} _____

g. S^{2-} _____   **h.** Al^{3+} _____   **i.** Fe^{3+} _____

j. Ba^{2+} _____   **k.** Cu^{2+} _____   **l.** N^{3-} _____

Answers   **a.** chloride   **b.** iron(II)   **c.** copper(I)
             d. silver   **e.** oxide   **f.** calcium
             g. sulfide   **h.** aluminum   **i.** iron(III)
             j. barium   **k.** copper(II)   **l.** nitride

♦ Learning Exercise 6.3C

Write the symbol for each of the following ions:

Name	Symbol
a. chromium(III) ion	_____
b. cobalt(II) ion	_____
c. zinc ion	_____
d. lead(IV) ion	_____
e. gold(I) ion	_____
f. silver ion	_____
g. potassium ion	_____
h. nickel(III) ion	_____

Answers   **a.** Cr^{3+} **b.** Co^{2+} **c.** Zn^{2+}
d. Pb^{4+} **e.** Au^+ **f.** Ag^+
g. K^+ **h.** Ni^{3+}

Some Metals That Form More Than One Positive Ion

Element	Possible Ions	Name of Ion
Bismuth	Bi^{3+}	Bismuth(III)
	Bi^{5+}	Bismuth(V)
Chromium	Cr^{2+}	Chromium(II)
	Cr^{3+}	Chromium(III)
Cobalt	Co^{2+}	Cobalt(II)
	Co^{3+}	Cobalt(III)
Copper	Cu^+	Copper(I)
	Cu^{2+}	Copper(II)
Gold	Au^+	Gold(I)
	Au^{3+}	Gold(III)
Iron	Fe^{2+}	Iron(II)
	Fe^{3+}	Iron(III)
Lead	Pb^{2+}	Lead(II)
	Pb^{4+}	Lead(IV)
Manganese	Mn^{2+}	Manganese(II)
	Mn^{3+}	Manganese(III)
Mercury	Hg_2^{2+}	Mercury(I)*
	Hg^{2+}	Mercury(II)
Nickel	Ni^{2+}	Nickel(II)
	Ni^{3+}	Nickel(III)
Tin	Sn^{2+}	Tin(II)
	Sn^{4+}	Tin(IV)

*Mercury(I) ions form an ion pair with a 2+ charge.

⚲ CORE CHEMISTRY SKILL

Naming Ionic Compounds

♦ Learning Exercise 6.3D

Write the symbols for the ions and the correct name for each of the following ionic compounds:

Formula	Ions		Name
a. $CrCl_2$	_____	_____	_____
b. $SnBr_4$	_____	_____	_____
c. Na_3P	_____	_____	_____
d. Ni_2O_3	_____	_____	_____
e. CuO	_____	_____	_____
f. Mg_3N_2	_____	_____	_____

Answers   **a.** Cr^{2+}, Cl^-, chromium(II) chloride   **b.** Sn^{4+}, Br^-, tin(IV) bromide
c. Na^+, P^{3-}, sodium phosphide   **d.** Ni^{3+}, O^{2-}, nickel(III) oxide
e. Cu^{2+}, O^{2-}, copper(II) oxide   **f.** Mg^{2+}, N^{3-}, magnesium nitride

Guide to Writing Formulas from the Name of an Ionic Compound	
STEP 1	Identify the cation and anion.
STEP 2	Balance the charges.
STEP 3	Write the formula, cation first, using subscripts from the charge balance.

♦ **Learning Exercise 6.3E**

Write the symbols for the ions and the correct formula for each of the following ionic compounds:

Compound	Cation	Anion	Formula of Compound
Aluminum sulfide			
Lead(II) chloride			
Barium oxide			
Gold(III) bromide			
Silver oxide			

Answers

Compound	Cation	Anion	Formula of Compound
Aluminum sulfide	Al^{3+}	S^{2-}	Al_2S_3
Lead(II) chloride	Pb^{2+}	Cl^-	$PbCl_2$
Barium oxide	Ba^{2+}	O^{2-}	BaO
Gold(III) bromide	Au^{3+}	Br^-	$AuBr_3$
Silver oxide	Ag^+	O^{2-}	Ag_2O

6.4   Polyatomic Ions

Learning Goal: Write the name and formula for an ionic compound containing a polyatomic ion.

Fertilizer
NH_4NO_3

- A polyatomic ion is a group of nonmetal atoms that carries an electrical charge, usually negative, 1−, 2−, or 3−. The polyatomic ion NH_4^+, ammonium ion, has a positive charge.

- Polyatomic ions cannot exist alone but are combined with an ion of the opposite charge.

- The names of ionic compounds containing polyatomic anions often end with *ate* or *ite*.

- When there is more than one polyatomic ion in the formula for a compound, the entire polyatomic ion formula is enclosed in parentheses and the subscript written outside the parentheses.

NH_4^+ NO_3^-
Ammonium ion   Nitrate ion

Credit: Pearson Education / Pearson Science

Study Note

By learning the most common polyatomic ions such as nitrate NO_3^-, carbonate CO_3^{2-}, sulfate SO_4^{2-}, and phosphate PO_4^{3-}, you can derive their related polyatomic ions. For example, the nitrite ion, NO_2^-, has one oxygen atom less than the nitrate ion: NO_3^-, nitrate, and NO_2^-, nitrite.

Names and Formulas of Some Common Polyatomic Ions

Nonmetal	Formula of Ion*	Name of Ion
Hydrogen	OH^-	Hydroxide
Nitrogen	NH_4^+	Ammonium
	NO_3^-	**Nitrate**
	NO_2^-	Nitrite
Chlorine	ClO_4^-	Perchlorate
	ClO_3^-	**Chlorate**
	ClO_2^-	Chlorite
	ClO^-	Hypochlorite
Carbon	**CO_3^{2-}**	**Carbonate**
	HCO_3^-	Hydrogen carbonate (or bicarbonate)
	CN^-	Cyanide
	$C_2H_3O_2^-$	Acetate
Sulfur	**SO_4^{2-}**	**Sulfate**
	HSO_4^-	Hydrogen sulfate (or bisulfate)
	SO_3^{2-}	Sulfite
	HSO_3^-	Hydrogen sulfite (or bisulfite)
Phosphorus	**PO_4^{3-}**	**Phosphate**
	HPO_4^{2-}	Hydrogen phosphate
	$H_2PO_4^-$	Dihydrogen phosphate
	PO_3^{3-}	Phosphite

*Formulas and names in bold type indicate the most common polyatomic ion for that element.

♦ **Learning Exercise 6.4A**

Write the polyatomic ion (symbol and charge) for each of the following:

a. sulfate ion _____   **b.** hydroxide ion _____

c. carbonate ion _____   **d.** sulfite ion _____

e. ammonium ion _____   **f.** phosphate ion _____

g. nitrate ion _____   **h.** nitrite ion _____

Answers
a. SO_4^{2-} **b.** OH^- **c.** CO_3^{2-}
d. SO_3^{2-} **e.** NH_4^+ **f.** PO_4^{3-}
g. NO_3^- **h.** NO_2^-

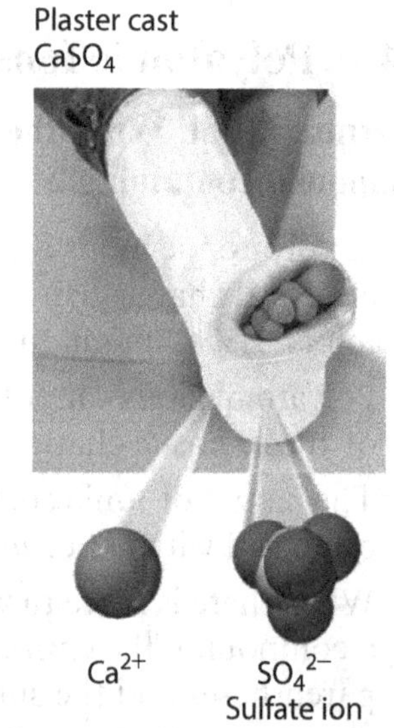

Plaster cast
$CaSO_4$

Ca^{2+} SO_4^{2-}
Sulfate ion
The sulfate ion in $CaSO_4$ is a poly-atomic ion, which is a group of atoms that have an ionic charge.

Credit: Pearson Education / Pearson Science

Guide to Writing Formulas with Polyatomic Ions	
STEP 1	Identify the cation and polyatomic ion (anion).
STEP 2	Balance the charges.
STEP 3	Write the formula, cation first, using the subscripts from charge balance.

♦ **Learning Exercise 6.4B**

Write the formulas for the ions and the correct formula for each of the following compounds:

Compound	Cation	Anion	Formula
Sodium phosphate			
Iron(II) hydroxide			
Ammonium carbonate			
Silver bicarbonate			
Chromium(III) sulfate			
Lead(II) nitrate			
Potassium sulfite			
Barium phosphate			

Answers

Compound	Cation	Anion	Formula
Sodium phosphate	Na^+	PO_4^{3-}	Na_3PO_4
Iron(II) hydroxide	Fe^{2+}	OH^-	$Fe(OH)_2$
Ammonium carbonate	NH_4^+	CO_3^{2-}	$(NH_4)_2CO_3$
Silver bicarbonate	Ag^+	HCO_3^-	$AgHCO_3$
Chromium(III) sulfate	Cr^{3+}	SO_4^{2-}	$Cr_2(SO_4)_3$
Lead(II) nitrate	Pb^{2+}	NO_3^-	$Pb(NO_3)_2$
Potassium sulfite	K^+	SO_3^{2-}	K_2SO_3
Barium phosphate	Ba^{2+}	PO_4^{3-}	$Ba_3(PO_4)_2$

Guide to Naming Ionic Compounds with Polyatomic Ions	
STEP 1	Identify the cation and polyatomic ion (anion).
STEP 2	Name the cation using a Roman numeral, if needed.
STEP 3	Name the polyatomic ion.
STEP 4	Write the name for the compound, cation first and polyatomic ion second.

◆ **Learning Exercise 6.4C**

Write the symbols for the ions and the correct name for each of the following ionic compounds:

Formula	Ions		Name
a. $Ba(NO_3)_2$	_____	_____	_____
b. $Fe_2(SO_4)_3$	_____	_____	_____
c. Na_3PO_3	_____	_____	_____
d. $Al(ClO_3)_3$	_____	_____	_____
e. NH_4NO_2	_____	_____	_____
f. $Cr(OH)_2$	_____	_____	_____

Answers

a. Ba^{2+}, NO_3^-, barium nitrate
c. Na^+, PO_3^{3-}, sodium phosphite
e. NH_4^+, NO_2^-, ammonium nitrite

b. Fe^{3+}, SO_4^{2-}, iron(III) sulfate
d. Al^{3+}, ClO_3^-, aluminum chlorate
f. Cr^{2+}, OH^-, chromium(II) hydroxide

6.5 Molecular Compounds: Sharing Electrons

Learning Goal: Given the formula of a molecular compound, write its correct name; given the name of a molecular compound, write its formula.

- Molecular compounds are composed of nonmetals bonded together to give discrete units called *molecules*.
- The name of a molecular compound is written using the element name of the first nonmetal and the first syllable of the second nonmetal name followed by the suffix *ide*.
- Prefixes are used to indicate the number of atoms of each nonmetal in the formula: mono (1), di (2), tri (3), tetra (4), penta (5), hexa (6), hepta (7), octa (8), nona (9), deca (10).
- In the name of a molecular compound, the prefix *mono* is usually omitted, as in NO, nitrogen oxide. Traditionally, however, CO is named carbon monoxide.
- If the vowels *o* and *o* or *a* and *o* appear together, the first vowel is omitted.
- The formula of a molecular compound is written using the symbols of the nonmetals in the name followed by subscripts determined from the prefixes.

Key Terms for Sections 6.1 to 6.5

Match each of the following key terms with the correct description:

a. ion	**b.** cation	**c.** anion
d. polyatomic ion	**e.** ionic bond	**f.** covalent bond

1. _____ a positively charged ion

2. _____ a negatively charged ion

3. _____ the attraction between atoms formed by sharing electrons

4. _____ a group of covalently bonded nonmetal atoms that has an overall charge

5. _____ an atom or group of atoms having an electrical charge

6. _____ the attraction between positively charged and negatively charged ions

Answers   **1.** b   **2.** c   **3.** f   **4.** d   **5.** a   **6.** e

Guide to Naming Molecular Compounds	
STEP 1	Name the first nonmetal by its element name.
STEP 2	Name the second nonmetal by using the first syllable of its element name followed by *ide*.
STEP 3	Add prefixes to indicate the number of atoms (subscripts).

♦ **Learning Exercise 6.5A**

▶ **CORE CHEMISTRY SKILL**

Writing the Names and Formulas for Molecular Compounds

Name the following molecular compounds:

a. CS_2 _____

b. PCl_3 _____

c. CO _____

d. SO_3 _____

e. N_2O_4 _____

f. CCl_4 _____

g. P_4S_6 _____

h. IF_7 _____

i. ClO_2 _____

j. S_2O _____

Answers
  a. carbon disulfide
  c. carbon monoxide
  e. dinitrogen tetroxide
  g. tetraphosphorus hexasulfide
  i. chlorine dioxide

  b. phosphorus trichloride
  d. sulfur trioxide
  f. carbon tetrachloride
  h. iodine heptafluoride
  j. disulfur oxide

Guide to Writing Formulas for Molecular Compounds	
STEP 1	Write the symbols in the order of the elements in the name.
STEP 2	Write any prefixes as subscripts.

♦ **Learning Exercise 6.5B**

Write the formula for each of the following molecular compounds:

a. dinitrogen oxide _____

b. silicon tetrabromide _____

c. bromine pentafluoride _____

d. carbon dioxide _____

e. sulfur hexafluoride _____

f. oxygen difluoride _____

g. phosphorus octasulfide _____

h. selenium hexafluoride _____

i. iodine chloride _____

j. xenon trioxide _____

Answers   a. N_2O b. $SiBr_4$ c. BrF_5 d. CO_2 e. SF_6
          f. OF_2 g. PS_8 h. SeF_6 i. ICl   j. XeO_3

Summary of Writing Formulas and Names

- In ionic compounds containing two different elements, the first takes its element name. The ending of the name of the second element is replaced by *ide*. For example, $BaCl_2$ is named *barium chloride*. If the metal forms two or more positive ions, a Roman numeral is added to its name to indicate the ionic charge in the compound. For example, $FeCl_3$ is named *iron(III) chloride*.

- In ionic compounds with three or more elements, a group of atoms is named as a polyatomic ion. The names of negative polyatomic ions end in *ate* or *ite*, except for hydroxide and cyanide. For example, Na_2SO_4 is named *sodium sulfate*. No prefixes are used.
- When a polyatomic ion occurs two or more times in a formula, its formula is placed inside parentheses, and the number of ions is shown as a subscript after the parentheses $Ca(NO_3)_2$. If the subscript is 1, it is not written.
- In naming molecular compounds, a prefix before the name of one or both nonmetal names indicates the numerical value of a subscript. For example, N_2O_3 is named *dinitrogen trioxide*.

♦ **Learning Exercise 6.5C**

Indicate the type of compound (ionic or molecular). Then give the formula and name for each.

Formula	Ionic or Molecular?	Name
$MgCl_2$		
		nitrogen trichloride
		potassium sulfate
Li_2O		
CBr_4		
Na_3PO_4		
		dihydrogen sulfide
		calcium hydrogen carbonate (bicarbonate)

Answers

Formula	Ionic or Molecular?	Name
$MgCl_2$	ionic	magnesium chloride
NCl_3	molecular	nitrogen trichloride
K_2SO_4	ionic	potassium sulfate
Li_2O	ionic	lithium oxide
CBr_4	molecular	carbon tetrabromide
Na_3PO_4	ionic	sodium phosphate
H_2S	molecular	dihydrogen sulfide
$Ca(HCO_3)_2$	ionic	calcium hydrogen carbonate (bicarbonate)

Checklist for Chapter 6

You are ready to take the Practice Test for Chapter 6. Be sure you have accomplished the following learning goals for this chapter. If not, review the section listed at the end of the goal. Then apply your new skills and understanding to the Practice Test.

After studying Chapter 6, I can successfully:

_____ Illustrate the octet rule for the formation of ions. (6.1)

_____ Write the formulas for ionic compounds containing the ions of metals and nonmetals of representative elements. (6.2)

_____ Use charge balance to write an ionic formula. (6.2)

_____ Write the name for an ionic compound. (6.3)

_____ Write the formula for a compound containing a polyatomic ion. (6.4)

_____ Write the names and formulas for molecular compounds. (6.5)

Practice Test for Chapter 6

The chapter sections to review are shown in parentheses at the end of each question.

For questions 1 through 4, consider an atom of phosphorus: (6.1)

1. Phosphorus is in Group
   A. 2A (2)   **B.** 3A (13)   **C.** 5A (15)   **D.** 7A (17)   **E.** 8A (18)

2. How many valence electrons does an atom of phosphorus have?
   A. 2   **B.** 3   **C.** 5   **D.** 8   **E.** 15

3. To achieve an octet, a phosphorus atom in an ionic compound will
   A. lose 1 electron   **B.** lose 2 electrons   **C.** lose 5 electrons
   D. gain 2 electrons   **E.** gain 3 electrons

4. A phosphide ion has a charge of
   A. 1+   **B.** 2+   **C.** 5+   **D.** 2−   **E.** 3−

5. To achieve an octet, a calcium atom (6.1)
   A. loses 1 electron   **B.** loses 2 electrons   **C.** loses 3 electrons
   D. gains 1 electron   **E.** gains 2 electrons

6. The correct ionic charge for a calcium ion is (6.1)
   A. 1+   **B.** 2+   **C.** 1−   **D.** 2−   **E.** 3−

7. Another name for a positive ion is (6.1)
   A. anion   **B.** cation   **C.** proton   **D.** positron   **E.** sodium

8. To achieve an octet, a chlorine atom (6.1)
   A. loses 1 electron   **B.** loses 2 electrons   **C.** loses 3 electrons
   D. gains 1 electron   **E.** gains 2 electrons

9. The correct ionic charge for a silver ion is (6.1)
   A. 1+   **B.** 2+   **C.** 1−   **D.** 2−   **E.** 3−

10. The correct ionic charge for a fluoride ion is (6.1)
    A. 1+   **B.** 2+   **C.** 1−   **D.** 2−   **E.** 3−

11. When the elements magnesium and sulfur react, (6.2)
    A. an ionic compound forms
    B. a molecular compound forms
    C. no reaction occurs
    D. the two repel each other and will not combine
    E. none of the above

12. An ionic bond typically occurs between (6.2)
    A. two different nonmetals
    B. two of the same type of nonmetals
    C. two noble gases
    D. two different metals
    E. a metal and a nonmetal

13. The formula for a compound between sodium and sulfur is (6.2)
    A. SoS   **B.** NaS   **C.** Na_2S **D.** NaS_2 **E.** Na_2SO_4

14. The formula for a compound between aluminum and oxygen is (6.2)
    A. AlO   **B.** Al_2O **C.** AlO_3 **D.** Al_2O_3 **E.** Al_3O_2

15. The formula for a compound between barium and sulfur is (6.2)
    A. BaS   **B.** Ba_2S **C.** BaS_2 **D.** Ba_2S_2 **E.** $BaSO_4$

16. The correct formula for iron(III) fluoride is (6.2, 6.3)
    A. FeF   **B.** FeF_2 **C.** Fe_2F **D.** Fe_3F **E.** FeF_3

17. The correct formula for copper(II) chloride is (6.2, 6.3)
    A. CoCl   **B.** CuCl   **C.** $CoCl_2$ **D.** $CuCl_2$ **E.** Cu_2Cl

18. The correct formula for silver oxide is (6.2, 6.3)
    A. AgO   **B.** Ag_2O **C.** AgO_2 **D.** Ag_3O_2 **E.** Ag_3O

19. The correct ionic charge for a phosphate ion is (6.4)
    A. 1+   **B.** 2+   **C.** 1−   **D.** 2−   **E.** 3−

20. The correct ionic charge for a sulfate ion is (6.4)
    A. 1+   **B.** 2+   **C.** 1−   **D.** 2−   **E.** 3−

21. The correct formula for ammonium sulfate is (6.2, 6.4)
    A. AmS   **B.** $AmSO_4$ **C.** $(NH_4)_2S$ **D.** NH_4SO_4 **E.** $(NH_4)_2SO_4$

22. The correct formula for lithium phosphate is (6.2, 6.4)
    A. $LiPO_4$ **B.** Li_2PO_4 **C.** Li_3PO_4 **D.** $Li_2(PO_4)_3$ **E.** $Li_3(PO_4)_2$

23. The correct formula for magnesium carbonate, used as a dietary magnesium supplement, is (6.2, 6.4)
    A. $MgCO_3$ **B.** Mg_2CO_3 **C.** $Mg(CO_3)_2$ **D.** MgCO   **E.** $Mg_2(CO_3)_3$

24. The correct formula for copper(I) sulfate, used as a fungicide, is (6.3, 6.4)
    A. $CuSO_3$ **B.** $CuSO_4$ **C.** Cu_2SO_3 **D.** $Cu(SO_4)_2$ **E.** Cu_2SO_4

25. The name of $Al_2(HPO_4)_3$ is (6.3, 6.4)
    A. aluminum hydrogen phosphite
    B. aluminum hydrogen phosphate
    C. aluminum hydrogen phosphorus
    D. aluminum hydrogen phosphorus oxide
    E. trialuminum dihydrogen phosphate

26. The name of CoS is (6.2, 6.3)
    A. copper sulfide
    B. cobalt(II) sulfate
    C. cobalt(I) sulfide
    D. cobalt sulfide
    E. cobalt(II) sulfide

27. The name of MnCl₂, used as a supplement in intravenous solutions, is (6.2, 6.3)
   A. magnesium chloride   **B.** manganese(II) chlorine   **C.** manganese(II) chloride
   D. manganese chlorine   **E.** manganese(III) chloride

28. The name of ZnCO₃ is (6.3, 6.4)
   A. zinc(III) carbonate   **B.** zinc(II) carbonate   **C.** zinc bicarbonate
   D. zinc carbon trioxide   **E.** zinc carbonate

29. The name of Al₂O₃ is (6.2)
   A. aluminum oxide   **B.** aluminum(II) oxide   **C.** aluminum trioxide
   D. dialuminum trioxide   **E.** aluminum oxygenate

30. The name of Cr₂(SO₃)₃ is (6.3, 6.4)
   A. chromium sulfite   **B.** dichromium trisulfite   **C.** chromium(III) sulfite
   D. chromium(III) sulfate   **E.** chromium sulfate

31. The name of PH₃ is (6.5)
   A. potassium trihydrogen   **B.** phosphorus trihydride   **C.** phosphorus trihydrogen
   D. potassium trihydride   **E.** phosphorus trihydrogenide

32. The name of NCl₃ is (6.5)
   A. nitrogen chloride   **B.** nitrogen trichloride   **C.** trinitrogen chloride
   D. nitrogen chlorine three   **E.** nitrogen trichlorine

33. The name of CO is (6.5)
   A. carbon monoxide   **B.** carbonic oxide   **C.** carbon oxide
   D. carbonious oxide   **E.** carboxide

Answers to the Practice Test

1. C	**2.** C	**3.** E	**4.** E	**5.** B
6. B	**7.** B	**8.** D	**9.** A	**10.** C
11. A	**12.** E	**13.** C	**14.** D	**15.** A
16. E	**17.** D	**18.** B	**19.** E	**20.** D
21. E	**22.** C	**23.** A	**24.** E	**25.** B
26. E	**27.** C	**28.** E	**29.** A	**30.** C
31. B	**32.** B	**33.** A		

Selected Answers and Solutions to Text Problems

6.1 Atoms with one, two, or three valence electrons will lose those electrons to acquire a noble gas electron configuration.
    a. Li loses 1 e^-.
    b. Ca loses 2 e^-.
    c. Ga loses 3 e^-.
    d. Cs loses 1 e^-.
    e. Ba loses 2 e^-.

6.3 Atoms form ions by losing or gaining valence electrons to achieve the same electron configuration as the nearest noble gas. Elements in Groups 1A (1), 2A (2), and 3A (13) lose valence electrons, whereas elements in Groups 5A (15), 6A (16), and 7A (17) gain valence electrons to complete octets.
    a. Sr loses 2 e^-.
    b. P gains 3 e^-.
    c. Elements in Group 7A (17) gain 1 e^-.
    d. Na loses 1 e^-.
    e. Br gains 1 e^-.

6.5   **a.** The element with 3 protons is lithium. In a lithium ion with 2 electrons, the ionic charge would be 1+, $(3+) + (2-) = 1+$. The lithium ion is written as Li^+.
    b. The element with 9 protons is fluorine. In a fluorine ion with 10 electrons, the ionic charge would be 1−, $(9+) + (10-) = 1-$. The fluoride ion is written as F^-.
    c. The element with 12 protons is magnesium. In a magnesium ion with 10 electrons, the ionic charge would be 2+, $(12+) + (10-) = 2+$. The magnesium ion is written as Mg^{2+}.
    d. The element with 26 protons is iron. In an iron ion with 23 electrons, the ionic charge would be 3+, $(26+) + (23-) = 3+$. This iron ion is written as Fe^{3+}.

6.7   **a.** Cu has an atomic number of 29, which means it has 29 protons. In a neutral atom, the number of electrons equals the number of protons. A copper ion with a charge of 2+ has lost 2 e^- to have $29 - 2 = 27$ electrons. $\therefore$ in a Cu^{2+} ion, there are 29 protons and 27 electrons.
    b. Se has an atomic number of 34, which means it has 34 protons. In a neutral atom, the number of electrons equals the number of protons. A selenium ion with a charge of 2− has gained 2 e^- to have $34 + 2 = 36$ electrons. $\therefore$ in a Se^{2-} ion, there are 34 protons and 36 electrons.
    c. Br has an atomic number of 35, which means it has 35 protons. In a neutral atom, the number of electrons equals the number of protons. A bromine ion with a charge of 1− has gained 1 e^- to have $35 + 1 = 36$ electrons. $\therefore$ in a Br^- ion, there are 35 protons and 36 electrons.
    d. Fe has an atomic number of 26, which means it has 26 protons. In a neutral atom, the number of electrons equals the number of protons. An iron ion with a charge of 3+ has lost 3 e^- to have $26 - 3 = 23$ electrons. $\therefore$ in a Fe^{3+} ion, there are 26 protons and 23 electrons.

6.9   **a.** Chlorine in Group 7A (17) gains one electron to form chloride ion, Cl^-.
    b. Cesium in Group 1A (1) loses one electron to form cesium ion, Cs^+.
    c. Nitrogen in Group 5A (15) gains three electrons to form nitride ion, N^{3-}.
    d. Radium in Group 2A (2) loses two electrons to form radium ion, Ra^{2+}.

6.11   **a.** Li^+ is called the lithium ion.
    b. Ca^{2+} is called the calcium ion.
    c. Ga^{3+} is called the gallium ion.
    d. P^{3-} is called the phosphide ion.

6.13 **a.** There are 8 protons and 10 electrons in the O^{2-} ion; $(8+) + (10-) = 2-$.
 b. There are 19 protons and 18 electrons in the K^+ ion; $(19+) + (18-) = 1+$.
 c. There are 53 protons and 54 electrons in the I^- ion; $(53+) + (54-) = 1-$.
 d. There are 11 protons and 10 electrons in the Na^+ ion; $(11+) + (10-) = 1+$.

6.15 **a** (Li and Cl) and **c** (K and O) will form ionic compounds.

6.17 **a.** Na^+ and $O^{2-} \rightarrow Na_2O$ **b.** Al^{3+} and $Br^- \rightarrow AlBr_3$
 c. Ba^{2+} and $N^{3-} \rightarrow Ba_3N_2$ **d.** Mg^{2+} and $F^- \rightarrow MgF_2$
 e. Al^{3+} and $S^{2-} \rightarrow Al_2S_3$

6.19 **a.** Ions: K^+ and $S^{2-} \rightarrow K_2S$ Check: $2(1+) + 1(2-) = 0$
 b. Ions: Na^+ and $N^{3-} \rightarrow Na_3N$ Check: $3(1+) + 1(3-) = 0$
 c. Ions: Al^{3+} and $I^- \rightarrow AlI_3$ Check: $1(3+) + 3(1-) = 0$
 d. Ions: Ga^{3+} and $O^{2-} \rightarrow Ga_2O_3$ Check: $2(3+) + 3(2-) = 0$

6.21 **a.** aluminum oxide   **b.** calcium chloride
 c. sodium oxide   **d.** magnesium phosphide
 e. potassium iodide   **f.** barium fluoride

6.23 **a.** iron(II)   **b.** copper(II)
 c. zinc   **d.** lead(IV)
 e. chromium(III)   **f.** manganese(II)

6.25 **a.** Ions: Sn^{2+} and $Cl^- \rightarrow$ tin(II) chloride
 b. Ions: Fe^{2+} and $O^{2-} \rightarrow$ iron(II) oxide
 c. Ions: Cu^+ and $S^{2-} \rightarrow$ copper(I) sulfide
 d. Ions: Cu^{2+} and $S^{2-} \rightarrow$ copper(II) sulfide
 e. Ions: Cd^{2+} and $Br^- \rightarrow$ cadmium bromide
 f. Ions: Hg^{2+} and $Cl^- \rightarrow$ mercury(II) chloride

6.27 **a.** Au^{3+} **b.** Fe^{3+}
 c. Pb^{4+} **d.** Sn^{2+}

6.29 **a.** Ions: Mg^{2+} and $Cl^- \rightarrow MgCl_2$ **b.** Ions: Na^+ and $S^{2-} \rightarrow Na_2S$
 c. Ions: Cu^+ and $O^{2-} \rightarrow Cu_2O$ **d.** Ions: Zn^{2+} and $P^{3-} \rightarrow Zn_3P_2$
 e. Ions: Au^{3+} and $N^{3-} \rightarrow AuN$ **f.** Ions: Co^{3+} and $F^- \rightarrow CoF_3$

6.31 **a.** Ions: Co^{3+} and $Cl^- \rightarrow CoCl_3$ **b.** Ions: Pb^{4+} and $O^{2-} \rightarrow PbO_2$
 c. Ions: Ag^+ and $I^- \rightarrow AgI$ **d.** Ions: Ca^{2+} and $N^{3-} \rightarrow Ca_3N_2$
 e. Ions: Cu^+ and $P^{3-} \rightarrow Cu_3P$ **f.** Ions: Cr^{2+} and $Cl^- \rightarrow CrCl_2$

6.33 **a.** Ions: K^+ and $P^{3-} \rightarrow K_3P$ **b.** Ions: Cu^{2+} and $Cl^- \rightarrow CuCl_2$
 c. Ions: Fe^{3+} and $Br^- \rightarrow FeBr_3$ **d.** Ions: Mg^{2+} and $O^{2-} \rightarrow MgO$

6.35 **a.** HCO_3^- **b.** NH_4^+
 c. PO_3^{3-} **d.** ClO_3^-

6.37 **a.** sulfate   **b.** carbonate
 c. hydrogen sulfite (or bisulfite)   **d.** nitrate

6.39

	NO$_2^-$	CO$_3^{2-}$	HSO$_4^-$	PO$_4^{3-}$
Li$^+$	LiNO$_2$ Lithium nitrite	Li$_2$CO$_3$ Lithium carbonate	LiHSO$_4$ Lithium hydrogen sulfate	Li$_3$PO$_4$ Lithium phosphate
Cu^{2+}	Cu(NO$_2$)$_2$ Copper(II) nitrite	CuCO$_3$ Copper(II) carbonate	Cu(HSO$_4$)$_2$ Copper(II) hydrogen sulfate	Cu$_3$(PO$_4$)$_2$ Copper(II) phosphate
Ba^{2+}	Ba(NO$_2$)$_2$ Barium nitrite	BaCO$_3$ Barium carbonate	Ba(HSO$_4$)$_2$ Barium hydrogen sulfate	Ba$_3$(PO$_4$)$_2$ Barium phosphate

6.41 **a.** Ions: Ba^{2+} and OH$^-$ → Ba(OH)$_2$ **b.** Ions: Na$^+$ and HSO$_4^-$ → NaHSO$_4$
c. Ions: Fe^{2+} and NO$_2^-$ → Fe(NO$_2$)$_2$ **d.** Ions: Zn^{2+} and PO$_4^{3-}$ → Zn$_3$(PO$_4$)$_2$
e. Ions: Fe^{3+} and CO$_3^{2-}$ → Fe$_2$(CO$_3$)$_3$

6.43 **a.** CO$_3^{2-}$, sodium carbonate **b.** NH$_4^+$, ammonium sulfide
c. PO$_4^{3-}$, barium phosphate **d.** NO$_2^-$, tin(II) nitrite

6.45 **a.** Ions: Zn^{2+} + C$_2$H$_3$O$_2^-$ → zinc acetate
b. Ions: Mg^{2+} and PO$_4^{3-}$ → magnesium phosphate
c. Ions: NH$_4^+$ + Cl$^-$ → ammonium chloride
d. Ions: Na$^+$ and HCO$_3^-$ → sodium bicarbonate or sodium hydrogen carbonate

6.47 When naming molecular compounds, prefixes are used to indicate the number of each atom as shown in the subscripts of the formula. The first nonmetal is named by its elemental name; the second nonmetal is named by using its elemental name with the ending changed to *ide*.
a. 1 P and 3 Br → phosphorus tribromide **b.** 2 Cl and 1 O → dichlorine oxide
c. 1 C and 4 Br → carbon tetrabromide **d.** 1 H and 1 F → hydrogen fluoride
e. 1 N and 3 F → nitrogen trifluoride

6.49 When naming molecular compounds, prefixes are used to indicate the number of each atom as shown in the subscripts of the formula. The first nonmetal is named by its elemental name; the second nonmetal is named by using its elemental name with the ending changed to *ide*.
a. 2 N and 3 O → dinitrogen trioxide
b. 2 Si and 6 Br → disilicon hexabromide
c. 4 P and 3 S → tetraphosphorus trisulfide
d. 1 P and 5 Cl → phosphorus pentachloride
e. 1 Se and 6 F → selenium hexafluoride

6.51 **a.** 1 C and 4 Cl → CCl$_4$ **b.** 1 C and 1 O → CO
c. 1 P and 3 Cl → PCl$_3$ **d.** 2 N and 4 O → N$_2$O$_4$

6.53 **a.** 1 O and 2 F → OF$_2$ **b.** 1 B and 3 Cl → BCl$_3$
c. 2 N and 3 O → N$_2$O$_3$ **d.** 1 S and 6 F → SF$_6$

6.55 **a.** Ions: Al^{3+} and SO$_4^{2-}$ → aluminum sulfate
b. Ions: Ca^{2+} and CO$_3^{2-}$ → calcium carbonate
c. 2 N and 1 O → dinitrogen oxide
d. Ions: Mg^{2+} + OH$^-$ → magnesium hydroxide

6.57 **a.** Ions: Mg^{2+} + SO$_4^{2-}$ → MgSO$_4$ **b.** Ions: Sn^{2+} + F$^-$ → SnF$_2$
c. Ions: Al^{3+} + OH$^-$ → Al(OH)$_3$

6.59 **a.** ionic **b.** ionic **c.** ionic

6.61 **a.** By losing two valence electrons from the third energy level, magnesium achieves an octet in the second energy level.

b. The magnesium ion Mg^{2+} has the same electron configuration as the noble gas Ne $(1s^2 2s^2 2p^6)$.

c. Group 1A (1) and 2A (2) elements achieve octets by losing electrons when they form compounds. Group 8A (18) elements already have a stable octet of valence electrons (or two electrons for helium), so they are not normally found in compounds.

6.63 **a.** The element with 15 protons is phosphorus. In an ion of phosphorus with 18 electrons, the ionic charge would be $3-$, $(15+) + (18-) = 3-$. The phosphide ion is written P^{3-}.

b. The element with 8 protons is oxygen. Since there are also 8 electrons, this is an oxygen (O) atom.

c. The element with 30 protons is zinc. In an ion of zinc with 28 electrons, the ionic charge would be $2+$, $(30+) + (28-) = 2+$. The zinc ion is written Zn^{2+}.

d. The element with 26 protons is iron. In an ion of iron with 23 electrons, the ionic charge would be $3+$, $(26+) + (23-) = 3+$. This iron ion is written Fe^{3+}.

6.65 **a.** X is in Group 1A (1); Y is in Group 6A (16).

b. ionic

c. X^+, Y^{2-}

d. X_2Y

e. X^+ and $S^{2-} \rightarrow X_2S$

f. 1 Y and 2 Cl $\rightarrow YCl_2$

g. molecular

6.67

Electron Configurations		Cation	Anion	Formula of Compound	Name of Compound
$1s^2 2s^2 2p^6 3s^2$	$1s^2 2s^2 2p^3$	Mg^{2+}	N^{3-}	Mg_3N_2	Magnesium nitride
$1s^2 2s^2 2p^6 3s^2 3p^6 4s^1$	$1s^2 2s^2 2p^4$	K^+	O^{2-}	K_2O	Potassium oxide
$1s^2 2s^2 2p^6 3s^2 3p^1$	$1s^2 2s^2 2p^6 3s^2 3p^5$	Al^{3+}	Cl^-	$AlCl_3$	Aluminum chloride

6.69 **a.** N^{3-} is called the nitride ion.

b. Mg^{2+} is called the magnesium ion.

c. O^{2-} is called the oxide ion.

d. Al^{3+} is called the aluminum ion.

6.71 **a.** An element that forms an ion with a $2+$ charge would be in Group 2A (2).

b. The Lewis symbol for an element in Group 2A (2) is $\dot{\ddot{X}}$

c. Mg is the Group 2A (2) element in Period 3.

d. Ions: X^{2+} and $N^{3-} \rightarrow X_3N_2$

6.73 **a.** Tin(IV) is Sn^{4+}.

b. The Sn^{4+} ion has 50 protons and $50 - 4 = 46$ electrons.

c. Ions: Sn^{4+} and $O^{2-} \rightarrow SnO_2$

d. Ions: Sn^{4+} and $PO_4^{3-} \rightarrow Sn_3(PO_4)_4$

6.75 **a.** Ions: Sn^{2+} and $S^{2-} \rightarrow SnS$ **b.** Ions: Pb^{4+} and $O^{2-} \rightarrow PbO_2$

c. Ions: Ag^+ and $Cl^- \rightarrow AgCl$ **d.** Ions: Ca^{2+} and $N^{3-} \rightarrow Ca_3N_2$

e. Ions: Cu^+ and $P^{3-} \rightarrow Cu_3P$ **f.** Ions: Cr^{2+} and $Br^- \rightarrow CrBr_2$

6.77 **a.** 1 N and 3 Cl $\rightarrow$ nitrogen trichloride

b. 2 N and 3 S $\rightarrow$ dinitrogen trisulfide

c. 2 N and 1 O $\rightarrow$ dinitrogen oxide

d. 1 I and 1 F $\rightarrow$ iodine fluoride

    e. 1 B and 3 F → boron trifluoride

    f. 2 P and 5 O → diphosphorus pentoxide

6.79 **a.** 1 C and 1 S → CS   **b.** 2 P and 5 O → P_2O_5

        c. 2 H and 1 S → H_2S **d.** 1 S and 2 Cl → SCl_2

6.81 **a.** Ionic, ions are Fe^{3+} and Cl^- → iron(III) chloride

        b. Ionic, ions are Na^+ and SO_4^{2-} → sodium sulfate

        c. Molecular, 1 N and 2 O → nitrogen dioxide

        d. Ionic, ions are Rb^+ and S^{2-} → rubidium sulfide

        e. Molecular, 1 P and 5 F → phosphorus pentafluoride

        f. Molecular, 1 C and 4 F → carbon tetrafluoride

6.83 **a.** Ions: Sn^{2+} and CO_3^{2-} → $SnCO_3$ **b.** Ions: Li^+ and P^{3-} → Li_3P

        c. Molecular, 1 Si and 4 Cl → $SiCl_4$ **d.** Ions: Mn^{3+} and O^{2-} → Mn_2O_3

        e. Molecular, 4 P and 3 Se → P_4Se_3 **f.** Ions: Ca^{2+} and Br^- → $CaBr_2$

6.85

Atom or Ion	Number of Protons	Number of Electrons	Electrons Lost / Gained
K^+	$19\ p^+$	$18\ e^-$	$1\ e^-$ lost
Mg^{2+}	$12\ p^+$	$10\ e^-$	$2\ e^-$ lost
O^{2-}	$8\ p^+$	$10\ e^-$	$2\ e^-$ gained
Al^{3+}	$13\ p^+$	$10\ e^-$	$3\ e^-$ lost

6.87 **a.** X as a X^{3+} ion would be in Group 3A (13).

        b. X as a X^{2-} ion would be in Group 6A (16).

        c. X as a X^{2+} ion would be in Group 2A (2).

6.89 Compounds with a metal and nonmetal are classified as ionic; compounds with two nonmetals are molecular.

        a. Ionic, ions are Li^+ and HPO_4^{2-} → lithium hydrogen phosphate

        b. Molecular, 1 Cl and 3 F → chlorine trifluoride

        c. Ionic, ions are Mg^{2+} and ClO_2^- → magnesium chlorite

        d. Molecular, 1 N and 3 F → nitrogen trifluoride

        e. Ionic, ions are Ca^{2+} and HSO_4^- → calcium bisulfate or calcium hydrogen sulfate

        f. Ionic, ions are K^+ and ClO_4^- → potassium perchlorate

        g. Ionic, ions are Au^{3+} and SO_3^{2-} → gold(III) sulfite

Chemical Quantities

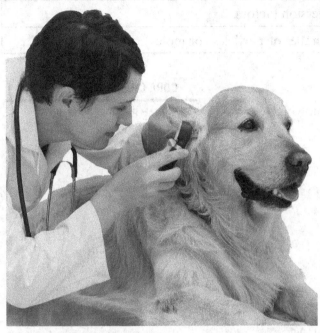

Credit: wavebreakmedia / Shutterstock

Max, a six-year-old dog, was taken to his veterinarian, Chris, two months ago for an examination. After a sample for a blood chemistry profile determined that Max had an elevated white blood cell count, his veterinarian treated him with 375-mg tablets of Clavamox, a broad-spectrum antibiotic that combines amoxicillin and clavulanic acid. A few days later, the lab tests were repeated and indicated a normal white blood cell count, revealing that the infection had cleared. Max's electrolytes, which are also indicators of good health, were all in the normal ranges. The electrolyte chloride, Cl^-, was in the normal range of 0.106 to 0.118 mol/L, and the electrolyte sodium, Na^+, was in the normal range of 0.144 to 0.160 mol/L. If amoxicillin has a formula of $C_{16}H_{19}N_3O_5S$ and a molar mass of 365 g/mol, how many moles of amoxicillin are in 250 mg of amoxicillin?

CHAPTER READINESS*

✿ Key Math Skills

- Calculating Percentages (1.4)
- Solving Equations (1.4)
- Converting Between Standard Numbers and Scientific Notation (1.5)
- Rounding Off (2.3)

🔑 Core Chemistry Skills

- Counting Significant Figures (2.2)
- Using Significant Figures in Calculations (2.3)
- Writing Conversion Factors from Equalities (2.5)
- Using Conversion Factors (2.6)

*These Key Math Skills and Core Chemistry Skills from previous chapters are listed here for your review as you proceed to the new material in this chapter.

LOOKING AHEAD

7.1 The Mole
7.2 Molar Mass

7.3 Calculations Using Molar Mass
7.4 Mass Percent Composition

7.5 Empirical Formulas
7.6 Molecular Formulas

7.1 The Mole

Learning Goal: Use Avogadro's number to calculate the number of particles in a given number of moles. Calculate the number of moles of an element in a given number of moles of a compound.

- A mole of any compound contains Avogadro's number, 6.022×10^{23}, of particles.
  1 mol = 6.022×10^{23} particles (atoms, molecules, ions, formula units)
- Avogadro's number provides conversion factors between moles and the number of particles:

$$\frac{6.022 \times 10^{23} \text{ particles}}{1 \text{ mol}} \quad \text{and} \quad \frac{1 \text{ mol}}{6.022 \times 10^{23} \text{ particles}}$$

Guide to Converting the Moles (or Particles) of a Substance to Particles (or Moles)	
STEP 1	State the given and needed quantities.
STEP 2	Write a plan to convert moles (or particles) to particles (or moles).
STEP 3	Use Avogadro's number to write conversion factors.
STEP 4	Set up the problem to calculate the number of particles (or moles).

♦ **Learning Exercise 7.1A**

Use Avogadro's number to calculate each of the following:

⟫ **CORE CHEMISTRY SKILL**

Converting Particles to Moles

a. number of Ca atoms in 3.00 mol of Ca

b. number of Zn atoms in 0.250 mol of Zn

c. number of SO_2 molecules in 0.118 mol of SO_2

d. number of moles of Ag in 4.88×10^{23} atoms of Ag

e. number of moles of NH_3 in 7.52×10^{23} molecules of NH_3

Answers
a. 1.81×10^{24} atoms of Ca
c. 7.11×10^{22} molecules of SO_2
e. 1.25 mol of NH_3

b. 1.51×10^{23} atoms of Zn
d. 0.810 mol of Ag

Study Note

The subscripts in the formula of a compound indicate the number of moles of each element in one mol of that compound. For example, consider the formula Mg_3N_2:

1 mol of Mg_3N_2 = 3 mol of Mg atoms and 2 mol of N atoms

Some conversion factors for the moles of elements can be written as

$$\frac{3 \text{ mol Mg atoms}}{1 \text{ mol } Mg_3N_2} \text{ and } \frac{1 \text{ mol } Mg_3N_2}{3 \text{ mol Mg atoms}} \qquad \frac{2 \text{ mol N atoms}}{1 \text{ mol } Mg_3N_2} \text{ and } \frac{1 \text{ mol } Mg_3N_2}{2 \text{ mol N atoms}}$$

Guide to Calculating Moles of a Compound or Element	
STEP 1	State the given and needed quantities.
STEP 2	Write a plan to convert moles of a compound to moles of an element (or moles of an element to moles of a compound).
STEP 3	Write the equalities and conversion factors using subscripts.
STEP 4	Set up the problem to calculate the moles of an element (or moles of a compound).

♦ **Learning Exercise 7.1B**

Vitamin C (ascorbic acid) has the formula $C_6H_8O_6$.

  a. How many moles of C are in 2.0 mol of vitamin C?

  b. How many moles of H are in 5.0 mol of vitamin C?

  c. How many moles of O are in 1.5 mol of vitamin C?

Answers   **a.** 12 mol of C   **b.** 40. mol of H   **c.** 9.0 mol of O

♦ **Learning Exercise 7.1C**

For the compound ibuprofen $(C_{13}H_{18}O_2)$ used in Advil and Motrin, calculate each of the following:

  a. moles of C in 2.20 mol of ibuprofen

  b. moles of H in 0.073 mol of ibuprofen

  c. moles of O in 0.750 mol of ibuprofen

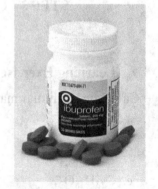

  d. moles of ibuprofen that contain 0.150 mol of H

Ibuprofen is an
anti-inflammatory.

*Credit: Pearson Education /
Pearson Science*

Answers   **a.** 28.6 mol of C   **b.** 1.31 mol of H
              c. 1.50 mol of O   **d.** 8.33×10^{-3} mol of ibuprofen

7.2   Molar Mass

Learning Goal: Given the chemical formula of a substance, calculate its molar mass.

- The molar mass (g/mol) of an element is numerically equal to its atomic mass in grams.
- The molar mass (g/mol) of a compound is the sum of the mass (grams) for each element as indicated by the subscripts in the formula.

Guide to Calculating Molar Mass	
STEP 1	Obtain the molar mass of each element.
STEP 2	Multiply each molar mass by the number of moles (subscript) in the formula.
STEP 3	Calculate the molar mass by adding the masses of the elements.

Example: What is the molar mass of silver nitrate, $AgNO_3$?

Analyze the Problem	**Given**	**Need**	**Connect**
	formula $AgNO_3$	molar mass of $AgNO_3$	periodic table

Solution: **STEP 1**   **Obtain the molar mass of each element.**

$$\frac{107.9 \text{ g Ag}}{1 \text{ mol Ag}} \quad \frac{14.01 \text{ g N}}{1 \text{ mol N}} \quad \frac{16.00 \text{ g O}}{1 \text{ mol O}}$$

STEP 2   **Multiply each molar mass by the number of moles (subscript) in the formula.**

$$1 \text{ mol Ag} \times \frac{107.9 \text{ g Ag}}{1 \text{ mol Ag}} = 107.9 \text{ g of Ag}$$

$$1 \text{ mol N} \times \frac{14.01 \text{ g N}}{1 \text{ mol N}} = 14.01 \text{ g of N}$$

$$3 \text{ mol O} \times \frac{16.00 \text{ g O}}{1 \text{ mol O}} = 48.00 \text{ g of O}$$

▶ **CORE CHEMISTRY SKILL**

Calculating Molar Mass

STEP 3   **Calculate the molar mass by adding the masses of the elements.**

$$\text{Molar mass of } AgNO_3 = (107.9 \text{ g} + 14.01 \text{ g} + 48.00 \text{ g}) = 169.9 \text{ g}$$

♦ **Learning Exercise 7.2**

Calculate the molar mass for each of the following:

a. K_2O

b. $AlCl_3$

c. $C_{13}H_{18}O_2$

d. C_4H_{10}

e. $Ca(NO_3)_2$

f. Mg_3N_2

g. $FeCO_3$

h. $(NH_4)_3PO_4$

$C_{12}H_{22}O_{11}$

1 mol of sucrose contains 342.3 g of sucrose (table sugar).

Credit: Pearson Education / Pearson Science

Answers   **a.** 94.20 g   **b.** 133.33 g   **c.** 206.3 g   **d.** 58.12 g
             e. 164.10 g   **f.** 100.95 g   **g.** 115.86 g   **h.** 149.10 g

7.3   Calculations Using Molar Mass

Learning Goal: Given the number of moles (or grams) of a substance, calculate the grams (or moles). Given the grams of a compound, calculate the grams of one of the elements.

- The molar mass is a conversion factor to change a given quantity in moles to grams or grams to moles.
- The two conversion factors for the molar mass of NaOH (40.01 g/mol) are:

$$\frac{40.01 \text{ g NaOH}}{1 \text{ mol NaOH}} \quad \text{and} \quad \frac{1 \text{ mol NaOH}}{40.01 \text{ g NaOH}}$$

Guide to Calculating the Moles (or Grams) of a Substance from Grams (or Moles)	
STEP 1	State the given and needed quantities.
STEP 2	Write a plan to convert moles to grams (or grams to moles).
STEP 3	Determine the molar mass and write conversion factors.
STEP 4	Set up the problem to convert moles to grams (or grams to moles).

Example: What is the mass, in grams, of 0.254 mol of Na_2CO_3?

Solution: **STEP 1**   **State the given and needed quantities.**

Analyze the Problem	Given	Need	Connect
	0.254 mol of Na_2CO_3	grams of Na_2CO_3	molar mass

STEP 2   **Write a plan to convert moles to grams.**

$$\text{moles of } Na_2CO_3 \xrightarrow{\text{Molar mass}} \text{grams of } Na_2CO_3$$

STEP 3   **Determine the molar mass and write conversion factors.**

$$1 \text{ mol of } Na_2CO_3 = 105.99 \text{ g of } Na_2CO_3$$

$$\frac{105.99 \text{ g } Na_2CO_3}{1 \text{ mol } Na_2CO_3} \quad \text{and} \quad \frac{1 \text{ mol } Na_2CO_3}{105.99 \text{ g } Na_2CO_3}$$

STEP 4   **Set up the problem to convert moles to grams.**

$$0.254 \text{ mol } Na_2CO_3 \times \frac{105.99 \text{ g } Na_2CO_3}{1 \text{ mol } Na_2CO_3} = 26.9 \text{ g of } Na_2CO_3$$

♦ **Learning Exercise 7.3A**

Calculate the mass, in grams, for each of the following:

a. 0.75 mol of S

b. 3.18 mol of K_2SO_4

🔧 **CORE CHEMISTRY SKILL**

Using Molar Mass as a Conversion Factor

c. 2.50 mol of NH_4Cl **d.** 4.08 mol of $FeCl_3$

e. 2.28 mol of PCl_3 **f.** 0.815 mol of $Mg(NO_3)_2$

Answers   **a.** 24 g   **b.** 554 g   **c.** 134 g   **d.** 662 g   **e.** 313 g   **f.** 121 g

◆ Learning Exercise 7.3B

Calculate the number of moles in each of the following quantities:

a. 108 g of C_5H_{12} **b.** 6.12 g of CO_2

c. 38.7 g of $CaBr_2$ **d.** 236 g of Cl_2

e. 128 g of $Mg(OH)_2$ **f.** 172 g of Al_2O_3

g. The methane used to heat a home has a formula of CH_4. If 725 g of methane is used in one day, how many moles of methane are used during this time?

h. A vitamin tablet contains 18 mg of iron. If there are 100 tablets in a bottle, how many moles of iron are contained in the vitamins in the bottle?

Answers   **a.** 1.50 mol   **b.** 0.139 mol   **c.** 0.194 mol   **d.** 3.33 mol
                      e. 2.19 mol   **f.** 1.69 mol   **g.** 45.2 mol   **h.** 0.032 mol

◆ Learning Exercise 7.3C

Octane, C_8H_{18}, is a component of gasoline.

a. How many moles of octane are in 34.7 g of octane?

b. How many grams of octane are in 0.737 mol of octane?

c. How many grams of C are in 135 g of octane?

d. How many grams of H are in 0.654 g of octane?

Answers   **a.** 0.304 mol   **b.** 84.2 g   **c.** 114 g   **d.** 0.104 g

7.4 Mass Percent Composition

Learning Goal: Given the formula of a compound, calculate the mass percent composition.

• Mass percent composition is the percent by mass of each element in a compound.

Guide to Calculating Mass Percent Composition from Molar Mass	
STEP 1	Determine the total mass of each element in the molar mass of a formula.
STEP 2	Divide the total mass of each element by the molar mass and multiply by 100%.

Example: What is the mass percent composition of NaOH?

Analyze the Problem	**Given**	**Need**	**Connect**
	NaOH	mass percent composition: %Na, %O, %H	periodic table

Solution: **STEP 1** Determine the total mass of each element in the molar mass of a formula.

$$1 \text{ mol Na} \times \frac{22.99 \text{ g Na}}{1 \text{ mol Na}} = 22.99 \text{ g of Na}$$

$$1 \text{ mol O} \times \frac{16.00 \text{ g O}}{1 \text{ mol O}} = 16.00 \text{ g of O}$$

$$1 \text{ mol H} \times \frac{1.008 \text{ g H}}{1 \text{ mol H}} = \underline{1.008 \text{ g of H}}$$

$$\text{Molar mass of NaOH} = 40.00 \text{ g of NaOH}$$

STEP 2 Divide the total mass of each element by the molar mass and multiply by 100%.

$$\frac{22.99 \text{ g Na}}{40.00 \text{ g NaOH}} \times 100\% = 57.48\% \text{ Na}$$

$$\frac{16.00 \text{ g O}}{40.00 \text{ g NaOH}} \times 100\% = 40.00\% \text{ O}$$

$$\frac{1.008 \text{ g H}}{40.00 \text{ g NaOH}} \times 100\% = 2.520\% \text{ H}$$

When we add the mass percents of all the elements the sum is 100.00%. In some cases because of rounding off, the sum of the mass percents may not total exactly 100%.

♦ **Learning Exercise 7.4**

Calculate the mass percent composition of each of the following:

a. C_3H_6O

b. $BaSO_4$

c. $Ca(NO_3)_2$

d. $NaOCl$

e. $(NH_4)_3PO_4$

⬧ **CORE CHEMISTRY SKILL**

Calculating Mass Percent Composition

Answers   **a.** C 62.04%, H 10.41%, O 27.55%
              b. Ba 58.83%, S 13.74%, O 27.42%
              c. Ca 24.42%, N 17.07%, O 58.50%
              d. Na 30.88%, O 21.49%, Cl 47.62%
              e. N 28.19%, H 8.115%, P 20.77%, O 42.92%

7.5   Empirical Formulas

Learning Goal: From the mass percent composition, calculate the empirical formula for a compound.

- An empirical (simplest) formula gives the lowest whole-number ratio of atoms in a compound.

Guide to Calculating an Empirical Formula	
STEP 1	Calculate the moles of each element.
STEP 2	Divide by the smallest number of moles.
STEP 3	Use the lowest whole-number ratio of moles as subscripts.

Example: What is the empirical formula of a compound containing 27.3% C and 72.7% O by mass?

Analyze the Problem	Given	Need	Connect
	27.3% C, 72.7% O	empirical formula	molar mass

Solution: **STEP 1**   **Calculate the moles of each element.** If we assume we have *exactly* 100 g of the compound, there would be 27.3 g of C and 72.7 g of O.

$$27.3 \; \cancel{g \, C} \times \frac{1 \; mol \; C}{12.01 \; \cancel{g \, C}} = 2.27 \; mol \; of \; C \qquad 72.7 \; \cancel{g \, O} \times \frac{1 \; mol \; O}{16.00 \; \cancel{g \, O}} = 4.54 \; mol \; of \; C$$

STEP 2   **Divide by the smallest number of moles.**

$$\frac{2.27 \; mol \; C}{2.27} = 1.00 \; mol \; of \; C \qquad \frac{4.54 \; mol \; O}{2.27} = 2.00 \; mol \; of \; O$$

STEP 3   **Use the lowest whole-number ratio of moles as subscripts.**

$$Empirical \; formula = C_{1.00}O_{2.00} \longrightarrow CO_2$$

Study Note

When the number of moles is a decimal that is within 0.1 of a whole number, round off to the whole number. For example, 1.98 mol is rounded off to 2 mol.

However, if a decimal is obtained that is greater than 0.1 or less than 0.9, it is not rounded off. Instead, multiply by a small integer to obtain a whole number. Then all other subscripts must be increased by multiplying by the same small integer.

Examples of Multipliers That Convert Decimal Values to Whole Numbers

Decimal	Multiplier	Example			Whole Number
0.20	5	1.20×5	=		6
0.25	4	2.25×4	=		9
0.33	3	1.33×3	=		4
0.50	2	2.50×2	=		5
0.67	3	1.67×3	=		5

◆ **Learning Exercise 7.5**

CORE CHEMISTRY SKILL

Calculate the empirical formula for each of the following compounds: Calculating an Empirical Formula

a. a compound that is 75.7% by mass tin (Sn) and 24.3% by mass fluorine (F)

b. a compound that is 20.2% by mass magnesium (Mg), 26.6% sulfur (S), and 53.2% oxygen (O)

c. a 4.58-g sample of a compound that contains 2.00 g of phosphorus (P) and the rest is oxygen (O)

d. a compound that contains 1.000 g of C, 0.168 g of H, and 0.890 g of O

Answers   **a.** SnF_2 **b.** $MgSO_4$ **c.** P_2O_5 **d.** $C_3H_6O_2$

7.6   Molecular Formulas

Learning Goal: Determine the molecular formula of a substance from the empirical formula and molar mass.

- A molecular formula gives the actual number of atoms of each element in the compound.
- A molecular formula is related to the empirical formula by a small integer such as 1, 2, or 3.
- The molecular formula is obtained using the empirical formula mass and molar mass of the compound.

Key Terms for Sections 7.1 to 7.6

Match each of the following key terms with the correct description:

a. mole
b. Avogadro's number
c. empirical formula
d. molar mass
e. molecular formula

1. _____ the formula that gives the actual number of atoms of each element in a compound

2. _____ the number of items in 1 mol is equal to 6.022×10^{23}

3. _____ the formula that gives the smallest whole-number ratio of atoms in a formula

4. _____ the amount of a substance that contains 6.022×10^{23} particles of that substance

5. _____ the mass in grams of an element or compound that is equal numerically to its atomic mass or sum of atomic masses as indicated by the formula

Answers   1. e   2. b   3. c   4. a   5. d

Guide to Calculating a Molecular Formula from an Empirical Formula	
STEP 1	Obtain the empirical formula and calculate the empirical formula mass.
STEP 2	Divide the molar mass by the empirical formula mass to obtain a small integer.
STEP 3	Multiply the empirical formula by the small integer to obtain the molecular formula.

Example: Lactic acid has the empirical formula CH_2O. If the experimental molar mass is 90. g, what is the molecular formula of lactic acid?

Analyze the Problem	Given	Need	Connect
	empirical formula CH_2O, molar mass 90. g	molecular formula	empirical formula mass

Solution: **STEP 1**   **Obtain the empirical formula and calculate the empirical formula mass.**

Empirical formula mass $= (1 \times 12.01) + (2 \times 1.008) + (1 \times 16.00) = 30.0$

STEP 2   **Divide the molar mass by the empirical formula mass to obtain a small integer.**

$$\frac{\text{Molar mass of lactic acid}}{\text{Empirical formula mass of } CH_2O} = \frac{90. \text{ g}}{30.03 \text{ g}} = 3 \,(\text{small integer})$$

STEP 3   **Multiply the empirical formula by the small integer to obtain the molecular formula.**

Molecular formula $= 3 \times (CH_2O) = C_3H_6O_3$

♦ Learning Exercise 7.6A

CORE CHEMISTRY SKILL

Calculating a Molecular Formula

Calculate the molecular formula for each compound from its empirical formula and approximate molar mass:

a. NS, 184 g/mol

b. KCO_2, 166 g/mol

c. HIO_3, 176 g/mol

d. $NaPO_3$, 306 g/mol

e. $C_7H_8O_3$, 420 g/mol

f. C_3H_5O, 228 g/mol

Answers

a. N_4S_4 **b.** $K_2C_2O_4$ **c.** HIO_3
d. $Na_3P_3O_9$ **e.** $C_{21}H_{24}O_9$ **f.** $C_{12}H_{20}O_4$

♦ Learning Exercise 7.6B

Calculate the molecular formula for each of the following:

a. pyrogallol with a composition of C 57.14%, H 4.80%, and O 38.06% and an experimental molar mass of 126 g

b. borazine with a composition of B 40.31%, H 7.51%, and N 52.18% and an experimental molar mass of 80. g

c. a substance used to tan leather with a composition of K 26.58%, Cr 35.36%, and O 38.07% and an experimental molar mass of 294 g

d. pyrocatechol, a topical anesthetic with a composition of C 65.44%, H 5.49%, and O 29.06% and an experimental molar mass of 110. g

Answers **a.** $C_6H_6O_3$ **b.** $B_3H_6N_3$ **c.** $K_2Cr_2O_7$ **d.** $C_6H_6O_2$

Checklist for Chapter 7

You are ready to take the Practice Test for Chapter 7. Be sure you have accomplished the following learning goals for this chapter. If not, review the section listed at the end of the goal. Then apply your new skills and understanding to the Practice Test.

After studying Chapter 7, I can successfully:

_____ Calculate the number of particles in a given number of moles of a substance. (7.1)

_____ Calculate the molar mass given the chemical formula of a substance. (7.2)

_____ Convert the grams of a substance to moles, and moles to grams. (7.3)

_____ Convert the number of grams of a substance to the number of particles and number of particles to the number of grams. (7.3)

_____ Calculate the mass percent composition of a compound. (7.4)

_____ Calculate the empirical formula for a compound. (7.5)

_____ Calculate the molecular formula for a compound. (7.6)

Practice Test for Chapter 7

The chapter sections to review are shown in parentheses at the end of each question.

1. The number of carbon atoms in 2.0 mol of C is (7.1)
   A. 6.0   **B.** 12   **C.** 6.0×10^{23} **D.** 3.0×10^{23} **E.** 1.2×10^{24}

2. The number of molecules in 0.25 mol of NH_3 is (7.1)
   A. 0.25   **B.** 2.5×10^{23} **C.** 4.2×10^{-25} **D.** 1.5×10^{23} **E.** 2.43×10^{24}

3. The number of moles in 8.8×10^{24} molecules of SO_3 is (7.1)
   A. 0.068   **B.** 2.6   **C.** 1.5   **D.** 15   **E.** 53

4. The moles of O in 2.0 mol of $Al(OH)_3$ is (7.1)
   A. 1   **B.** 2   **C.** 3   **D.** 4   **E.** 6

5. The number of atoms of P in 1.00 mol of Be_3P_2 is (7.1)
   A. 3.01×10^{23} **B.** 6.02×10^{23} **C.** 1.20×10^{23} **D.** 9.03×10^{23} **E.** 1.20×10^{24}

6. What is the molar mass of Li_2SO_4? (7.2)
   A. 55.01 g   **B.** 62.10 g   **C.** 103.01 g   **D.** 109.95 g   **E.** 103.11 g

7. What is the molar mass of $Ca(NO_3)_2$? (7.2)
   A. 102.09 g   **B.** 116.10 g   **C.** 164.10 g   **D.** 204.18 g   **E.** 314.20 g

8. The number of grams in 0.600 mol of Cl_2 is (7.3)
   A. 71.0 g   **B.** 21.3 g   **C.** 42.5 g   **D.** 84.5 g   **E.** 4.30 g

9. How many grams are in 4.00 mol of NH_3? (7.3)
   A. 4.00 g   **B.** 17.0 g   **C.** 34.0 g   **D.** 68.1 g   **E.** 0.240 g

10. How many grams are in 0.450 mol of Na_2SO_4? (7.3)
    A. 6.22 g   **B.** 32.0 g   **C.** 53.6 g   **D.** 63.9 g   **E.** 142 g

11. The number of moles of aluminum in 54.0 g of Al is (7.3)
    A. 0.500 mol   **B.** 1.00 mol   **C.** 2.00 mol   **D.** 3.00 mol   **E.** 4.00 mol

12. How many moles is 8.00 g of NaOH? (7.3)
    A. 0.100 mol   **B.** 0.200 mol   **C.** 0.400 mol   **D.** 2.00 mol   **E.** 4.00 mol

13. The number of moles of water in 3.60 g of H_2O is (7.3)
  A. 0.0500 mol   **B.** 0.100 mol   **C.** 0.200 mol   **D.** 0.300 mol   **E.** 0.400 mol

14. What is the number of moles in 2.20 g of CO_2? (7.3)
  A. 2.00 mol   **B.** 1.00 mol   **C.** 0.200 mol   **D.** 0.0500 mol   **E.** 0.0100 mol

15. 20.0 g of $CaCl_2$ = _____ mol of $CaCl_2$. (7.3)
  A. 0.132 mol   **B.** 0.180 mol   **C.** 0.241 mol   **D.** 0.265 mol   **E.** 22.0 mol

16. The mass percent of C in C_3H_8 is (7.4)
  A. 18.29%   **B.** 27.32%   **C.** 59.84%   **D.** 81.71%   **E.** 92.24%

17. What is the empirical formula of a compound that is 42.9% C and 57.1% O? (7.5)
  A. CO   **B.** CO_2 **C.** C_2O **D.** C_2O_3 **E.** CO_3

18. What is the empirical formula of a compound that contains 3.88 g of Cl and 6.12 g of O? (7.5)
  A. ClO   **B.** Cl_2O_7 **C.** Cl_3O **D.** Cl_2O_3 **E.** ClO_4

19. What is the molecular formula of a compound that has an empirical formula of CH_2 and an experimental molar mass of 70. g? (7.6)
  A. CH_2 **B.** C_2H_4 **C.** C_3H_6 **D.** C_4H_8 **E.** C_5H_{10}

20. What is the molecular formula of a compound that has an empirical formula of P_2O_3 and an experimental molar mass of 220. g? (7.6)
  A. P_2O_3 **B.** P_2O_5 **C.** P_4O_6 **D.** P_4O_8 **E.** P_4O_{10}

For questions 21 through 25, consider the compound glucose that has a mass percent composition of C 40.0%, H 6.72%, and O 53.3%. In a 100.-g sample there are (7.3)

21. A. 0.300 mol of C   **B.** 3.00 mol of C   **C.** 3.33 mol of C   **D.** 0.333 mol of C   **E.** 1.00 mol of C

22. A. 0.667 mol of H   **B.** 6.67 mol of H   **C.** 1.49 mol of H   **D.** 0.149 mol of H   **E.** 1.00 mol of H

23. A. 0.974 mol of O   **B.** 9.74 mol of O   **C.** 3.33 mol of O   **D.** 0.333 mol of O   **E.** 1.00 mol of O

24. The empirical formula of glucose is (7.5)
  A. CHO   **B.** CHO_2 **C.** C_2HO_2 **D.** CH_2O **E.** C_2HO

25. If the experimental molar mass of glucose is 180. g, the molecular formula is (7.6)
  A. CHO   **B.** CH_2O **C.** $C_6H_6O_6$ **D.** $C_6H_{12}O_6$ **E.** $C_{12}H_6O_6$

Answers to the Practice Test

1. E	**2.** D	**3.** D	**4.** E	**5.** E
6. D	**7.** C	**8.** C	**9.** D	**10.** D
11. C	**12.** B	**13.** C	**14.** D	**15.** B
16. D	**17.** A	**18.** B	**19.** E	**20.** C
21. C	**22.** B	**23.** C	**24.** D	**25.** D

Selected Answers and Solutions to Text Problems

7.1 One mole contains 6.022×10^{23} atoms of an element, molecules of a molecular substance, or formula units of an ionic substance.

7.3 **a.** $0.500 \text{ mol C} \times \dfrac{6.022 \times 10^{23} \text{ atoms C}}{1 \text{ mol C}} = 3.01 \times 10^{23}$ atoms of C (3 SFs)

b. $1.28 \text{ mol SO}_2 \times \dfrac{6.022 \times 10^{23} \text{ molecules SO}_2}{1 \text{ mol SO}_2} = 7.71 \times 10^{23}$ molecules of SO_2 (3 SFs)

c. $5.22 \times 10^{22} \text{ atoms Fe} \times \dfrac{1 \text{ mol Fe}}{6.022 \times 10^{23} \text{ atoms Fe}} = 0.0867$ mol of Fe (3 SFs)

d. $8.50 \times 10^{24} \text{ atoms C}_2\text{H}_6\text{O} \times \dfrac{1 \text{ mol C}_2\text{H}_6\text{O}}{6.022 \times 10^{23} \text{ atoms C}_2\text{H}_6\text{O}} = 14.1$ mol of C_2H_6O (3 SFs)

7.5 1 mol of H_3PO_4 contains 3 mol of H atoms, 1 mol of P atoms, and 4 mol of O atoms.

a. $2.00 \text{ mol H}_3\text{PO}_4 \times \dfrac{3 \text{ mol H}}{1 \text{ mol H}_3\text{PO}_4} = 6.00$ mol of H (3 SFs)

b. $2.00 \text{ mol H}_3\text{PO}_4 \times \dfrac{4 \text{ mol O}}{1 \text{ mol H}_3\text{PO}_4} = 8.00$ mol of O (3 SFs)

c. $2.00 \text{ mol H}_3\text{PO}_4 \times \dfrac{1 \text{ mol P}}{1 \text{ mol H}_3\text{PO}_4} \times \dfrac{6.022 \times 10^{23} \text{ atoms P}}{1 \text{ mol P}} = 1.20 \times 10^{24}$ atoms of P (3 SFs)

d. $2.00 \text{ mol H}_3\text{PO}_4 \times \dfrac{4 \text{ mol O}}{1 \text{ mol H}_3\text{PO}_4} \times \dfrac{6.022 \times 10^{23} \text{ atoms O}}{1 \text{ mol O}} = 4.82 \times 10^{24}$ atoms of O (3 SFs)

7.7 The subscripts indicate the moles of each element in one mole of that compound.

a. $1.5 \text{ mol quinine} \times \dfrac{24 \text{ mol H}}{1 \text{ mol quinine}} = 36$ mol of H (2 SFs)

b. $5.0 \text{ mol quinine} \times \dfrac{20 \text{ mol C}}{1 \text{ mol quinine}} = 1.0 \times 10^2$ mol of C (2 SFs)

c. $0.020 \text{ mol quinine} \times \dfrac{2 \text{ mol N}}{1 \text{ mol quinine}} = 0.040$ mol of N (2 SFs)

7.9 **a.** $2.30 \text{ mol naproxen} \times \dfrac{14 \text{ mol C}}{1 \text{ mol naproxen}} = 32.2$ mol of C (3 SFs)

b. $0.444 \text{ mol naproxen} \times \dfrac{14 \text{ mol H}}{1 \text{ mol naproxen}} = 6.22$ mol of H (3 SFs)

c. $0.0765 \text{ mol naproxen} \times \dfrac{3 \text{ mol O}}{1 \text{ mol naproxen}} = 0.230$ mol of O (3 SFs)

7.11 **a.** $2 \text{ mol Cl} \times \dfrac{35.45 \text{ g Cl}}{1 \text{ mol Cl}} = 70.90$ g of Cl

$\therefore$ Molar mass of Cl_2 = 70.90 g

b. $3 \text{ mol C} \times \dfrac{12.01 \text{ g C}}{1 \text{ mol C}} = 36.03 \text{ g of C}$

$6 \text{ mol H} \times \dfrac{1.008 \text{ g H}}{1 \text{ mol H}} = 6.048 \text{ g of H}$

$3 \text{ mol O} \times \dfrac{16.00 \text{ g O}}{1 \text{ mol O}} = 48.00 \text{ g of O}$

3 mol of C   = 36.03 g of C

6 mol of H   = 6.048 g of H

3 mol of O   = 48.00 g of O

∴ Molar mass of $C_3H_6O_3$ = 90.08 g

c. $3 \text{ mol Mg} \times \dfrac{24.31 \text{ g Mg}}{1 \text{ mol Mg}} = 72.93 \text{ g of Mg}$

$2 \text{ mol P} \times \dfrac{30.97 \text{ g P}}{1 \text{ mol P}} = 61.94 \text{ g of P}$

$8 \text{ mol O} \times \dfrac{16.00 \text{ g O}}{1 \text{ mol O}} = 128.0 \text{ g of O}$

3 mol of Mg   = 72.93 g of Mg

2 mol of P   = 61.94 g of P

8 mol of O   = 128.0 g of O

∴ Molar mass of $Mg_3(PO_4)_2$ = 262.9 g

7.13   **a.** $1 \text{ mol Al} \times \dfrac{26.98 \text{ g Al}}{1 \text{ mol Al}} = 26.98 \text{ g of Al}$

$3 \text{ mol F} \times \dfrac{19.00 \text{ g F}}{1 \text{ mol F}} = 57.00 \text{ g of F}$

1 mol of Al   = 26.98 g of Al

3 mol of F   = 57.00 g of F

∴ Molar mass of AlF_3 = 83.98 g

b. $2 \text{ mol C} \times \dfrac{12.01 \text{ g C}}{1 \text{ mol C}} = 24.02 \text{ g of C}$

$4 \text{ mol H} \times \dfrac{1.008 \text{ g H}}{1 \text{ mol H}} = 4.032 \text{ g of H}$

$2 \text{ mol Cl} \times \dfrac{35.45 \text{ g Cl}}{1 \text{ mol Cl}} = 70.90 \text{ g of Cl}$

2 mol of C   = 24.02 g of C

4 mol of H   = 4.032 g of H

2 mol of Cl   = 70.90 g of Cl

∴ Molar mass of $C_2H_4Cl_2$ = 98.95 g

c. $1 \text{ mol Sn} \times \dfrac{118.7 \text{ g Sn}}{1 \text{ mol Sn}} = 118.7 \text{ g of Sn}$

$2 \text{ mol F} \times \dfrac{19.00 \text{ g F}}{1 \text{ mol F}} = 38.00 \text{ g of F}$

1 mol of Sn   = 118.7 g of Sn

2 mol of F   = 38.00 g of F

∴ Molar mass of SnF_2 = 156.7 g

7.15 a. $1 \text{ mol Na} \times \dfrac{22.99 \text{ g Na}}{1 \text{ mol Na}} = 22.99 \text{ g of Na}$

$1 \text{ mol Cl} \times \dfrac{35.45 \text{ g Cl}}{1 \text{ mol Cl}} = 35.45 \text{ g of Cl}$

$$
\begin{array}{ll}
1 \text{ mol of Na} & = 22.99 \text{ g of Na} \\
1 \text{ mol of Cl} & = 35.45 \text{ g of Cl} \\
\hline
\therefore \text{ Molar mass of NaCl} & = 58.44 \text{ g}
\end{array}
$$

b. $2 \text{ mol Fe} \times \dfrac{55.85 \text{ g Fe}}{1 \text{ mol Fe}} = 111.7 \text{ g of Fe}$

$3 \text{ mol O} \times \dfrac{16.00 \text{ g O}}{1 \text{ mol O}} = 48.00 \text{ g of O}$

$$
\begin{array}{ll}
2 \text{ mol of Fe} & = 111.7 \text{ g of Fe} \\
3 \text{ mol of O} & = 48.00 \text{ g of O} \\
\hline
\therefore \text{ Molar mass of Fe}_2\text{O}_3 & = 159.7 \text{ g}
\end{array}
$$

c. $19 \text{ mol C} \times \dfrac{12.01 \text{ g C}}{1 \text{ mol C}} = 228.2 \text{ g of C}$

$20 \text{ mol H} \times \dfrac{1.008 \text{ g H}}{1 \text{ mol H}} = 20.16 \text{ g of H}$

$1 \text{ mol F} \times \dfrac{19.00 \text{ g F}}{1 \text{ mol F}} = 19.00 \text{ g of F}$

$1 \text{ mol N} \times \dfrac{14.01 \text{ g N}}{1 \text{ mol N}} = 14.01 \text{ g of N}$

$3 \text{ mol O} \times \dfrac{16.00 \text{ g O}}{1 \text{ mol O}} = 48.00 \text{ g of O}$

$$
\begin{array}{ll}
19 \text{ mol of C} & = 228.2 \text{ g of C} \\
20 \text{ mol of H} & = 20.16 \text{ g of H} \\
1 \text{ mol of F} & = 19.00 \text{ g of F} \\
1 \text{ mol of N} & = 14.01 \text{ g of N} \\
3 \text{ mol of O} & = 48.00 \text{ g of O} \\
\hline
\therefore \text{ Molar mass of C}_{19}\text{H}_{20}\text{FNO}_3 & = 329.4 \text{ g}
\end{array}
$$

7.17 a. $2 \text{ mol Al} \times \dfrac{26.98 \text{ g Al}}{1 \text{ mol Al}} = 53.96 \text{ g of Al}$

$3 \text{ mol S} \times \dfrac{32.07 \text{ g S}}{1 \text{ mol S}} = 96.21 \text{ g of S}$

$12 \text{ mol O} \times \dfrac{16.00 \text{ g O}}{1 \text{ mol O}} = 192.0 \text{ g of O}$

$$
\begin{array}{ll}
2 \text{ mol of Al} & = 53.96 \text{ g of Al} \\
3 \text{ mol of S} & = 96.21 \text{ g of S} \\
12 \text{ mol of O} & = 192.0 \text{ g of O} \\
\hline
\therefore \text{ Molar mass of Al}_2(\text{SO}_4)_3 & = 342.2 \text{ g}
\end{array}
$$

b. $1 \ \cancel{\text{mol K}} \times \dfrac{39.10 \text{ g K}}{1 \ \cancel{\text{mol K}}} = 39.10 \text{ g of K}$

$4 \ \cancel{\text{mol C}} \times \dfrac{12.01 \text{ g C}}{1 \ \cancel{\text{mol C}}} = 48.04 \text{ g of C}$

$5 \ \cancel{\text{mol H}} \times \dfrac{1.008 \text{ g H}}{1 \ \cancel{\text{mol H}}} = 5.040 \text{ g of H}$

$6 \ \cancel{\text{mol O}} \times \dfrac{16.00 \text{ g O}}{1 \ \cancel{\text{mol O}}} = 96.00 \text{ g of O}$

1 mol of K	=	39.10 g of K
4 mol of C	=	48.04 g of C
5 mol of H	=	5.040 g of H
6 mol of O	=	96.00 g of O

∴ Molar mass of $KC_4H_5O_6$ = 188.18 g

c. $16 \ \cancel{\text{mol C}} \times \dfrac{12.01 \text{ g C}}{1 \ \cancel{\text{mol C}}} = 192.2 \text{ g of C}$

$19 \ \cancel{\text{mol H}} \times \dfrac{1.008 \text{ g H}}{1 \ \cancel{\text{mol H}}} = 19.15 \text{ g of H}$

$3 \ \cancel{\text{mol N}} \times \dfrac{14.01 \text{ g N}}{1 \ \cancel{\text{mol N}}} = 42.03 \text{ g of N}$

$5 \ \cancel{\text{mol O}} \times \dfrac{16.00 \text{ g O}}{1 \ \cancel{\text{mol O}}} = 80.00 \text{ g of O}$

$1 \ \cancel{\text{mol S}} \times \dfrac{32.07 \text{ g S}}{1 \ \cancel{\text{mol S}}} = 32.07 \text{ g of S}$

16 mol of C	=	192.2 g of C
19 mol of H	=	19.15 g of H
3 mol of N	=	42.03 g of N
5 mol of O	=	80.00 g of O
1 mol of S	=	32.07 g of S

∴ Molar mass of $C_{16}H_{19}N_3O_5S$ = 365.5 g

7.19 a. $8 \ \cancel{\text{mol C}} \times \dfrac{12.01 \text{ g C}}{1 \ \cancel{\text{mol C}}} = 96.08 \text{ g of C}$

$9 \ \cancel{\text{mol H}} \times \dfrac{1.008 \text{ g H}}{1 \ \cancel{\text{mol H}}} = 9.072 \text{ g of H}$

$1 \ \cancel{\text{mol N}} \times \dfrac{14.01 \text{ g N}}{1 \ \cancel{\text{mol N}}} = 14.01 \text{ g of N}$

$2 \ \cancel{\text{mol O}} \times \dfrac{16.00 \text{ g O}}{1 \ \cancel{\text{mol O}}} = 32.00 \text{ g of O}$

8 mol of C	=	96.08 g of C
9 mol of H	=	9.072 g of H
1 mol of N	=	14.01 g of N
2 mol of O	=	32.00 g of O

∴ Molar mass of $C_8H_9NO_2$ = 151.16 g

b. $3 \text{ mol Ca} \times \dfrac{40.08 \text{ g Ca}}{1 \text{ mol Ca}} = 120.2 \text{ g of Ca}$

$12 \text{ mol C} \times \dfrac{12.01 \text{ g C}}{1 \text{ mol C}} = 144.1 \text{ g of C}$

$10 \text{ mol H} \times \dfrac{1.008 \text{ g H}}{1 \text{ mol H}} = 10.08 \text{ g of H}$

$14 \text{ mol O} \times \dfrac{16.00 \text{ g O}}{1 \text{ mol O}} = 224.0 \text{ g of O}$

3 mol of Ca	= 120.2 g of Ca
12 mol of C	= 144.1 g of C
10 mol of H	= 10.08 g of H
14 mol of O	= 224.0 g of O

$\therefore$ Molar mass of $Ca_3(C_6H_5O_7)_2 = 498.4 \text{ g}$

c. $17 \text{ mol C} \times \dfrac{12.01 \text{ g C}}{1 \text{ mol C}} = 204.2 \text{ g of C}$

$18 \text{ mol H} \times \dfrac{1.008 \text{ g H}}{1 \text{ mol H}} = 18.14 \text{ g of H}$

$1 \text{ mol F} \times \dfrac{19.00 \text{ g F}}{1 \text{ mol F}} = 19.00 \text{ g of F}$

$3 \text{ mol N} \times \dfrac{14.01 \text{ g N}}{1 \text{ mol N}} = 42.03 \text{ g of N}$

$3 \text{ mol O} \times \dfrac{16.00 \text{ g O}}{1 \text{ mol O}} = 48.00 \text{ g of O}$

17 mol of C	= 204.2 g of C
18 mol of H	= 18.14 g of H
1 mol of F	= 19.00 g of F
3 mol of N	= 42.03 g of N
3 mol of O	= 48.00 g of O

$\therefore$ Molar mass of $C_{17}H_{18}FN_3O_3 = 331.4 \text{ g}$

7.21 a. $1.50 \text{ mol Na} \times \dfrac{22.99 \text{ g Na}}{1 \text{ mol Na}} = 34.5 \text{ g of Na (3 SFs)}$

b. $2.80 \text{ mol Ca} \times \dfrac{40.08 \text{ g Ca}}{1 \text{ mol Ca}} = 112 \text{ g of Ca (3 SFs)}$

c. Molar mass of CO_2

$= 1 \text{ mol of C} + 2 \text{ mol of O} = 12.01 \text{ g} + 2(16.00 \text{ g}) = 44.01 \text{ g}$

$0.125 \text{ mol CO}_2 \times \dfrac{44.01 \text{ g CO}_2}{1 \text{ mol CO}_2} = 5.50 \text{ g of CO}_2 \text{ (3 SFs)}$

d. Molar mass of Na_2CO_3

$= 2 \text{ mol of Na} + 1 \text{ mol of C} + 3 \text{ mol of O} = 2(22.99 \text{ g}) + 12.01 \text{ g} + 3(16.00 \text{ g})$

$= 105.99 \text{ g}$

$0.0485 \text{ mol Na}_2\text{CO}_3 \times \dfrac{105.99 \text{ g Na}_2\text{CO}_3}{1 \text{ mol Na}_2\text{CO}_3} = 5.14 \text{ g of Na}_2\text{CO}_3 \text{ (3 SFs)}$

e. Molar mass of PCl_3

$= 1 \text{ mol of } P + 3 \text{ mol of } Cl = 30.97 \text{ g} + 3(35.45 \text{ g}) = 137.32 \text{ g}$

$$7.14 \times 10^2 \text{ mol } PCl_3 \times \frac{137.32 \text{ g } PCl_3}{1 \text{ mol } PCl_3} = 9.80 \times 10^4 \text{ g of } PCl_3 \text{ (3 SFs)}$$

7.23 a. $0.150 \text{ mol } Ne \times \dfrac{20.18 \text{ g Ne}}{1 \text{ mol } Ne} = 3.03 \text{ g of Ne}$

b. Molar mass of I_2

$= 2 \text{ mol of } I = 2(126.9 \text{ g}) = 253.8 \text{ g}$

$$0.150 \text{ mol } I_2 \times \frac{253.8 \text{ g } I_2}{1 \text{ mol } I_2} = 38.1 \text{ g of } I_2 \text{ (3 SFs)}$$

c. Molar mass of Na_2O

$= 2 \text{ mol of Na} + 1 \text{ mol of O} = 2(22.99 \text{ g}) + 16.00 \text{ g} = 61.98 \text{ g}$

$$0.150 \text{ mol } Na_2O \times \frac{61.98 \text{ g } Na_2O}{1 \text{ mol } Na_2O} = 9.30 \text{ g of } Na_2O \text{ (3 SFs)}$$

d. Molar mass of $Ca(NO_3)_2$

$= 1 \text{ mol of Ca} + 2 \text{ mol of N} + 6 \text{ mol of O} = 40.08 \text{ g} + 2(14.01 \text{ g}) + 6(16.00 \text{ g}) = 164.10 \text{ g}$

$$0.150 \text{ mol } Ca(NO_3)_2 \times \frac{164.10 \text{ g } Ca(NO_3)_2}{1 \text{ mol } Ca(NO_3)_2} = 24.6 \text{ g of } Ca(NO_3)_2 \text{ (3 SFs)}$$

e. Molar mass of C_6H_{14}

$= 6 \text{ mol of C} + 14 \text{ mol of H} = 6(12.01 \text{ g}) + 14(1.008 \text{ g}) = 86.17 \text{ g}$

$$0.150 \text{ mol } C_6H_{14} \times \frac{86.17 \text{ g } C_6H_{14}}{1 \text{ mol } C_6H_{14}} = 12.9 \text{ g of } C_6H_{14} \text{ (3 SFs)}$$

7.25 a. $82.0 \text{ g } Ag \times \dfrac{1 \text{ mol Ag}}{107.9 \text{ g } Ag} = 0.760 \text{ mol of Ag (3 SFs)}$

b. $0.288 \text{ g } C \times \dfrac{1 \text{ mol C}}{12.01 \text{ g } C} = 0.0240 \text{ mol of C (3 SFs)}$

c. Molar mass of NH_3

$= 1 \text{ mol of N} + 3 \text{ mol of H} = 14.01 \text{ g} + 3(1.008 \text{ g}) = 17.03 \text{ g}$

$$15.0 \text{ g } NH_3 \times \frac{1 \text{ mol } NH_3}{17.03 \text{ g } NH_3} = 0.881 \text{ mol of } NH_3 \text{ (3 SFs)}$$

d. Molar mass of CH_4

$= 1 \text{ mol of C} + 4 \text{ mol of H} = 12.01 \text{ g} + 4(1.008 \text{ g}) = 16.04 \text{ g}$

$$7.25 \text{ g } CH_4 \times \frac{1 \text{ mol } CH_4}{16.04 \text{ g } CH_4} = 0.452 \text{ mol of } CH_4 \text{ (3 SFs)}$$

e. Molar mass of Fe_2O_3

$= 2 \text{ mol of Fe} + 3 \text{ mol of O} = 2(55.85 \text{ g}) + 3(16.00 \text{ g}) = 159.7 \text{ g}$

$$245 \text{ g } Fe_2O_3 \times \frac{1 \text{ mol } Fe_2O_3}{159.7 \text{ g } Fe_2O_3} = 1.53 \text{ mol of } Fe_2O_3 \text{ (3 SFs)}$$

7.27 **a.** $25.0 \text{ g He} \times \dfrac{1 \text{ mol He}}{4.003 \text{ g He}} = 6.25 \text{ mol of He (3 SFs)}$

    b. Molar mass of O_2

       $= 2 \text{ mol of O} = 2(16.00 \text{ g}) = 32.00 \text{ g}$

       $25.0 \text{ g O}_2 \times \dfrac{1 \text{ mol O}_2}{32.00 \text{ g O}_2} = 0.781 \text{ mol of O}_2 \text{ (3 SFs)}$

    c. Molar mass of $Al(OH)_3$

       $= 1 \text{ mol of Al} + 3 \text{ mol of O} + 3 \text{ mol of H} = 26.98 \text{ g} + 3(16.00 \text{ g}) + 3(1.008 \text{ g})$

       $= 78.00 \text{ g}$

       $25.0 \text{ g Al(OH)}_3 \times \dfrac{1 \text{ mol Al(OH)}_3}{78.00 \text{ g Al(OH)}_3} = 0.321 \text{ mol of Al(OH)}_3 \text{ (3 SFs)}$

    d. Molar mass of Ga_2S_3

       $= 2 \text{ mol of Ga} + 3 \text{ mol of S} = 2(69.72 \text{ g}) + 3(32.07 \text{ g}) = 235.7 \text{ g}$

       $25.0 \text{ g Ga}_2\text{S}_3 \times \dfrac{1 \text{ mol Ga}_2\text{S}_3}{235.7 \text{ g Ga}_2\text{S}_3} = 0.106 \text{ mol of Ga}_2\text{S}_3 \text{ (3 SFs)}$

    e. Molar mass of C_4H_{10}

       $= 4 \text{ mol of C} + 10 \text{ mol of H} = 4(12.01 \text{ g}) + 10(1.008 \text{ g}) = 58.12 \text{ g}$

       $25.0 \text{ g C}_4\text{H}_{10} \times \dfrac{1 \text{ mol C}_4\text{H}_{10}}{58.12 \text{ g C}_4\text{H}_{10}} = 0.430 \text{ mol of C}_4\text{H}_{10} \text{ (3 SFs)}$

7.29 **a.** Molar mass of CO_2

       $= 1 \text{ mol of C} + 2 \text{ mol of O} = 12.01 \text{ g} + 2(16.00 \text{ g}) = 44.01 \text{ g}$

      $0.688 \text{ g CO}_2 \times \dfrac{1 \text{ mol CO}_2}{44.01 \text{ g CO}_2} \times \dfrac{1 \text{ mol C}}{1 \text{ mol CO}_2} \times \dfrac{12.01 \text{ g C}}{1 \text{ mol C}} = 0.188 \text{ g of C (3 SFs)}$

    b. Molar mass of C_3H_6

       $= 3 \text{ mol of C} + 6 \text{ mol of H} = 3(12.01 \text{ g}) + 6(1.008 \text{ g}) = 42.08 \text{ g}$

      $275 \text{ g C}_3\text{H}_6 \times \dfrac{1 \text{ mol C}_3\text{H}_6}{42.08 \text{ g C}_3\text{H}_6} \times \dfrac{3 \text{ mol C}}{1 \text{ mol C}_3\text{H}_6} \times \dfrac{12.01 \text{ g C}}{1 \text{ mol C}} = 235 \text{ g of C (3 SFs)}$

    c. Molar mass of C_2H_6O

       $= 2 \text{ mol of C} + 6 \text{ mol of H} + 1 \text{ mol of O} = 2(12.01 \text{ g}) + 6(1.008 \text{ g}) + 16.00 \text{ g} = 46.07 \text{ g}$

      $1.84 \text{ g C}_2\text{H}_6\text{O} \times \dfrac{1 \text{ mol C}_2\text{H}_6\text{O}}{46.07 \text{ g C}_2\text{H}_6\text{O}} \times \dfrac{2 \text{ mol C}}{1 \text{ mol C}_2\text{H}_6\text{O}} \times \dfrac{12.01 \text{ g C}}{1 \text{ mol C}} = 0.959 \text{ g of C (3 SFs)}$

    d. Molar mass of $C_8H_{16}O_2$

       $= 8 \text{ mol of C} + 16 \text{ mol of H} + 2 \text{ mol of O} = 8(12.01 \text{ g}) + 16(1.008 \text{ g}) + 2(16.00 \text{ g})$

       $= 144.2 \text{ g}$

      $73.4 \text{ g C}_8\text{H}_{16}\text{O}_2 \times \dfrac{1 \text{ mol C}_8\text{H}_{16}\text{O}_2}{144.2 \text{ g C}_8\text{H}_{16}\text{O}_2} \times \dfrac{8 \text{ mol C}}{1 \text{ mol C}_8\text{H}_{16}\text{O}_2} \times \dfrac{12.01 \text{ g C}}{1 \text{ mol C}} = 48.9 \text{ g of C (3 SFs)}$

7.31 Molar mass of C_3H_8
= 3 mol of C + 8 mol of H = 3(12.01 g) + 8(1.008 g) = 44.09 g

a. $34.0 \text{ g } C_3H_8 \times \dfrac{1 \text{ mol } C_3H_8}{44.09 \text{ g } C_3H_8} \times \dfrac{8 \text{ mol H}}{1 \text{ mol } C_3H_8} = 6.17 \text{ mol of H (3 SFs)}$

b. $1.50 \text{ mol } C_3H_8 \times \dfrac{3 \text{ mol C}}{1 \text{ mol } C_3H_8} \times \dfrac{12.01 \text{ g C}}{1 \text{ mol C}} = 54.0 \text{ g of C (3 SFs)}$

c. $34.0 \text{ g } C_3H_8 \times \dfrac{1 \text{ mol } C_3H_8}{44.09 \text{ g } C_3H_8} \times \dfrac{3 \text{ mol C}}{1 \text{ mol } C_3H_8} \times \dfrac{12.01 \text{ g C}}{1 \text{ mol C}} = 27.8 \text{ g of C (3 SFs)}$

d. $0.254 \text{ g } C_3H_8 \times \dfrac{1 \text{ mol } C_3H_8}{44.09 \text{ g } C_3H_8} \times \dfrac{8 \text{ mol H}}{1 \text{ mol } C_3H_8} \times \dfrac{1.008 \text{ g H}}{1 \text{ mol H}}$

$= 0.0465 \text{ g of H (3 SFs)}$

7.33 a. Molar mass of $MgSO_4$
= 1 mol of Mg + 1 mol of S + 4 mol of O = 24.31 g + 32.07 g + 4(16.00 g)
= 120.38 g

$5.00 \text{ mol } MgSO_4 \times \dfrac{120.38 \text{ g } MgSO_4}{1 \text{ mol } MgSO_4} = 602 \text{ g of } MgSO_4 \text{ (3 SFs)}$

b. Molar mass of KI
= 1 mol of K + 1 mol of I = 39.10 g + 126.9 g = 166.0 g

$0.450 \text{ mol KI} \times \dfrac{166.0 \text{ g KI}}{1 \text{ mol KI}} = 74.7 \text{ g of KI (3 SFs)}$

7.35 Molar mass of N_2O
= 2 mol of N + 1 mol of O = 2(14.01 g) + 16.00 g = 44.02 g

a. $1.50 \text{ mol } N_2O \times \dfrac{44.02 \text{ g } N_2O}{1 \text{ mol } N_2O} = 66.0 \text{ g of } N_2O \text{ (3 SFs)}$

b. $34.0 \text{ g } N_2O \times \dfrac{1 \text{ mol } N_2O}{44.02 \text{ g } N_2O} = 0.772 \text{ mol of } N_2O \text{ (3 SFs)}$

c. $34.0 \text{ g } N_2O \times \dfrac{1 \text{ mol } N_2O}{44.02 \text{ g } N_2O} \times \dfrac{2 \text{ mol N}}{1 \text{ mol } N_2O} \times \dfrac{14.01 \text{ g N}}{1 \text{ mol N}} = 21.6 \text{ g of N (3 SFs)}$

7.37 Mass percent of an element $= \dfrac{\text{mass of an element}}{\text{total mass of the compound}} \times 100\%$

a. Mass of compound = 4.68 g Si + 5.32 g O = 10.00 g of compound

Mass % Si $= \dfrac{4.68 \text{ g Si}}{10.00 \text{ g compound}} \times 100\% = 46.8\% \text{ Si (3 SFs)}$

Mass % O $= \dfrac{5.32 \text{ g O}}{10.00 \text{ g compound}} \times 100\% = 53.2\% \text{ O (3 SFs)}$

b. Mass of compound = 5.72 g C + 1.28 g H = 7.00 g of compound

Mass % C $= \dfrac{5.72 \text{ g C}}{7.00 \text{ g compound}} \times 100\% = 81.7\% \text{ C (3 SFs)}$

Mass % H $= \dfrac{1.28 \text{ g H}}{7.00 \text{ g compound}} \times 100\% = 18.3\% \text{ H (3 SFs)}$

c. Mass of compound = 16.1 g Na + 22.5 g S + 11.3 g O = 49.9 g of compound

$$\text{Mass \% Na} = \frac{16.1 \text{ g Na}}{49.9 \text{ g compound}} \times 100\% = 32.3\% \text{ Na (3 SFs)}$$

$$\text{Mass \% S} = \frac{22.5 \text{ g S}}{49.9 \text{ g compound}} \times 100\% = 45.1\% \text{ S (3 SFs)}$$

$$\text{Mass \% O} = \frac{11.3 \text{ g O}}{49.9 \text{ g compound}} \times 100\% = 22.6\% \text{ O (3 SFs)}$$

d. Mass of compound = 6.22 g C + 1.04 g H + 4.14 g O = 11.4 g of compound

$$\text{Mass \% C} = \frac{6.22 \text{ g C}}{11.4 \text{ g compound}} \times 100\% = 54.6\% \text{ C (3 SFs)}$$

$$\text{Mass \% H} = \frac{1.04 \text{ g H}}{11.4 \text{ g compound}} \times 100\% = 9.12\% \text{ H (3 SFs)}$$

$$\text{Mass \% O} = \frac{4.14 \text{ g O}}{11.4 \text{ g compound}} \times 100\% = 36.3\% \text{ O (3 SFs)}$$

7.39 a. $1 \text{ mol Mg} \times \dfrac{24.31 \text{ g Mg}}{1 \text{ mol Mg}} = 24.31 \text{ g of Mg}$

$2 \text{ mol F} \times \dfrac{19.00 \text{ g F}}{1 \text{ mol F}} = 38.00 \text{ g of F}$

$\therefore$ Molar mass of MgF_2 = 62.31 g

$$\text{Mass \% Mg} = \frac{24.31 \text{ g Mg}}{62.31 \text{ g MgF}_2} \times 100\% = 39.01\% \text{ Mg (4 SFs)}$$

$$\text{Mass \% F} = \frac{38.00 \text{ g F}}{62.31 \text{ g MgF}_2} \times 100\% = 60.99\% \text{ F (4 SFs)}$$

b. $1 \text{ mol Ca} \times \dfrac{40.08 \text{ g Ca}}{1 \text{ mol Ca}} = 40.08 \text{ g of Ca}$

$2 \text{ mol O} \times \dfrac{16.00 \text{ g O}}{1 \text{ mol O}} = 32.00 \text{ g of O}$

$2 \text{ mol H} \times \dfrac{1.008 \text{ g H}}{1 \text{ mol H}} = 2.016 \text{ g of H}$

$\therefore$ Molar mass of $Ca(OH)_2$ = 74.10 g

$$\text{Mass \% Ca} = \frac{40.08 \text{ g Ca}}{74.10 \text{ g Ca(OH)}_2} \times 100\% = 54.09\% \text{ Ca (4 SFs)}$$

$$\text{Mass \% O} = \frac{32.00 \text{ g O}}{74.10 \text{ g Ca(OH)}_2} \times 100\% = 43.18\% \text{ O (4 SFs)}$$

$$\text{Mass \% H} = \frac{2.016 \text{ g H}}{74.10 \text{ g Ca(OH)}_2} \times 100\% = 2.721\% \text{ H (4 SFs)}$$

c. $4 \text{ mol C} \times \dfrac{12.01 \text{ g C}}{1 \text{ mol C}} = 48.04 \text{ g of C}$

$8 \text{ mol H} \times \dfrac{1.008 \text{ g H}}{1 \text{ mol H}} = 8.064 \text{ g of H}$

$4 \text{ mol O} \times \dfrac{16.00 \text{ g O}}{1 \text{ mol O}} = 64.00 \text{ g of O}$

$\therefore$ Molar mass of $C_4H_8O_4 = 120.10$ g

$\text{Mass \% C} = \dfrac{48.04 \text{ g C}}{120.10 \text{ g } C_4H_8O_4} \times 100\% = 40.00\% \text{ C (4 SFs)}$

$\text{Mass \% H} = \dfrac{8.064 \text{ g H}}{120.10 \text{ g } C_4H_8O_4} \times 100\% = 6.714\% \text{ H (4 SFs)}$

$\text{Mass \% O} = \dfrac{64.00 \text{ g O}}{120.10 \text{ g } C_4H_8O_4} \times 100\% = 53.29\% \text{ O (4 SFs)}$

d. $3 \text{ mol N} \times \dfrac{14.01 \text{ g N}}{1 \text{ mol N}} = 42.03 \text{ g of N}$

$12 \text{ mol H} \times \dfrac{1.008 \text{ g H}}{1 \text{ mol H}} = 12.10 \text{ g of H}$

$1 \text{ mol P} \times \dfrac{30.97 \text{ g P}}{1 \text{ mol P}} = 30.97 \text{ g of P}$

$4 \text{ mol O} \times \dfrac{16.00 \text{ g O}}{1 \text{ mol O}} = 64.00 \text{ g of O}$

$\therefore$ Molar mass of $(NH_4)_3PO_4 = 149.10$ g

$\text{Mass \% N} = \dfrac{42.03 \text{ g N}}{149.10 \text{ g } (NH_4)_3PO_4} \times 100\% = 28.19\% \text{ N (4 SFs)}$

$\text{Mass \% H} = \dfrac{12.10 \text{ g H}}{149.10 \text{ g } (NH_4)_3PO_4} \times 100\% = 8.115\% \text{ H (4 SFs)}$

$\text{Mass \% P} = \dfrac{30.97 \text{ g P}}{149.10 \text{ g } (NH_4)_3PO_4} \times 100\% = 20.77\% \text{ P (4 SFs)}$

$\text{Mass \% O} = \dfrac{64.00 \text{ g O}}{149.10 \text{ g } (NH_4)_3PO_4} \times 100\% = 42.92\% \text{ O (4 SFs)}$

e. $17 \text{ mol C} \times \dfrac{12.01 \text{ g C}}{1 \text{ mol C}} = 204.2 \text{ g of C}$

$19 \text{ mol H} \times \dfrac{1.008 \text{ g H}}{1 \text{ mol H}} = 19.15 \text{ g of H}$

$1 \text{ mol N} \times \dfrac{14.01 \text{ g N}}{1 \text{ mol N}} = 14.01 \text{ g of N}$

$3 \text{ mol O} \times \dfrac{16.00 \text{ g O}}{1 \text{ mol O}} = 48.00 \text{ g of O}$

$\therefore$ Molar mass of $C_{17}H_{19}NO_3 = 285.4$ g

$$\text{Mass \% C} = \frac{204.2 \text{ g C}}{285.4 \text{ g C}_{17}\text{H}_{19}\text{NO}_3} \times 100\% = 71.55\% \text{ C (4 SFs)}$$

$$\text{Mass \% H} = \frac{19.15 \text{ g H}}{285.4 \text{ g C}_{17}\text{H}_{19}\text{NO}_3} \times 100\% = 6.710\% \text{ H (4 SFs)}$$

$$\text{Mass \% N} = \frac{14.01 \text{ g N}}{285.4 \text{ g C}_{17}\text{H}_{19}\text{NO}_3} \times 100\% = 4.909\% \text{ N (4 SFs)}$$

$$\text{Mass \% O} = \frac{48.00 \text{ g O}}{285.4 \text{ g C}_{17}\text{H}_{19}\text{NO}_3} \times 100\% = 16.82\% \text{ O (4 SFs)}$$

7.41 **a.** $2 \text{ mol N} \times \dfrac{14.01 \text{ g N}}{1 \text{ mol N}} = 28.02 \text{ g of N}$

$5 \text{ mol O} \times \dfrac{16.00 \text{ g O}}{1 \text{ mol O}} = 80.00 \text{ g of O}$

$\therefore$ Molar mass of $N_2O_5 = 108.02$ g

$$\text{Mass \% N} = \frac{28.02 \text{ g N}}{108.02 \text{ g N}_2\text{O}_5} \times 100\% = 25.94\% \text{ N (4 SFs)}$$

b. $1 \text{ mol N} \times \dfrac{14.01 \text{ g N}}{1 \text{ mol N}} = 14.01 \text{ g of N}$

$4 \text{ mol H} \times \dfrac{1.008 \text{ g H}}{1 \text{ mol H}} = 4.032 \text{ g of H}$

$1 \text{ mol Cl} \times \dfrac{35.45 \text{ g Cl}}{1 \text{ mol Cl}} = 35.45 \text{ g of Cl}$

$\therefore$ Molar mass of $NH_4Cl = 53.49$ g

$$\text{Mass \% N} = \frac{14.01 \text{ g N}}{53.49 \text{ g NH}_4\text{Cl}} \times 100\% = 26.19\% \text{ N (4 SFs)}$$

c. $2 \text{ mol C} \times \dfrac{12.01 \text{ g C}}{1 \text{ mol C}} = 24.02 \text{ g of C}$

$8 \text{ mol H} \times \dfrac{1.008 \text{ g H}}{1 \text{ mol H}} = 8.064 \text{ g of H}$

$2 \text{ mol N} \times \dfrac{14.01 \text{ g N}}{1 \text{ mol N}} = 28.02 \text{ g of N}$

$\therefore$ Molar mass of $C_2H_8N_2 = 60.10$ g

$$\text{Mass \% N} = \frac{28.02 \text{ g N}}{60.10 \text{ g C}_2\text{H}_8\text{N}_2} \times 100\% = 46.62\% \text{ N (4 SFs)}$$

d. $9 \text{ mol C} \times \dfrac{12.01 \text{ g C}}{1 \text{ mol C}} = 108.1 \text{ g of C}$

$15 \text{ mol H} \times \dfrac{1.008 \text{ g H}}{1 \text{ mol H}} = 15.12 \text{ g of H}$

$5 \text{ mol N} \times \dfrac{14.01 \text{ g N}}{1 \text{ mol N}} = 70.05 \text{ g of N}$

$1 \text{ mol O} \times \dfrac{16.00 \text{ g O}}{1 \text{ mol O}} = 16.00 \text{ g of O}$

$\therefore$ Molar mass of $C_9H_{15}N_5O = 209.3$ g

$$\text{Mass \% N} = \frac{70.05 \text{ g N}}{209.3 \text{ g } C_9H_{15}N_5O} \times 100\% = 33.47\% \text{ N (4 SFs)}$$

e. $14 \text{ mol C} \times \dfrac{12.01 \text{ g C}}{1 \text{ mol C}} = 168.1 \text{ g of C}$

$22 \text{ mol H} \times \dfrac{1.008 \text{ g H}}{1 \text{ mol H}} = 22.18 \text{ g of H}$

$2 \text{ mol N} \times \dfrac{14.01 \text{ g N}}{1 \text{ mol N}} = 28.02 \text{ g of N}$

$1 \text{ mol O} \times \dfrac{16.00 \text{ g O}}{1 \text{ mol O}} = 16.00 \text{ g of O}$

$\therefore$ Molar mass of $C_{14}H_{22}N_2O = 234.3 \text{ g}$

$$\text{Mass \% N} = \frac{28.02 \text{ g N}}{234.3 \text{ g } C_{14}H_{22}N_2O} \times 100\% = 11.96\% \text{ N (4 SFs)}$$

7.43 a. $3.57 \text{ g N} \times \dfrac{1 \text{ mol N}}{14.01 \text{ g N}} = 0.255 \text{ mol of N}$

$2.04 \text{ g O} \times \dfrac{1 \text{ mol O}}{16.00 \text{ g O}} = 0.128 \text{ mol of O (smaller number of moles)}$

$\dfrac{0.255 \text{ mol N}}{0.128} = 1.99 \text{ mol of N}$ $\dfrac{0.128 \text{ mol O}}{0.128} = 1.00 \text{ mol of O}$

$\therefore$ Empirical formula $= N_{1.99}O_{1.00} \rightarrow N_2O$

b. $7.00 \text{ g C} \times \dfrac{1 \text{ mol C}}{12.01 \text{ g C}} = 0.583 \text{ mol of C (smaller number of moles)}$

$1.75 \text{ g H} \times \dfrac{1 \text{ mol H}}{1.008 \text{ g H}} = 1.74 \text{ mol of H}$

$\dfrac{0.583 \text{ mol C}}{0.583} = 1.00 \text{ mol of C}$ $\dfrac{1.74 \text{ mol H}}{0.583} = 2.98 \text{ mol of H}$

$\therefore$ Empirical formula $= C_{1.00}H_{2.98} \rightarrow CH_3$

c. $0.175 \text{ g H} \times \dfrac{1 \text{ mol H}}{1.008 \text{ g H}} = 0.174 \text{ mol of H (smallest number of moles)}$

$2.44 \text{ g N} \times \dfrac{1 \text{ mol N}}{14.01 \text{ g N}} = 0.174 \text{ mol of N}$

$8.38 \text{ g O} \times \dfrac{1 \text{ mol O}}{16.00 \text{ g O}} = 0.524 \text{ mol of O}$

$\dfrac{0.174 \text{ mol H}}{0.174} = 1.00 \text{ mol of H}$ $\dfrac{0.174 \text{ mol N}}{0.174} = 1.00 \text{ mol of N}$

$\dfrac{0.524 \text{ mol O}}{0.174} = 3.01 \text{ mol of O}$

$\therefore$ Empirical formula $= H_{1.00}N_{1.00}O_{3.01} \rightarrow HNO_3$

d. $2.06 \text{ g Ca} \times \dfrac{1 \text{ mol Ca}}{40.08 \text{ g Ca}} = 0.0514 \text{ mol of Ca}$

$2.66 \text{ g Cr} \times \dfrac{1 \text{ mol Cr}}{52.00 \text{ g Cr}} = 0.0512 \text{ mol of Cr (smallest number of moles)}$

$3.28 \text{ g O} \times \dfrac{1 \text{ mol O}}{16.00 \text{ g O}} = 0.205 \text{ mol of O}$

$\dfrac{0.0514 \text{ mol Ca}}{0.0512} = 1.00 \text{ mol of Ca}$ $\qquad \dfrac{0.0512 \text{ mol Cr}}{0.0512} = 1.00 \text{ mol of Cr}$

$\dfrac{0.205 \text{ mol O}}{0.0512} = 4.00 \text{ mol of O}$

$\therefore$ Empirical formula $= Ca_{1.00}Cr_{1.00}O_{4.00} \rightarrow CaCrO_4$

7.45 **a.** In exactly 100. g of compound, there are 70.9 g of K and 29.1 g of S.

$70.9 \text{ g K} \times \dfrac{1 \text{ mol K}}{39.10 \text{ g K}} = 1.81 \text{ mol of K}$

$29.1 \text{ g S} \times \dfrac{1 \text{ mol S}}{32.07 \text{ g S}} = 0.907 \text{ mol of S (smaller number of moles)}$

$\dfrac{1.81 \text{ mol K}}{0.907} = 2.00 \text{ mol of K}$ $\qquad \dfrac{0.907 \text{ mol S}}{0.907} = 1.00 \text{ mol of S}$

$\therefore$ Empirical formula $= K_{2.00}S_{1.00} \rightarrow K_2S$

b. In exactly 100. g of compound, there are 55.0 g of Ga and 45.0 g of F.

$55.0 \text{ g Ga} \times \dfrac{1 \text{ mol Ga}}{69.72 \text{ g Ga}} = 0.789 \text{ mol of Ga (smaller number of moles)}$

$45.0 \text{ g F} \times \dfrac{1 \text{ mol F}}{19.00 \text{ g F}} = 2.37 \text{ mol of F}$

$\dfrac{0.789 \text{ mol Ga}}{0.789} = 1.00 \text{ mol of Ga}$ $\qquad \dfrac{2.37 \text{ mol F}}{0.789} = 3.00 \text{ mol of F}$

$\therefore$ Empirical formula $= Ga_{1.00}F_{3.00} \rightarrow GaF_3$

c. In exactly 100. g of compound, there are 31.0 g of B and 69.0 g of O.

$31.0 \text{ g B} \times \dfrac{1 \text{ mol B}}{10.81 \text{ g B}} = 2.87 \text{ mol of B (smaller number of moles)}$

$69.0 \text{ g O} \times \dfrac{1 \text{ mol O}}{16.00 \text{ g O}} = 4.31 \text{ mol of O}$

$\dfrac{2.87 \text{ mol B}}{2.87} = 1.00 \text{ mol of B}$ $\qquad \dfrac{4.31 \text{ mol O}}{2.87} = 1.50 \text{ mol of O}$

$\therefore$ Empirical formula $= B_{1.00}O_{1.50} \rightarrow B_{(1.00 \times 2)}O_{(1.50 \times 2)} = B_{2.00}O_{3.00} \rightarrow B_2O_3$

d. In exactly 100. g of compound, there are 18.8 g of Li, 16.3 g of C, and 64.9 g of O.

$$18.8 \text{ g Li} \times \frac{1 \text{ mol Li}}{6.941 \text{ g Li}} = 2.71 \text{ mol of Li}$$

$$16.3 \text{ g C} \times \frac{1 \text{ mol C}}{12.01 \text{ g C}} = 1.36 \text{ mol of C (smallest number of moles)}$$

$$64.9 \text{ g O} \times \frac{1 \text{ mol O}}{16.00 \text{ g O}} = 4.06 \text{ mol of O}$$

$$\frac{2.71 \text{ mol Li}}{1.36} = 1.99 \text{ mol of Li} \qquad \frac{1.36 \text{ mol C}}{1.36} = 1.00 \text{ mol of C}$$

$$\frac{4.06 \text{ mol O}}{1.36} = 2.99 \text{ mol of O}$$

$\therefore$ Empirical formula $= \text{Li}_{1.99}\text{C}_{1.00}\text{O}_{2.99} \rightarrow \text{Li}_2\text{CO}_3$

e. In exactly 100. g of compound, there are 51.7 g of C, 6.95 g of H, and 41.3 g of O.

$$51.7 \text{ g C} \times \frac{1 \text{ mol C}}{12.01 \text{ g C}} = 4.30 \text{ mol of C}$$

$$6.95 \text{ g H} \times \frac{1 \text{ mol H}}{1.008 \text{ g H}} = 6.89 \text{ mol of H}$$

$$41.3 \text{ g O} \times \frac{1 \text{ mol O}}{16.00 \text{ g O}} = 2.58 \text{ mol of O (smallest number of moles)}$$

$$\frac{4.30 \text{ mol C}}{2.58} = 1.67 \text{ mol of C} \qquad \frac{6.89 \text{ mol H}}{2.58} = 2.67 \text{ mol of H}$$

$$\frac{2.58 \text{ mol O}}{2.58} = 1.00 \text{ mol of O}$$

$\therefore$ Empirical formula $= \text{C}_{1.67}\text{H}_{2.67}\text{O}_{1.00} \rightarrow \text{C}_{(1.67\times3)}\text{H}_{(2.67\times3)}\text{O}_{(1.00\times3)}$

$$= \text{C}_{5.01}\text{H}_{8.01}\text{O}_{3.00} \rightarrow \text{C}_5\text{H}_8\text{O}_3$$

7.47 **a.** $\text{H}_2\text{O}_2 \rightarrow \text{H}_{(2\div2)}\text{O}_{(2\div2)} = \text{HO}$ (empirical formula)
b. $\text{C}_{18}\text{H}_{12} \rightarrow \text{C}_{(18\div6)}\text{H}_{(12\div6)} = \text{C}_3\text{H}_2$ (empirical formula)
c. $\text{C}_{10}\text{H}_{16}\text{O}_2 \rightarrow \text{C}_{(10\div2)}\text{H}_{(16\div2)}\text{O}_{(2\div2)} = \text{C}_5\text{H}_8\text{O}$ (empirical formula)
d. $\text{C}_9\text{H}_{18}\text{N}_6 \rightarrow \text{C}_{(9\div3)}\text{H}_{(18\div3)}\text{N}_{(6\div3)} = \text{C}_3\text{H}_6\text{N}_2$ (empirical formula)
e. $\text{C}_2\text{H}_4\text{N}_2\text{O}_2 \rightarrow \text{C}_{(2\div2)}\text{H}_{(4\div2)}\text{N}_{(2\div2)}\text{O}_{(2\div2)} = \text{CH}_2\text{NO}$ (empirical formula)

7.49 Empirical formula mass of $\text{CH}_2\text{O} = 12.01 \text{ g} + 2(1.008 \text{ g}) + 16.00 \text{ g} = 30.03 \text{ g}$

$$\text{Small integer} = \frac{\text{molar mass of fructose}}{\text{empirical formula mass of CH}_2\text{O}} = \frac{180 \text{ g}}{30.03 \text{ g}} = 6$$

$\therefore$ Molecular formula of fructose $= \text{C}_{(1\times6)}\text{H}_{(2\times6)}\text{O}_{(1\times6)} = \text{C}_6\text{H}_{12}\text{O}_6$

7.51 Empirical formula mass of $\text{CH} = 12.01 \text{ g} + 1.008 \text{ g} = 13.02 \text{ g}$

For benzene: $\quad \text{Small integer} = \dfrac{\text{molar mass of benzene}}{\text{empirical formula mass of CH}} = \dfrac{78 \text{ g}}{13.02 \text{ g}} = 6$

$\therefore$ Molecular formula of benzene $= \text{C}_{(1\times6)}\text{H}_{(1\times6)} = \text{C}_6\text{H}_6$

For acetylene: $\quad \text{Small integer} = \dfrac{\text{molar mass of acetylene}}{\text{empirical formula mass of CH}} = \dfrac{26 \text{ g}}{13.02 \text{ g}} = 2$

$\therefore$ Molecular formula of acetylene $= \text{C}_{(1\times2)}\text{H}_{(1\times2)} = \text{C}_2\text{H}_2$

7.53 In exactly 100. g of mevalonic acid, there are 48.64 g of C, 8.16 g of H, and 43.20 g of O.

$$48.64 \text{ g C} \times \frac{1 \text{ mol C}}{12.01 \text{ g C}} = 4.050 \text{ mol of C}$$

$$8.16 \text{ g H} \times \frac{1 \text{ mol H}}{1.008 \text{ g H}} = 8.10 \text{ mol of H}$$

$$43.20 \text{ g O} \times \frac{1 \text{ mol O}}{16.00 \text{ g O}} = 2.70 \text{ mol of O (smallest number of moles)}$$

$$\frac{4.050 \text{ mol C}}{2.70} = 1.50 \text{ mol of C} \qquad \frac{8.10 \text{ mol H}}{2.70} = 3.00 \text{ mol of H}$$

$$\frac{2.70 \text{ mol O}}{2.70} = 1.00 \text{ mol of O}$$

∴ Empirical formula $= C_{1.50}H_{3.00}O_{1.00} \rightarrow C_{(1.50 \times 2)}H_{(3.00 \times 2)}O_{(1.00 \times 2)}$
$$= C_{3.00}H_{6.00}O_{2.00} \rightarrow C_3H_6O_2$$

Empirical formula mass of $C_3H_6O_2 = 3(12.01 \text{ g}) + 6(1.008 \text{ g}) + 2(16.00 \text{ g}) = 74.08 \text{ g}$

$$\text{Small integer} = \frac{\text{molar mass of mevalonic acid}}{\text{empirical formula mass of } C_3H_6O_2} = \frac{148 \text{ g}}{74.08 \text{ g}} = 2$$

∴ Molecular formula of mevalonic acid $= C_{(3 \times 2)}H_{(6 \times 2)}O_{(2 \times 2)} = C_6H_{12}O_4$

7.55 In exactly 100. g of vanillic acid, there are 57.14 g of C, 4.80 g of H, and 38.06 g of O.

$$57.14 \text{ g C} \times \frac{1 \text{ mol C}}{12.01 \text{ g C}} = 4.758 \text{ mol of C}$$

$$4.80 \text{ g H} \times \frac{1 \text{ mol H}}{1.008 \text{ g H}} = 4.76 \text{ mol of H}$$

$$38.06 \text{ g O} \times \frac{1 \text{ mol O}}{16.00 \text{ g O}} = 2.379 \text{ mol of O (smallest number of moles)}$$

$$\frac{4.758 \text{ mol C}}{2.379} = 2.000 \text{ mol of C} \qquad \frac{4.76 \text{ mol H}}{2.379} = 2.00 \text{ mol of H}$$

$$\frac{2.379 \text{ mol O}}{2.379} = 1.000 \text{ mol of O}$$

∴ Empirical formula $= C_{2.000}H_{2.00}O_{1.000} \rightarrow C_2H_2O$

Empirical formula mass of $C_2H_2O = 2(12.01 \text{ g}) + 2(1.008 \text{ g}) + 16.00 \text{ g} = 42.04 \text{ g}$

$$\text{Small integer} = \frac{\text{molar mass of vanillic acid}}{\text{empirical formula mass of } C_2H_2O} = \frac{168 \text{ g}}{42.04 \text{ g}} = 4$$

∴ Molecular formula of vanillic acid $= C_{(2 \times 4)}H_{(2 \times 4)}O_{(1 \times 4)} = C_8H_8O_4$

7.57 In exactly 100. g of nicotine, there are 74.0 g of C, 8.70 g of H, and 17.3 g of N.

$$74.0 \text{ g C} \times \frac{1 \text{ mol C}}{12.01 \text{ g C}} = 6.16 \text{ mol of C}$$

$$8.70 \text{ g H} \times \frac{1 \text{ mol H}}{1.008 \text{ g H}} = 8.63 \text{ mol of H}$$

$$17.3 \text{ g N} \times \frac{1 \text{ mol N}}{14.01 \text{ g N}} = 1.23 \text{ mol of N (smallest number of moles)}$$

$$\frac{6.16 \text{ mol C}}{1.23} = 5.00 \text{ mol of C} \qquad \frac{8.63 \text{ mol H}}{1.23} = 7.02 \text{ mol of H}$$

$$\frac{1.23 \text{ mol N}}{1.23} = 1.00 \text{ mol of N}$$

∴ Empirical formula $= C_{5.00}H_{7.02}N_{1.00} \rightarrow C_5H_7N$

Empirical formula mass of $C_5H_7N = 5(12.01 \text{ g}) + 7(1.008 \text{ g}) + 14.01 \text{ g} = 81.12 \text{ g}$

$$\text{Small integer} = \frac{\text{molar mass of nicotine}}{\text{empirical formula mass of } C_5H_7N} = \frac{162 \text{ g}}{81.12 \text{ g}} = 2$$

∴ Molecular formula of nicotine $= C_{(5 \times 2)}H_{(7 \times 2)}N_{(1 \times 2)} = C_{10}H_{14}N_2$

7.59 **a.** Molar mass of clavulanic acid
= 8 mol of C + 9 mol of H + 1 mol of N + 5 mol of O
$= 8(12.01 \text{ g}) + 9(1.008 \text{ g}) + 14.01 \text{ g} + 5(16.00 \text{ g}) = 199.16 \text{ g (5 SFs)}$

b. Mass % C $= \dfrac{96.08 \text{ g C}}{199.16 \text{ g C}_8\text{H}_9\text{NO}_5} \times 100\% = 48.24\% \text{ C (4 SFs)}$

c. $29 \text{ kg} \times \dfrac{2.5 \text{ mg clavulanic acid}}{1 \text{ kg}} \times \dfrac{1 \text{ g clavulanic acid}}{1000 \text{ mg clavulanic acid}} \times \dfrac{1 \text{ mol clavulanic acid}}{199.16 \text{ g clavulanic acid}}$

$= 3.6 \times 10^{-4} \text{ mol of clavulanic acid (2 SFs)}$

7.61 **a.** $C_{10}H_8N_2O_2S_2 \rightarrow C_{(10 \div 2)}H_{(8 \div 2)}N_{(2 \div 2)}O_{(2 \div 2)}S_{(2 \div 2)} = C_5H_4NOS$ (empirical formula)

b. $10 \text{ mol C} \times \dfrac{12.01 \text{ g C}}{1 \text{ mol C}} = 120.1 \text{ g of C}$

$8 \text{ mol H} \times \dfrac{1.008 \text{ g H}}{1 \text{ mol H}} = 8.064 \text{ g of H}$

$2 \text{ mol N} \times \dfrac{14.01 \text{ g N}}{1 \text{ mol N}} = 28.02 \text{ g of N}$

$2 \text{ mol O} \times \dfrac{16.00 \text{ g O}}{1 \text{ mol O}} = 32.00 \text{ g of O}$

$2 \text{ mol S} \times \dfrac{32.07 \text{ g S}}{1 \text{ mol S}} = 64.14 \text{ g of S}$

∴ Molar mass of $C_{10}H_8N_2O_2S_2 = 252.3 \text{ g}$

c. Mass % O $= \dfrac{32.00 \text{ g O}}{252.3 \text{ g C}_{10}\text{H}_8\text{N}_2\text{O}_2\text{S}_2} \times 100\% = 12.68\% \text{ O (4 SFs)}$

d. $25.0 \text{ g C}_{10}\text{H}_8\text{N}_2\text{O}_2\text{S}_2 \times \dfrac{1 \text{ mol C}_{10}\text{H}_8\text{N}_2\text{O}_2\text{S}_2}{252.3 \text{ g C}_{10}\text{H}_8\text{N}_2\text{O}_2\text{S}_2} \times \dfrac{10 \text{ mol C}}{1 \text{ mol C}_{10}\text{H}_8\text{N}_2\text{O}_2\text{S}_2}$

$\times \dfrac{12.01 \text{ g C}}{1 \text{ mol C}}$

$= 11.9 \text{ g of C (3 SFs)}$

e. $25.0 \text{ g } \overline{C_{10}H_8N_2O_2S_2} \times \dfrac{1 \text{ mol } C_{10}H_8N_2O_2S_2}{252.3 \text{ g } \overline{C_{10}H_8N_2O_2S_2}}$

$= 0.0991 \text{ mol of } C_{10}H_8N_2O_2S_2 \text{ (3 SFs)}$

7.63 (1) a. molecular formula $= S_2Cl_2$

    b. $S_2Cl_2 \rightarrow S_{(2 \div 2)}Cl_{(2 \div 2)} = SCl$ (empirical formula)

    c. $2 \text{ mol S} \times \dfrac{32.07 \text{ g S}}{1 \text{ mol S}} = 64.14 \text{ g of S}$

         $2 \text{ mol Cl} \times \dfrac{35.45 \text{ g Cl}}{1 \text{ mol Cl}} = 70.90 \text{ g of Cl}$

         $\therefore$ Molar mass of $S_2Cl_2 = 135.04 \text{ g}$

    d. Mass % S $= \dfrac{64.14 \text{ g S}}{135.04 \text{ g } S_2Cl_2} \times 100\% = 47.50\% \text{ S (4 SFs)}$

         Mass % Cl $= \dfrac{70.90 \text{ g Cl}}{135.04 \text{ g } S_2Cl_2} \times 100\% = 52.50\% \text{ Cl (4 SFs)}$

  (2) a. molecular formula $= C_6H_6$

    b. $C_6H_6 \rightarrow C_{(6 \div 6)}H_{(6 \div 6)} = CH$ (empirical formula)

    c. $6 \text{ mol C} \times \dfrac{12.01 \text{ g C}}{1 \text{ mol C}} = 72.06 \text{ g of C}$

         $6 \text{ mol H} \times \dfrac{1.008 \text{ g H}}{1 \text{ mol H}} = 6.048 \text{ g of H}$

         $\therefore$ Molar mass of $C_6H_6 = 78.11 \text{ g}$

    d. Mass % C $= \dfrac{72.06 \text{ g C}}{78.11 \text{ g } C_6H_6} \times 100\% = 92.25\% \text{ C (4 SFs)}$

         Mass % H $= \dfrac{6.048 \text{ g H}}{78.11 \text{ g } C_6H_6} \times 100\% = 7.743\% \text{ H (4 SFs)}$

7.65 a. $1 \text{ mol Zn} \times \dfrac{65.41 \text{ g Zn}}{1 \text{ mol Zn}} = 65.41 \text{ g of Zn}$

      $1 \text{ mol S} \times \dfrac{32.07 \text{ g S}}{1 \text{ mol S}} = 32.07 \text{ g of S}$

      $4 \text{ mol O} \times \dfrac{16.00 \text{ g O}}{1 \text{ mol O}} = 64.00 \text{ g of O}$

      1 mol of Zn $= 65.41 \text{ g of Zn}$

      1 mol of S $= 32.07 \text{ g of S}$

      4 mol of O $= 64.00 \text{ g of O}$

      $\therefore$ Molar mass of $ZnSO_4 = 161.48 \text{ g}$

b. $1 \text{ mol Ca} \times \dfrac{40.08 \text{ g Ca}}{1 \text{ mol Ca}} = 40.08 \text{ g of Ca}$

$2 \text{ mol I} \times \dfrac{126.9 \text{ g I}}{1 \text{ mol I}} = 253.8 \text{ g of I}$

$6 \text{ mol O} \times \dfrac{16.00 \text{ g O}}{1 \text{ mol O}} = 96.00 \text{ g of O}$

$$
\begin{aligned}
1 \text{ mol of Ca} &= 40.08 \text{ g of Ca} \\
2 \text{ mol of I} &= 253.8 \text{ g of I} \\
6 \text{ mol of O} &= \underline{96.00 \text{ g of O}} \\
\therefore \text{ Molar mass of Ca(IO}_3)_2 &= 389.9 \text{ g}
\end{aligned}
$$

c. $5 \text{ mol C} \times \dfrac{12.01 \text{ g C}}{1 \text{ mol C}} = 60.05 \text{ g of C}$

$8 \text{ mol H} \times \dfrac{1.008 \text{ g H}}{1 \text{ mol H}} = 8.064 \text{ g of H}$

$1 \text{ mol N} \times \dfrac{14.01 \text{ g N}}{1 \text{ mol N}} = 14.01 \text{ g of N}$

$1 \text{ mol Na} \times \dfrac{22.99 \text{ g Na}}{1 \text{ mol Na}} = 22.99 \text{ g of Na}$

$4 \text{ mol O} \times \dfrac{16.00 \text{ g O}}{1 \text{ mol O}} = 64.00 \text{ g of O}$

$$
\begin{aligned}
5 \text{ mol of C} &= 60.05 \text{ g of C} \\
8 \text{ mol of H} &= 8.064 \text{ g of H} \\
1 \text{ mol of N} &= 14.01 \text{ g of N} \\
1 \text{ mol of Na} &= 22.99 \text{ g of Na} \\
4 \text{ mol of O} &= \underline{64.00 \text{ g of O}} \\
\therefore \text{ Molar mass of C}_5\text{H}_8\text{NNaO}_4 &= 169.11 \text{ g}
\end{aligned}
$$

d. $6 \text{ mol C} \times \dfrac{12.01 \text{ g C}}{1 \text{ mol C}} = 72.06 \text{ g of C}$

$12 \text{ mol H} \times \dfrac{1.008 \text{ g H}}{1 \text{ mol H}} = 12.10 \text{ g of H}$

$2 \text{ mol O} \times \dfrac{16.00 \text{ g O}}{1 \text{ mol O}} = 32.00 \text{ g of O}$

$$
\begin{aligned}
6 \text{ mol of C} &= 72.06 \text{ g of C} \\
12 \text{ mol of H} &= 12.10 \text{ g of H} \\
2 \text{ mol of O} &= \underline{32.00 \text{ g of O}} \\
\therefore \text{ Molar mass of C}_6\text{H}_{12}\text{O}_2 &= 116.16 \text{ g}
\end{aligned}
$$

7.67 a. Molar mass of H_2O

$= 2$ mol of H $+ 1$ mol of O $= 2(1.008 \text{ g}) + 16.00 \text{ g} = 18.02 \text{ g}$

$50.0 \text{ g } H_2O \times \dfrac{1 \text{ mol } H_2O}{18.02 \text{ g } H_2O} \times \dfrac{1 \text{ mol } O}{1 \text{ mol } H_2O} \times \dfrac{16.00 \text{ g } O}{1 \text{ mol } O} = 44.4 \text{ g of O (3 SFs)}$

b. Molar mass of CO_2

$= 1$ mol of C $+ 2$ mol of O $= 12.01 \text{ g} + 2(16.00 \text{ g}) = 44.01 \text{ g}$

$17.5 \text{ g } CO_2 \times \dfrac{1 \text{ mol } CO_2}{44.01 \text{ g } CO_2} \times \dfrac{2 \text{ mol } O}{1 \text{ mol } CO_2} \times \dfrac{16.00 \text{ g } O}{1 \text{ mol } O} = 12.7 \text{ g of O (3 SFs)}$

c. Molar mass of $C_7H_6O_2$

$= 7$ mol of C $+ 6$ mol of H $+ 2$ mol of O $= 7(12.01 \text{ g}) + 6(1.008 \text{ g}) + 2(16.00 \text{ g}) = 122.12 \text{ g}$

$48 \text{ g } C_7H_6O_2 \times \dfrac{1 \text{ mol } C_7H_6O_2}{122.12 \text{ g } C_7H_6O_2} \times \dfrac{2 \text{ mol } O}{1 \text{ mol } C_7H_6O_2} \times \dfrac{16.00 \text{ g } O}{1 \text{ mol } O} = 13 \text{ g of O (2 SFs)}$

7.69 Mass percent of an element $= \dfrac{\text{mass of an element}}{\text{total mass of the compound}} \times 100\%$

a. Mass of compound $= 3.85 \text{ g Ca} + 3.65 \text{ g F} = 7.50 \text{ g of compound}$

Mass % Ca $= \dfrac{3.85 \text{ g Ca}}{7.50 \text{ g compound}} \times 100\% = 51.3\% \text{ Ca (3 SFs)}$

Mass % F $= \dfrac{3.65 \text{ g F}}{7.50 \text{ g compound}} \times 100\% = 48.7\% \text{ F (3 SFs)}$

b. Mass of compound $= 0.389 \text{ g Na} + 0.271 \text{ g O} = 0.660 \text{ g of compound}$

Mass % Na $= \dfrac{0.389 \text{ g Na}}{0.660 \text{ g compound}} \times 100\% = 58.9\% \text{ Na (3 SFs)}$

Mass % O $= \dfrac{0.271 \text{ g O}}{0.660 \text{ g compound}} \times 100\% = 41.1\% \text{ O (3 SFs)}$

c. Mass of compound $= 12.4 \text{ g K} + 17.4 \text{ g Mn} + 20.3 \text{ g O} = 50.1 \text{ g of compound}$

Mass % K $= \dfrac{12.4 \text{ g K}}{50.1 \text{ g compound}} \times 100\% = 24.8\% \text{ K (3 SFs)}$

Mass % Mn $= \dfrac{17.4 \text{ g Mn}}{50.1 \text{ g compound}} \times 100\% = 34.7\% \text{ Mn (3 SFs)}$

Mass % O $= \dfrac{20.3 \text{ g O}}{50.1 \text{ g compound}} \times 100\% = 40.5\% \text{ O (3 SFs)}$

7.71 a. $2 \text{ mol K} \times \dfrac{39.10 \text{ g K}}{1 \text{ mol K}} = 78.20 \text{ g of K}$

$1 \text{ mol Cr} \times \dfrac{52.00 \text{ g Cr}}{1 \text{ mol Cr}} = 52.00 \text{ g of Cr}$

$4 \text{ mol O} \times \dfrac{16.00 \text{ g O}}{1 \text{ mol O}} = 64.00 \text{ g of O}$

$\therefore$ Molar mass of $K_2CrO_4 = 194.20 \text{ g}$

$$\text{Mass \% K} = \frac{78.20 \text{ g K}}{194.20 \text{ g K}_2\text{CrO}_4} \times 100\% = 40.27\% \text{ K (4 SFs)}$$

$$\text{Mass \% Cr} = \frac{52.00 \text{ g Cr}}{194.20 \text{ g K}_2\text{CrO}_4} \times 100\% = 26.78\% \text{ Cr (4 SFs)}$$

$$\text{Mass \% O} = \frac{64.00 \text{ g O}}{194.20 \text{ g K}_2\text{CrO}_4} \times 100\% = 32.96\% \text{ O (4 SFs)}$$

b. $1 \text{ mol Al} \times \dfrac{26.98 \text{ g Al}}{1 \text{ mol Al}} = 26.98 \text{ g of Al}$

$3 \text{ mol H} \times \dfrac{1.008 \text{ g H}}{1 \text{ mol H}} = 3.024 \text{ g of H}$

$3 \text{ mol C} \times \dfrac{12.01 \text{ g C}}{1 \text{ mol C}} = 36.03 \text{ g of C}$

$9 \text{ mol O} \times \dfrac{16.00 \text{ g O}}{1 \text{ mol O}} = 144.0 \text{ g of O}$

$\therefore$ Molar mass of $Al(HCO_3)_3 = \overline{210.0 \text{ g}}$

$$\text{Mass \% Al} = \frac{26.98 \text{ g Al}}{210.0 \text{ g Al(HCO}_3)_3} \times 100\% = 12.85\% \text{ Al (4 SFs)}$$

$$\text{Mass \% H} = \frac{3.024 \text{ g H}}{210.0 \text{ g Al(HCO}_3)_3} \times 100\% = 1.440\% \text{ H (4 SFs)}$$

$$\text{Mass \% C} = \frac{36.03 \text{ g C}}{210.0 \text{ g Al(HCO}_3)_3} \times 100\% = 17.16\% \text{ C (4 SFs)}$$

$$\text{Mass \% O} = \frac{144.0 \text{ g O}}{210.0 \text{ g Al(HCO}_3)_3} \times 100\% = 68.57\% \text{ O (4 SFs)}$$

c. $6 \text{ mol C} \times \dfrac{12.01 \text{ g C}}{1 \text{ mol C}} = 72.06 \text{ g of C}$

$12 \text{ mol H} \times \dfrac{1.008 \text{ g H}}{1 \text{ mol H}} = 12.10 \text{ g of H}$

$6 \text{ mol O} \times \dfrac{16.00 \text{ g O}}{1 \text{ mol O}} = 96.00 \text{ g of O}$

$\therefore$ Molar mass of $C_6H_{12}O_6 = \overline{180.16 \text{ g}}$

$$\text{Mass \% C} = \frac{72.06 \text{ g C}}{180.16 \text{ g C}_6\text{H}_{12}\text{O}_6} \times 100\% = 40.00\% \text{ C (4 SFs)}$$

$$\text{Mass \% H} = \frac{12.10 \text{ g H}}{180.16 \text{ g C}_6\text{H}_{12}\text{O}_6} \times 100\% = 6.716\% \text{ H (4 SFs)}$$

$$\text{Mass \% O} = \frac{96.00 \text{ g O}}{180.16 \text{ g C}_6\text{H}_{12}\text{O}_6} \times 100\% = 53.29\% \text{ O (4 SFs)}$$

7.73 a. $9 \text{ mol C} \times \dfrac{12.01 \text{ g C}}{1 \text{ mol C}} = 108.1 \text{ g of C}$

$8 \text{ mol H} \times \dfrac{1.008 \text{ g H}}{1 \text{ mol H}} = 8.064 \text{ g of H}$

$4 \text{ mol O} \times \dfrac{16.00 \text{ g O}}{1 \text{ mol O}} = 64.00 \text{ g of O}$

$\therefore$ Molar mass of $C_9H_8O_4 = \overline{180.2 \text{ g}}$

$$\text{Mass \% C} = \frac{108.1 \text{ g C}}{180.2 \text{ g C}_9\text{H}_8\text{O}_4} \times 100\% = 59.99\% \text{ C (4 SFs)}$$

$$\text{Mass \% H} = \frac{8.064 \text{ g H}}{180.2 \text{ g C}_9\text{H}_8\text{O}_4} \times 100\% = 4.475\% \text{ H (4 SFs)}$$

$$\text{Mass \% O} = \frac{64.00 \text{ g O}}{180.2 \text{ g C}_9\text{H}_8\text{O}_4} \times 100\% = 35.52\% \text{ O (4 SFs)}$$

b. 5.0×10^{24} atoms C $\times \dfrac{1 \text{ mol C}}{6.022 \times 10^{23} \text{ atoms C}} \times \dfrac{1 \text{ mol C}_9\text{H}_8\text{O}_4}{9 \text{ mol C}}$

$= 0.92$ mol of $\text{C}_9\text{H}_8\text{O}_4$ (aspirin) (2 SFs)

c. 7.50 g $\text{C}_9\text{H}_8\text{O}_4 \times \dfrac{1 \text{ mol C}_9\text{H}_8\text{O}_4}{180.2 \text{ g C}_9\text{H}_8\text{O}_4} \times \dfrac{4 \text{ mol O}}{1 \text{ mol C}_9\text{H}_8\text{O}_4} \times \dfrac{16.00 \text{ g O}}{1 \text{ mol O}}$

$= 2.66$ g of O (3 SFs)

d. 2.50 g H $\times \dfrac{1 \text{ mol H}}{1.008 \text{ g H}} \times \dfrac{1 \text{ mol C}_9\text{H}_8\text{O}_4}{8 \text{ mol H}} \times \dfrac{6.022 \times 10^{23} \text{ molecules C}_9\text{H}_8\text{O}_4}{1 \text{ mol C}_9\text{H}_8\text{O}_4}$

$= 1.87 \times 10^{23}$ molecules of $\text{C}_9\text{H}_8\text{O}_4$ (3 SFs)

7.75 a. 0.250 mol $\text{Mn}_2\text{O}_3 \times \dfrac{3 \text{ mol O}}{1 \text{ mol Mn}_2\text{O}_3} = 0.750$ mol of O

Molar mass of MnO_2

$= 1$ mol of Mn $+ 2$ mol of O $= 54.94 \text{ g} + 2(16.00 \text{ g}) = 86.94$ g

20.0 g $\text{MnO}_2 \times \dfrac{1 \text{ mol MnO}_2}{86.94 \text{ g MnO}_2} \times \dfrac{2 \text{ mol O}}{1 \text{ mol MnO}_2} = 0.460$ mol of O

$\therefore$ Total moles of O $= 0.750$ mol O $+ 0.460$ mol O $= 1.210$ mol of O (4 SFs)

b. 0.250 mol $\text{Mn}_2\text{O}_3 \times \dfrac{2 \text{ mol Mn}}{1 \text{ mol Mn}_2\text{O}_3} \times \dfrac{54.94 \text{ g Mn}}{1 \text{ mol Mn}} = 27.5$ g of Mn

20.0 g $\text{MnO}_2 \times \dfrac{1 \text{ mol MnO}_2}{86.94 \text{ g MnO}_2} \times \dfrac{1 \text{ mol Mn}}{1 \text{ mol MnO}_2} \times \dfrac{54.94 \text{ g Mn}}{1 \text{ mol Mn}} = 12.6$ g of Mn

$\therefore$ Total grams of Mn $= 27.5$ g Mn $+ 12.6$ g Mn $= 40.1$ g of Mn (3 SFs)

7.77 a. $\text{C}_5\text{H}_5\text{N}_5 \rightarrow \text{C}_{(5 \div 5)}\text{H}_{(5 \div 5)}\text{N}_{(5 \div 5)} = \text{CHN}$ (empirical formula)

b. $\text{FeC}_2\text{O}_4 \rightarrow \text{Fe}_{(1 \div 1)}\text{C}_{(2 \div 1)}\text{O}_{(4 \div 1)} = \text{FeC}_2\text{O}_4$ (empirical formula)

c. $\text{C}_{16}\text{H}_{16}\text{N}_4 \rightarrow \text{C}_{(16 \div 4)}\text{H}_{(16 \div 4)}\text{N}_{(4 \div 4)} = \text{C}_4\text{H}_4\text{N}$ (empirical formula)

d. $\text{C}_6\text{H}_{14}\text{N}_2\text{O}_2 \rightarrow \text{C}_{(6 \div 2)}\text{H}_{(14 \div 2)}\text{N}_{(2 \div 2)}\text{O}_{(2 \div 2)} = \text{C}_3\text{H}_7\text{NO}$ (empirical formula)

7.79 a. 2.20 g S $\times \dfrac{1 \text{ mol S}}{32.07 \text{ g S}} = 0.0686$ mol of S (smaller number of moles)

7.81 g F $\times \dfrac{1 \text{ mol F}}{19.00 \text{ g F}} = 0.411$ mol of F

$\dfrac{0.0686 \text{ mol S}}{0.0686} = 1.00$ mol of S $\qquad \dfrac{0.411 \text{ mol F}}{0.0686} = 5.99$ mol of F

$\therefore$ Empirical formula $= \text{S}_{1.00}\text{F}_{5.99} \rightarrow \text{SF}_6$

b. $6.35 \text{ g Ag} \times \dfrac{1 \text{ mol Ag}}{107.9 \text{ g Ag}} = 0.0589 \text{ mol of Ag (smallest number of moles)}$

$0.825 \text{ g N} \times \dfrac{1 \text{ mol N}}{14.01 \text{ g N}} = 0.0589 \text{ mol of N}$

$2.83 \text{ g O} \times \dfrac{1 \text{ mol O}}{16.00 \text{ g O}} = 0.177 \text{ mol of O}$

$\dfrac{0.0589 \text{ mol Ag}}{0.0589} = 1.00 \text{ mol of Ag} \qquad \dfrac{0.0589 \text{ mol N}}{0.0589} = 1.00 \text{ mol of N}$

$\dfrac{0.177 \text{ mol O}}{0.0589} = 3.01 \text{ mol of O}$

∴ Empirical formula $= Ag_{1.00}N_{1.00}O_{3.01} \rightarrow AgNO_3$

c. In exactly 100. g of compound, there are 43.6 g P and 56.4 g O.

$43.6 \text{ g P} \times \dfrac{1 \text{ mol P}}{30.97 \text{ g P}} = 1.41 \text{ mol of P (smaller number of moles)}$

$56.4 \text{ g O} \times \dfrac{1 \text{ mol O}}{16.00 \text{ g O}} = 3.53 \text{ mol of O}$

$\dfrac{1.41 \text{ mol P}}{1.41} = 1.00 \text{ mol of P} \qquad \dfrac{3.53 \text{ mol O}}{1.41} = 2.50 \text{ mol of O}$

∴ Empirical formula $= P_{1.00}O_{2.50} \rightarrow P_{(1.00 \times 2)}O_{(2.50 \times 2)} \rightarrow P_2O_5$

d. In exactly 100. g of compound, there are 22.1 g Al, 25.4 g P, and 52.5 g O.

$22.1 \text{ g Al} \times \dfrac{1 \text{ mol Al}}{26.98 \text{ g Al}} = 0.819 \text{ mol of Al (smallest number of moles)}$

$25.4 \text{ g P} \times \dfrac{1 \text{ mol P}}{30.97 \text{ g P}} = 0.820 \text{ mol of P}$

$52.5 \text{ g O} \times \dfrac{1 \text{ mol O}}{16.00 \text{ g O}} = 3.28 \text{ mol of O}$

$\dfrac{0.819 \text{ mol Al}}{0.819} = 1.00 \text{ mol of Al} \qquad \dfrac{0.820 \text{ mol P}}{0.819} = 1.00 \text{ mol of P}$

$\dfrac{3.28 \text{ mol O}}{0.819} = 4.00 \text{ mol of O}$

∴ Empirical formula $= Al_{1.00}P_{1.00}O_{4.00} \rightarrow AlPO_4$

7.81 a. In exactly 100. g of oleic acid, there are 76.54 g of C, 12.13 g of H, and 11.33 g of O.

$76.54 \text{ g C} \times \dfrac{1 \text{ mol C}}{12.01 \text{ g C}} = 6.373 \text{ mol of C}$

$12.13 \text{ g H} \times \dfrac{1 \text{ mol H}}{1.008 \text{ g H}} = 12.03 \text{ mol of H}$

$11.33 \text{ g O} \times \dfrac{1 \text{ mol O}}{16.00 \text{ g O}} = 0.7081 \text{ mol of O (smallest number of moles)}$

$\dfrac{6.373 \text{ mol C}}{0.7081} = 9.000 \text{ mol of C} \qquad \dfrac{12.03 \text{ mol H}}{0.7081} = 16.99 \text{ mol of H}$

$\dfrac{0.7081 \text{ mol O}}{0.7081} = 1.000 \text{ mol of O}$

∴ Empirical formula $= C_{9.000}H_{16.99}O_{1.000} \rightarrow C_9H_{17}O$

Empirical formula mass of $C_9H_{17}O = 9(12.01\text{ g}) + 17(1.008\text{ g}) + 16.00\text{ g} = 141.2\text{ g}$

$$\text{Small integer} = \frac{\text{molar mass of oleic acid}}{\text{empirical formula mass of } C_9H_{17}O} = \frac{282\text{ g}}{141.2\text{ g}} = 2$$

$\therefore$ Molecular formula of oleic acid $= C_{(9\times2)}H_{(17\times2)}O_{(1\times2)} = C_{18}H_{34}O_2$

b. $3.00\text{ mL } C_{18}H_{34}O_2 \times \dfrac{0.895\text{ g } C_{18}H_{34}O_2}{1\text{ mL } C_{18}H_{34}O_2} \times \dfrac{1\text{ mol } C_{18}H_{34}O_2}{282\text{ g } C_{18}H_{34}O_2}$

$\times \dfrac{6.022\times10^{23}\text{ molecules } C_{18}H_{34}O_2}{1\text{ mol } C_{18}H_{34}O_2}$

$= 5.73\times10^{21}$ molecules of $C_{18}H_{34}O_2$ (3 SFs)

7.83 In exactly 100. g of succinic acid, there are 40.7 g of C, 5.12 g of H, and 54.2 g of O.

$40.7\text{ g C} \times \dfrac{1\text{ mol C}}{12.01\text{ g C}} = 3.39$ mol of C

$5.12\text{ g H} \times \dfrac{1\text{ mol H}}{1.008\text{ g H}} = 5.08$ mol of H

$54.2\text{ g O} \times \dfrac{1\text{ mol O}}{16.00\text{ g O}} = 3.39$ mol of O (smallest number of moles)

$\dfrac{3.39\text{ mol C}}{3.39} = 1.00$ mol of C $\dfrac{5.08\text{ mol H}}{3.39} = 1.50$ mol of H

$\dfrac{3.39\text{ mol O}}{3.39} = 1.00$ mol of O

$\therefore$ Empirical formula $= C_{1.00}H_{1.50}O_{1.00} \rightarrow C_{(1.00\times2)}H_{(1.50\times2)}O_{(1.00\times2)}$

$= C_{2.00}H_{3.00}O_{2.00} \rightarrow C_2H_3O_2$

Empirical formula mass of $C_2H_3O_2 = 2(12.01\text{ g}) + 3(1.008\text{ g}) + 2(16.0\text{ g}) = 59.04\text{ g}$

$$\text{Small integer} = \frac{\text{molar mass of succinic acid}}{\text{empirical formula mass of } C_2H_3O_2} = \frac{118\text{ g}}{59.04\text{ g}} = 2$$

$\therefore$ Molecular formula of succinic acid $= C_{(2\times2)}H_{(3\times2)}O_{(2\times2)} = C_4H_6O_4$

7.85 In the sample of compound, there are 1.65×10^{23} atoms of C, 0.552 g of H, and 4.39 g of O.

$1.65\times10^{23}\text{ atoms C} \times \dfrac{1\text{ mol C}}{6.022\times10^{23}\text{ atoms C}} = 0.274$ mol of C (smallest number of moles)

$0.552\text{ g H} \times \dfrac{1\text{ mol H}}{1.008\text{ g H}} = 0.548$ mol of H

$4.39\text{ g O} \times \dfrac{1\text{ mol O}}{16.00\text{ g O}} = 0.274$ mol of O

$\dfrac{0.274\text{ mol C}}{0.274} = 1.00$ mol of C $\dfrac{0.548\text{ mol H}}{0.274} = 2.00$ mol of H

$\dfrac{0.274\text{ mol O}}{0.274} = 1.00$ mol of O

$\therefore$ Empirical formula $= C_{1.00}H_{2.00}O_{1.00} \rightarrow CH_2O$

Since 1 mol of compound contains 4 mol of O,

$\therefore$ Molecular formula $= C_{(1\times4)}H_{(2\times4)}O_{(1\times4)} = C_4H_8O_4$

Molar mass of compound ($C_4H_8O_4$)
$= 4$ mol of C $+ 8$ mol of H $+ 4$ mol of O
$= 4(12.01\ g) + 8(1.008\ g) + 4(16.00\ g) = 120.10\ g$

7.87 a. Molar mass of NaF
$= 1$ mol of Na $+ 1$ mol of F $= 22.99\ g + 19.00\ g = 41.99\ g$

$$1\ \text{tube} \times \frac{119\ \text{g toothpaste}}{1\ \text{tube}} \times \frac{0.240\ \text{g NaF}}{100\ \text{g toothpaste}} \times \frac{1\ \text{mol NaF}}{41.99\ \text{g NaF}}$$

$= 0.00680$ mol of NaF (3 SFs)

b. $1\ \text{tube} \times \dfrac{119\ \text{g toothpaste}}{1\ \text{tube}} \times \dfrac{0.240\ \text{g NaF}}{100\ \text{g toothpaste}} \times \dfrac{1\ \text{mol NaF}}{41.99\ \text{g NaF}} \times \dfrac{1\ \text{mol F}^-\ \text{ions}}{1\ \text{mol NaF}}$

$$\times\ \frac{6.022 \times 10^{23}\ \text{F}^-\ \text{ions}}{1\ \text{mol F}^-\ \text{ions}} = 4.10 \times 10^{21}\ \text{F}^-\ \text{ions (3 SFs)}$$

c. $1.50\ \text{g toothpaste} \times \dfrac{0.240\ \text{g NaF}}{100\ \text{g toothpaste}} \times \dfrac{1\ \text{mol NaF}}{41.99\ \text{g NaF}} \times \dfrac{1\ \text{mol Na}^+}{1\ \text{mol NaF}} \times \dfrac{22.99\ \text{g Na}^+}{1\ \text{mol Na}^+}$

$= 0.00197$ g of Na$^+$ ions (3 SFs)

d. $12\ \text{mol C} \times \dfrac{12.01\ \text{g C}}{1\ \text{mol C}} = 144.1\ \text{g of C}$

$7\ \text{mol H} \times \dfrac{1.008\ \text{g H}}{1\ \text{mol H}} = 7.056\ \text{g of H}$

$3\ \text{mol Cl} \times \dfrac{35.45\ \text{g Cl}}{1\ \text{mol Cl}} = 106.4\ \text{g of Cl}$

$2\ \text{mol O} \times \dfrac{16.00\ \text{g O}}{1\ \text{mol O}} = 32.00\ \text{g of O}$

$\therefore$ Molar mass of $C_{12}H_7Cl_3O_2 = 289.6\ g$

$1\ \text{tube} \times \dfrac{119\ \text{g toothpaste}}{1\ \text{tube}} \times \dfrac{0.30\ \text{g}\ C_{12}H_7Cl_3O_2}{100\ \text{g toothpaste}} \times \dfrac{1\ \text{mol}\ C_{12}H_7Cl_3O_2}{289.6\ \text{g}\ C_{12}H_7Cl_3O_2}$

$$\times\ \frac{6.022 \times 10^{23}\ \text{molecules}\ C_{12}H_7Cl_3O_2}{1\ \text{mol}\ C_{12}H_7Cl_3O_2}$$

$= 7.4 \times 10^{20}$ molecules of $C_{12}H_7Cl_3O_2$ (triclosan) (2 SFs)

e. Mass % C $= \dfrac{144.1\ \text{g C}}{289.6\ \text{g}\ C_{12}H_7Cl_3O_2} \times 100\% = 49.76\%$ C (4 SFs)

Mass % H $= \dfrac{7.056\ \text{g H}}{289.6\ \text{g}\ C_{12}H_7Cl_3O_2} \times 100\% = 2.436\%$ H (4 SFs)

Mass % Cl $= \dfrac{106.4\ \text{g Cl}}{289.6\ \text{g}\ C_{12}H_7Cl_3O_2} \times 100\% = 36.74\%$ Cl (4 SFs)

Mass % O $= \dfrac{32.00\ \text{g O}}{289.6\ \text{g}\ C_{12}H_7Cl_3O_2} \times 100\% = 11.05\%$ O (4 SFs)

7.89 In exactly 100. g of iron(III) chromate, there are 24.3 g of Fe, 33.9 g of Cr, and 41.8 g of O.

$$24.3 \text{ g Fe} \times \frac{1 \text{ mol Fe}}{55.85 \text{ g Fe}} = 0.435 \text{ mol of Fe (smallest number of moles)}$$

$$33.9 \text{ g Cr} \times \frac{1 \text{ mol Cr}}{52.00 \text{ g Cr}} = 0.652 \text{ mol of Cr}$$

$$41.8 \text{ g O} \times \frac{1 \text{ mol O}}{16.00 \text{ g O}} = 2.61 \text{ mol of O}$$

$$\frac{0.435 \text{ mol Fe}}{0.435} = 1.00 \text{ mol of Fe} \qquad \frac{0.652 \text{ mol Cr}}{0.435} = 1.50 \text{ mol of Cr}$$

$$\frac{2.61 \text{ mol O}}{0.435} = 6.00 \text{ mol of O}$$

$\therefore$ Empirical formula $= Fe_{1.00}Cr_{1.50}O_{6.00} \rightarrow Fe_{(1.00 \times 2)}Cr_{(1.50 \times 2)}O_{(6.00 \times 2)}$

$$= Fe_{2.00}Cr_{3.00}O_{12.0} \rightarrow Fe_2Cr_3O_{12}$$

Empirical formula mass of $Fe_2Cr_3O_{12} = 2(55.85 \text{ g}) + 3(52.00 \text{ g}) + 12(16.00 \text{ g}) = 459.7 \text{ g}$

$$\text{Small integer} = \frac{\text{molar mass of compound}}{\text{empirical formula mass of } Fe_2Cr_3O_{12}} = \frac{460 \text{ g}}{459.7 \text{ g}} = 1$$

$\therefore$ Molecular formula $= Fe_{(2 \times 1)}Cr_{(3 \times 1)}O_{(12 \times 1)} = Fe_2Cr_3O_{12}$

Selected Answers to Combining Ideas from Chapters 4 to 7

CI.7 **a.** X is a metal; elements in Group 2A (2) are metals.
   b. Y is a nonmetal; elements in Group 7A (17) are nonmetals.
   c. X^{2+}, Y^-
   d. $X = 1s^2 2s^2 2p^6 3s^2$ $Y = 1s^2 2s^2 2p^6 3s^2 3p^5$
   e. $MgCl_2$, magnesium chloride

CI.9 **a.** In exactly 100. g of oxalic acid, there are 26.7 g of C, 2.24 g of H, and 71.1 g of O.

$$26.7 \text{ g C} \times \frac{1 \text{ mol C}}{12.01 \text{ g C}} = 2.22 \text{ mol of C}$$

$$2.24 \text{ g H} \times \frac{1 \text{ mol H}}{1.008 \text{ g H}} = 2.22 \text{ mol of H (smallest number of moles)}$$

$$71.1 \text{ g O} \times \frac{1 \text{ mol O}}{16.00 \text{ g O}} = 4.44 \text{ mol of O}$$

$$\frac{2.22 \text{ mol C}}{2.22} = 1.00 \text{ mol of C} \qquad \frac{2.22 \text{ mol H}}{2.22} = 1.00 \text{ mol of H}$$

$$\frac{4.44 \text{ mol O}}{2.22} = 2.00 \text{ mol of O}$$

∴ Empirical formula $= C_{1.00}H_{1.00}O_{2.00} \rightarrow CHO_2$

   b. Empirical formula mass of $CHO_2 = 12.01 + 1.008 + 2(16.00) = 45.02$ g

$$\text{Small integer} = \frac{\text{molar mass of oxalic acid}}{\text{empirical formula mass of } CHO_2} = \frac{90. \text{ g}}{45.02 \text{ g}} = 2$$

∴ Molecular formula $= C_{(1\times2)}H_{(1\times2)}O_{(2\times2)} = C_2H_2O_4$

   c. $160 \text{ lb} \times \dfrac{1 \text{ kg}}{2.205 \text{ lb}} \times \dfrac{375 \text{ mg oxalic acid}}{1 \text{ kg}} \times \dfrac{1 \text{ g oxalic acid}}{1000 \text{ mg oxalic acid}}$
   $= 27$ g of oxalic acid (2 SFs)

   d. $160 \text{ lb} \times \dfrac{1 \text{ kg}}{2.205 \text{ lb}} \times \dfrac{375 \text{ mg oxalic acid}}{1 \text{ kg}} \times \dfrac{1 \text{ g oxalic acid}}{1000 \text{ mg oxalic acid}}$
   $\times \dfrac{100 \text{ g rhubarb leaves}}{0.5 \text{ g oxalic acid}} \times \dfrac{1 \text{ kg}}{1000 \text{ g}}$
   $= 5$ kg of rhubarb leaves (1 SF)

CI.11 **a.** $C_{16}H_{28}N_2O_4 \rightarrow C_{(16\div2)}H_{(28\div2)}N_{(2\div2)}O_{(4\div2)} = C_8H_{14}NO_2$ (empirical formula)

   b. $16 \text{ mol C} \times \dfrac{12.01 \text{ g C}}{1 \text{ mol C}} = 192.2 \text{ g of C}$

   $28 \text{ mol H} \times \dfrac{1.008 \text{ g H}}{1 \text{ mol H}} = 28.22 \text{ g of H}$

   $2 \text{ mol N} \times \dfrac{14.01 \text{ g N}}{1 \text{ mol N}} = 28.02 \text{ g of N}$

   $4 \text{ mol O} \times \dfrac{16.00 \text{ g O}}{1 \text{ mol O}} = 64.00 \text{ g of O}$

∴ Molar mass of $C_{16}H_{28}N_2O_4 = 312.4$ g

$$\text{Mass \% C} = \frac{192.2 \text{ g C}}{312.4 \text{ g C}_{16}\text{H}_{28}\text{N}_2\text{O}_4} \times 100\% = 61.52\% \text{ C (4 SFs)}$$

$$\text{Mass \% H} = \frac{28.22 \text{ g H}}{312.4 \text{ g C}_{16}\text{H}_{28}\text{N}_2\text{O}_4} \times 100\% = 9.033\% \text{ H (4 SFs)}$$

$$\text{Mass \% N} = \frac{28.02 \text{ g N}}{312.4 \text{ g C}_{16}\text{H}_{28}\text{N}_2\text{O}_4} \times 100\% = 8.969\% \text{ N (4 SFs)}$$

$$\text{Mass \% O} = \frac{64.00 \text{ g O}}{312.4 \text{ g C}_{16}\text{H}_{28}\text{N}_2\text{O}_4} \times 100\% = 20.49\% \text{ O (4 SFs)}$$

c. The formula of shikimic acid is $C_7H_{10}O_5$.

d. Molar mass of shikimic acid $(C_7H_{10}O_5)$
$$= 7(12.01 \text{ g}) + 10(1.008 \text{ g}) + 5(16.00 \text{ g}) = 174.15 \text{ g (5 SFs)}$$

$$1.3 \text{ g } C_7H_{10}O_5 \times \frac{1 \text{ mol } C_7H_{10}O_5}{174.15 \text{ g } C_7H_{10}O_5} = 0.0075 \text{ mol of } C_7H_{10}O_5 \text{ (shikimic acid) (2 SFs)}$$

e. $154 \text{ g star anise} \times \dfrac{0.13 \text{ g } C_7H_{10}O_5}{2.6 \text{ g star anise}} \times \dfrac{75 \text{ mg Tamiflu}}{0.13 \text{ g } C_7H_{10}O_5} \times \dfrac{1 \text{ capsule}}{75 \text{ mg Tamiflu}}$

$= 59 \text{ capsules of Tamiflu (2 SFs)}$

f. $1 \text{ capsule} \times \dfrac{75 \text{ mg } C_{16}H_{28}N_2O_4}{1 \text{ capsule}} \times \dfrac{1 \text{ g } C_{16}H_{28}N_2O_4}{1000 \text{ mg } C_{16}H_{28}N_2O_4} \times \dfrac{1 \text{ mol } C_{16}H_{28}N_2O_4}{312.4 \text{ g } C_{16}H_{28}N_2O_4}$

$\times \dfrac{16 \text{ mol C}}{1 \text{ mol } C_{16}H_{28}N_2O_4} \times \dfrac{12.01 \text{ g C}}{1 \text{ mol C}} = 0.046 \text{ g of C (2 SFs)}$

g. $500\,000 \text{ people} \times \dfrac{2 \text{ capsules}}{1 \text{ day } 1 \text{ person}} \times 5 \text{ days} \times \dfrac{75 \text{ mg Tamiflu}}{1 \text{ capsule}}$

$\times \dfrac{1 \text{ g Tamiflu}}{1000 \text{ mg Tamiflu}} \times \dfrac{1 \text{ kg Tamiflu}}{1000 \text{ g Tamiflu}}$

$= 4 \times 10^2 \text{ kg of Tamiflu (1 SF)}$

Chemical Reactions

Since Natalie's diagnosis of mild emphysema due to second-hand cigarette smoke, she has been working with Angela, an exercise physiologist. Natalie's workout started with low-intensity exercises that use smaller muscles. Now that Natalie has increased the oxygen concentration in her blood, Angela has added some exercises that use larger muscles, which need more O_2. Today Natalie had a stress test and an ECG to assess and evaluate her progress. Her results show that her heart is functioning more efficiently and that she needs less oxygen. Her doctor told her that she has begun to reverse the progression of emphysema. During exercise, aqueous stearic acid, $C_{18}H_{36}O_2$, undergoes reaction with O_2 gas to form CO_2 gas and liquid H_2O. Write a balanced chemical equation for the reaction.

Credit: Javier Larrea/AGE Fotostock

CHAPTER READINESS*

Core Chemistry Skills

- Writing Ionic Formulas (6.2)
- Naming Ionic Compounds (6.3)

- Writing the Names and Formulas for Molecular Compounds (6.5)

*These Core Chemistry Skills from previous chapters are listed here for your review as you proceed to the new material in this chapter.

LOOKING AHEAD

8.1 Equations for Chemical Reactions

Learning Goal: Identify a balanced chemical equation; determine the number of atoms in the reactants and products.

- A chemical change occurs when a substance is converted into one or more new substances.
- Chemical change may be indicated by a change in color, formation of a gas (bubbles), formation of a solid, and/or heat produced or heat absorbed.
- When new substances form, a chemical reaction has taken place.
- A chemical equation shows the formulas of the reactants on the left side of the arrow and the formulas of the products on the right side.
- In a balanced equation, numbers called *coefficients*, which appear in front of the symbols or formulas, provide the same number of atoms for each kind of element on the reactant and product sides.
- Each formula in an equation is followed by an abbreviation, in parentheses, that gives the physical state of the substance: solid (*s*), liquid (*l*), or gas (*g*), and, if dissolved in water, an aqueous solution (*aq*).
- The Greek letter delta (Δ) over the arrow in an equation represents the application of heat to the reaction.

♦ **Learning Exercise 8.1**

Indicate the number of atoms of each element on the reactant side and on the product side for each of the following balanced equations:

a. $CaCO_3(s) \longrightarrow CaO(s) + CO_2(g)$

Element	Atoms on Reactant Side	Atoms on Product Side
Ca		
C		
O		

$2H_2(g) + O_2(g) \longrightarrow 2H_2O(g)$

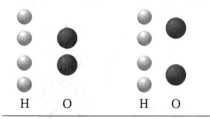

H   O   H   O

Reactant atoms   =   Product atoms

When there are the same numbers of each type of atom in the reactants as in the products, the equation is balanced.

b. $2Na(s) + H_2O(l) \longrightarrow Na_2O(s) + H_2(g)$

Element	Atoms on Reactant Side	Atoms on Product Side
Na		
H		
O		

c. $C_5H_{12}(g) + 8O_2(g) \xrightarrow{\Delta} 5CO_2(g) + 6H_2O(g)$

Element	Atoms on Reactant Side	Atoms on Product Side
C		
H		
O		

d. $2AgNO_3(aq) + K_2S(aq) \longrightarrow Ag_2S(s) + 2KNO_3(aq)$

Element	Atoms on Reactant Side	Atoms on Product Side
Ag		
N		
O		
K		
S		

e. $2Al(OH)_3(s) + 3H_2SO_4(aq) \longrightarrow 6H_2O(l) + Al_2(SO_4)_3(aq)$

Element	Atoms on Reactant Side	Atoms on Product Side
Al		
O		
H		
S		

Answers

a. $CaCO_3(s) \longrightarrow CaO(s) + CO_2(g)$

Element	Atoms on Reactant Side	Atoms on Product Side
Ca	1	1
C	1	1
O	3	3

b. $2Na(s) + H_2O(l) \longrightarrow Na_2O(s) + H_2(g)$

Element	Atoms on Reactant Side	Atoms on Product Side
Na	2	2
H	2	2
O	1	1

c. $C_5H_{12}(g) + 8O_2(g) \xrightarrow{\Delta} 5CO_2(g) + 6H_2O(g)$

Element	Atoms on Reactant Side	Atoms on Product Side
C	5	5
H	12	12
O	16	16

d. $2AgNO_3(aq) + K_2S(aq) \longrightarrow Ag_2S(s) + 2KNO_3(aq)$

Element	Atoms on Reactant Side	Atoms on Product Side
Ag	2	2
N	2	2
O	6	6
K	2	2
S	1	1

e. $2Al(OH)_3(s) + 3H_2SO_4(aq) \longrightarrow 6H_2O(l) + Al_2(SO_4)_3(aq)$

Element	Atoms on Reactant Side	Atoms on Product Side
Al	2	2
O	18	18
H	12	12
S	3	3

8.2   Balancing a Chemical Equation

Learning Goal: Write a balanced chemical equation from the formulas of the reactants and products for a chemical reaction.

- A chemical equation is balanced by writing coefficients, small whole numbers, in front of the formulas to equalize the atoms of each of the elements in the reactants and the products.

Guide to Writing and Balancing a Chemical Equation	
STEP 1	Write an equation using the correct formulas for the reactants and products.
STEP 2	Count the atoms of each element in the reactants and products.
STEP 3	Use coefficients to balance each element.
STEP 4	Check the final equation to confirm it is balanced.

Example: Balance the following equation:

$$N_2(g) + H_2(g) \longrightarrow NH_3(g)$$

CORE CHEMISTRY SKILL

Balancing a Chemical Equation

Analyze the Problem	Given	Need	Connect
	reactants, products	balanced equation	equal numbers of atoms in reactants and products

Solution:

STEP 1 **Write an equation using the correct formulas for the reactants and products.**

$$N_2(g) + H_2(g) \longrightarrow NH_3(g)$$

STEP 2 **Count the atoms of each element in the reactants and products.**

$$N_2(g) + H_2(g) \longrightarrow NH_3(g)$$
2N   2H   1N, 3H

STEP 3 **Use coefficients to balance each element.** Balance the N atoms by placing a coefficient of 2 in front of NH_3. (This increases the number of H atoms, too.) Recheck the number of N atoms and the number of H atoms.

$$N_2(g) + H_2(g) \longrightarrow 2NH_3(g)$$
2N   2H   2N, 6H

Balance the H atoms by placing a coefficient of 3 in front of H_2.

$$N_2(g) + 3H_2(g) \longrightarrow 2NH_3(g)$$

STEP 4 **Check the final equation to confirm it is balanced.**

$$N_2(g) + 3H_2(g) \longrightarrow 2NH_3(g)$$
2N   6H   2N, 6H   Balanced

♦ **Learning Exercise 8.2**

Balance each of the following equations by placing coefficients in front of the formulas as needed:

a. ____ $MgO(s) \longrightarrow$ ____ $Mg(s) +$ ____ $O_2(g)$

b. ____ $Zn(s) +$ ____ $HCl(aq) \longrightarrow$ ____ $H_2(g) +$ ____ $ZnCl_2(aq)$

c. ____ $Al(s) +$ ____ $CuSO_4(aq) \longrightarrow$ ____ $Cu(s) +$ ____ $Al_2(SO_4)_3(aq)$

d. ____ $Al_2S_3(s) +$ ____ $H_2O(l) \longrightarrow$ ____ $Al(OH)_3(s) +$ ____ $H_2S(g)$

e. ____ $BaCl_2(aq) +$ ____ $Na_2SO_4(aq) \longrightarrow$ ____ $BaSO_4(s) +$ ____ $NaCl(aq)$

f. ____ $CO(g) +$ ____ $Fe_2O_3(s) \longrightarrow$ ____ $Fe(s) +$ ____ $CO_2(g)$

g. ___ K(s) + ___ H₂O(l) ⟶ ___ H₂(g) + ___ K₂O(s)

h. ___ Fe(OH)₃(s) ⟶ ___ H₂O(l) + ___ Fe₂O₃(s)

i. Galactose, $C_6H_{12}O_6(aq)$, is found in milk. In the body, galactose reacts with O_2 gas to form CO_2 gas and liquid water. Write a balanced chemical equation for the reaction.

j. Maltose, $C_{12}H_{22}O_{11}(aq)$, is found in corn syrup. In the body, maltose reacts with O_2 gas to form CO_2 gas and liquid water. Write a balanced chemical equation for the reaction.

Answers

a. $2MgO(s) \longrightarrow 2Mg(s) + O_2(g)$

b. $Zn(s) + 2HCl(aq) \longrightarrow H_2(g) + ZnCl_2(aq)$

c. $2Al(s) + 3CuSO_4(aq) \longrightarrow 3Cu(s) + Al_2(SO_4)_3(aq)$

d. $Al_2S_3(s) + 6H_2O(l) \longrightarrow 2Al(OH)_3(s) + 3H_2S(g)$

e. $BaCl_2(aq) + Na_2SO_4(aq) \longrightarrow BaSO_4(s) + 2NaCl(aq)$

f. $3CO(g) + Fe_2O_3(s) \longrightarrow 2Fe(s) + 3CO_2(g)$

g. $2K(s) + H_2O(l) \longrightarrow H_2(g) + K_2O(s)$

h. $2Fe(OH)_3(s) \longrightarrow 3H_2O(l) + Fe_2O_3(s)$

i. $C_6H_{12}O_6(aq) + 6O_2(g) \longrightarrow 6CO_2(g) + 6H_2O(l)$

j. $C_{12}H_{22}O_{11}(aq) + 12O_2(g) \longrightarrow 12CO_2(g) + 11H_2O(l)$

8.3   Types of Chemical Reactions

Learning Goal: Identify a chemical reaction as a combination, decomposition, single replacement, double replacement, or combustion.

- Reactions are classified as combination, decomposition, single replacement, double replacement, or combustion.

- In a *combination* reaction, two or more reactants form one product:

  A + B ⟶ AB

- In a *decomposition* reaction, a reactant splits into two or more simpler products:

  AB ⟶ A + B

- In a *single replacement* reaction, an uncombined element takes the place of an element in a compound:

  A + BC ⟶ AC + B

- In a *double replacement* reaction, the positive ions in the reactants switch places:

$$AB + CD \longrightarrow AD + CB$$

- In a *combustion* reaction, a compound of carbon and hydrogen reacts with oxygen to form CO_2, H_2O, and energy.

$$C_XH_Y(g) + ZO_2(g) \xrightarrow{\Delta} XCO_2(g) + \frac{Y}{2}H_2O(g) + energy$$

♦ **Learning Exercise 8.3A**

🜂 **CORE CHEMISTRY SKILL**

Match each of the following reactions with the type of reaction:

Classifying Types of Chemical Reactions

    a. combination   **b.** decomposition

    c. single replacement   **d.** double replacement   **e.** combustion

1. _____ $N_2(g) + 3H_2(g) \longrightarrow 2NH_3(g)$

2. _____ $BaCl_2(aq) + K_2CO_3(aq) \longrightarrow BaCO_3(s) + 2KCl(aq)$

3. _____ $2H_2O_2(aq) \longrightarrow 2H_2O(l) + O_2(g)$

4. _____ $CuO(s) + H_2(g) \longrightarrow Cu(s) + H_2O(l)$

5. _____ $N_2(g) + 2O_2(g) \longrightarrow 2NO_2(g)$

6. _____ $C_2H_4(g) + 3O_2(g) \xrightarrow{\Delta} 2CO_2(g) + 2H_2O(g) + energy$

7. _____ $PbCO_3(s) \longrightarrow PbO(s) + CO_2(g)$

8. _____ $2Al(s) + Fe_2O_3(s) \longrightarrow 2Fe(s) + Al_2O_3(s)$

Answers	**1.** a	**2.** d	**3.** b	**4.** c
	5. a	**6.** e	**7.** b	**8.** c

In a combustion reaction, a candle burns using the oxygen in the air.

Credit: Sergiy Zavgorodny / Shutterstock

♦ **Learning Exercise 8.3B**

One way to remove tarnish from silver is to place the silver object on a piece of aluminum foil and add boiling water and some baking soda. The unbalanced chemical equation is the following:

$$Al(s) + Ag_2S(s) \longrightarrow Ag(s) + Al_2S_3(s)$$

a. Write the balanced chemical equation for the reaction.

b. What type of reaction takes place?

Answers   **a.** $2Al(s) + 3Ag_2S(s) \longrightarrow 6Ag(s) + Al_2S_3(s)$

            b. single replacement

♦ **Learning Exercise 8.3C**

Octane, C_8H_{18}, a liquid compound in gasoline, burns in oxygen to produce the gases carbon dioxide and water, and energy.

a. Write the balanced chemical equation for the reaction.

b. What type of reaction takes place?

Answers   **a.** $2C_8H_{18}(l) + 25O_2(g) \xrightarrow{\Delta} 16CO_2(g) + 18H_2O(g) + \text{energy}$
b. combustion

♦ **Learning Exercise 8.3D**

Predict the products of each of the following reactions and balance:

a. decomposition: $Al_2O_3(s) \longrightarrow$

b. single replacement: $Mg(s) + H_3PO_4(aq) \longrightarrow$

c. combustion: $C_6H_{12}O_2(l) + O_2(g) \xrightarrow{\Delta}$

d. combination: $Ca(s) + N_2(g) \longrightarrow$

Answers   **a.** decomposition: $2Al_2O_3(s) \longrightarrow 4Al(s) + 3O_2(g)$
b. single replacement: $6Mg(s) + 2H_3PO_4(aq) \longrightarrow 3H_2(g) + 2Mg_3(PO_4)_2(s)$
c. combustion: $C_6H_{12}O_2(l) + 8O_2(g) \xrightarrow{\Delta} 6CO_2(g) + 6H_2O(g)$
d. combination: $3Ca(s) + N_2(g) \longrightarrow Ca_3N_2(s)$

8.4   Oxidation–Reduction Reactions

Learning Goal: Define the terms oxidation and reduction; identify the reactants oxidized and reduced.

- In an oxidation–reduction reaction, there is a loss and gain of electrons. In an oxidation, electrons are lost. In a reduction, electrons are gained.
- An oxidation must always be accompanied by a reduction. The number of electrons lost in the oxidation reaction is equal to the number of electrons gained in the reduction reaction.
- In biological systems, the term *oxidation* describes the gain of oxygen or the loss of hydrogen. The term *reduction* is used to describe a loss of oxygen or a gain of hydrogen.

Key Terms for Sections 8.1 to 8.4

Match each of the following key terms with the correct description:

    a. combustion reaction   **b.** chemical equation   **c.** combination reaction

    d. decomposition reaction   **e.** single replacement reaction   **f.** oxidation–reduction reaction

1. _____ a reaction in which a single reactant splits into two or more simpler substances

2. _____ a reaction in which one element replaces a different element in a compound

3. _____ a reaction in which a carbon-containing compound burns in oxygen from the air to produce carbon dioxide, water, and energy

4. _____ a reaction in which reactants combine to form a single product

5. _____ a shorthand method of writing a chemical reaction with the formulas of the reactants written on the left side of an arrow and the formulas of the products written on the right side

6. _____ a reaction in which an equal number of electrons is lost from one reactant and gained by another reactant

Answers   **1.** d   **2.** e   **3.** a   **4.** c   **5.** b   **6.** f

♦ **Learning Exercise 8.4**

CORE CHEMISTRY SKILL

For each of the following reactions, indicate whether the underlined element is *oxidized* or *reduced*.

Identifying Oxidized and Reduced Substances

a. $\underline{Fe}^{3+}(aq) + e^- \longrightarrow Fe^{2+}(aq)$ Fe^{3+} is   _____

b. $2\underline{Cl}^-(aq) \longrightarrow Cl_2(g) + 2\,e^-$ Cl^- is   _____

c. $\underline{Ca}O(s) + H_2(g) \longrightarrow Ca(s) + H_2O(l)$ Ca^{2+} is   _____

d. $4\underline{Al}(s) + 3O_2(g) \longrightarrow 2Al_2O_3(s)$ Al is   _____

e. $\underline{Cu}Cl_2(aq) + Zn(s) \longrightarrow ZnCl_2(aq) + Cu(s)$ Cu^{2+} is   _____

Answers

    a. Fe^{3+} is reduced to Fe^{2+}.   **b.** Cl^- is oxidized to Cl_2.

    c. Ca^{2+} is reduced to Ca.   **d.** Al is oxidized to Al^{3+}.

    e. Cu^{2+} is reduced to Cu.

Checklist for Chapter 8

You are ready to take the Practice Test for Chapter 8. Be sure you have accomplished the following learning goals for this chapter. If not, review the section listed at the end of the goal. Then apply your new skills and understanding to the Practice Test.

After studying Chapter 8, I can successfully:

_____ State a chemical equation in words, and calculate the total number of atoms of each element in the reactants and products. (8.1)

_____ Write a balanced equation for a chemical reaction from the formulas of the reactants and products. (8.2)

_____ Identify a reaction as a combination, decomposition, single replacement, double replacement, or combustion reaction. (8.3)

_____ Identify the substance oxidized and the substance reduced in an oxidation–reduction reaction. (8.4)

Practice Test for Chapter 8

The chapter sections to review are shown in parentheses at the end of each question.

For each of the following balanced chemical equations in questions 1 through 3, indicate the number of oxygen atoms on the reactant and product sides: (8.1)

A. 2   **B.** 4   **C.** 6   **D.** 7   **E.** 8

1. _____ $4Cr(s) + 3O_2(g) \longrightarrow 2Cr_2O_3(s)$

2. _____ $2N_2O(g) + 3O_2(g) \longrightarrow 2N_2O_4(g)$

3. _____ $C_2H_6O(g) + 3O_2(g) \xrightarrow{\Delta} 2CO_2(g) + 3H_2O(g)$

*For questions 4 through 7, state if the equation is balanced (**B**) or not balanced (**NB**):* (8.1, 8.2)

4. _____ $Mg(s) + O_2(g) \longrightarrow 2MgO(s)$

5. _____ $I_2(s) + 2Cl_2(g) \longrightarrow ICl_5(s)$

6. _____ $MgSO_4(aq) + K_2S(aq) \longrightarrow K_2SO_4(aq) + MgS(s)$

7. _____ $B_2H_6(l) + 3O_2(g) \longrightarrow B_2O_3(s) + 3H_2O(l)$

*For each of the unbalanced equations in questions 8 through 12, balance the equation and indicate the correct coefficient for the component in the equation written in **boldface type**.* (8.2)

A. 1   **B.** 2   **C.** 3   **D.** 4   **E.** 5

8. _____ $Sn(s) + \mathbf{Cl_2}(g) \longrightarrow SnCl_4(s)$

9. _____ $Al(s) + H_2O(l) \longrightarrow \mathbf{H_2}(g) + Al_2O_3(s)$

10. _____ $C_3H_8(g) + \mathbf{O_2}(g) \xrightarrow{\Delta} CO_2(g) + H_2O(g)$

11. _____ $\mathbf{NH_3}(g) + O_2(g) \longrightarrow N_2(g) + H_2O(g)$

12. _____ $N_2O(g) \longrightarrow N_2(g) + \mathbf{O_2}(g)$

For questions 13 through 17, classify each reaction as one of the following: (8.3)

    A. combination   **B.** decomposition   **C.** single replacement

    D. double replacement   **E.** combustion

13. _____ $S(s) + O_2(g) \longrightarrow SO_2(g)$

14. _____ $C_4H_8(g) + 6O_2(g) \xrightarrow{\Delta} 4CO_2(g) + 4H_2O(g)$

15. _____ $CaCO_3(s) \longrightarrow CaO(s) + CO_2(g)$

16. _____ $Mg(s) + 2AgNO_3(aq) \longrightarrow Mg(NO_3)_2(aq) + 2Ag(s)$

17. _____ $Na_2S(aq) + Pb(NO_3)_2(aq) \longrightarrow PbS(s) + 2NaNO_3(aq)$

For questions 18 through 22, identify each reaction as an oxidation (O) *or a reduction* (R). (8.4)

18. $Ca \longrightarrow Ca^{2+} + 2\,e^-$ ____   **19.** $Fe^{3+} + 3\,e^- \longrightarrow Fe$ ____

20. $Al^{3+} + 3\,e^- \longrightarrow Al$ ____   **21.** $Sn^{2+} \longrightarrow Sn^{4+} + 2\,e^-$ ____

22. $Br_2 + 2\,e^- \longrightarrow 2Br^-$ ____

23. Identify the reactant that is oxidized in the following equation: (8.4)

    $Mg(s) + CuCl_2(aq) \longrightarrow MgCl_2(aq) + Cu(s)$

    A. Mg   **B.** Cu^{2+} **C.** Cl^- **D.** Mg^{2+} **E.** Cu

24. Identify the reactant that is reduced in the following equation: (8.4)

    $3C(s) + 2Fe_2O_3(s) \longrightarrow 3CO_2(g) + 4Fe(s)$

    A. C   **B.** Fe^{3+} **C.** O^{2-} **D.** Fe   **E.** CO_2

Answers to the Practice Test

1. C	**2.** E	**3.** D	**4.** NB	**5.** NB
6. B	**7.** B	**8.** B	**9.** C	**10.** E
11. D	**12.** A	**13.** A	**14.** E	**15.** B
16. C	**17.** D	**18.** O	**19.** R	**20.** R
21. O	**22.** R	**23.** A	**24.** B	

Selected Answers and Solutions to Text Problems

8.1 **a.** reactants/products: 7 O atoms
b. reactants/products: 4 O atoms
c. reactants/products: 10 O atoms
d. reactants/products: 4 O atoms

8.3 An equation is balanced when there are an equal number of atoms of each element on the reactant side and on the product side.
a. not balanced ($2\,O \neq 3\,O$)   **b.** not balanced ($2\,Cl \neq 3\,Cl$)
c. balanced   **d.** balanced

8.5 **a.** reactants/products: 2 Na atoms, 2 Cl atoms
b. reactants/products: 1 P atom, 3 Cl atoms, 6 H atoms
c. reactants/products: 4 P atoms, 16 O atoms, 12 H atoms

8.7 Place coefficients in front of formulas until you make the atoms of each element equal on each side of the equation.
a. $N_2(g) + O_2(g) \longrightarrow 2NO(g)$ **b.** $2HgO(s) \xrightarrow{\Delta} 2Hg(l) + O_2(g)$
c. $4Fe(s) + 3O_2(g) \longrightarrow 2Fe_2O_3(s)$ **d.** $2Na(s) + Cl_2(g) \longrightarrow 2NaCl(s)$
e. $2Cu_2O(s) + O_2(g) \longrightarrow 4CuO(s)$

8.9 **a.** $Mg(s) + 2AgNO_3(aq) \longrightarrow 2Ag(s) + Mg(NO_3)_2(aq)$
b. $CuCO_3(s) \longrightarrow CuO(s) + CO_2(g)$
c. $2Al(s) + 3CuSO_4(aq) \longrightarrow 3Cu(s) + Al_2(SO_4)_3(aq)$
d. $Pb(NO_3)_2(aq) + 2NaCl(aq) \longrightarrow PbCl_2(s) + 2NaNO_3(aq)$
e. $2Al(s) + 6HCl(aq) \longrightarrow 3H_2(g) + 2AlCl_3(aq)$

8.11 **a.** $Fe_2O_3(s) + 3CO(g) \longrightarrow 2Fe(s) + 3CO_2(g)$
b. $2Li_3N(s) \longrightarrow 6Li(s) + N_2(g)$
c. $2Al(s) + 6HBr(aq) \longrightarrow 3H_2(g) + 2AlBr_3(aq)$
d. $3Ba(OH)_2(aq) + 2Na_3PO_4(aq) \longrightarrow Ba_3(PO_4)_2(s) + 6NaOH(aq)$
e. $As_4S_6(s) + 9O_2(g) \longrightarrow As_4O_6(s) + 6SO_2(g)$

8.13 **a.** $2Li(s) + 2H_2O(l) \longrightarrow H_2(g) + 2LiOH(aq)$
b. $2P(s) + 5Cl_2(g) \longrightarrow 2PCl_5(s)$
c. $FeO(s) + CO(g) \longrightarrow Fe(s) + CO_2(g)$
d. $2C_5H_{10}(l) + 15O_2(g) \xrightarrow{\Delta} 10CO_2(g) + 10H_2O(g)$
e. $3H_2S(g) + 2FeCl_3(s) \longrightarrow Fe_2S_3(s) + 6HCl(g)$

8.15 $NH_4NO_3(s) \xrightarrow{\Delta} 2H_2O(g) + N_2O(g)$

8.17 $2C_3H_7NO_2(aq) + 6O_2(g) \longrightarrow 5CO_2(g) + 5H_2O(l) + CH_4N_2O(aq)$

8.19 **a.** This is a decomposition reaction because a single reactant splits into two simpler substances.
b. This is a single replacement reaction because one element in the reacting compound (I in BaI_2) is replaced by the other reactant (Br in Br_2).
c. This is a combustion reaction because a carbon-containing fuel burns in oxygen to produce carbon dioxide and water.
d. This is a double replacement reaction because the positive ions in the two reactants exchange places to form two products.
e. This is a combination reaction because atoms of two different elements bond to form one product.

8.21 **a.** combination   **b.** single replacement   **c.** decomposition
       d. double replacement   **e.** combustion

8.23 **a.** Since this is a combination reaction, the product is a single compound.
$$Mg(s) + Cl_2(g) \longrightarrow MgCl_2(s)$$
       b. Since this is a decomposition reaction, the single reactant breaks down into more than one simpler substance.
$$2HBr(g) \longrightarrow H_2(g) + Br_2(l)$$
       c. Since this is a single replacement reaction, the uncombined metal element takes the place of the metal element in the compound.
$$Mg(s) + Zn(NO_3)_2(aq) \longrightarrow Zn(s) + Mg(NO_3)_2(aq)$$
       d. Since this is a double replacement reaction, the positive ions in the reacting compounds switch places.
$$K_2S(aq) + Pb(NO_3)_2(aq) \longrightarrow PbS(s) + 2KNO_3(aq)$$
       e. Since this is a combustion reaction, a fuel burns in oxygen to produce carbon dioxide and water.
$$2C_2H_6(g) + 7O_2(g) \xrightarrow{\Delta} 4CO_2(g) + 6H_2O(g)$$

8.25 Oxidation is the loss of electrons; reduction is the gain of electrons.
       a. Na^+ gains an electron to form Na; this is a reduction.
       b. Ni loses electrons to form Ni^{2+}; this is an oxidation.
       c. Cr^{3+} gains electrons to form Cr; this is a reduction.
       d. $2H^+$ gain electrons to form H_2; this is a reduction.

8.27 An oxidized substance has lost electrons; a reduced substance has gained electrons.
       a. Zn loses electrons and is oxidized. Cl_2 gains electrons and is reduced.
       b. Br^- (in NaBr) loses electrons and is oxidized. Cl_2 gains electrons and is reduced.
       c. O^{2-} (in PbO) loses electrons and is oxidized. Pb^{2+} (in PbO) gains electrons and is reduced.
       d. Sn^{2+} loses electrons and is oxidized. Fe^{3+} gains electrons and is reduced.

8.29 **a.** Fe^{3+} gains an electron to form Fe^{2+}; this is a reduction.
       b. Fe^{2+} loses an electron to form Fe^{3+}; this is an oxidation.

8.31 Linoleic acid gains hydrogen atoms and is reduced.

8.33 **a.** $C_6H_{12}O_6(aq) + 6O_2(g) \longrightarrow 6CO_2(g) + 6H_2O(l)$
       b. $6CO_2(g) + 6H_2O(l) \longrightarrow C_6H_{12}O_6(aq) + 6O_2(g)$

8.35 **a.** 1, 1, 2 combination
       b. 2, 2, 1 decomposition

8.37 **a.** reactants NO and O_2; product NO_2
       b. $2NO(g) + O_2(g) \longrightarrow 2NO_2(g)$
       c. combination

8.39 **a.** reactant NI_3; products N_2 and I_2
       b. $2NI_3(g) \longrightarrow N_2(g) + 3I_2(g)$
       c. decomposition

8.41 **a.** reactants Cl_2 and O_2; product OCl_2
       b. $2Cl_2(g) + O_2(g) \longrightarrow 2OCl_2(g)$
       c. combination

8.43 **a.** Atoms of a metal and a nonmetal forming an ionic compound is a combination reaction.
       b. When a compound of hydrogen and carbon reacts with oxygen, it is a combustion reaction.
       c. When calcium carbonate is heated to produce calcium oxide and carbon dioxide, it is a decomposition reaction.
       d. Zinc replacing copper in $Cu(NO_3)_2$ is a single replacement reaction.

8.45 **a.** $NH_3(g) + HCl(g) \longrightarrow NH_4Cl(s)$ combination
    b. $C_4H_8(g) + 6O_2(g) \xrightarrow{\Delta} 4CO_2(g) + 4H_2O(g)$ combustion
    c. $2Sb(s) + 3Cl_2(g) \longrightarrow 2SbCl_3(s)$ combination
    d. $2NI_3(s) \longrightarrow N_2(g) + 3I_2(g)$ decomposition
    e. $2KBr(aq) + Cl_2(aq) \longrightarrow 2KCl(aq) + Br_2(l)$ single replacement
    f. $2Fe(s) + 3H_2SO_4(aq) \longrightarrow 3H_2(g) + Fe_2(SO_4)_3(aq)$ single replacement
    g. $Al_2(SO_4)_3(aq) + 6NaOH(aq) \longrightarrow 2Al(OH)_3(s)$ double replacement
                                          $+ 3Na_2SO_4(aq)$

8.47 **a.** Since this is a single replacement reaction, the uncombined metal element takes the place of the metal element in the compound.
       $Zn(s) + 2HCl(aq) \longrightarrow H_2(g) + ZnCl_2(aq)$
    b. Since this is a decomposition reaction, the single reactant breaks down into more than one simpler substance.
       $BaCO_3(s) \xrightarrow{\Delta} BaO(s) + CO_2(g)$
    c. Since this is a double replacement reaction, the positive ions in the reacting compounds switch places.
       $NaOH(aq) + HCl(aq) \longrightarrow H_2O(l) + NaCl(aq)$
    d. Since this is a combination reaction, the product is a single compound.
       $2Al(s) + 3F_2(g) \longrightarrow 2AlF_3(s)$

8.49 **a.** $4Na(s) + O_2(g) \longrightarrow 2Na_2O(s)$ combination
    b. $NaCl(aq) + AgNO_3(aq) \longrightarrow AgCl(s) + NaNO_3(aq)$ double replacement
    c. $C_2H_6O(l) + 3O_2(g) \xrightarrow{\Delta} 2CO_2(g) + 3H_2O(g)$ combustion

8.51 **a.** N_2 is oxidized; O_2 is reduced.
    b. The C (in CO) is oxidized; H_2 is reduced.
    c. Mg is oxidized; Br_2 is reduced.

8.53 **a.** $K_2O(s) + H_2O(g) \longrightarrow 2KOH(aq)$ combination
    b. $2C_8H_{18}(l) + 25O_2(g) \xrightarrow{\Delta} 16CO_2(g) + 18H_2O(g)$ combustion
    c. $2Fe(OH)_3(s) \longrightarrow Fe_2O_3(s) + 3H_2O(g)$ decomposition
    d. $CuS(s) + 2HCl(aq) \longrightarrow CuCl_2(aq) + H_2S(g)$ double replacement

8.55 **a.** $Fe_3O_4(s) + 4H_2(g) \longrightarrow 3Fe(s) + 4H_2O(g)$
    b. $2C_4H_{10}(g) + 13O_2(g) \xrightarrow{\Delta} 8CO_2(g) + 10H_2O(g)$
    c. $4Al(s) + 3O_2(g) \longrightarrow 2Al_2O_3(s)$
    d. $2NaOH(aq) + ZnSO_4(aq) \longrightarrow Zn(OH)_2(s) + Na_2SO_4(aq)$

8.57 **a.** $3Pb(NO_3)_2(aq) + 2Na_3PO_4(aq) \longrightarrow Pb_3(PO_4)_2(s) + 6NaNO_3(aq)$ double replacement
    b. $4Ga(s) + 3O_2(g) \xrightarrow{\Delta} 2Ga_2O_3(s)$ combination
    c. $2NaNO_3(s) \xrightarrow{\Delta} 2NaNO_2(s) + O_2(g)$ decomposition

8.59 **a.** reactants: X and Y_2; product: XY_3 **b.** $2X + 3Y_2 \longrightarrow 2XY_3$
    c. combination

9

Chemical Quantities in Reactions

Credit: Bart Coenders/E+/Getty Images

Lance, an environmental scientist, has returned to a farm where he found small amounts of Sevin, a pesticide, in the soil and water. His recent tests indicate that the Sevin is now within acceptable range in both the soil and the water in the lake near the farm.

Lance specializes in measuring levels of contaminants in soil and water caused by manufacturing processes or farming. In his work, Lance also acts as an advisor, consultant, and safety officer to ensure that soil and water on the farm are safe.

Sevin has a molecular formula $C_{12}H_{11}NO_2$ with a molar mass of 201.2. It is an insectide applied to fruit and vegetable crops as well as to livestock. If 15 g of Sevin is mixed in 1.0 L of water to spray farm buildings, how many moles of Sevin are used?

CHAPTER READINESS*

⚙ **Key Math Skills**

- Calculating Percentages (1.4)

🧪 **Core Chemistry Skills**

- Counting Significant Figures (2.2)
- Using Significant Figures in Calculations (2.3)

- Writing Conversion Factors from Equalities (2.5)
- Using Conversion Factors (2.6)
- Using Energy Units (3.4)
- Calculating Molar Mass (7.2)
- Using Molar Mass as a Conversion Factor (7.3)
- Balancing a Chemical Equation (8.2)
- Classifying Types of Chemical Reactions (8.3)

*These Key Math Skills and Core Chemistry Skills from previous chapters are listed here for your review as you proceed to the new material in this chapter.

LOOKING AHEAD

9.1   Conservation of Mass

Learning Goal: Calculate the total mass of reactants and the total mass of products in a balanced chemical equation.

- The law of conservation of mass states that mass is neither lost nor gained during a chemical reaction.

♦ Learning Exercise 9.1

Calculate the total mass of the reactants and the products for the following equation:

$$2Na(s) + 2H_2O(l) \longrightarrow H_2(g) + 2NaOH(aq)$$

	Reactants		**Products**
Equation	_____	$\longrightarrow$	_____
Moles	_____	$\longrightarrow$	_____
Mass	_____	$\longrightarrow$	_____
Total Mass	_____	$=$	_____

Answer

	Reactants		**Products**
Equation	$2Na(s) + 2H_2O(l)$	$\longrightarrow$	$H_2(g) + 2NaOH(aq)$
Moles	2 mol of Na + 2 mol of H_2O	$\longrightarrow$	1 mol of H_2 + 2 mol of NaOH
Mass	45.98 g of Na + 36.04 g of H_2O	$\longrightarrow$	2.016 g of H_2 + 80.00 g of NaOH
Total Mass	82.02 g of reactants	$=$	82.02 g of products

9.2   Calculating Moles Using Mole–Mole Factors

Learning Goal: Use a mole–mole factor from a balanced chemical equation to calculate the number of moles of another substance in the reaction.

- The coefficients in a balanced chemical equation indicate the moles of reactants and products in a reaction.
- Using the coefficients, mole–mole conversion factors can be written to relate any two substances in an equation.
- For the reaction of oxygen forming ozone, $3O_2(g) \longrightarrow 2O_3(g)$, the mole–mole factors are the following:

$$\frac{3 \text{ mol } O_2}{2 \text{ mol } O_3} \text{ and } \frac{2 \text{ mol } O_3}{3 \text{ mol } O_2}$$

♦ Learning Exercise 9.2A

Write all of the possible mole–mole factors for the following equation:

$$N_2(g) + O_2(g) \longrightarrow 2NO(g)$$

Answers

For N_2 and O_2:

$$\frac{1 \text{ mol } N_2}{1 \text{ mol } O_2} \text{ and } \frac{1 \text{ mol } O_2}{1 \text{ mol } N_2}$$

For N_2 and NO:

$$\frac{2 \text{ mol NO}}{1 \text{ mol } N_2} \text{ and } \frac{1 \text{ mol } N_2}{2 \text{ mol NO}}$$

For O_2 and NO:

$$\frac{2 \text{ mol NO}}{1 \text{ mol } O_2} \text{ and } \frac{1 \text{ mol } O_2}{2 \text{ mol NO}}$$

Guide to Calculating the Quantities of Reactants and Products in a Chemical Reaction	
STEP 1	State the given and needed quantities (moles or grams).
STEP 2	Write a plan to convert the given to the needed quantity (moles or grams).
STEP 3	Use coefficients to write mole–mole factors; write molar masses if needed.
STEP 4	Set up the problem to give the needed quantity (moles or grams).

Study Note

The appropriate mole–mole factor is used to change the number of given moles to the number of needed moles.

Example: For the equation $N_2(g) + O_2(g) \longrightarrow 2NO(g)$, calculate the number of moles of NO produced from 3.0 mol of N_2.

Solution: **STEP 1 State the given and needed quantities (moles).**

Analyze the Problem	Given	Need	Connect
	3.0 mol of N_2	moles of NO	mole–mole factor

STEP 2 Write a plan to convert the given to the needed quantity (moles).

moles of N_2 | Mole–mole factor | moles of NO

STEP 3 Use coefficients to write mole–mole factors; write molar masses if needed.

$$\frac{2 \text{ mol NO}}{1 \text{ mol } N_2} \text{ and } \frac{1 \text{ mol } N_2}{2 \text{ mol NO}}$$

STEP 4 Set up the problem to give the needed quantity (moles).

$$3.0 \text{ mol } N_2 \times \frac{\overset{\text{Exact}}{2 \text{ mol NO}}}{1 \text{ mol } N_2} = 6.0 \text{ mol of NO}$$

Two SFs   Exact   Two SFs

♦ **Learning Exercise 9.2B**

Use the balanced chemical equation to answer the following questions:

> CORE CHEMISTRY SKILL
>
> Using Mole–Mole Factors

$$C_3H_8(g) + 5O_2(g) \xrightarrow{\Delta} 3CO_2(g) + 4H_2O(g)$$

a. How many moles of O_2 are needed to react with 2.00 mol of C_3H_8?

b. How many moles of CO_2 are produced when 4.00 mol of O_2 react?

c. How many moles of C_3H_8 react with 3.00 mol of O_2?

d. How many moles of H_2O are produced from 0.500 mol of C_3H_8?

Answers   **a.** 10.0 mol of O_2 **b.** 2.40 mol of CO_2 **c.** 0.600 mol of C_3H_8
              d. 2.00 mol of H_2O

9.3   Mass Calculations for Reactions

Learning Goal: Given the mass in grams of a substance in a reaction, calculate the mass in grams of another substance in the reaction.

• The grams of a substance in an equation are converted to grams of another substance using their molar masses and mole–mole factors.

Substance A   Substance B

grams of A   Molar mass A   moles of A   Mole–mole factor B/A   moles of B   Molar mass B   grams of B

Example: How many grams of water are produced when 45.0 g of O_2 reacts with sufficient NH_3?

Solution: **STEP 1   State the given and needed quantities (grams).**

Analyze the Problem	Given	Need	Connect
	45.0 g of O_2	grams of H_2O	molar masses, mole–mole factor
Equation	$4NH_3(g) + 3O_2(g) \longrightarrow 2N_2(g) + 6H_2O(g)$		

STEP 2   Write a plan to convert the given to the needed quantity (grams).

grams of O_2 Molar mass   moles of O_2 Mole–mole factor   moles of H_2O Molar mass   grams of H_2O

STEP 3   Use coefficients to write relationships and mole–mole factors; write molar masses.

1 mol of O_2 = 32.00 g of O_2 1 mol of H_2O = 18.02 g of H_2O

$$\frac{32.00 \text{ g } O_2}{1 \text{ mol } O_2} \text{ and } \frac{1 \text{ mol } O_2}{32.00 \text{ g } O_2} \qquad \frac{18.02 \text{ g } H_2O}{1 \text{ mol } H_2O} \text{ and } \frac{1 \text{ mol } H_2O}{18.02 \text{ g } H_2O}$$

3 mol of O_2 = 6 mol of H_2O

$$\frac{3 \text{ mol } O_2}{6 \text{ mol } H_2O} \text{ and } \frac{6 \text{ mol } H_2O}{3 \text{ mol } O_2}$$

STEP 4   Set up the problem to give the needed quantity (grams).

$$45.0 \text{ g } O_2 \times \frac{1 \text{ mol } O_2}{32.00 \text{ g } O_2} \times \frac{6 \text{ mol } H_2O}{3 \text{ mol } O_2} \times \frac{18.02 \text{ g } H_2O}{1 \text{ mol } H_2O} = 50.7 \text{ g of } H_2O$$

♦ **Learning Exercise 9.3**

Use the balanced chemical equation to answer the questions below:

$$2C_2H_6(g) + 7O_2(g) \xrightarrow{\Delta} 4CO_2(g) + 6H_2O(g)$$

a. How many grams of O_2 are needed to react with 120. g of C_2H_6?

b. How many grams of C_2H_6 are needed to react with 115 g of O_2?

c. How many grams of C_2H_6 react if 2.00 g of CO_2 is produced?

d. How many grams of CO_2 are produced when 60.0 g of C_2H_6 reacts with sufficient oxygen?

e. How many grams of H_2O are produced when 82.5 g of O_2 reacts with sufficient C_2H_6?

Answers   **a.** 447 g of O_2 **b.** 30.9 g of C_2H_6 **c.** 0.683 g of C_2H_6
d. 176 g of CO_2 **e.** 39.8 g of H_2O

9.4 Limiting Reactants

Learning Goal: Identify a limiting reactant when given the quantities of two reactants; calculate the amount of product formed from the limiting reactant.

- In a limiting reactant problem, the availability of one of the reactants limits the amount of product.
- The reactant that is used up is the limiting reactant; the reactant that remains is the excess reactant.
- The limiting reactant produces the smaller number of moles of product.

Guide to Calculating the Quantity (Moles or Grams) of Product from a Limiting Reactant	
STEP 1	State the given and needed quantity (moles or grams).
STEP 2	Write a plan to convert the quantity (moles or grams) of each reactant to the quantity (moles or grams) of product.
STEP 3	Write the mole–mole factors and molar masses (if needed).
STEP 4	Calculate the quantity (moles or grams) of product from each reactant and select the smaller quantity (moles or grams) as the limiting reactant.

Example: The amount of product possible from a mixture of two reactants is determined by calculating the grams of product each will produce. The limiting reactant produces the smaller amount of product. In the reaction $S(l) + 3F_2(g) \longrightarrow SF_6(g)$, how many grams of SF_6 are possible when 16.1 g of S is mixed with 45.6 g of F_2?

Solution: **STEP 1   State the given and needed quantity (grams).**

Analyze the Problem	Given	Need	Connect
	16.1 g of S, 45.6 g of F_2	grams of SF_6	molar masses, mole–mole factors
Equation	$S(l) + 3F_2(g) \longrightarrow SF_6(g)$		

STEP 2   Write a plan to convert the quantity (grams) of each reactant to quantity (grams) of product.

grams of S $\longrightarrow$ moles of S $\longrightarrow$ moles of SF_6 $\longrightarrow$ grams of SF_6

grams of F_2 $\longrightarrow$ moles of F_2 $\longrightarrow$ moles of SF_6 $\longrightarrow$ grams of SF_6

STEP 3   Write the mole–mole factors and molar masses.

$$1 \text{ mol of S} = 1 \text{ mol of SF}_6 \qquad 3 \text{ mol of F}_2 = 1 \text{ mol of SF}_6$$

$$\frac{1 \text{ mol S}}{1 \text{ mol SF}_6} \text{ and } \frac{1 \text{ mol SF}_6}{1 \text{ mol S}} \qquad \frac{3 \text{ mol F}_2}{1 \text{ mol SF}_6} \text{ and } \frac{1 \text{ mol SF}_6}{3 \text{ mol F}_2}$$

$$1 \text{ mol of F}_2 = 38.00 \text{ g of F}_2 \qquad 1 \text{ mol of SF}_6 = 146.1 \text{ g of SF}_6$$

$$\frac{38.00 \text{ g F}_2}{1 \text{ mol F}_2} \text{ and } \frac{1 \text{ mol F}_2}{38.00 \text{ g F}_2} \qquad \frac{146.1 \text{ g SF}_6}{1 \text{ mol SF}_6} \text{ and } \frac{1 \text{ mol SF}_6}{146.1 \text{ g SF}_6}$$

STEP 4   Calculate the quantity (grams) of product from each reactant and select the smaller quantity (grams) as the limiting reactant.

$$16.1 \text{ g S} \times \frac{1 \text{ mol S}}{32.07 \text{ g S}} \times \frac{1 \text{ mol SF}_6}{1 \text{ mol S}} \times \frac{146.1 \text{ g SF}_6}{1 \text{ mol SF}_6} = 73.3 \text{ g of SF}_6$$

$$45.6 \text{ g F}_2 \times \frac{1 \text{ mol F}_2}{38.00 \text{ g F}_2} \times \frac{1 \text{ mol SF}_6}{3 \text{ mol F}_2} \times \frac{146.1 \text{ g SF}_6}{1 \text{ mol SF}_6} = 58.4 \text{ g of SF}_6 \text{ from the limiting reactant } F_2 \text{ (smaller amount of product)}$$

◆ **Learning Exercise 9.4A**

Use the balanced chemical equation to answer the following:

How many grams of Co_2S_3 can be produced from the reaction of 250. g of Co and 250. g of S?

$$2Co(s) + 3S(s) \longrightarrow Co_2S_3(s)$$

⬥ **CORE CHEMISTRY SKILL**

Calculating Quantity of Product from a Limiting Reactant

Answer   454 g

♦ **Learning Exercise 9.4B**

Use the balanced chemical equation to answer the following:

How many grams of NO_2 can be produced from the reaction of 32.0 g of NO and 24.0 g of O_2?

$$2NO(g) + O_2(g) \longrightarrow 2NO_2(g)$$

Answer   49.1 g

9.5   Percent Yield

Learning Goal: Given the actual quantity of product, determine the percent yield for a reaction.

- In most reactions, some product is lost due to incomplete reactions or side reactions. The amount of product collected and measured is called the actual yield.
- Theoretical yield is the maximum amount of product calculated for a given amount of a reactant.
- Percent yield is the ratio of the actual amount (yield) of product obtained to the theoretical yield.

$$\text{Percent yield}\,(\%) = \frac{\text{actual yield}}{\text{theoretical yield}} \times 100\%$$

Study Note

The percent yield is the ratio of the actual yield obtained to the theoretical yield, which is calculated for a given amount of starting reactant. If we calculate that a reaction can theoretically produce 43.1 g of NH_3, but the actual yield is 26.0 g of NH_3, the percent yield is:

$$\frac{26.0 \text{ g } NH_3 \,(\text{actual mass produced})}{43.1 \text{ g } NH_3 \,(\text{theoretical mass})} \times 100\% = 60.3\%$$

	Guide to Calculating Percent Yield
STEP 1	State the given and needed quantities.
STEP 2	Write a plan to calculate the theoretical yield and the percent yield.
STEP 3	Write the molar masses and the mole–mole factor from the balanced equation.
STEP 4	Calculate the percent yield by dividing the actual yield (given) by the theoretical yield and multiplying the result by 100%.

♦ **Learning Exercise 9.5A**

CORE CHEMISTRY SKILL

Calculating Percent Yield

Use the balanced chemical equation to answer the following:

$$2H_2S(g) + 3O_2(g) \longrightarrow 2SO_2(g) + 2H_2O(g)$$

a. If 60.0 g of H_2S reacts with sufficient oxygen and produces 45.5 g of SO_2, what is the percent yield of SO_2?

b. If 25.0 g of O_2 reacting with sufficient H_2S produces 18.6 g of SO_2, what is the percent yield of SO_2?

Answers **a.** 40.3% **b.** 55.7%

♦ **Learning Exercise 9.5B**

Use the balanced chemical equation to answer the following:

$$\underset{\text{Ethane}}{2C_2H_6(g)} + 7O_2(g) \xrightarrow{\Delta} 4CO_2(g) + 6H_2O(g)$$

a. If 125 g of C_2H_6 reacting with sufficient oxygen produces 175 g of CO_2, what is the percent yield of CO_2?

b. When 35.0 g of O_2 reacts with sufficient ethane to produce 12.5 g of H_2O, what is the percent yield of H_2O?

Answers **a.** 47.8% **b.** 74.0%

♦ **Learning Exercise 9.5C**

Valine, $C_5H_{11}NO_2$, an amino acid, reacts in the body according to the following reaction:

$$\underset{\text{Valine}}{2C_5H_{11}NO_2(aq)} + 12O_2(g) \longrightarrow 9CO_2(g) + 9H_2O(l) + \underset{\text{Urea}}{CH_4N_2O(aq)}$$

a. How many grams of O_2 will react with 15.0 g of valine?

b. How many grams of CO_2 are formed when 100. g of valine and 200. g of O_2 react?

c. If 45.0 g of valine reacts to give 6.75 g of urea, what is the percent yield of urea?

Answers   **a.** 24.6 g   **b.** 169 g   **c.** 58.7%

♦   **Learning Exercise 9.5D**

Galactose, $C_6H_{12}O_6(aq)$, found in milk products, reacts in the body by the following reaction:

$$C_6H_{12}O_6(aq) + 6O_2(g) \longrightarrow 6CO_2(g) + 6H_2O(l)$$

a. How many grams of O_2 will react with 46.4 g of galactose?

b. How many grams of CO_2 are formed when 77.0 g of galactose and 100. g of O_2 react?

c. If 23.3 g of galactose reacts to give 26.3 g of CO_2, what is the percent yield of CO_2?

Answers   **a.** 49.4 g   **b.** 113 g   **c.** 76.9%

9.6   Energy in Chemical Reactions

Learning Goal: Given the heat of reaction (enthalpy change), calculate the loss or gain of heat for an exothermic or endothermic reaction.

- The heat of reaction is the energy released or absorbed when the reaction occurs.
- The heat of reaction or enthalpy change, ΔH, is the energy difference between the reactants and the products.
- In exothermic reactions, the energy of the products is lower than that of the reactants, which means that energy is released. The heat of reaction for an exothermic reaction is written as a ΔH value with a negative sign $(-)$ or by writing the heat of reaction as a product.

- In endothermic reactions, the energy of the products is higher than that of the reactants, which means that energy is absorbed. The heat of reaction for an endothermic reaction is written as a ΔH value with a positive sign ($+$) or by writing the heat of reaction as a reactant.

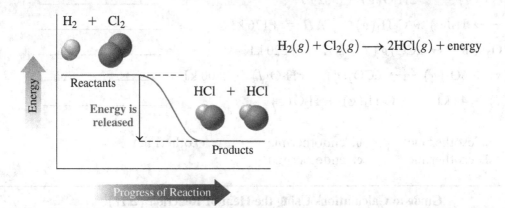

$$H_2(g) + Cl_2(g) \longrightarrow 2HCl(g) + energy$$

- The value of ΔH refers to the heat change, in kilojoules, for the number of moles of each substance in the balanced chemical equation.
- Heat factors can be written for each substance: heat/mol of substance and mol of substance/heat.
- Hess's law states that the heat of a reaction can be absorbed or released in a single step or in several steps.

Key Terms for Sections 9.1 to 9.6

Match each of the following key terms with the correct description:

a. actual yield
b. exothermic reaction
c. endothermic reaction
d. theoretical yield
e. mole–mole factor
f. Hess's law
g. conservation of mass

1. _____ a reaction in which the energy of the reactants is less than that of the products

2. _____ the maximum amount of product that a reaction can produce from a given amounts of reactants

3. _____ a reaction in which the energy of the reactants is greater than that of the products

4. _____ the law that states that there is no change in the mass of substances reacting

5. _____ a factor that relates the number of moles of two compounds derived from the coefficients in a balanced equation

6. _____ the amount of product produced by a reaction

7. _____ the law that states that the heat of a reaction can be absorbed or released in a single step or in several steps

Answers   1. c   2. d   3. b   4. g   5. e   6. a   7. f

♦ **Learning Exercise 9.6A**

Indicate whether each of the following is an endothermic or exothermic reaction:

a. $2H_2(g) + O_2(g) \longrightarrow 2H_2O(g) + 582\,kJ$ _____

b. $C_2H_4(g) \longrightarrow H_2(g) + C_2H_2(g)$ $\Delta H = +176\,kJ$ _____

c. $2C(s) + O_2(g) \longrightarrow 2CO(g)$ $\Delta H = -220\,kJ$ _____

d. $C_6H_{12}O_6(s) + 6O_2(g) \longrightarrow 6CO_2(g) + 6H_2O(l) + 2800\,kJ$ _____

e. $C_2H_5OH(l) + 44\,kJ \longrightarrow C_2H_4(g) + H_2O(g)$ _____

Answers   **a.** exothermic   **b.** endothermic   **c.** exothermic
      d. exothermic   **e.** endothermic

Guide to Calculations Using the Heat of Reaction (ΔH)	
STEP 1	State the given and needed quantities.
STEP 2	Write a plan using the heat of reaction and any molar mass needed.
STEP 3	Write the conversion factors including heat of reaction.
STEP 4	Set up the problem to calculate the heat.

♦ **Learning Exercise 9.6B**   CORE CHEMISTRY SKILL

Use the balanced chemical equation to answer the following: Using the Heat of Reaction

$C_6H_{12}O_6(s) + 6O_2(g) \longrightarrow 6CO_2(g) + 6H_2O(l)$ $\Delta H = -2800\,kJ$

How much heat, in kilojoules, is produced when 50.0 g of CO_2 is formed?

Answer   530 kJ

♦ **Learning Exercise 9.6C**

For each of the following reactions, calculate the number of kilojoules absorbed or released:

$C_2H_4(g) \longrightarrow H_2(g) + C_2H_2(g)$ $\Delta H = +176\,kJ$

a. 3.50 mol of C_2H_2 is produced.

b. 75.0 g of C_2H_4 reacts.

Answers   **a.** 616 kJ is absorbed   **b.** 471 kJ is absorbed

♦ **Learning Exercise 9.6D**

For each of the following reactions, calculate the number of kilojoules absorbed or released:

$$2H_2(g) + O_2(g) \longrightarrow 2H_2O(g) \qquad \Delta H = -582 \text{ kJ}$$

a. 0.820 mol of H_2 reacts.

b. 2.25 g of H_2O is produced.

Answers   **a.** 239 kJ released   **b.** 36.4 kJ released

♦ **Learning Exercise 9.6E**

Calculate the enthalpy change for the reaction $2CH_4(g) + 3O_2(g) \longrightarrow 2CO(g) + 4H_2O(l)$ from the following equations:

$$CH_4(g) + 2O_2(g) \longrightarrow CO_2(g) + 2H_2O(l) \qquad \Delta H = -890 \text{ kJ}$$
$$2CO(g) + O_2(g) \longrightarrow 2CO_2(g) \qquad \Delta H = -566 \text{ kJ}$$

Answer $\Delta H = -1214 \text{ kJ}$

Checklist for Chapter 9

You are ready to take the Practice Test for Chapter 9. Be sure you have accomplished the following learning goals for this chapter. If not, review the section listed at the end of the goal. Then apply your new skills and understanding to the Practice Test.

After studying Chapter 9, I can successfully:

_____ Calculate the total mass of reactants and products in a balanced chemical equation. (9.1)

_____ Use a mole–mole factor from a balanced chemical equation to calculate the number of moles of another substance in the reaction. (9.2)

_____ Calculate the mass in grams of a substance in an equation using mole–mole factors and molar masses. (9.3)

_____ Given the quantities of two reactants, determine the limiting reactant and calculate the amount of product formed. (9.4)

_____ Calculate the percent yield for a reaction given the actual yield of a product. (9.5)

_____ Given the heat of reaction, describe a reaction as exothermic or endothermic. (9.6)

_____ Given the mass of reactants or products, calculate the heat, in kilojoules, released or absorbed by a chemical reaction. (9.6)

_____ Calculate the overall energy change for a reaction from the energy changes for that reaction. (9.6)

Practice Test for Chapter 9

The chapter sections to review are shown in parentheses at the end of each question.

1. For the balanced reaction $2Ag_2O(s) + C(s) \longrightarrow CO_2(g) + 4Ag(s)$, the total mass of the reactants and the products are (9.1):
   A. reactants 243.8 g, products 135.9 g **B.** reactants 243.8 g, products 243.8 g
   C. reactants 3 g, products 5 g **D.** reactants 475.6 g, products 475.6 g
   E. reactants 151.9 g, products 151.9 g

For questions 2 through 8, use the reaction $C_2H_6O(g) + 3O_2(g) \xrightarrow{\Delta} 2CO_2(g) + 3H_2O(g)$
(9.2, 9.3, 9.4)   Ethanol

2. How many moles of oxygen are needed to react with 0.360 mol of ethanol?
   A. 0.120 mol **B.** 1.08 mol **C.** 1.20 mol **D.** 3.00 mol **E.** 6.00 mol

3. How many moles of oxygen would be needed to produce 1.00 mol of CO_2?
   A. 0.670 mol **B.** 1.00 mol **C.** 1.50 mol **D.** 2.00 mol **E.** 3.00 mol

4. How many moles of water are produced when 12.0 mol of oxygen reacts?
   A. 3.00 mol **B.** 6.00 mol **C.** 8.00 mol **D.** 12.0 mol **E.** 36.0 mol

5. How many grams of carbon dioxide are produced when 92.0 g of ethanol reacts?
   A. 22.0 g **B.** 44.0 g **C.** 88.0 g **D.** 92.0 g **E.** 176 g

6. How many grams of water will be produced if 23.0 g of ethanol reacts?
   A. 54.0 g **B.** 27.0 g **C.** 18.0 g **D.** 9.00 g **E.** 6.00 g

7. How many grams of water will be produced if 138 g of ethanol and 160. g of O_2 react?
   A. 3.60 g **B.** 18.0 g **C.** 54.1 g **D.** 90.1 g **E.** 162 g

8. How many grams of water will be produced if 230. g of ethanol and 384 g of O_2 react?
   A. 270 g **B.** 216 g **C.** 54.1 g **D.** 36.0 g **E.** 18.0 g

For questions 9 and 10, use the reaction $C_3H_8(g) + 5O_2(g) \xrightarrow{\Delta} 3CO_2(g) + 4H_2O(g)$ (9.4)

9. If 50.0 g of C_3H_8 and 150. g of O_2 react, the limiting reactant is:
   A. C_3H_8 **B.** O_2 **C.** both **D.** neither **E.** CO_2

10. If 50.0 g of C_3H_8 and 150. g of O_2 react, the mass of CO_2 formed is:
   A. 150. g **B.** 206 g **C.** 200. g **D.** 124 g **E.** 132 g

For questions 11 and 12, use the reaction: $N_2(g) + 3H_2(g) \longrightarrow 2NH_3(g)$ (9.4, 9.5)

11. When 25.0 g of N_2 reacts, what is the theoretical yield of NH_3?
   A. 7.60 g **B.** 15.2 g **C.** 28.02 g **D.** 30.4 g **E.** 60.8 g

12. When 25.0 g of N_2 reacts, the actual yield of NH_3 is 20.8 g. What is the percent yield of NH_3?
   A. 68.4% **B.** 20.8% **C.** 25.0% **D.** 82.5% **E.** 100%

13. In an experiment using the equation below, 36.0 g of NO is produced from 20.0 g of N_2. What is the percent yield of NO?
   $N_2(g) + O_2(g) \xrightarrow{\Delta} 2NO(g)$ (9.5)
   A. 42.9% **B.** 55.6% **C.** 71.4% **D.** 84.1% **E.** 93.4%

For questions 14 through 17, use the following equation and heat of reaction: (9.6)

$$C_2H_6O(l) + 3O_2(g) \longrightarrow 2CO_2(g) + 3H_2O(l) \qquad \Delta H = -1420 \text{ kJ}$$

14. The reaction is
    A. thermic   B. endothermic   C. exothermic   D. nonthermic   E. subthermic

15. How much heat, in kilojoules, is released when 0.500 mol of C_2H_6O reacts?
    A. 710 kJ   B. 1420 kJ   C. 2130 kJ   D. 2840 kJ   E. 4260 kJ

16. How many grams of CO_2 are produced if 355 kJ of heat are released?
    A. 11.0 g   B. 22.0 g   C. 33.0 g   D. 44.0 g   E. 88.0 g

17. How much heat, in kilojoules, is released when 16.0 g of O_2 reacts?
    A. 4260 kJ   B. 1420 kJ   C. 710 kJ   D. 237 kJ   E. 118 kJ

18. Calculate the enthalpy change for the reaction

    $$2NOCl(g) \longrightarrow N_2(g) + O_2(g) + Cl_2(g)$$

    from the following equations:

    $$N_2(g) + O_2(g) \longrightarrow 2NO(g) \qquad \Delta H = +181 \text{ kJ}$$

    $$2NO(g) + Cl_2(g) \longrightarrow 2NOCl(g) \qquad \Delta H = -77 \text{ kJ}$$

    A. +154 kJ   B. −154 kJ   C. +104 kJ   D. −104   E. −358 kJ

Answers to the Practice Test

1. D	2. B	3. C	4. D	5. E
6. B	7. D	8. B	9. B	10. D
11. D	12. A	13. D	14. C	15. A
16. B	17. D	18. D		

Selected Answers and Solutions to Text Problems

9.1 **a.** Reactants: 2 mol of SO_2 and 1 mol of O_2 = 2 mol $(64.07\ g/mol)$ + 1 mol $(32.00\ g/mol)$
= 128.14 g + 32.00 g = 160.14 g of reactants
Products: 2 mol of SO_3 = 2 mol $(80.07\ g/mol)$ = 160.14 g of products

b. Reactants: 4 mol of P and 5 mol of O_2 = 4 mol $(30.97\ g/mol)$ + 5 mol $(32.00\ g/mol)$
= 123.88 g + 160.00 g = 283.88 g of reactants
Products: 2 mol of P_2O_5 = 2 mol $(141.94\ g/mol)$ = 283.88 g of products

9.3 **a.** $\dfrac{2\ mol\ SO_2}{1\ mol\ O_2}$ and $\dfrac{1\ mol\ O_2}{2\ mol\ SO_2}$; $\dfrac{2\ mol\ SO_2}{2\ mol\ SO_3}$ and $\dfrac{2\ mol\ SO_3}{2\ mol\ SO_2}$; $\dfrac{1\ mol\ O_2}{2\ mol\ SO_3}$ and $\dfrac{2\ mol\ SO_3}{1\ mol\ O_2}$

b. $\dfrac{4\ mol\ P}{5\ mol\ O_2}$ and $\dfrac{5\ mol\ O_2}{4\ mol\ P}$; $\dfrac{4\ mol\ P}{2\ mol\ P_2O_5}$ and $\dfrac{2\ mol\ P_2O_5}{4\ mol\ P}$; $\dfrac{5\ mol\ O_2}{2\ mol\ P_2O_5}$ and $\dfrac{2\ mol\ P_2O_5}{5\ mol\ O_2}$

9.5 **a.** mol SO_2 $\times$ $\dfrac{2\ mol\ SO_3}{2\ mol\ SO_2}$ = mol of SO_3

b. mol P $\times$ $\dfrac{5\ mol\ O_2}{4\ mol\ P}$ = mol of O_2

9.7 **a.** 2.6 mol H_2 $\times$ $\dfrac{1\ mol\ O_2}{2\ mol\ H_2}$ = 1.3 mol of O_2 (2 SFs)

b. 5.0 mol O_2 $\times$ $\dfrac{2\ mol\ H_2}{1\ mol\ O_2}$ = 10. mol of H_2 (2 SFs)

c. 2.5 mol O_2 $\times$ $\dfrac{2\ mol\ H_2O}{1\ mol\ O_2}$ = 5.0 mol of H_2O (2 SFs)

9.9 **a.** 0.500 mol SO_2 $\times$ $\dfrac{5\ mol\ C}{2\ mol\ SO_2}$ = 1.25 mol of C (3 SFs)

b. 1.2 mol C $\times$ $\dfrac{4\ mol\ CO}{5\ mol\ C}$ = 0.96 mol of CO (2 SFs)

c. 0.50 mol CS_2 $\times$ $\dfrac{2\ mol\ SO_2}{1\ mol\ CS_2}$ = 1.0 mol of SO_2 (2 SFs)

d. 2.5 mol C $\times$ $\dfrac{1\ mol\ CS_2}{5\ mol\ C}$ = 0.50 mol of CS_2 (2 SFs)

9.11 **a.** 57.5 g Na $\times$ $\dfrac{1\ mol\ Na}{22.99\ g\ Na}$ $\times$ $\dfrac{2\ mol\ Na_2O}{4\ mol\ Na}$ $\times$ $\dfrac{61.98\ g\ Na_2O}{1\ mol\ Na_2O}$ = 77.5 g of Na_2O (3 SFs)

b. 18.0 g Na $\times$ $\dfrac{1\ mol\ Na}{22.99\ g\ Na}$ $\times$ $\dfrac{1\ mol\ O_2}{4\ mol\ Na}$ $\times$ $\dfrac{32.00\ g\ O_2}{1\ mol\ O_2}$ = 6.26 g of O_2 (3 SFs)

c. 75.0 g Na_2O $\times$ $\dfrac{1\ mol\ Na_2O}{61.98\ g\ Na_2O}$ $\times$ $\dfrac{1\ mol\ O_2}{2\ mol\ Na_2O}$ $\times$ $\dfrac{32.00\ g\ O_2}{1\ mol\ O_2}$ = 19.4 g of O_2 (3 SFs)

9.13 **a.** 13.6 g NH_3 $\times$ $\dfrac{1\ mol\ NH_3}{17.03\ g\ NH_3}$ $\times$ $\dfrac{3\ mol\ O_2}{4\ mol\ NH_3}$ $\times$ $\dfrac{32.00\ g\ O_2}{1\ mol\ O_2}$ = 19.2 g of O_2 (3 SFs)

b. 6.50 g O_2 $\times$ $\dfrac{1\ mol\ O_2}{32.00\ g\ O_2}$ $\times$ $\dfrac{2\ mol\ N_2}{3\ mol\ O_2}$ $\times$ $\dfrac{28.02\ g\ N_2}{1\ mol\ N_2}$ = 3.79 g of N_2 (3 SFs)

c. 34.0 g NH_3 $\times$ $\dfrac{1\ mol\ NH_3}{17.03\ g\ NH_3}$ $\times$ $\dfrac{6\ mol\ H_2O}{4\ mol\ NH_3}$ $\times$ $\dfrac{18.02\ g\ H_2O}{1\ mol\ H_2O}$ = 54.0 g of H_2O (3 SFs)

9.15 a. $28.0 \text{ g NO}_2 \times \dfrac{1 \text{ mol NO}_2}{46.01 \text{ g NO}_2} \times \dfrac{1 \text{ mol H}_2\text{O}}{3 \text{ mol NO}_2} \times \dfrac{18.02 \text{ g H}_2\text{O}}{1 \text{ mol H}_2\text{O}} = 3.66 \text{ g of H}_2\text{O (3 SFs)}$

b. $15.8 \text{ g H}_2\text{O} \times \dfrac{1 \text{ mol H}_2\text{O}}{18.02 \text{ g H}_2\text{O}} \times \dfrac{1 \text{ mol NO}}{1 \text{ mol H}_2\text{O}} \times \dfrac{30.01 \text{ g NO}}{1 \text{ mol NO}} = 26.3 \text{ g of NO (3 SFs)}$

c. $8.25 \text{ g NO}_2 \times \dfrac{1 \text{ mol NO}_2}{46.01 \text{ g NO}_2} \times \dfrac{2 \text{ mol HNO}_3}{3 \text{ mol NO}_2} \times \dfrac{63.02 \text{ g HNO}_3}{1 \text{ mol HNO}_3} = 7.53 \text{ g of HNO}_3 \text{ (3 SFs)}$

9.17 a. $2\text{PbS}(s) + 3\text{O}_2(g) \longrightarrow 2\text{PbO}(s) + 2\text{SO}_2(g)$

b. $29.9 \text{ g PbS} \times \dfrac{1 \text{ mol PbS}}{239.3 \text{ g PbS}} \times \dfrac{3 \text{ mol O}_2}{2 \text{ mol PbS}} \times \dfrac{32.00 \text{ g O}_2}{1 \text{ mol O}_2} = 6.00 \text{ g of O}_2 \text{ (3 SFs)}$

c. $65.0 \text{ g PbS} \times \dfrac{1 \text{ mol PbS}}{239.3 \text{ g PbS}} \times \dfrac{2 \text{ mol SO}_2}{2 \text{ mol PbS}} \times \dfrac{64.07 \text{ g SO}_2}{1 \text{ mol SO}_2} = 17.4 \text{ g of SO}_2 \text{ (3 SFs)}$

d. $128 \text{ g PbO} \times \dfrac{1 \text{ mol PbO}}{223.2 \text{ g PbO}} \times \dfrac{2 \text{ mol PbS}}{2 \text{ mol PbO}} \times \dfrac{239.3 \text{ g PbS}}{1 \text{ mol PbS}} = 137 \text{ g of PbS (3 SFs)}$

9.19 a. The limiting factor is the number of drivers: with only eight drivers available, only eight taxis can be used to pick up passengers.

b. The limiting factor is the number of taxis: only seven taxis are in working condition to be driven.

9.21 a. $3.0 \text{ mol N}_2 \times \dfrac{2 \text{ mol NH}_3}{1 \text{ mol N}_2} = 6.0 \text{ mol of NH}_3$

$5.0 \text{ mol H}_2 \times \dfrac{2 \text{ mol NH}_3}{3 \text{ mol H}_2} = 3.3 \text{ mol of NH}_3 \text{ (smaller amount of product)}$

The limiting reactant is 5.0 mol of H_2. (2 SFs)

b. $8.0 \text{ mol N}_2 \times \dfrac{2 \text{ mol NH}_3}{1 \text{ mol N}_2} = 16 \text{ mol of NH}_3$

$4.0 \text{ mol H}_2 \times \dfrac{2 \text{ mol NH}_3}{3 \text{ mol H}_2} = 2.7 \text{ mol of NH}_3 \text{ (smaller amount of product)}$

The limiting reactant is 4.0 mol of H_2. (2 SFs)

c. $3.0 \text{ mol N}_2 \times \dfrac{2 \text{ mol NH}_3}{1 \text{ mol N}_2} = 6.0 \text{ mol of NH}_3 \text{ (smaller amount of product)}$

$12.0 \text{ mol H}_2 \times \dfrac{2 \text{ mol NH}_3}{3 \text{ mol H}_2} = 8.0 \text{ mol of NH}_3$

The limiting reactant is 3.0 mol of N_2. (2 SFs)

9.23 a. $20.0 \text{ g Al} \times \dfrac{1 \text{ mol Al}}{26.98 \text{ g Al}} \times \dfrac{2 \text{ mol AlCl}_3}{2 \text{ mol Al}} \times \dfrac{133.33 \text{ g AlCl}_3}{1 \text{ mol AlCl}_3} = 98.8 \text{ g of AlCl}_3$

$20.0 \text{ g Cl}_2 \times \dfrac{1 \text{ mol Cl}_2}{70.90 \text{ g Cl}_2} \times \dfrac{2 \text{ mol AlCl}_3}{3 \text{ mol Cl}_2} \times \dfrac{133.33 \text{ g AlCl}_3}{1 \text{ mol AlCl}_3} = 25.1 \text{ g of AlCl}_3$ (smaller amount of product)

∴ Cl_2 is the limiting reactant, and 25.1 g of $AlCl_3$ (3 SFs) can be produced.

b. $20.0 \text{ g NH}_3 \times \dfrac{1 \text{ mol NH}_3}{17.03 \text{ g NH}_3} \times \dfrac{6 \text{ mol H}_2\text{O}}{4 \text{ mol NH}_3} \times \dfrac{18.02 \text{ g H}_2\text{O}}{1 \text{ mol H}_2\text{O}} = 31.7 \text{ g of H}_2\text{O}$

$20.0 \text{ g O}_2 \times \dfrac{1 \text{ mol O}_2}{32.00 \text{ g O}_2} \times \dfrac{6 \text{ mol H}_2\text{O}}{5 \text{ mol O}_2} \times \dfrac{18.02 \text{ g H}_2\text{O}}{1 \text{ mol H}_2\text{O}} = 13.5 \text{ g of H}_2\text{O}$
(smaller amount of product)

∴ O_2 is the limiting reactant, and 13.5 g of H_2O (3 SFs) can be produced.

c. $20.0 \text{ g CS}_2 \times \dfrac{1 \text{ mol CS}_2}{76.15 \text{ g CS}_2} \times \dfrac{2 \text{ mol SO}_2}{1 \text{ mol CS}_2} \times \dfrac{64.07 \text{ g SO}_2}{1 \text{ mol SO}_2} = 33.7 \text{ g of SO}_2$

$20.0 \text{ g O}_2 \times \dfrac{1 \text{ mol O}_2}{32.00 \text{ g O}_2} \times \dfrac{2 \text{ mol SO}_2}{3 \text{ mol O}_2} \times \dfrac{64.07 \text{ g SO}_2}{1 \text{ mol SO}_2} = 26.7 \text{ g of SO}_2$
(smaller amount of product)

∴ O_2 is the limiting reactant, and 26.7 g of SO_2 (3 SFs) can be produced.

9.25 a. $25.0 \text{ g SO}_2 \times \dfrac{1 \text{ mol SO}_2}{64.07 \text{ g SO}_2} \times \dfrac{2 \text{ mol SO}_3}{2 \text{ mol SO}_2} \times \dfrac{80.07 \text{ g SO}_3}{1 \text{ mol SO}_3} = 31.2 \text{ g of SO}_3$
(smaller amount of product)

$40.0 \text{ g O}_2 \times \dfrac{1 \text{ mol O}_2}{32.00 \text{ g O}_2} \times \dfrac{2 \text{ mol SO}_3}{1 \text{ mol O}_2} \times \dfrac{80.07 \text{ g SO}_3}{1 \text{ mol SO}_3} = 200. \text{ g of SO}_3$

∴ SO_2 is the limiting reactant, and 31.2 g of SO_3 (3 SFs) can be produced.

b. $25.0 \text{ g Fe} \times \dfrac{1 \text{ mol Fe}}{55.85 \text{ g Fe}} \times \dfrac{1 \text{ mol Fe}_3\text{O}_4}{3 \text{ mol Fe}} \times \dfrac{231.6 \text{ g Fe}_3\text{O}_4}{1 \text{ mol Fe}_3\text{O}_4} = 34.6 \text{ g of Fe}_3\text{O}_4$
(smaller amount of product)

$40.0 \text{ g H}_2\text{O} \times \dfrac{1 \text{ mol H}_2\text{O}}{18.02 \text{ g H}_2\text{O}} \times \dfrac{1 \text{ mol Fe}_3\text{O}_4}{4 \text{ mol H}_2\text{O}} \times \dfrac{231.6 \text{ g Fe}_3\text{O}_4}{1 \text{ mol Fe}_3\text{O}_4} = 129 \text{ g of Fe}_3\text{O}_4$

∴ Fe is the limiting reactant, and 34.6 g of Fe_3O_4 (3 SFs) can be produced.

c. $25.0 \text{ g C}_7\text{H}_{16} \times \dfrac{1 \text{ mol C}_7\text{H}_{16}}{100.20 \text{ g C}_7\text{H}_{16}} \times \dfrac{7 \text{ mol CO}_2}{1 \text{ mol C}_7\text{H}_{16}} \times \dfrac{44.01 \text{ g CO}_2}{1 \text{ mol CO}_2} = 76.9 \text{ g of CO}_2$

$40.0 \text{ g O}_2 \times \dfrac{1 \text{ mol O}_2}{32.00 \text{ g O}_2} \times \dfrac{7 \text{ mol CO}_2}{11 \text{ mol O}_2} \times \dfrac{44.01 \text{ g CO}_2}{1 \text{ mol CO}_2} = 35.0 \text{ g of CO}_2$
(smaller amount of product)

∴ O_2 is the limiting reactant, and 35.0 g of CO_2 (3 SFs) can be produced.

9.27 a. Theoretical yield of CS_2:

$40.0 \text{ g C} \times \dfrac{1 \text{ mol C}}{12.01 \text{ g C}} \times \dfrac{1 \text{ mol CS}_2}{5 \text{ mol C}} \times \dfrac{76.15 \text{ g CS}_2}{1 \text{ mol CS}_2} = 50.7 \text{ g of CS}_2$

Percent yield: $\dfrac{36.0 \text{ g CS}_2 \text{ (actual)}}{50.7 \text{ g CS}_2 \text{ (theoretical)}} \times 100\% = 71.0\%$ (3 SFs)

b. Theoretical yield of CS_2:

$32.0 \text{ g SO}_2 \times \dfrac{1 \text{ mol SO}_2}{64.07 \text{ g SO}_2} \times \dfrac{1 \text{ mol CS}_2}{2 \text{ mol SO}_2} \times \dfrac{76.15 \text{ g CS}_2}{1 \text{ mol CS}_2} = 19.0 \text{ g of CS}_2$

Percent yield: $\dfrac{12.0 \text{ g CS}_2 \text{ (actual)}}{19.0 \text{ g CS}_2 \text{ (theoretical)}} \times 100\% = 63.2\%$ (3 SFs)

9.29 Theoretical yield of Al_2O_3:

$$50.0 \text{ g Al} \times \frac{1 \text{ mol Al}}{26.98 \text{ g Al}} \times \frac{2 \text{ mol Al}_2O_3}{4 \text{ mol Al}} \times \frac{101.96 \text{ g Al}_2O_3}{1 \text{ mol Al}_2O_3} = 94.5 \text{ g of Al}_2O_3$$

Use the percent yield to convert theoretical to actual:

$$94.5 \text{ g Al}_2O_3 \times \frac{75.0 \text{ g Al}_2O_3}{100 \text{ g Al}_2O_3} = 70.9 \text{ g of Al}_2O_3 \text{ (actual) (3 SFs)}$$

9.31 Theoretical yield of CO_2:

$$30.0 \text{ g C} \times \frac{1 \text{ mol C}}{12.01 \text{ g C}} \times \frac{2 \text{ mol CO}}{3 \text{ mol C}} \times \frac{28.01 \text{ g CO}}{1 \text{ mol CO}} = 46.6 \text{ g of CO}$$

Percent yield: $\dfrac{28.2 \text{ g CO (actual)}}{46.6 \text{ g CO (theoretical)}} \times 100\% = 60.5\% \text{ (3 SFs)}$

9.33 In exothermic reactions, the energy of the products is lower than that of the reactants.

9.35 **a.** An exothermic reaction releases energy.
b. An endothermic reaction has a higher energy level for the products than the reactants.
c. Metabolism is an exothermic reaction, providing energy for the body.

9.37 **a.** Heat is released, which makes the reaction exothermic with $\Delta H = -802 \text{ kJ}$.
b. Heat is absorbed, which makes the reaction endothermic with $\Delta H = +65.3 \text{ kJ}$.
c. Heat is released, which makes the reaction exothermic with $\Delta H = -850 \text{ kJ}$.

9.39 **a.** $125 \text{ g Cl}_2 \times \dfrac{1 \text{ mol Cl}_2}{70.90 \text{ g Cl}_2} \times \dfrac{-657 \text{ kJ}}{2 \text{ mol Cl}_2} = -579 \text{ kJ (3 SFs)}$

∴ 579 kJ are released.

b. $278 \text{ g PCl}_5 \times \dfrac{1 \text{ mol PCl}_5}{208.2 \text{ g PCl}_5} \times \dfrac{+67 \text{ kJ}}{1 \text{ mol PCl}_5} = +89 \text{ kJ (2 SFs)}$

∴ 89 kJ are absorbed.

9.41 We see that the reactants and products are contained in Equation 1 and Equation 2. By combining, rearranging, and multiplying the substances, including the ΔH, in these two equations, we can obtain the equation we need. Using Hess's law, the sum of the ΔH values involved in each will give us the ΔH value for the reaction.

same	$N_2(g) + O_2(g) \longrightarrow 2NO(g)$	$\Delta H = +180 \text{ kJ}$
reversed	$2NO(g) + O_2(g) \longrightarrow 2NO_2(g)$	$\Delta H = -112 \text{ kJ}$
	$N_2(g) + 2O_2(g) \longrightarrow 2NO_2(g)$	$\Delta H = +68 \text{ kJ}$

9.43 We see that the reactants and products are contained in Equation 1 and Equation 2. By combining, rearranging, and multiplying the substances, including the ΔH, in these two equations, we can obtain the equation we need. Using Hess's law, the sum of the ΔH values involved in each will give us the ΔH value for the reaction.

same	$S(s) + \frac{3}{2}O_2(g) \longrightarrow SO_3(g)$	$\Delta H = -396 \text{ kJ}$
reversed	$SO_3(g) \longrightarrow SO_2(g) + \frac{1}{2}O_2(g)$	$\Delta H = +90 \text{ kJ}$
	$S(s) + O_2(g) \longrightarrow SO_2(g)$	$\Delta H = -306 \text{ kJ}$

9.45 a. Molar mass of $C_{11}H_7O_2Cl$

$= 11(12.01 \text{ g}) + 7(1.008 \text{ g}) + 2(16.00 \text{ g}) + 35.45 \text{ g} = 206.6 \text{ g}$ (4 SFs)

Molar mass of $C_{10}H_8O$

$= 10(12.01 \text{ g}) + 8(1.008 \text{ g}) + 16.00 \text{ g} = 144.2 \text{ g}$ (4 SFs)

$2.2 \times 10^2 \text{ kg } C_{10}H_8O \times \dfrac{1000 \text{ g } C_{10}H_8O}{1 \text{ kg } C_{10}H_8O} \times \dfrac{1 \text{ mol } C_{10}H_8O}{144.2 \text{ g } C_{10}H_8O} \times \dfrac{1 \text{ mol } C_{11}H_7O_2Cl}{1 \text{ mol } C_{10}H_8O}$

$\times \dfrac{206.6 \text{ g } C_{11}H_7O_2Cl}{1 \text{ mol } C_{11}H_7O_2Cl} \times \dfrac{1 \text{ kg } C_{11}H_7O_2Cl}{1000 \text{ g } C_{11}H_7O_2Cl} = 3.2 \times 10^2 \text{ kg of } C_{11}H_7O_2Cl$ (2 SFs)

b. Molar mass of $COCl_2$

$= 12.01 \text{ g} + 16.00 \text{ g} + 2(35.45 \text{ g}) = 98.91 \text{ g}$ (4 SFs)

$100. \text{ g } C_{10}H_8O \times \dfrac{1 \text{ mol } C_{10}H_8O}{144.2 \text{ g } C_{10}H_8O} \times \dfrac{1 \text{ mol } C_{11}H_7O_2Cl}{1 \text{ mol } C_{10}H_8O} \times \dfrac{206.6 \text{ g } C_{11}H_7O_2Cl}{1 \text{ mol } C_{11}H_7O_2Cl}$

$= 143 \text{ g of } C_{11}H_7O_2Cl$ (smaller amount of product)

$100. \text{ g } COCl_2 \times \dfrac{1 \text{ mol } COCl_2}{98.91 \text{ g } COCl_2} \times \dfrac{1 \text{ mol } C_{11}H_7O_2Cl}{1 \text{ mol } COCl_2} \times \dfrac{206.6 \text{ g } C_{11}H_7O_2Cl}{1 \text{ mol } C_{11}H_7O_2Cl}$

$= 209 \text{ g of } C_{11}H_7O_2Cl$

$\therefore C_{10}H_8O$ is the limiting reactant and 143 g of $C_{11}H_7O_2Cl$ (3 SFs) can be produced.

c. Percent yield: $\dfrac{115 \text{ g } C_{11}H_7O_2Cl \text{ (actual)}}{143 \text{ g } C_{11}H_7O_2Cl \text{ (theoretical)}} \times 100\% = 80.4\%$ (3 SFs)

9.47 a. $2NO(g) + O_2(g) \longrightarrow 2NO_2(g)$

b. NO is the limiting reactant.

9.49 a. $N_2(g) + 3H_2(g) \longrightarrow 2NH_3(g)$

b. diagram A; N_2 is the excess reactant.

9.51 a. $2NI_3(s) \longrightarrow N_2(g) + 3I_2(g)$

b. Theoretical yield: from the $6NI_3$ we could obtain

$6NI_3 \times \dfrac{1N_2}{2NI_3} = 3N_2$ $6NI_3 \times \dfrac{3I_2}{2NI_3} = 9I_2$

Actual yield: in the actual products, we obtain $2N_2$ and $6I_2$.

Percent yield: $\dfrac{2N_2 \text{ and } 6I_2 \text{ (actual)}}{3N_2 \text{ and } 9I_2 \text{ (theoretical)}} \times 100\%$

$= \dfrac{2 (N_2 \text{ and } 3I_2) \text{ (actual)}}{3 (N_2 \text{ and } 3I_2) \text{ (theoretical)}} \times 100\% = 67\%$

9.53

FeS	O₂	FeO	SO₂
2.0 mol	3.0 mol	2.0 mol	2.0 mol
4.6 mol	6.9 mol	4.6 mol	4.6 mol

9.55 a. $4.00 \text{ mol HF} \times \dfrac{2 \text{ mol NH}_3}{6 \text{ mol HF}} = 1.33 \text{ mol of NH}_3$ (3 SFs)

$4.00 \text{ mol HF} \times \dfrac{5 \text{ mol F}_2}{6 \text{ mol HF}} = 3.33 \text{ mol of F}_2$ (3 SFs)

b. $25.5 \text{ g } NH_3 \times \dfrac{1 \text{ mol } NH_3}{17.03 \text{ g } NH_3} \times \dfrac{5 \text{ mol } F_2}{2 \text{ mol } NH_3} \times \dfrac{38.00 \text{ g } F_2}{1 \text{ mol } F_2} = 142 \text{ g of } F_2 \text{ (3 SFs)}$

c. $3.40 \text{ g } NH_3 \times \dfrac{1 \text{ mol } NH_3}{17.03 \text{ g } NH_3} \times \dfrac{1 \text{ mol } N_2F_4}{2 \text{ mol } NH_3} \times \dfrac{104.02 \text{ g } N_2F_4}{1 \text{ mol } N_2F_4} = 10.4 \text{ g of } N_2F_4 \text{ (3 SFs)}$

9.57 a. $3.00 \text{ mol } H_2O \times \dfrac{2 \text{ mol } H_2O_2}{2 \text{ mol } H_2O} = 3.00 \text{ mol of } H_2O_2 \text{ (3 SFs)}$

b. $36.5 \text{ g } O_2 \times \dfrac{1 \text{ mol } O_2}{32.00 \text{ g } O_2} \times \dfrac{2 \text{ mol } H_2O_2}{1 \text{ mol } O_2} \times \dfrac{34.02 \text{ g } H_2O_2}{1 \text{ mol } H_2O_2} = 77.6 \text{ g of } H_2O_2 \text{ (3 SFs)}$

c. $12.2 \text{ g } H_2O_2 \times \dfrac{1 \text{ mol } H_2O_2}{34.02 \text{ g } H_2O_2} \times \dfrac{2 \text{ mol } H_2O}{2 \text{ mol } H_2O_2} \times \dfrac{18.02 \text{ g } H_2O}{1 \text{ mol } H_2O} = 6.46 \text{ g of } H_2O \text{ (3 SFs)}$

9.59 $12.8 \text{ g } Na \times \dfrac{1 \text{ mol } Na}{22.99 \text{ g } Na} \times \dfrac{2 \text{ mol } NaCl}{2 \text{ mol } Na} \times \dfrac{58.44 \text{ g } NaCl}{1 \text{ mol } NaCl} = 32.5 \text{ g of } NaCl$

$10.2 \text{ g } Cl_2 \times \dfrac{1 \text{ mol } Cl_2}{70.90 \text{ g } Cl_2} \times \dfrac{2 \text{ mol } NaCl}{1 \text{ mol } Cl_2} \times \dfrac{58.44 \text{ g } NaCl}{1 \text{ mol } NaCl} = 16.8 \text{ g of } NaCl$

(smaller amount of product)

∴ Cl_2 is the limiting reactant, and 16.8 g of NaCl (3 SFs) can be produced.

9.61 a. $4.00 \text{ mol } H_2O \times \dfrac{1 \text{ mol } C_5H_{12}}{6 \text{ mol } H_2O} = 0.667 \text{ mol of } C_5H_{12} \text{ (3 SFs)}$

b. $32.0 \text{ g } O_2 \times \dfrac{1 \text{ mol } O_2}{32.00 \text{ g } O_2} \times \dfrac{5 \text{ mol } CO_2}{8 \text{ mol } O_2} \times \dfrac{44.01 \text{ g } CO_2}{1 \text{ mol } CO_2} = 27.5 \text{ g of } CO_2 \text{ (3 SFs)}$

c. $44.5 \text{ g } C_5H_{12} \times \dfrac{1 \text{ mol } C_5H_{12}}{72.15 \text{ g } C_5H_{12}} \times \dfrac{5 \text{ mol } CO_2}{1 \text{ mol } C_5H_{12}} \times \dfrac{44.01 \text{ g } CO_2}{1 \text{ mol } CO_2} = 136 \text{ g of } CO_2$

$108 \text{ g } O_2 \times \dfrac{1 \text{ mol } O_2}{32.00 \text{ g } O_2} \times \dfrac{5 \text{ mol } CO_2}{8 \text{ mol } O_2} \times \dfrac{44.01 \text{ g } CO_2}{1 \text{ mol } CO_2} = 92.8 \text{ g of } CO_2$

(smaller amount of product)

∴ O_2 is the limiting reactant, and 92.8 g of CO_2 (3 SFs) can be produced.

9.63 a. Theoretical yield of CO_2:

$22.0 \text{ g } C_2H_2 \times \dfrac{1 \text{ mol } C_2H_2}{26.04 \text{ g } C_2H_2} \times \dfrac{4 \text{ mol } CO_2}{2 \text{ mol } C_2H_2} \times \dfrac{44.01 \text{ g } CO_2}{1 \text{ mol } CO_2} = 74.4 \text{ g of } CO_2 \text{ (3 SFs)}$

b. Percent yield: $\dfrac{64.0 \text{ g } CO_2 \text{ (actual)}}{74.4 \text{ g } CO_2 \text{ (theoretical)}} \times 100\% = 86.0\% \text{ (3 SFs)}$

9.65 Theoretical yield of C_2H_6:

$28.0 \text{ g } C_2H_2 \times \dfrac{1 \text{ mol } C_2H_2}{26.04 \text{ g } C_2H_2} \times \dfrac{1 \text{ mol } C_2H_6}{1 \text{ mol } C_2H_2} \times \dfrac{30.07 \text{ g } C_2H_6}{1 \text{ mol } C_2H_6} = 32.3 \text{ g of } C_2H_6$

Percent yield: $\dfrac{24.5 \text{ g } C_2H_6 \text{ (actual)}}{32.3 \text{ g } C_2H_6 \text{ (theoretical)}} \times 100\% = 75.9\% \text{ (3 SFs)}$

9.67 a. Use the limiting reactant to calculate the theoretical yield of NH_3:

$$50.0 \text{ g } N_2 \times \frac{1 \text{ mol } N_2}{28.02 \text{ g } N_2} \times \frac{2 \text{ mol } NH_3}{1 \text{ mol } N_2} \times \frac{17.03 \text{ g } NH_3}{1 \text{ mol } NH_3} = 60.8 \text{ g of } NH_3$$
(smaller amount of product)

$$20.0 \text{ g } H_2 \times \frac{1 \text{ mol } H_2}{2.016 \text{ g } H_2} \times \frac{2 \text{ mol } NH_3}{3 \text{ mol } H_2} \times \frac{17.03 \text{ g } NH_3}{1 \text{ mol } NH_3} = 113 \text{ g of } NH_3$$

$\therefore N_2$ is the limiting reactant, and 60.8 g of NH_3 (3 SFs) can be produced (theoretical yield).

b. Use the percent yield to calculate the actual yield of NH_3:

$$60.8 \text{ g } NH_3 \text{ (theoretical)} \times \frac{62.0 \text{ g } NH_3 \text{ (actual)}}{100 \text{ g } NH_3 \text{ (theoretical)}} = 37.7 \text{ g of } NH_3 \text{ (3 SFs)}$$

9.69 a. $3.00 \text{ g } NO \times \dfrac{1 \text{ mol } NO}{30.01 \text{ g } NO} \times \dfrac{+90.2 \text{ kJ}}{2 \text{ mol } NO} = +4.51 \text{ kJ (3 SFs)}$

$\therefore$ 4.51 kJ are required.

b. The equation for the decomposition of NO will be the reverse of the equation for the formation of NO.

$$2NO(g) \longrightarrow N_2(g) + O_2(g) + 90.2 \text{ kJ}$$

c. $5.00 \text{ g } NO \times \dfrac{1 \text{ mol } NO}{30.01 \text{ g } NO} \times \dfrac{-90.2 \text{ kJ}}{2 \text{ mol } NO} = -7.51 \text{ kJ (3 SFs)}$

$\therefore$ 7.51 kJ are released.

9.71 a. Heat is released, which makes the reaction exothermic.
b. Heat is absorbed, which makes the reaction endothermic.

9.73 a. $4.50 \text{ mol } Cr \times \dfrac{3 \text{ mol } O_2}{4 \text{ mol } Cr} = 3.38 \text{ mol of } O_2 \text{ (3 SFs)}$

b. $24.8 \text{ g } Cr \times \dfrac{1 \text{ mol } Cr}{52.00 \text{ g } Cr} \times \dfrac{2 \text{ mol } Cr_2O_3}{4 \text{ mol } Cr} \times \dfrac{152.0 \text{ g } Cr_2O_3}{1 \text{ mol } Cr_2O_3} = 36.2 \text{ g of } Cr_2O_3 \text{ (3 SFs)}$

c. $26.0 \text{ g } Cr \times \dfrac{1 \text{ mol } Cr}{52.00 \text{ g } Cr} \times \dfrac{2 \text{ mol } Cr_2O_3}{4 \text{ mol } Cr} \times \dfrac{152.0 \text{ g } Cr_2O_3}{1 \text{ mol } Cr_2O_3} = 38.0 \text{ g of } Cr_2O_3$

$$8.00 \text{ g } O_2 \times \frac{1 \text{ mol } O_2}{32.00 \text{ g } O_2} \times \frac{2 \text{ mol } Cr_2O_3}{3 \text{ mol } O_2} \times \frac{152.0 \text{ g } Cr_2O_3}{1 \text{ mol } Cr_2O_3} = 25.3 \text{ g of } Cr_2O_3$$
(smaller amount of product)

$\therefore O_2$ is the limiting reactant, and 25.3 g of Cr_2O_3 (3 SFs) can be produced.

d. $74.0 \text{ g } Cr \times \dfrac{1 \text{ mol } Cr}{52.00 \text{ g } Cr} \times \dfrac{2 \text{ mol } Cr_2O_3}{4 \text{ mol } Cr} \times \dfrac{152.0 \text{ g } Cr_2O_3}{1 \text{ mol } Cr_2O_3} = 108 \text{ g of } Cr_2O_3$
(smaller amount of product)

$$62.0 \text{ g } O_2 \times \frac{1 \text{ mol } O_2}{32.00 \text{ g } O_2} \times \frac{2 \text{ mol } Cr_2O_3}{3 \text{ mol } O_2} \times \frac{152.0 \text{ g } Cr_2O_3}{1 \text{ mol } Cr_2O_3} = 196 \text{ g of } Cr_2O_3$$

$\therefore$ Cr is the limiting reactant, and 108 g of Cr_2O_3 (3 SFs) can be produced (theoretical yield).

Percent yield: $\dfrac{87.3 \text{ g } Cr_2O_3 \text{ (actual)}}{108 \text{ g } Cr_2O_3 \text{ (theoretical)}} \times 100\% = 80.8\% \text{ (3 SFs)}$

9.75 **a.** $0.225 \text{ mol } C_3H_4 \times \dfrac{4 \text{ mol } O_2}{1 \text{ mol } C_3H_4} = 0.900 \text{ mol of } O_2 \text{ (3 SFs)}$

b. $64.0 \text{ g } O_2 \times \dfrac{1 \text{ mol } O_2}{32.00 \text{ g } O_2} \times \dfrac{2 \text{ mol } H_2O}{4 \text{ mol } O_2} \times \dfrac{18.02 \text{ g } H_2O}{1 \text{ mol } H_2O} = 18.0 \text{ g of } H_2O \text{ (3 SFs)}$

c. $78.0 \text{ g } C_3H_4 \times \dfrac{1 \text{ mol } C_3H_4}{40.06 \text{ g } C_3H_4} \times \dfrac{3 \text{ mol } CO_2}{1 \text{ mol } C_3H_4} \times \dfrac{44.01 \text{ g } CO_2}{1 \text{ mol } CO_2} = 257 \text{ g of } CO_2 \text{ (3 SFs)}$

d. Percent yield: $\dfrac{186 \text{ g } CO_2 \text{ (actual)}}{257 \text{ g } CO_2 \text{ (theoretical)}} \times 100\% = 72.4\% \text{ (3 SFs)}$

9.77

FeS	O₂	FeO	SO₂
26 g	14 g	21 g	19 g
10.9 g	5.95 g	8.90 g	7.94 g

9.79 **a.** The heat of reaction (ΔH) is positive, so the reaction is endothermic.

b. $1.5 \text{ mol } SO_3 \times \dfrac{+790 \text{ kJ}}{2 \text{ mol } SO_3} = +590 \text{ kJ (2 SFs)}$

∴ 590 kJ are required.

c. $150 \text{ g } O_2 \times \dfrac{1 \text{ mol } O_2}{32.00 \text{ g } O_2} \times \dfrac{+790 \text{ kJ}}{3 \text{ mol } O_2} = +1200 \text{ kJ (2 SFs)}$

∴ 1200 kJ are required.

9.81 We see that the reactants and products are contained in Equation 1, Equation 2, and Equation 3. By combining, rearranging, and multiplying the substances, including the ΔH, in these three equations, we can obtain the equation we need. Using Hess's law, the sum of the ΔH values involved in each will give us the ΔH value for the reaction.

$$\text{same} \qquad \tfrac{1}{2}H_2(g) + \tfrac{1}{2}Cl_2(g) \longrightarrow HCl(g) \qquad\qquad \Delta H = -92 \text{ kJ}$$

$$\text{reversed} \div 2 \quad NH_4Cl(s) \longrightarrow \tfrac{1}{2}N_2(g) + 2H_2(g) + \tfrac{1}{2}Cl_2(g) \quad \Delta H = \tfrac{1}{2}(+631 \text{ kJ})$$

$$\text{same} \div 2 \qquad \tfrac{1}{2}N_2(g) + \tfrac{3}{2}H_2(g) \longrightarrow NH_3(g) \qquad\qquad \Delta H = \tfrac{1}{2}(-296 \text{ kJ})$$

$$NH_4Cl(s) \longrightarrow NH_3(g) + HCl(g) \qquad\qquad \Delta H = +76 \text{ kJ}$$

10

Bonding and Properties of Solids and Liquids

Credit: Jose Luis Pelaez Inc/Alamy

Lisa, a histologist, prepares thin tissue samples and mounts them on slides for review by a pathologist. She then adds dyes that stain the structures in the tissue to make them visible. Recently, a dermatologist excised a lesion from the calf of a 37-year-old woman. When Lisa received the sample, she froze it in liquid nitrogen, which has a heat of vaporization of 198 J/g at the boiling point of nitrogen, which is −196 °C. Then she cut thin slices of the frozen tissue and mounted them on slides. The pathologist reported that the lesion contained a variety of abnormal growth patterns that led to a diagnosis of Rosai–Dorfman disease. Because this disease is benign, there is minimal risk that it will progress and will instead probably regress. How many joules are needed to convert 25 g of liquid N_2 to vapor at −196 °C?

CHAPTER READINESS*

✿ Key Math Skills

- Using Positive and Negative Numbers in Calculations (1.4)
- Solving Equations (1.4)

⚗ Core Chemistry Skills

- Using the Heat Equation (3.5)
- Writing Electron Configurations (5.4)
- Drawing Lewis Symbols (5.6)

*These Key Math Skills and Core Chemistry Skills from previous chapters are listed here for your review as you proceed to the new material in this chapter.

LOOKING AHEAD

10.1 Lewis Structures for Molecules and Polyatomic Ions

Learning Goal: Draw the Lewis structures for molecular compounds or polyatomic ions with single and multiple bonds.

- In a covalent bond, atoms of nonmetals share electrons to achieve an octet (two for H).

- In a double bond, two pairs of electrons are shared between the same two atoms to complete octets.

- In a triple bond, three pairs of electrons are shared to complete octets.
- Bonding pairs of electrons are shared between atoms.
- Nonbonding electrons (lone pairs) are not shared between atoms.

♦ **Learning Exercise 10.1A**

Determine the total number of valence electrons for each of the following:

a. N_2O _____   **b.** HSCN _____   **c.** PH_4^+ _____   **d.** CBr_4 _____

e. ClO_4^- _____   **f.** BrCl _____   **g.** H_2CO _____   **h.** OF_2 _____

Answers   **a.** 16   **b.** 16   **c.** 8   **d.** 32   **e.** 32   **f.** 14   **g.** 12   **h.** 20

Guide to Drawing Lewis Structures	
STEP 1	Determine the arrangement of atoms.
STEP 2	Determine the total number of valence electrons.
STEP 3	Attach each bonded atom to the central atom with a pair of electrons.
STEP 4	Place the remaining electrons using single or multiple bonds to complete octets (two for H).

Example: Draw the Lewis structure for SCl_2.

Solution:

Analyze the Problem	Given	Need	Connect
	SCl_2	Lewis structure	total valence electrons

STEP 1   **Determine the arrangement of atoms.**

     Cl   S   Cl

STEP 2   **Determine the total number of valence electrons.**

     1 S atom $\times 6\,e^-$ $= 6\,e^-$
     2 Cl atoms $\times 7\,e^-$ $= \underline{14\,e^-}$
     Total valence electrons for $SCl_2 = 20\,e^-$

STEP 3   **Attach each bonded atom to the central atom with a pair of electrons.** Four electrons are used to form two single bonds to connect the S atom and each Cl atom. Each bonding pair can be represented by a bond line.

     Cl:S:Cl   or   Cl—S—Cl $20\,e^- - 4\,e^-$
                                  $= 16$ remaining valence electrons (8 pairs)

STEP 4   **Place the remaining electrons using single or multiple bonds to complete octets.** The remaining 16 electrons are placed as lone pairs around the S and Cl atoms to complete octets for all the atoms.

     :Cl:S:Cl: or   :Cl—S̈—Cl:

♦ **Learning Exercise 10.1B**

Draw the Lewis structure for each of the following:

a. H_2

b. NCl_3

c. HCl

d. Cl_2

e. H_2S

f. CBr_4

Answers

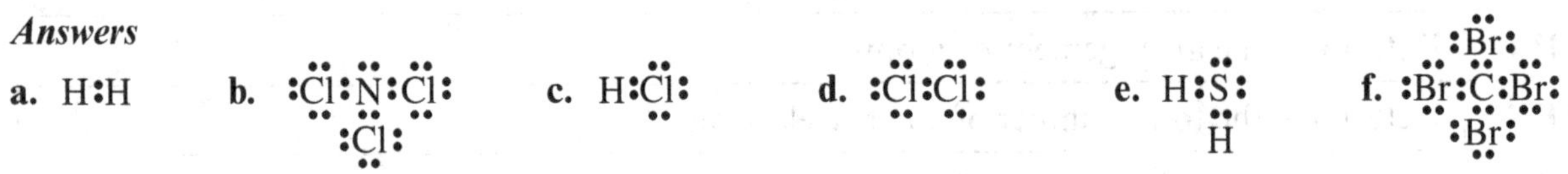

a. H:H

b. :C̈l:N̈:C̈l:
   ..C̈l:..

c. H:C̈l:

d. :C̈l:C̈l:

e. H:S̈:
      H

f. :B̈r:
   :B̈r:C:B̈r:
      :B̈r:

Study Note

A Lewis structure for the polyatomic ion ClO_2^- can be drawn as follows:

STEP 1 **Determine the arrangement of atoms.**

$$[O \quad Cl \quad O]^-$$

STEP 2 **Determine the total number of valence electrons.**

1 Cl atom $\times$ 7 e^-	$= 7\ e^-$
2 O atoms $\times$ 6 e^-	$= 12\ e^-$
Ionic charge (negative)	$= \underline{1\ e^-}$
Total valence electrons for ClO_2^-	$= 20\ e^-$

STEP 3 **Attach each bonded atom to the central atom with a pair of electrons.** Each bonding pair can be represented by a bond line.

$$[O:Cl:O]^- \quad or \quad [O{-}Cl{-}O]^-$$

STEP 4 **Place the remaining electrons using single or multiple bonds to complete octets.**

20 valence e^- $-$ 4 bonding e^- $=$ 16 remaining valence electrons (8 pairs)

Thus, 8 lone pairs of electrons are placed on the Cl and O atoms.

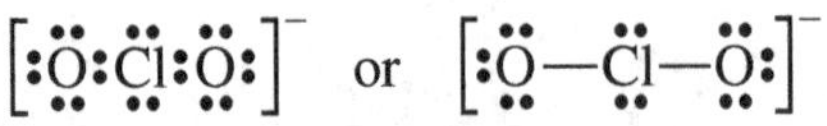

$$\left[:\ddot{O}:\ddot{C}l:\ddot{O}:\right]^- \quad or \quad \left[:\ddot{O}{-}\ddot{C}l{-}\ddot{O}:\right]^-$$

Study Note

A Lewis structure for CO_2 can be drawn with double bonds as follows:

STEP 1 **Determine the arrangement of atoms.**

O   C   O

STEP 2 **Determine the total number of valence electrons.**

$$1 \text{ C atom} \times 4 \, e^- \qquad = 4 \, e^-$$
$$2 \text{ O atoms} \times 6 \, e^- \qquad = \underline{12 \, e^-}$$
$$\text{Total valence electrons for } CO_2 = 16 \, e^-$$

STEP 3 **Attach each bonded atom to the central atom with a pair of electrons.** Each bonding pair can be represented by a bond line.

O—C—O $16 \, e^- - 4 \, e^- = 12$ remaining valence electrons (6 pairs)

STEP 4 **Place the remaining electrons using single or multiple bonds to complete octets.**

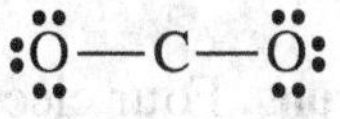

:Ö—C—Ö:

Because the octet for C is not complete, a lone pair from each O atom is shared with the central C atom to form double bonds.

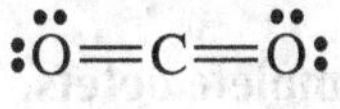

:Ö=C=Ö:

♦ **Learning Exercise 10.1C**

Draw the Lewis structures for each of the following molecules or ions:

a. CS_2

b. PH_4^+

c. H_2CCH_2

d. HONO

Answers   **a.** 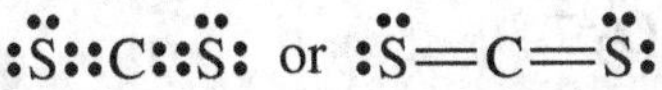 :S::C::S: or :S=C=S:

b. $\left[\text{H:}\overset{\text{H}}{\underset{\text{H}}{\text{P}}}\text{:H} \right]^+$ or $\left[\text{H—}\overset{\text{H}}{\underset{\text{H}}{\text{P}}}\text{—H} \right]^+$

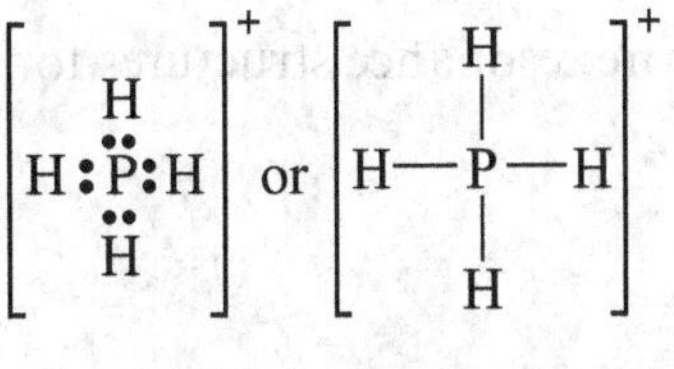

c. H:C::C:H or H—C=C—H (with H H above and H H below)

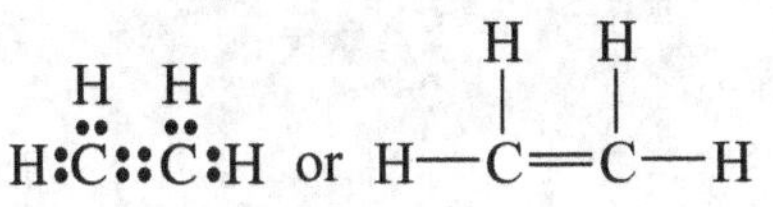

d. 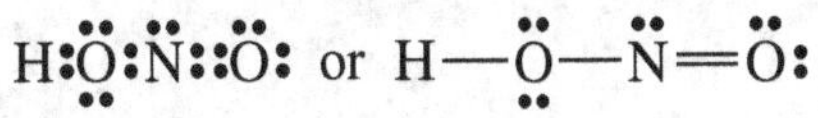 H:Ö:N::Ö: or H—Ö—N=Ö:

10.2   Resonance Structures

Learning Goal: Draw Lewis structures for molecules or polyatomic ions that have two or more resonance structures.

- When two or more Lewis structures can be drawn for the same compound, they are called *resonance structures.*

Study Note

When a molecule or polyatomic ion contains multiple bonds, it may be possible to draw two or more Lewis structures.

Example: Draw two resonance structures for selenium dioxide, SeO₂.

Solution:

Analyze the Problem	Given	Need	Connect
	SeO₂	Lewis structure	total valence electrons

STEP 1 **Determine the arrangement of atoms.** In SeO₂, the Se atom is the central atom.

O   Se   O

STEP 2 **Determine the total number of valence electrons.**

1 Se atom × 6 e⁻ = 6 e⁻
2 O atoms × 6 e⁻ = 12 e⁻
Total valence electrons for SeO₂ = 18 e⁻

STEP 3 **Attach each bonded atom to the central atom with a pair of electrons.** Four electrons are used to form two single bonds to connect the Se atom and each O atom.

O—Se—O

STEP 4 **Place the remaining electrons using single or multiple bonds to complete octets.** The remaining 14 electrons are drawn as lone pairs, which complete the octets of the O atoms but not the Se atom.

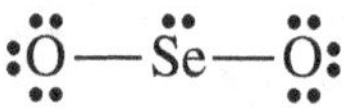

:Ö—Se—Ö:

To complete the octet for Se, a lone pair from one of the O atoms is shared with Se to form a double bond. Because the lone pair that is shared can come from either O atom, two resonance structures can be drawn with a double-headed arrow shown between them.

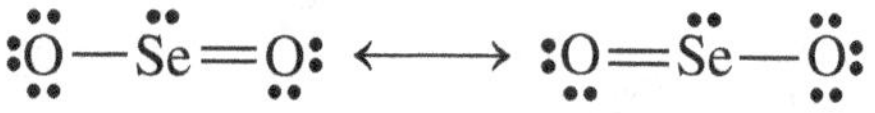

:Ö—Se=O: ⟷ :O=Se—Ö:

♦ **Learning Exercise 10.2**

Draw two or more resonance structures for each of the following:

CORE CHEMISTRY SKILL
Drawing Resonance Structures

a. SO₂

b. CO₃²⁻

c. NO₃⁻

Answers

a. :Ö—S̈=Ö: ⟷ :Ö=S̈—Ö:
b. [:Ö—C=Ö:]²⁻ ⟷ [:Ö—C̈—Ö:]²⁻ ⟷ [:Ö=C—Ö:]²⁻

c. [:Ö=N—Ö:]⁻ ⟷ [:Ö—N—Ö:]⁻ ⟷ [:Ö—N=Ö:]⁻

10.3   Shapes of Molecules and Polyatomic Ions (VSEPR Theory)

Learning Goal: Predict the three-dimensional structure of a molecule or a polyatomic ion.

- The VSEPR theory predicts the geometry of a molecule or polyatomic ion by placing the electron groups around a central atom as far apart as possible.
- One electron group may consist of the electrons in a lone pair, single, double, or triple bond.
- The atoms bonded to the central atom determine the three-dimensional geometry of a molecule or a polyatomic ion.
- A central atom with two electron groups has a linear geometry (180°); three electron groups give a trigonal planar geometry (120°); and four electron pairs give a tetrahedral geometry (109°).
- A linear molecule has a central atom bonded to two atoms and no lone pairs.

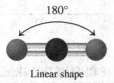

Linear shape

- A trigonal planar molecule has a central atom bonded to three atoms and no lone pairs. A bent molecule with a bond angle of 120° has a central atom bonded to two atoms and one lone pair.

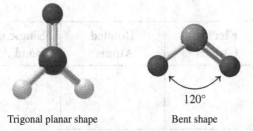

Trigonal planar shape   Bent shape

- A tetrahedral molecule has a central atom bonded to four atoms and no lone pairs. In a trigonal pyramidal molecule, a central atom is bonded to three atoms and one lone pair. In a bent molecule with a bond angle of 109°, a central atom is bonded to two atoms and two lone pairs.

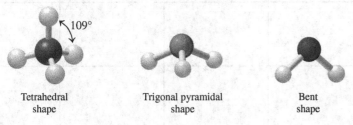

| Tetrahedral shape | Trigonal pyramidal shape | Bent shape |

Chapter 10

◆ **Learning Exercise 10.3A**

Match the shape of a molecule with the following descriptions:

    a. linear   **b.** trigonal planar   **c.** tetrahedral
    d. trigonal pyramidal   **e.** bent (120°)   **f.** bent (109°)

1. three electron groups with three bonded atoms _____

2. two electron groups with two bonded atoms _____

3. four electron groups with three bonded atoms _____

4. three electron groups with two bonded atoms _____

5. four electron groups with four bonded atoms _____

6. four electron groups with two bonded atoms _____

Answers   **1.** b   **2.** a   **3.** d   **4.** e   **5.** c   **6.** f

Guide to Predicting Shape (VSEPR Theory)	
STEP 1	Draw the Lewis structure.
STEP 2	Arrange the electron groups around the central atom to minimize repulsion.
STEP 3	Use the atoms bonded to the central atom to determine the shape.

◆ **Learning Exercise 10.3B**

For each molecule or ion, state the total number of valence electrons, draw the Lewis structure, state the number of electron groups and bonded atoms, and predict the shape and bond angles. When resonance structures are possible, only one Lewis structure is needed.

⮞ CORE CHEMISTRY SKILL

Predicting Shapes

Molecule or Ion	Valence Electrons	Lewis Structure	Electron Groups	Bonded Atoms	Shape and Bond Angle
CH_4					
PCl_3					
SO_3					
H_2S					
NO_2^-					

Answers

Molecule or Ion	Valence Electrons	Lewis Structure	Electron Groups	Bonded Atoms	Shape and Bond Angle
CH₄	8	H:C:H (with H above and below)	4	4	Tetrahedral, 109°
PCl₃	26	:Cl:P:Cl: :Cl:	4	3	Trigonal pyramidal, 109°
SO₃	24	:O:S::O: :O:	3	3	Trigonal planar, 120°
H₂S	8	H:S: H	4	2	Bent, 109°
NO₂⁻	18	[:O:N::O:]⁻	3	2	Bent, 120°

10.4 Electronegativity and Bond Polarity

Learning Goal: Use electronegativity to determine the polarity of a bond.

- Electronegativity values indicate the ability of an atom to attract electrons in a chemical bond.
- Electronegativity values are low for metals and high for nonmetals.

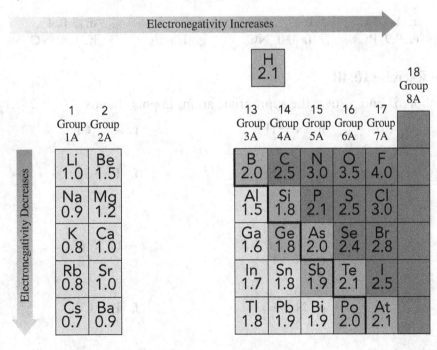

- When atoms sharing electrons have the same or similar electronegativity values, electrons are shared equally and the bond is *nonpolar covalent*.
- Electrons are shared unequally in *polar covalent* bonds because they are attracted to the more electronegative atom.
- An electronegativity difference of 0.0 to 0.4 indicates a nonpolar covalent bond, whereas a difference of 0.5 to 1.8 indicates a polar covalent bond.
- An electronegativity difference of 1.9 to 3.3 indicates a bond that is ionic.
- A polar bond has a charge separation, which is called a *dipole*; the positive end of the dipole is marked as δ^+ and the negative end as δ^-.

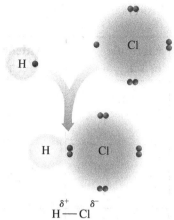

$\overset{\delta^+}{H}\!-\!\overset{\delta^-}{Cl}$
Electrons are shared unequally in the polar covalent bond of HCl.

♦ **Learning Exercise 10.4A**

CORE CHEMISTRY SKILL

Using the electronegativity values, determine the following:

Using Electronegativity

1. the electronegativity difference for each pair

2. the type of bonding as ionic (I), polar covalent (PC), or nonpolar covalent (NC)

Elements	Electronegativity Difference	Type of Bonding	Elements	Electronegativity Difference	Type of Bonding
a. H and O	_____	_____	b. N and S	_____	_____
c. Al and O	_____	_____	d. Li and F	_____	_____
e. H and Cl	_____	_____	f. Cl and Cl	_____	_____
g. S and F	_____	_____	h. H and C	_____	_____

Answers **a.** 1.4, PC   **b.** 0.5, PC   **c.** 2.0, I   **d.** 3.0, I
e. 0.9, PC   **f.** 0.0, NC   **g.** 1.5, PC   **h.** 0.4, NC

♦ **Learning Exercise 10.4B**

Write the symbols δ^+ and δ^- over the appropriate atoms in polar bonds.

a. H—O   **b.** N—H   **c.** C—Cl

d. O—F   **e.** N—F   **f.** P—Cl

Answers

a. $\overset{\delta^+}{H}\!-\!\overset{\delta^-}{O}$ **b.** $\overset{\delta^-}{N}\!-\!\overset{\delta^+}{H}$ **c.** $\overset{\delta^+}{C}\!-\!\overset{\delta^-}{Cl}$

d. $\overset{\delta^+}{O}\!-\!\overset{\delta^-}{F}$ **e.** $\overset{\delta^+}{N}\!-\!\overset{\delta^-}{F}$ **f.** $\overset{\delta^+}{P}\!-\!\overset{\delta^-}{Cl}$

10.5   Polarity of Molecules

Learning Goal: Use the three-dimensional structure of a molecule to classify it as polar or nonpolar.

- Nonpolar molecules can have polar bonds when the dipoles are in a symmetric arrangement and the bond dipoles cancel each other.
- In polar molecules, the bond dipoles do not cancel each other.

	Guide to Determining the Polarity of a Molecule
STEP 1	Determine if the bonds are polar covalent or nonpolar covalent.
STEP 2	If the bonds are polar covalent, draw the Lewis structure and determine if the dipoles cancel.

> **CORE CHEMISTRY SKILL**
>
> Identifying Polarity of Molecules

♦ **Learning Exercise 10.5**

Draw the bond dipoles in each of the following molecules and determine whether the molecule is polar or nonpolar:

a. CF_4

b. HCl

c. NH_3

d. OF_2

Answers

a. CF_4

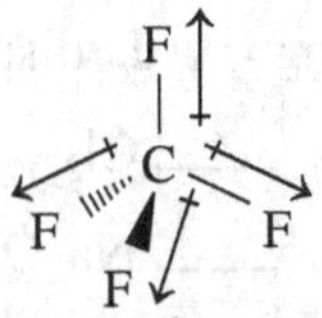

bond dipoles cancel, nonpolar

b. HBr   H—Br

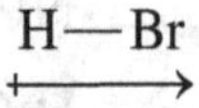

bond dipole does not cancel, polar

c. NH_3

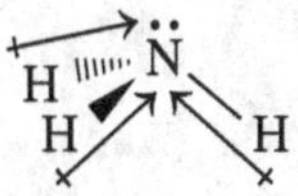

bond dipoles do not cancel, polar

d. SF_2

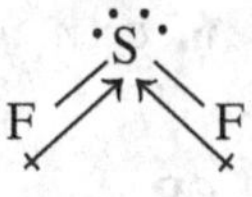

bond dipoles do not cancel, polar

10.6 Intermolecular Forces Between Atoms or Molecules

Learning Goal: Describe the intermolecular forces between ions, polar covalent molecules, and nonpolar covalent molecules.

• The intermolecular forces between particles in solids and liquids determine their melting and boiling points.

• Ionic solids have high melting points due to strong ionic interactions between positive and negative ions.

• In polar substances, dipole–dipole attractions occur between the positive end of one molecule and the negative end of another.

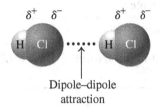

Dipole–dipole
attraction

• Hydrogen bonding, a particularly strong type of dipole–dipole attraction, occurs between partially positive hydrogen atoms and the strongly electronegative atoms N, O, or F.

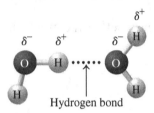

Hydrogen bond

• Dispersion forces occur when temporary dipoles form within the nonpolar molecules, causing weak attractions to other nonpolar molecules. Dispersion forces are the only intermolecular forces that occur in nonpolar molecules.

♦ **Learning Exercise 10.6A**

Match the major type of intermolecular force with each of the following:

⟩**CORE CHEMISTRY SKILL**

Identifying Intermolecular Forces

   a. ionic   **b.** dipole–dipole attractions   **c.** hydrogen bonding   **d.** dispersion forces

1. _____ KCl   **2.** _____ NF_3 **3.** _____ PF_3 **4.** _____ Cl_2

5. _____ HF   **6.** _____ H_2O **7.** _____ C_4H_{10} **8.** _____ Na_2O

Answers   **1.** a   **2.** b   **3.** b   **4.** d
                 5. c   **6.** c   **7.** d   **8.** a

♦ **Learning Exercise 10.6B**

Identify the substance that would have the higher boiling point in each pair and give the major type of intermolecular force:

a. KCl or HCl _____   **b.** Br_2 or HBr _____   **c.** H_2O or H_2S _____

d. C_2H_6 or C_8H_{18} _____   **e.** $MgCl_2$ or OCl_2 _____   **f.** NH_3 or AsH_3 _____

Answers   **a.** KCl (ionic)   **b.** HBr (dipole–dipole attractions)
          c. H_2O (hydrogen bonding)   **d.** C_8H_{18} (dispersion forces are stronger in the
                                             larger molecule)
          e. $MgCl_2$ (ionic)   **f.** NH_3 (hydrogen bonding)

10.7   Changes of State

Learning Goal: Describe the changes of state between solids, liquids, and gases; calculate the energy involved.

- The three states of matter are solid, liquid, and gas.
- A substance melts and freezes at its melting (freezing) point. During the process of melting or freezing, the temperature remains constant.
- The heat of fusion is the energy required to change 1 g of solid to liquid at the melting point. For ice to melt at 0 °C, 334 J/g is required. This is also the amount of heat lost when 1 g of water freezes at 0 °C.
- A substance boils and condenses at its boiling point. During the process of boiling or condensing, the temperature remains constant.
- The heat of vaporization is the energy required to change 1 g of liquid to 1 g of gas at the boiling point. When water boils at 100 °C, 2260 J/g is required. This is also the amount of heat released when water vapor condenses at 100 °C.
- Evaporation is a surface phenomenon while boiling occurs throughout the liquid.
- Sublimation is the change of state from a solid directly to a gas. Deposition is the opposite process.
- A heating or cooling curve illustrates the changes in temperature and state as heat is added to or removed from a substance. An example of a cooling curve for water is shown below.

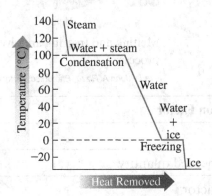

A cooling curve for water illustrates
the change in temperature and
changes of state as heat is removed.

- When a substance is heated or cooled, the energy gained or released is the total of the energy involved in temperature changes as well as the energy involved in changes of state.

Key Terms for Sections 10.1 to 10.7

Match each of the following key terms with the correct description:

a. polar covalent bond	**b.** multiple bond	**c.** VSEPR theory
d. nonpolar covalent bond	**e.** resonance	**f.** dipole–dipole attraction
g. deposition	**h.** dispersion forces	**i.** heat of fusion

1. _____ equal sharing of valence electrons by two atoms

2. _____ occurs when two or more Lewis structures can be drawn for the same compound

3. _____ the unequal attraction for shared electrons in a covalent bond

4. _____ the atoms in a molecule are arranged to minimize repulsion between electrons

5. _____ energy needed to change 1 g of solid to liquid at the melting point

6. _____ the sharing of two or three pairs of electrons by two atoms

7. _____ interaction between the positive end of a polar molecule and the negative end of another

8. _____ results from momentary polarization of nonpolar molecules

9. _____ change of a gas directly to a solid

Answers   **1.** d   **2.** e   **3.** a   **4.** c   **5.** i
            6. b   **7.** f   **8.** h   **9.** g

♦ **Learning Exercise 10.7A**

Identify each of the following as

    a. melting   **b.** freezing   **c.** sublimation   **d.** condensation

1. _____ A liquid changes to a solid.

2. _____ Ice forms on the surface of a lake in winter.

3. _____ Dry ice in an ice cream cart changes to a gas.

4. _____ Butter in a hot pan turns to liquid.

5. _____ A gas changes to a liquid.

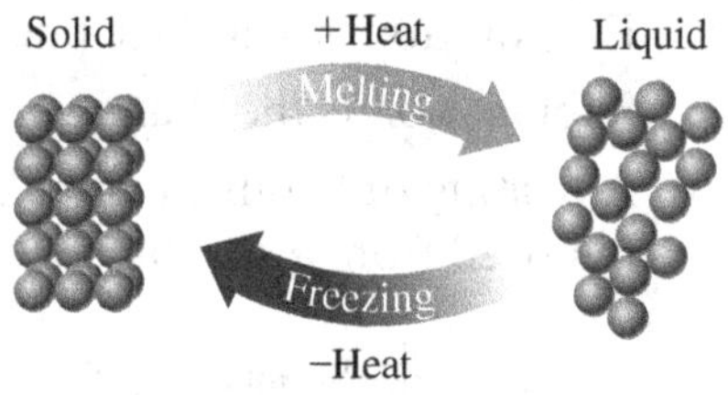

Solid   +Heat   Liquid

Melting

Freezing

−Heat

Melting and freezing are reversible processes.

Credit: John A. Rizzo / Photodisc / Getty Images

Answers   **1.** b   **2.** b   **3.** c   **4.** a   **5.** d

Guide to Using a Heat Conversion Factor	
STEP 1	State the given and needed quantities.
STEP 2	Write a plan to convert the given quantity to the needed quantity.
STEP 3	Write the heat conversion factor and any metric factor.
STEP 4	Set up the problem and calculate the needed quantity.

Study Note
The amount of heat needed or released during a change of state can be calculated using the heat conversion factor: Heat (joules) = mass (g) × heat conversion factor (J/g)

CORE CHEMISTRY SKILL

Calculating Heat for Change of State

♦ **Learning Exercise 10.7B**

Calculate each of the following when a substance melts or freezes:

a. How many joules of heat are needed to melt 24.0 g of ice at 0 °C?

b. How much heat, in kilojoules, is released when 325 g of water freezes at 0 °C?

Bonding and Properties of Solids and Liquids

c. How many grams of water can be converted to ice at 0 °C when 41.8 kJ of heat is released?

d. How many grams of ice would melt when 1680 J of heat were absorbed?

An ice bag is used to treat a sports injury.
Credit: wsphotos / Getty Images

Answers　　**a.** 8020 J　　**b.** 109 kJ　　**c.** 125 g　　**d.** 5.03 g

◆ Learning Exercise 10.7C

Calculate each of the following when a substance boils or condenses:

a. How many joules are needed to completely change 5.00 g of water to vapor at 100 °C?

b. How many kilojoules are released when 515 g of steam at 100 °C condenses to form liquid water at 100 °C?

c. How many grams of water can be converted to steam at 100 °C when 155 kJ of energy is absorbed?

d. How many grams of steam condense if 24 800 J is released at 100 °C?

Answers　　**a.** 11 300 J　　**b.** 1160 kJ　　**c.** 68.6 g　　**d.** 11.0 g

◆ Learning Exercise 10.7D

Draw each heating or cooling curve and indicate the portion that corresponds to a solid, liquid, or gas and the changes in state.

a. Draw a heating curve for water that begins at −20 °C and ends at 120 °C.

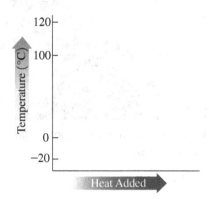

b. Draw a heating curve for bromine from −25 °C to 75 °C. Bromine has a melting point of −7 °C and a boiling point of 59 °C.

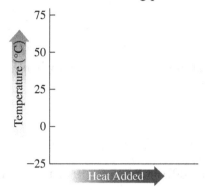

c. Draw a cooling curve for sodium from 1000 °C to 0 °C. Sodium has a freezing point of 98 °C and a boiling (condensation) point of 883 °C.

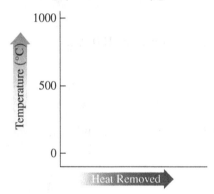

Answers

a.

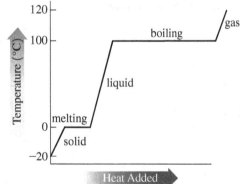

b.

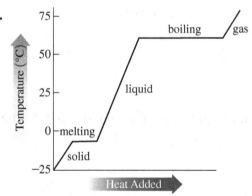

c.

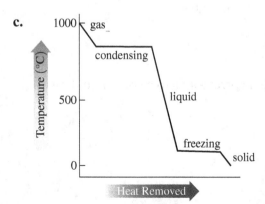

♦ **Learning Exercise 10.7E**

a. How many kilojoules are released when 35.0 g of water at 65.0 °C is cooled to 0.0 °C and changed to solid? (*Hint*: Two steps are needed.)

b. How many kilojoules are needed to melt 15.0 g of ice at 0. °C, heat the water to 100. °C, and convert the water to gas at 100. °C? (*Hint*: Three steps are needed.)

c. Calculate the number of kilojoules released when 25.0 g of steam at 100. °C condenses, then cools, and finally freezes at 0. °C. (*Hint*: Three steps are needed.)

Answers　　**a.** 21.2 kJ　　**b.** 45.2 kJ　　**c.** 75.3 kJ

Checklist for Chapter 10

You are ready to take the Practice Test for Chapter 10. Be sure you have accomplished the following learning goals for this chapter. If not, review the section listed at the end of the goal. Then apply your new skills and understanding to the Practice Test.

After studying Chapter 10, I can successfully:

_____ Draw a Lewis structure for a molecular compound or polyatomic ion with multiple bonds. (10.1)

_____ Draw resonance structures when two or more Lewis structures are possible. (10.2)

_____ Predict the shape and bond angles for a molecule or polyatomic ion. (10.3)

_____ Determine the electronegativity difference for two atoms in a compound. (10.4)

_____ Classify a bond as nonpolar covalent, polar covalent, or ionic. (10.4)

_____ Classify a molecule as polar or nonpolar. (10.5)

_____ Identify the intermolecular forces between particles in a liquid or solid and predict relative melting and boiling points. (10.6)

_____ Calculate the heat change when a specific amount of a substance changes temperature. (10.7)

_____ Calculate the heat change for a change of state for a specific amount of a substance. (10.7)

_____ Calculate the total heat for a combination of change of temperature and change of state for a specific amount of a substance. (10.7)

_____ Draw heating and cooling curves using the melting and boiling points of a substance. (10.7)

Practice Test for Chapter 10

The chapter sections to review are shown in parentheses at the end of each question.

1. The correct Lewis structure for PO_3^{3-} is (10.1)

A. $\left[\ddot{O}-P=\ddot{O} \right]^{3-}$ B. $\left[\ddot{O}-P-\ddot{O} \right]^{3-}$ C. $\left[\ddot{O}=P=\ddot{O} \right]^{3-}$ D. $\left[\ddot{O}-P-\ddot{O} \right]^{3-}$ E. $\left[\ddot{O}=P=\ddot{O} \right]^{3-}$

2. The resonance structures of NO_3^- are (10.1)

A. $\left[\ddot{O}=N-\ddot{O} \right]^{-}$ B. $\left[\ddot{O}=N=\ddot{O} \right]^{-}$ C. $\left[\ddot{O}-N-\ddot{O} \right]^{-}$ D. $\left[\ddot{O}-N=\ddot{O} \right]^{-}$ E. $\left[\ddot{O}=N-\ddot{O} \right]^{-}$

   A. A, B   **B.** A, C   **C.** A, D   **D.** C, D, E   **E.** A, C, D

3. A polar covalent bond occurs between (10.4)
   A. two different nonmetals
   B. two of the same type of nonmetals
   C. two noble gases
   D. two different metals
   E. a metal and a nonmetal

For questions 4 through 9, indicate the type of bonding expected to occur between the following elements: (10.4)

   ionic (I)   nonpolar covalent (NC)   polar covalent (PC)

   4. _____ silicon and oxygen   **5.** _____ barium and chlorine

   6. _____ aluminum and fluorine   **7.** _____ chlorine and chlorine

   8. _____ sulfur and oxygen   **9.** _____ nitrogen and oxygen

For questions 10 through 15, match each of the following shapes with one of the descriptions: (10.3)

   A. linear   **B.** trigonal planar   **C.** tetrahedral
   D. trigonal pyramidal **E.** bent (120°)   **F.** bent (109°)

10. _____ has a central atom with three electron groups and three bonded atoms

11. _____ has a central atom with two electron groups and two bonded atoms

12. _____ has a central atom with four electron groups and three bonded atoms

13. _____ has a central atom with three electron groups and two bonded atoms

14. _____ has a central atom with four electron groups and four bonded atoms

15. _____ has a central atom with four electron groups and two bonded atoms

For questions 16 through 20, determine the shape and bond angles for each of the following molecules: (10.3)

A. linear, 180° **B.** trigonal planar, 120° **C.** bent, 120°
D. tetrahedral, 109° **E.** trigonal pyramidal, 109° **F.** bent, 109°

16. PCl_3 **17.** CBr_4 **18.** H_2S **19.** BrO_2^- **20.** ClO_3^-

For questions 21 through 28, indicate the major type of intermolecular force (A to D) between particles of each of the following: (10.6)

A. ionic **B.** dipole–dipole attractions
C. hydrogen bonds **D.** dispersion forces

21. _____ CaO   **22.** _____ NCl_3

23. _____ SF_2 **24.** _____ Br_2

25. _____ HF   **26.** _____ NH_3

27. _____ C_3H_8 **28.** _____ BaF_2

For questions 29 through 33, match items A to E with one of the following: (10.7)

A. condensation **B.** evaporation **C.** heat of fusion
D. heat of vaporization **E.** boiling

29. _____ the energy required to convert 1 g of solid to liquid at the melting point

30. _____ the heat required to convert 1 g of liquid to gas at the boiling point

31. _____ the conversion of liquid molecules to gas at the surface of a liquid

32. _____ the conversion of water vapor to liquid water

33. _____ the formation of a gas within the liquid as well as on the surface

34. The number of joules needed to convert 15.0 g of ice to liquid at 0. °C is (10.7)
A. 73 J   **B.** 334 J   **C.** 2260 J   **D.** 4670 J   **E.** 5010 J

35. The total number of joules released when 2.0 g of water at 50. °C is cooled and frozen at 0. °C is (10.7)
A. 420 J   **B.** 670 J   **C.** 1100 J   **D.** 2100 J   **E.** 3400 J

36. What is the total number of kilojoules required to convert 25 g of ice at 0. °C to gas at 100. °C? (10.7)
A. 8.4 kJ   **B.** 19 kJ   **C.** 59 kJ   **D.** 67 kJ   **E.** 75 kJ

Use the heating curve for p-toluidine to answer questions 37 through 40: (10.7)

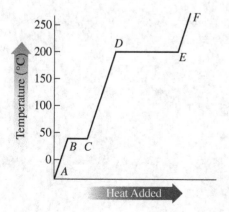

37. On the heating curve, segment *BC* indicates
    A. solid   **B.** melting   **C.** liquid   **D.** boiling   **E.** gas

38. On the heating curve, segment *CD* indicates
    A. solid   **B.** melting   **C.** liquid   **D.** boiling   **E.** gas

39. The boiling point of toluidine would be about
    A. 20 °C   **B.** 40 °C   **C.** 100 °C   **D.** 200 °C   **E.** 250 °C

40. On the heating curve, segment *EF* shows toluidine as
    A. solid   **B.** melting   **C.** liquid   **D.** boiling   **E.** gas

For questions 41 through 45, draw the Lewis structure for CO₂: (10.1, 10.2, 10.3, 10.5)

41. The number of valence electrons needed in the Lewis structure of CO_2 is
    A. 10   **B.** 12   **C.** 14   **D.** 16   **E.** 18

42. The number of single bonds is
    A. 0   **B.** 1   **C.** 2   **D.** 3   **E.** 4

43. The number of double bonds is
    A. 0   **B.** 1   **C.** 2   **D.** 3   **E.** 4

44. The molecule is
    A. nonpolar   **B.** polar   **C.** ionic

45. The shape of the molecule is
    A. linear   **B.** bent (120°)   **C.** bent (109°)
    D. trigonal planar   **E.** tetrahedral

Answers to the Practice Test

1. B	**2.** E	**3.** A	**4.** PC	**5.** I
6. I	**7.** NC	**8.** PC	**9.** PC	**10.** B
11. A	**12.** D	**13.** E	**14.** C	**15.** F
16. E	**17.** D	**18.** F	**19.** F	**20.** E
21. A	**22.** B	**23.** B	**24.** D	**25.** C
26. C	**27.** D	**28.** A	**29.** C	**30.** D
31. B	**32.** A	**33.** E	**34.** E	**35.** C
36. E	**37.** B	**38.** C	**39.** D	**40.** E
41. D	**42.** A	**43.** C	**44.** A	**45.** A

Selected Answers and Solutions to Text Problems

10.1 **a.** $2\,H(1\,e^-) + 1\,S(6\,e^-) = 2 + 6 = 8$ valence electrons

  b. $2\,I(7\,e^-) = 14$ valence electrons

  c. $1\,C(4\,e^-) + 4\,Cl(7\,e^-) = 4 + 28 = 32$ valence electrons

  d. $1\,O(6\,e^-) + 1\,H(1\,e^-) + 1\,e^-(\text{negative charge}) = 6 + 1 + 1 = 8$ valence electrons

10.3 **a.** $1\,H(1\,e^-) + 1\,F(7\,e^-) = 1 + 7 = 8$ valence electrons

  H$\ddot{\underset{\cdot\cdot}{\text{F}}}$:   or   H—$\ddot{\underset{\cdot\cdot}{\text{F}}}$:

  b. $1\,S(6\,e^-) + 2\,F(7\,e^-) = 6 + 14 = 20$ valence electrons

  :$\ddot{\text{F}}$:$\ddot{\text{S}}$:$\ddot{\text{F}}$:   or   :$\ddot{\text{F}}$—$\ddot{\text{S}}$—$\ddot{\text{F}}$:

  c. $1\,N(5\,e^-) + 3\,Br(7\,e^-) = 5 + 21 = 26$ valence electrons

$$
\begin{array}{ccc}
 & :\ddot{\text{Br}}: & \\
:\ddot{\text{Br}}:\text{N}:\ddot{\text{Br}}: & \text{or} & :\ddot{\text{Br}}—\overset{|}{\underset{|}{\text{N}}}—\ddot{\text{Br}}:
\end{array}
$$

  d. $1\,B(3\,e^-) + 4\,H(1\,e^-) + 1\,e^-(\text{negative charge}) = 3 + 4 + 1 = 8$ valence electrons

$$
\left[\begin{array}{c} \text{H} \\ \text{H}:\text{B}:\text{H} \\ \text{H} \end{array}\right]^- \quad \text{or} \quad \left[\begin{array}{c} \text{H} \\ | \\ \text{H}—\text{B}—\text{H} \\ | \\ \text{H} \end{array}\right]^-
$$

  e. $1\,C(4\,e^-) + 4\,H(1\,e^-) + 1\,O(6\,e^-) = 4 + 4 + 6 = 14$ valence electrons

$$
\begin{array}{ccc}
\text{H} & & \text{H} \\
\text{H}:\text{C}:\ddot{\text{O}}:\text{H} & \text{or} & \text{H}—\overset{|}{\underset{|}{\text{C}}}—\ddot{\text{O}}—\text{H} \\
\text{H} & & \text{H}
\end{array}
$$

  f. $2\,N(5\,e^-) + 4\,H(1\,e^-) = 10 + 4 = 14$ valence electrons

$$
\begin{array}{ccc}
\text{H} \quad \text{H} & & \text{H} \quad \text{H} \\
\text{H}:\ddot{\text{N}}:\ddot{\text{N}}:\text{H} & \text{or} & \text{H}—\overset{|}{\underset{}{\text{N}}}—\overset{|}{\underset{}{\text{N}}}—\text{H}
\end{array}
$$

10.5 If complete octets cannot be formed by using all the valence electrons, it is necessary to draw multiple bonds.

10.7 **a.** $1\,C(4\,e^-) + 1\,O(6\,e^-) = 4 + 6 = 10$ valence electrons

  :C$::$O: or   :C$\equiv$O:

  b. $1\,C(4\,e^-) + 1\,N(5\,e^-) + 1\,e^-(\text{negative charge}) = 4 + 5 + 1 = 10$ valence electrons

  $\big[$:C$::$N:$\big]^-$ or $\big[$:C$\equiv$N:$\big]^-$

  c. $2\,H(1\,e^-) + 1\,C(4\,e^-) + 1\,O(6\,e^-) = 2 + 4 + 6 = 12$ valence electrons

$$
\begin{array}{ccc}
:\ddot{\text{O}}: & & :\ddot{\text{O}}: \\
 & & \| \\
\text{H}:\text{C}:\text{H} & \text{or} & \text{H}—\text{C}—\text{H}
\end{array}
$$

10.9 Resonance occurs when we can draw two or more Lewis structures for the same molecule or ion.

10.11 a. $1 \, Cl(7 \, e^-) + 1 \, N(5 \, e^-) + 2 \, O(6 \, e^-) = 7 + 5 + 12 = 24$ valence electrons

$$
\ddot{:}\!\overset{\displaystyle :\!\ddot{O}\!:}{\underset{}{\overset{\|}{}}}\qquad\qquad
$$

:Cl—N—Ö: ⟷ :Cl—N=Ö:

b. $1 \, O(6 \, e^-) + 1 \, C(4 \, e^-) + 1 \, N(5 \, e^-) + 1 \, e^-$ (negative charge)
$= 6 + 4 + 5 + 1 = 16$ valence electrons

$$\left[:O\!\equiv\!C\!-\!\ddot{N}\!: \right]^- \longleftrightarrow \left[:\ddot{O}\!=\!C\!=\!\ddot{N}\!: \right]^- \longleftrightarrow \left[:\ddot{O}\!-\!C\!\equiv\!N\!: \right]^-$$

10.13 a. 6, tetrahedral; Four electron groups around a central atom have a tetrahedral electron-group geometry. With four bonded atoms, the shape of the molecule is tetrahedral.

b. 5, trigonal pyramidal; Four electron groups around a central atom have a tetrahedral electron-group geometry. With three bonded atoms (and one lone pair of electrons), the shape of the molecule is trigonal pyramidal.

c. 3, trigonal planar; Three electron groups around a central atom have a trigonal planar electron-group geometry. With three bonded atoms, the shape of the molecule is trigonal planar.

10.15 SeO_3 $1 \, Se(6 \, e^-) + 3 \, O(6 \, e^-) = 6 + 18 = 24$ valence electrons

$$
\ddot{:}\!\overset{\displaystyle :\!O\!:}{\underset{}{\overset{\|}{}}}
$$

:Ö—Se—Ö: (only 1 of the resonance structures is shown)

a. There are <u>three</u> electron groups around the central Se atom.
b. The electron-group geometry is <u>trigonal planar</u>.
c. The number of atoms attached to the central Se atom is <u>three</u>.
d. The shape of the molecule is <u>trigonal planar</u>.

10.17 In CF_4, the central atom C has four bonded atoms and no lone pairs of electrons, which gives CF_4 a tetrahedral shape. In NF_3, the central atom N has three bonded atoms and one lone pair of electrons, which gives NF_3 a trigonal pyramidal shape.

$$
\begin{array}{cc}
:\!\ddot{F}\!: & \\
\mid & \\
:\!\ddot{F}\!-\!C\!-\!\ddot{F}\!: \qquad :\!\ddot{F}\!-\!\ddot{N}\!-\!\ddot{F}\!: \\
\mid \qquad\qquad \mid \\
:\!\ddot{F}\!: \qquad\qquad :\!\ddot{F}\!:
\end{array}
$$

10.19 a. The central atom Ga has three electron groups bonded to three hydrogen atoms; GaH_3 has a trigonal planar shape.

$$
\begin{array}{c}
H \\
\mid \\
Ga \\
\diagup \quad \diagdown \\
H \qquad H
\end{array}
$$

b. The central O atom has four electron groups with two bonded atoms and two lone pairs of electrons, which gives OF_2 a bent shape ($109°$).

$$
\begin{array}{c}
:\!\ddot{F}\!: \\
\mid \\
:\!\ddot{F}\!-\!\ddot{O}\!:
\end{array}
$$

c. The central atom C has two electron groups bonded to two atoms; HCN is linear.

$$H—C≡N:$$

d. The central atom C has four electron groups bonded to four chlorine atoms; CCl_4 has a tetrahedral shape.

$$
\begin{array}{c}
:\ddot{C}l: \\
| \\
:\ddot{C}l—C—\ddot{C}l: \\
| \\
:\ddot{C}l:
\end{array}
$$

10.21 To find the total valence electrons for an ion, add the total valence electrons for each atom and add the number of electrons indicated by a negative charge, or subtract for a positive charge.

a. $1\,Al(3\,e^-) + 4\,H(1\,e^-) + 1\,e^-$ (negative charge) $= 3 + 4 + 1 = 8$ valence electrons

$$
\left[\begin{array}{c} H \\ \ddot{} \\ H:\!\ddot{A}l\!:H \\ \ddot{} \\ H \end{array}\right]^- \quad \text{or} \quad \left[\begin{array}{c} H \\ | \\ H—Al—H \\ | \\ H \end{array}\right]^- \quad
\begin{array}{l} \text{four electron groups around Al bonded to four atoms;} \\ \text{tetrahedral shape} \end{array}
$$

b. $1\,S(6\,e^-) + 4\,O(6\,e^-) + 2\,e^-$ (negative charge) $= 6 + 24 + 2 = 32$ valence electrons

$$
\left[\begin{array}{c} :\ddot{O}: \\ | \\ :\ddot{O}—S—\ddot{O}: \\ | \\ :\ddot{O}: \end{array}\right]^{2-} \quad \text{four electron groups around S bonded to four atoms; tetrahedral shape}
$$

c. $1\,N(5\,e^-) + 4\,H(1\,e^-) - 1\,e^-$ (positive charge) $= 5 + 4 - 1 = 8$ valence electrons

$$
\left[\begin{array}{c} H \\ | \\ H—N—H \\ | \\ H \end{array}\right]^+ \quad \text{four electron groups around N bonded to four atoms; tetrahedral shape}
$$

d. $1\,N(5\,e^-) + 2\,O(6\,e^-) - 1\,e^-$ (positive charge) $= 5 + 12 - 1 = 16$ valence electrons

$$\left[:\ddot{O}\!=\!N\!=\!\ddot{O}:\right]^+ \quad \text{two electron groups around N bonded to two atoms; linear shape}$$

10.23 a. The electronegativity values increase going from left to right across Period 2 from B to F.
b. The electronegativity values decrease going down Group 2A (2) from Mg to Ba.
c. The electronegativity values decrease going down Group 7A (17) from F to I.

10.25 a; A nonpolar covalent bond would result from an electronegativity difference from 0.0 to 0.4.

10.27 a. Electronegativity increases going up a group: K, Na, Li.
b. Electronegativity increases going left to right across a period: Na, P, Cl.
c. Electronegativity increases going across a period and at the top of a group: Ca, Se, O.

10.29 a. Si—Br electronegativity difference: 2.8 − 1.8 = 1.0, polar covalent
    b. Li—F electronegativity difference: 4.0 − 1.0 = 3.0, ionic
    c. Br—F electronegativity difference: 4.0 − 2.8 = 1.2, polar covalent
    d. I—I electronegativity difference: 2.5 − 2.5 = 0.0, nonpolar covalent
    e. N—P electronegativity difference: 3.0 − 2.1 = 0.9, polar covalent
    f. C—P electronegativity difference: 2.5 − 2.1 = 0.4, nonpolar covalent

10.31 A dipole arrow points from the atom with the lower electronegativity value (more positive) to the atom in the bond that has the higher electronegativity value (more negative).

    a. $\overset{\delta^+\quad\delta^-}{N-F}$ **b.** $\overset{\delta^+\quad\delta^-}{Si-Br}$

    c. $\overset{\delta^+\quad\delta^-}{C-O}$ **d.** $\overset{\delta^+\quad\delta^-}{P-Br}$

    e. $\overset{\delta^-\quad\delta^+}{N-P}$

10.33 F_2 is a nonpolar molecule because there is a nonpolar covalent bond between F atoms, which have identical electronegativity values. In HF, the bond is a polar bond because there is a large electronegativity difference, which makes HF a polar molecule.

    :F̈—F̈:   H—F̈:

    nonpolar   polar

10.35 a. The molecule CS_2 contains two polar C—S bonds whose dipoles point in opposite directions. As a result of this linear geometry, the dipoles cancel, which makes CS_2 a nonpolar molecule.

    :S̈=C=S̈:

    nonpolar

    b. The molecule NF_3 contains three polar covalent N—F bonds and a lone pair of electrons on the central N atom. This asymmetric trigonal pyramidal shape makes NF_3 a polar molecule.

    polar

    c. In the molecule CHF_3, there are three polar covalent C—F bonds and one nonpolar covalent C—H bond, which makes CHF_3 a polar molecule.

    polar

    d. In the molecule SO_3, there are three polar covalent S—O bonds arranged in a trigonal planar shape. As a result of this symmetric geometry, the dipoles cancel, which makes SO_3 a nonpolar molecule.

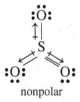

    nonpolar

10.37 In the linear molecule CO_2, the two $C-O$ dipoles cancel, resulting in a nonpolar molecule; in CO, there is only one dipole, making it a polar molecule.

$$:\overset{..}{O}=C=\overset{..}{O}: \qquad :C\equiv O:$$

nonpolar   polar

10.39 a. BrF is a polar molecule. An attraction between the positive end of one polar molecule and the negative end of another polar molecule is called a dipole–dipole attraction.
 b. An ionic bond is an attraction between a positive and negative ion, as in KCl.
 c. NF_3 is a polar molecule. An attraction between the positive end of one polar molecule and the negative end of another polar molecule is called a dipole–dipole attraction.
 d. Cl_2 is a nonpolar molecule. The weak attractions that occur between temporary dipoles in nonpolar molecules are called dispersion forces.

10.41 a. CH_3OH is a polar molecule. Hydrogen bonds are strong dipole–dipole attractions that occur between a partially positive hydrogen atom of one molecule and one of the strongly electronegative atoms N, O, or F in another, as is seen with CH_3OH molecules.
 b. CO is a polar molecule. Dipole–dipole attractions occur between dipoles in polar molecules.
 c. CF_4 is a nonpolar molecule. The weak attractions that occur between temporary dipoles in nonpolar molecules are called dispersion forces.
 d. CH_3-CH_3 is a nonpolar molecule. The weak attractions that occur between temporary dipoles in nonpolar molecules are called dispersion forces.

10.43 a. HF would have a higher boiling point than HBr; the hydrogen bonds in HF are stronger than the dipole–dipole attractions in HBr.
 b. NaF would have a higher boiling point than HF; the ionic bonds in NaF are stronger than the hydrogen bonds in HF.
 c. $MgBr_2$ would have a higher boiling point than PBr_3; the ionic bonds in $MgBr_2$ are stronger than the dipole–dipole attractions in PBr_3.
 d. CH_3OH would have a higher boiling point than CH_4; the hydrogen bonds in CH_3OH are stronger than the dispersion forces in CH_4.

10.45 a. $65.0 \text{ g ice} \times \dfrac{334 \text{ J}}{1 \text{ g ice}} = 21\,700 \text{ J (3 SFs) is absorbed}$

 b. $17.0 \text{ g benzene} \times \dfrac{128 \text{ J}}{1 \text{ g benzene}} = 2180 \text{ J (3 SFs) is absorbed}$

 c. $225 \text{ g acetic acid} \times \dfrac{300 \text{ J}}{1 \text{ g acetic acid}} \times \dfrac{1 \text{ kJ}}{1000 \text{ J}} = 87.8 \text{ kJ (3 SFs) is released}$

 d. $0.0500 \text{ kg water} \times \dfrac{1000 \text{ g water}}{1 \text{ kg water}} \times \dfrac{334 \text{ J}}{1 \text{ g water}} \times \dfrac{1 \text{ kJ}}{1000 \text{ J}} = 16.7 \text{ kJ (3 SFs) is released}$

10.47 a. $10.0 \text{ g water} \times \dfrac{2260 \text{ J}}{1 \text{ g water}} = 22\,600 \text{ J (3 SFs) is absorbed}$

 b. $50.0 \text{ g ethanol} \times \dfrac{841 \text{ J}}{1 \text{ g ethanol}} \times \dfrac{1 \text{ kJ}}{1000 \text{ J}} = 42.1 \text{ kJ (3 SFs) is absorbed}$

 c. $8.00 \text{ g acetic acid} \times \dfrac{390 \text{ J}}{1 \text{ g acetic acid}} = 3100 \text{ J (2 SFs) is released}$

 d. $0.175 \text{ kg ammonia} \times \dfrac{1000 \text{ g ammonia}}{1 \text{ kg ammonia}} \times \dfrac{1380 \text{ J}}{1 \text{ g ammonia}} \times \dfrac{1 \text{ kJ}}{1000 \text{ J}}$

   $= 242 \text{ kJ (3 SFs) is released}$

10.49 a. water 15 °C $\longrightarrow$ 72 °C: $\Delta T = 72\,°C - 15\,°C = 57\,°C$;

$$20.0\ \cancel{g} \times 57\ \cancel{°C} \times \frac{4.184\ J}{\cancel{g}\ \cancel{°C}}$$

$$= 4800\ J\ (2\ SFs)$$

b. Two calculations are needed:

(1) ice 0 °C $\longrightarrow$ water 0 °C: $50.0\ \cancel{g\ ice} \times \dfrac{334\ J}{1\ \cancel{g\ ice}} = 16\,700\ J\ (3\ SFs)$

(2) water 0 °C $\longrightarrow$ 65.0 °C: $\Delta T = 65.0\,°C - 0\,°C = 65.0\,°C$;

$$50.0\ \cancel{g} \times 65.0\ \cancel{°C} \times \frac{4.184\ J}{\cancel{g}\ \cancel{°C}} = 13\,600\ J\ (3\ SFs)$$

∴ Total heat needed $= 16\,700\ J + 13\,600\ J = 30\,300\ J\ (3\ SFs)$

c. Two calculations are needed:

(1) steam 100 °C $\longrightarrow$ water 100 °C: $15.0\ \cancel{g\ steam} \times \dfrac{2260\ J}{1\ \cancel{g\ steam}} \times \dfrac{1\ kJ}{1000\ \cancel{J}} = 33.9\ kJ\ (3\ SFs)$

(2) water 100 °C $\longrightarrow$ 0 °C: $15.0\ \cancel{g\ water} \times 100.\ \cancel{°C} \times \dfrac{4.184\ \cancel{J}}{\cancel{g}\ \cancel{°C}} \times \dfrac{1\ kJ}{1000\ \cancel{J}} = 6.28\ kJ\ (3\ SFs)$

∴ Total heat released $= 33.9\ kJ + 6.28\ kJ = 40.2\ kJ\ (3\ SFs)$

d. Three calculations are needed:

(1) ice 0 °C $\longrightarrow$ water 0 °C: $24.0\ \cancel{g\ ice} \times \dfrac{334\ \cancel{J}}{1\ \cancel{g\ ice}} \times \dfrac{1\ kJ}{1000\ \cancel{J}} = 8.02\ kJ\ (3\ SFs)$

(2) water 0 °C $\longrightarrow$ 100 °C: $24.0\ \cancel{g\ water} \times 100.\ \cancel{°C} \times \dfrac{4.184\ \cancel{J}}{\cancel{g}\ \cancel{°C}} \times \dfrac{1\ kJ}{1000\ \cancel{J}} = 10.0\ kJ\ (3\ SFs)$

(3) water 100 °C $\longrightarrow$ steam 100 °C: $24.0\ \cancel{g\ water} \times \dfrac{2260\ \cancel{J}}{1\ \cancel{g\ water}} \times \dfrac{1\ kJ}{1000\ \cancel{J}} = 54.2\ kJ\ (3\ SFs)$

∴ Total heat needed $= 8.02\ kJ + 10.0\ kJ + 54.2\ kJ = 72.2\ kJ\ (3\ SFs)$

10.51 Two calculations are needed:

(1) ice 0 °C $\longrightarrow$ water 0 °C: $275\ \cancel{g\ ice} \times \dfrac{334\ \cancel{J}}{1\ \cancel{g\ ice}} \times \dfrac{1\ kJ}{1000\ \cancel{J}} = 91.9\ kJ\ (3\ SFs)$

(2) water 0 °C $\longrightarrow$ 24.0 °C: $\Delta T = 24.0\,°C - 0\,°C = 24.0\,°C$;

$$275\ \cancel{g\ water} \times 24.0\ \cancel{°C} \times \frac{4.184\ \cancel{J}}{\cancel{g}\ \cancel{°C}} \times \frac{1\ kJ}{1000\ \cancel{J}} = 27.6\ kJ\ (3\ SFs)$$

∴ Total heat absorbed $= 91.9\ kJ + 27.6\ kJ = 119.5\ kJ\ (4\ SFs)$

10.53 $15.8\ \cancel{g\ N_2} \times \dfrac{198\ J}{1\ \cancel{g\ N_2}} = 3130\ J$ needed (3 SFs)

10.55 a. C—C electronegativity difference $2.5 - 2.5 = 0.0$; nonpolar covalent
    b. C—H electronegativity difference $2.5 - 2.1 = 0.4$; nonpolar covalent
    c. C—O electronegativity difference $3.5 - 2.5 = 1.0$; polar covalent

10.57 a. $1 C (4 e^-) + 3 O (6 e^-) + 2 e^- (\text{negative charge}) = 4 + 18 + 2 = 24$ valence electrons

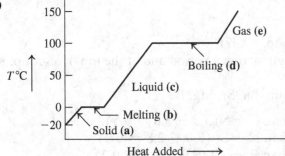

three electron groups around C bonded to three atoms; trigonal planar shape

b. The central atom C has three electron groups bonded to three atoms, which gives CO_3^{2-} a trigonal planar shape.

10.59 a. 2 valence electrons, 1 bonding pair, no lone pairs
b. 8 valence electrons, 1 bonding pair, 3 lone pairs
c. 14 valence electrons, 1 bonding pair, 6 lone pairs

10.61 a. 2; trigonal pyramidal shape, polar molecule
b. 1; bent shape (109°), polar molecule
c. 3; tetrahedral shape, nonpolar molecule

10.63 a. C—O (EN $3.5 - 2.5 = 1.0$) and N—O (EN $3.5 - 3.0 = 0.5$) are polar covalent bonds.
b. O—O (EN $3.5 - 3.5 = 0.0$) is a nonpolar covalent bond.
c. Ca—O (EN $3.5 - 1.0 = 2.5$) and K—O (EN $3.5 - 0.8 = 2.7$) are ionic bonds.
d. C—O, N—O, O—O

10.65 a. PH_3 is a nonpolar molecule. Dispersion forces occur between temporary dipoles in nonpolar molecules.
b. NO_2 is a polar molecule. Dipole–dipole attractions occur between dipoles in polar molecules.
c. Hydrogen bonds are strong dipole–dipole attractions that occur between a partially positive hydrogen atom of one molecule and one of the strongly electronegative atoms N, O, or F in another.
d. Dispersion forces occur between temporary dipoles in Ar atoms.

10.67 a. The heat from the skin is used to evaporate the water (perspiration). Therefore, the skin is cooled.
b. On a hot day, there are more liquid water molecules in the damp towels that have sufficient energy to become water vapor. Thus, water evaporates from the towels more readily on a hot day.
c. In a closed plastic bag, some water molecules evaporate, but they cannot escape and will condense back to liquid; the clothes will not dry.

10.69

10.71 a. The melting point of chloroform is about −60 °C.
b. The boiling point of chloroform is about 60 °C.
c. The diagonal line **A** represents the solid state as temperature increases. The horizontal line **B** represents the change from solid to liquid, or melting of the substance. The diagonal line **C** represents the liquid state as temperature increases. The horizontal line **D** represents the change from liquid to gas, or boiling of the liquid. The diagonal line **E** represents the gas state as temperature increases.
d. At −80 °C, it is solid; at −40 °C, it is liquid; at 25 °C, it is liquid; at 80 °C, it is gas.

10.73 a. $1\,H(1\,e^-) + 1\,N(5\,e^-) + 2\,O(6\,e^-) = 1 + 5 + 12 = 18$ valence electrons
b. $2\,C(4\,e^-) + 4\,H(1\,e^-) + 1\,O(6\,e^-) = 8 + 4 + 6 = 18$ valence electrons
c. $1\,P(5\,e^-) + 4\,H(1\,e^-) - 1\,e^-\,(\text{positive charge}) = 5 + 4 - 1 = 8$ valence electrons
d. $1\,S(6\,e^-) + 3\,O(6\,e^-) + 2\,e^-\,(\text{negative charge}) = 6 + 18 + 2 = 26$ valence electrons

10.75 a. $1\,B(3\,e^-) + 4\,F(7\,e^-) + 1\,e^-\,(\text{negative charge}) = 3 + 28 + 1 = 32$ valence electrons

b. $2\,Cl(7\,e^-) + 1\,O(6\,e^-) = 14 + 6 = 20$ valence electrons

c. $1\,N(5\,e^-) + 3\,H(1\,e^-) + 1\,O(6\,e^-) = 5 + 3 + 6 = 14$ valence electrons

d. $2\,C(4\,e^-) + 2\,H(1\,e^-) + 2\,Cl(7\,e^-) = 8 + 2 + 14 = 24$ valence electrons

10.77 a. $3\,N(5\,e^-) + 1\,e^-\,(\text{negative charge}) = 15 + 1 = 16$ valence electrons

b. $1\,N(5\,e^-) + 2\,O(6\,e^-) - 1\,e^-\,(\text{positive charge}) = 5 + 12 - 1 = 16$ valence electrons

c. $1\,H(1\,e^-) + 1\,C(4\,e^-) + 2\,O(6\,e^-) + 1\,e^-\,(\text{negative charge})$
$= 1 + 4 + 12 + 1 = 18$ valence electrons

10.79 a. Electronegativity increases going up a group: I, Cl, F
b. Electronegativity increases going left to right across a period and at the top of a group: K, Li, S, Cl
c. Electronegativity increases going up a group: Ba, Sr, Mg, Be

10.81 a. C—O $(3.5 - 2.5 = 1.0)$ is more polar than C—N$(3.5 - 3.0 = 0.5)$.
b. N—F $(4.0 - 3.0 = 1.0)$ is more polar than N—Br $(3.0 - 2.8 = 0.2)$.
c. S—Cl $(3.0 - 2.5 = 0.5)$ is more polar than Br—Cl $(3.0 - 2.8 = 0.2)$.
d. Br—I $(2.8 - 2.5 = 0.3)$ is more polar than Br—Cl $(3.0 - 2.8 = 0.2)$.
e. N—F $(4.0 - 3.0 = 1.0)$ is more polar than N—O $(3.5 - 3.0 = 0.5)$.

10.83 A dipole arrow points from the atom with the lower electronegativity value (more positive) to the atom in the bond that has the higher electronegativity value (more negative).

a. $\overset{\delta^+}{Si}\,\text{—}\,\overset{\delta^-}{Cl}$
$\longmapsto$

b. $\overset{\delta^+}{C}\,\text{—}\,\overset{\delta^-}{N}$
$\longmapsto$

c. $\overset{\delta^-}{F}\,\text{—}\,\overset{\delta^+}{Cl}$
$\longleftarrowtail$

d. $\overset{\delta^+}{C}\,\text{—}\,\overset{\delta^-}{F}$
$\longmapsto$

e. $\overset{\delta^+}{N}\,\text{—}\,\overset{\delta^-}{O}$
$\longmapsto$

10.85 a. Si—Cl electronegativity difference: $3.0 - 1.8 = 1.2$, polar covalent
b. C—C electronegativity difference: $2.5 - 2.5 = 0.0$, nonpolar covalent
c. Na—Cl electronegativity difference: $3.0 - 0.9 = 2.1$, ionic
d. C—H electronegativity difference: $2.5 - 2.1 = 0.4$, nonpolar covalent
e. F—F electronegativity difference: $4.0 - 4.0 = 0.0$, nonpolar covalent

10.87 a. $1\,N(5\,e^-) + 3\,F(7\,e^-) = 5 + 21 = 26$ valence electrons

$$:\!\ddot{F}\!-\!\ddot{N}\!-\!\ddot{F}\!:$$
$$|$$
$$:\!\ddot{F}\!:$$

The central atom N has four electron groups with three bonded atoms and one lone pair of electrons, which gives NF_3 a trigonal pyramidal shape.

b. $1\,Si(4\,e^-) + 4\,Br(7\,e^-) = 4 + 28 = 32$ valence electrons

$$:\!\ddot{Br}\!:$$
$$|$$
$$:\!\ddot{Br}\!-\!Si\!-\!\ddot{Br}\!:$$
$$|$$
$$:\!\ddot{Br}\!:$$

The central atom Si has four electron groups bonded to four bromine atoms; $SiBr_4$ has a tetrahedral shape.

c. $1\,C(4\,e^-) + 2\,Se(6\,e^-) = 4 + 12 = 16$ valence electrons

$$:\!\ddot{Se}\!=\!C\!=\!\ddot{Se}\!:$$

The central atom C has two electron groups bonded to two selenium atoms; CSe_2 has a linear shape.

d. $1\,S(6\,e^-) + 2\,O(6\,e^-) = 6 + 12 = 18$ valence electrons

$$:\!\ddot{O}\!=\!\ddot{S}\!-\!\ddot{O}\!: \;\longleftrightarrow\; :\!\ddot{O}\!-\!\ddot{S}\!=\!\ddot{O}\!:$$

The central atom S has three electron groups with two bonded atoms and one lone pair of electrons, which gives SO_2 a bent shape $(120°)$.

10.89 a. $1 \, Br(7 \, e^-) + 2 \, O(6 \, e^-) + 1 \, e^-(\text{negative charge}) = 7 + 12 + 1 = 20$ valence electrons

$$\left[\ddot{\underset{\cdot\cdot}{O}} - \ddot{Br} - \ddot{\underset{\cdot\cdot}{O}} \right]^{-}$$

The central atom Br has four electron groups with two bonded atoms and two lone pairs of electrons, which gives BrO_2^- a bent shape ($109°$).

b. $1 \, O(6 \, e^-) + 2 \, H(1 \, e^-) = 6 + 2 = 8$ valence electrons

$$\begin{array}{c} H \\ | \\ \ddot{\underset{\cdot\cdot}{O}} - H \end{array}$$

The central atom O has four electron groups with two bonded atoms and two lone pairs of electrons, which gives H_2O a bent shape ($109°$).

c. $1 \, C(4 \, e^-) + 4 \, Br(7 \, e^-) = 4 + 28 = 32$ valence electrons

$$\begin{array}{c} \ddot{\underset{\cdot\cdot}{Br}} \\ | \\ \ddot{\underset{\cdot\cdot}{Br}} - C - \ddot{\underset{\cdot\cdot}{Br}} \\ | \\ \ddot{\underset{\cdot\cdot}{Br}} \end{array}$$

The central atom C has four electron groups bonded to four bromine atoms, which gives CBr_4 a tetrahedral shape.

d. $1 \, P(5 \, e^-) + 3 \, O(6 \, e^-) + 3 \, e^-(\text{negative charge}) = 5 + 18 + 3 = 26$ valence electrons

$$\left[\begin{array}{c} \ddot{\underset{\cdot\cdot}{O}} \\ | \\ \ddot{\underset{\cdot\cdot}{O}} - P - \ddot{\underset{\cdot\cdot}{O}} \end{array} \right]^{3-}$$

The central atom P has four electron groups with three bonded atoms and one lone pair of electrons, which gives PO_3^{3-} a trigonal pyramidal shape.

10.91 a. A molecule that has a central atom with three bonded atoms and one lone pair will have a trigonal pyramidal shape. This asymmetric shape means the dipoles do not cancel, and the molecule will be polar.

b. A molecule that has a central atom with two bonded atoms and two lone pairs will have a bent shape ($109°$). This asymmetric shape means the dipoles do not cancel, and the molecule will be polar.

10.93 a. The molecule HBr contains only a single polar covalent H—Br bond (EN $2.8 - 2.1 = 0.7$); it is a polar molecule.

$$H - \ddot{Br}\!:$$

b. The molecule SiO_2 contains two polar covalent Si—O bonds (EN $3.5 - 1.8 = 1.7$) and has a linear shape. The two equal dipoles directed away from each other at $180°$ will cancel, resulting in a nonpolar molecule.

$$\ddot{\underset{\cdot\cdot}{O}} = Si = \ddot{\underset{\cdot\cdot}{O}}\!:$$

c. The molecule NCl_3 contains three nonpolar covalent N—Cl bonds (EN $3.0 - 3.0 = 0.0$) and a lone pair of electrons on the central N atom, which gives the molecule a trigonal pyramidal shape. Since the molecule contains only nonpolar bonds, NCl_3 is a nonpolar molecule.

$$\begin{array}{c} Cl^{\,\text{\tiny||||}}\ddot{N} \\ Cl \qquad \diagdown Cl \end{array}$$

d. The molecule CH_3Cl consists of a central atom, C, with three nonpolar covalent C—H bonds (EN $2.5 - 2.1 = 0.4$) and one polar covalent C—Cl bond (EN $3.0 - 2.5 = 0.5$). The molecule has a tetrahedral shape, but the single dipole makes CH_3Cl a polar molecule.

e. The molecule NI_3 contains three polar covalent N—I bonds (EN $3.0 - 2.5 = 0.5$) and a lone pair of electrons on the central N atom, which gives the molecule a trigonal pyramidal shape. This asymmetric shape does not allow the dipoles to cancel, which makes NI_3 a polar molecule.

f. The molecule H_2O contains two polar covalent O—H bonds (EN $3.5 - 2.1 = 1.4$) and two lone pairs of electrons on the central O atom. The molecule has a bent shape ($109°$) and the asymmetric geometry of the O—H dipoles makes H_2O a polar molecule.

10.95 a. NF_3 is a polar molecule. Dipole–dipole attractions (2) occur between dipoles in polar molecules.

b. ClF is a polar molecule (EN $4.0 - 3.0 = 1.0$). Dipole–dipole attractions (2) occur between dipoles in polar molecules.

c. Dispersion forces (4) occur between temporary dipoles in nonpolar Br_2 molecules.

d. Ionic bonds (1) are strong attractions between positive and negative ions, as in Cs_2O.

e. Dispersion forces (4) occur between temporary dipoles in nonpolar C_4H_{10} molecules.

f. CH_3OH is a polar molecule and contains the polar O—H bond. Hydrogen bonding (3) involves strong dipole–dipole attractions that occur between a partially positive hydrogen atom of one polar molecule and one of the strongly electronegative atoms F, O, or N in another, as is seen with CH_3OH molecules.

10.97 When water vapor condenses or liquid water freezes, heat is released, which warms the air.

10.99 Given 1540 J absorbed; heat of fusion for water $= 334$ J/g H_2O

Need grams of ice that will melt at 0 °C

Plan J $\longrightarrow$ g of H_2O $\dfrac{1 \text{ g } H_2O}{334 \text{ J}}$

Set-up $1540 \text{ J} \times \dfrac{1 \text{ g } H_2O}{334 \text{ J}} = 4.61$ g of H_2O (3 SFs)

10.101 Given 5.25 kJ removed; heat of fusion for acetic acid $= 192$ J/g acetic acid

Need grams of acetic acid that will freeze at its freezing point

Plan kJ $\longrightarrow$ J $\longrightarrow$ g of acetic acid $\dfrac{1000 \text{ J}}{1 \text{ kJ}}$ $\dfrac{1 \text{ g acetic acid}}{192 \text{ J}}$

Set-up $5.25 \text{ kJ} \times \dfrac{1000 \text{ J}}{1 \text{ kJ}} \times \dfrac{1 \text{ g acetic acid}}{192 \text{ J}} = 27.3$ g of acetic acid (3 SFs)

10.103 **a.** $3\,H(1\,e^-) + 1\,N(5\,e^-) + 1\,C(4\,e^-) + 1\,O(6\,e^-) = 3 + 5 + 4 + 6 = 18$ valence electrons

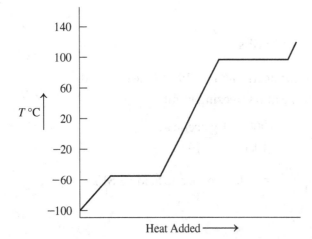

b. $1\,Cl(7\,e^-) + 2\,H(1\,e^-) + 2\,C(4\,e^-) + 1\,N(5\,e^-) = 7 + 2 + 8 + 5 = 22$ valence electrons

c. $2\,H(1\,e^-) + 2\,N(5\,e^-) = 2 + 10 = 12$ valence electrons

$$H-\overset{..}{N}=\overset{..}{N}-H$$

d. $1\,Cl(7\,e^-) + 2\,C(4\,e^-) + 2\,O(6\,e^-) + 3\,H(1\,e^-) = 7 + 8 + 12 + 3 = 30$ valence electrons

10.105 **a.** The central atom N has four electron groups with three bonded atoms and one lone pair of electrons, which gives NH_2Cl a trigonal pyramidal shape.

$$H-\overset{..}{N}-\overset{..}{\underset{..}{C}l}:$$
$$\;\;\;\;\;\;\;|$$
$$\;\;\;\;\;\;H$$

b. The central atom P has four electron pairs bonded to four H atoms; PH_4^+ has a tetrahedral shape.

c. The central atom C has two electron groups bonded to two atoms and no lone pairs, which gives SCN^- a linear shape.

$$\left[:S\equiv C-\overset{..}{\underset{..}{N}}:\right]^-$$

10.107

Copyright © 2017 Pearson Education, Inc.

a. Dibromomethane is a solid at -75 °C.

b. At -53 °C, solid dibromomethane melts (solid $\longrightarrow$ liquid).

c. Dibromomethane is a liquid at -18 °C.

d. At 110 °C, dibromomethane is a gas.

e. Both solid and liquid dibromomethane will be present at the melting temperature of -53 °C.

10.109 Two calculations are required:

(1) ice 0 °C $\longrightarrow$ water 0 °C:

$$\text{Heat} = \text{mass} \times \text{heat of fusion} = 45.0 \ \cancel{g} \times \frac{334 \ \text{J}}{\cancel{g}} = 1.50 \times 10^4 \ \text{J} \ (3 \ \text{SFs})$$

(2) water 8.0 °C $\longrightarrow$ 0.0 °C:

$$\text{Heat} = \text{mass} \times \Delta T \times SH$$

$$\therefore \text{Mass}_{\text{water}} = \frac{\text{heat}}{\Delta T \times SH} = \frac{1.50 \times 10^4 \ \cancel{J}}{(8.0 \ \cancel{°C}) \left(\dfrac{4.184 \ \cancel{J}}{1 \ \text{g} \ \cancel{°C}} \right)} = 450 \ \text{g of water} \ (2 \ \text{SFs})$$

10.111 Two calculations are required:

(1) water 25 °C $\longrightarrow$ 0 °C: $325 \ \cancel{g} \times 25 \ \cancel{°C} \times \dfrac{4.184 \ \cancel{J}}{\cancel{g} \ \cancel{°C}} \times \dfrac{1 \ \text{kJ}}{1000 \ \cancel{J}} = 34 \ \text{kJ}$

(2) water 0 °C $\longrightarrow$ ice 0 °C: $325 \ \cancel{g} \times \dfrac{334 \ \cancel{J}}{1 \ \cancel{g}} \times \dfrac{1 \ \text{kJ}}{1000 \ \cancel{J}} = 109 \ \text{kJ}$

$\therefore$ Total heat removed $= 34 \ \text{kJ} + 109 \ \text{kJ} = 143 \ \text{kJ removed} \ (3 \ \text{SFs})$

Selected Answers to Combining Ideas from Chapters 8 to 10

CI.13 a. $8.56 \text{ g Cu} \times \dfrac{1 \text{ mol Cu}}{63.55 \text{ g Cu}} \times \dfrac{6.022 \times 10^{23} \text{ Cu atoms}}{1 \text{ mol Cu}} = 8.11 \times 10^{22}$ atoms of Cu (3 SFs)

b. $2Cu(s) + O_2(g) \longrightarrow 2CuO(s)$

c. The reaction of the elements copper and oxygen to form copper(II) oxide is a combination reaction.

d. $8.56 \text{ g Cu} \times \dfrac{1 \text{ mol Cu}}{63.55 \text{ g Cu}} \times \dfrac{1 \text{ mol O}_2}{2 \text{ mol Cu}} \times \dfrac{32.00 \text{ g O}_2}{1 \text{ mol O}_2} = 2.16$ g of O_2 (3 SFs)

e. From part **d**, we know that 8.56 g of Cu reacts with 2.16 g of O_2
∴ there is excess O_2; Cu is the limiting reactant.
Molar mass of CuO = 1 mol of Cu + 1 mol of O = 63.55 g + 16.00 g = 79.55 g

$8.56 \text{ g Cu} \times \dfrac{1 \text{ mol Cu}}{63.55 \text{ g Cu}} \times \dfrac{2 \text{ mol CuO}}{2 \text{ mol Cu}} \times \dfrac{79.55 \text{ g CuO}}{1 \text{ mol CuO}} = 10.7$ g of CuO (3 SFs)

f. Theoretical yield of CuO (from part **e**) = 10.7 g of CuO
Use the percent yield to convert theoretical to actual:

$10.7 \text{ g CuO} \times \dfrac{85.0 \text{ g CuO}}{100 \text{ g CuO}} = 9.10$ g of CuO (actual) (3 SFs)

CI.15 a. The formula of sodium hypochlorite is NaClO.
Molar mass of NaClO
$= 1$ mol of Na + 1 mol of Cl + 1 mol of O $= 22.99 \text{ g} + 35.45 \text{ g} + 16.00 \text{ g} = 74.44$ g

b. $1 \text{ Cl}(7 \, e^-) + 1 \text{ O}(6 \, e^-) + 1 \, e^- (\text{negative charge}) = 7 + 6 + 1 = 14$ valence electrons

$$\left[:\ddot{\underset{..}{\text{Cl}}} - \ddot{\underset{..}{\text{O}}}: \right]^-$$

c. **Given** gallons of bleach   **Need** number of ClO^- ions
Plan gal → qt → mL → g of solution → g of NaClO → mol of NaClO → mol of ClO^-
→ number of ClO^- ions

Set-up $1.00 \text{ gal} \times \dfrac{4 \text{ qt}}{1 \text{ gal}} \times \dfrac{946.4 \text{ mL}}{1 \text{ qt}} \times \dfrac{1.08 \text{ g solution}}{1 \text{ mL solution}} \times \dfrac{5.25 \text{ g NaClO}}{100 \text{ g solution}}$

$\times \dfrac{1 \text{ mol NaClO}}{74.44 \text{ g NaClO}} \times \dfrac{1 \text{ mol ClO}^-}{1 \text{ mol NaClO}} \times \dfrac{6.022 \times 10^{23} \text{ ClO}^- \text{ ions}}{1 \text{ mol ClO}^-}$

$= 1.74 \times 10^{24} \text{ ClO}^-$ ions (3 SFs)

d. $2NaOH(aq) + Cl_2(g) \longrightarrow NaClO(aq) + NaCl(aq) + H_2O(l)$

e. **Given** gallons of bleach   **Need** grams of NaOH
Plan gal → qt → mL → g of solution → g of NaClO → mol of NaClO → mol of NaOH
→ g of NaOH

Set-up $1.00 \text{ gal} \times \dfrac{4 \text{ qt}}{1 \text{ gal}} \times \dfrac{946.4 \text{ mL}}{1 \text{ qt}} \times \dfrac{1.08 \text{ g solution}}{1 \text{ mL solution}} \times \dfrac{5.25 \text{ g NaClO}}{100 \text{ g solution}}$

$\times \dfrac{1 \text{ mol NaClO}}{74.44 \text{ g NaClO}} \times \dfrac{2 \text{ mol NaOH}}{1 \text{ mol NaClO}} \times \dfrac{40.00 \text{ g NaOH}}{1 \text{ mol NaOH}}$

$= 231$ g of NaOH (3 SFs)

f. $165 \text{ g Cl}_2 \times \dfrac{1 \text{ mol Cl}_2}{70.90 \text{ g Cl}_2} \times \dfrac{1 \text{ mol NaClO}}{1 \text{ mol Cl}_2} \times \dfrac{74.44 \text{ g NaClO}}{1 \text{ mol NaClO}} = 173$ g of NaClO (smaller amount of product)

$275 \text{ g NaOH} \times \dfrac{1 \text{ mol NaOH}}{40.00 \text{ g NaOH}} \times \dfrac{1 \text{ mol NaClO}}{2 \text{ mol NaOH}} \times \dfrac{74.44 \text{ g NaClO}}{1 \text{ mol NaClO}} = 256$ g of NaClO

∴ Cl_2 is the limiting reactant and 173 g of NaClO can be produced (theoretical).

∴ Percent yield: $\dfrac{162 \text{ g NaClO (actual)}}{173 \text{ g NaClO (theoretical)}} \times 100\% = 93.6\%$ (3 SFs)

CI.17 a. $3 \text{ Cl}(7\,e^-) + 2 \text{ C}(4\,e^-) + 2 \text{ O}(6\,e^-) + 3 \text{ H}(1\,e^-) = 21 + 8 + 12 + 3$
$$= 44 \text{ valence electrons}$$

Chloral hydrate

$3 \text{ Cl}(7\,e^-) + 2 \text{ C}(4\,e^-) + 1 \text{ O}(6\,e^-) + 1 \text{ H}(1\,e^-) = 21 + 8 + 6 + 1$
$$= 36 \text{ valence electrons}$$

Chloral

b. Chloral hydrate, $C_2H_3O_2Cl_3 \rightarrow C_{(2\div1)}H_{(3\div1)}O_{(2\div1)}Cl_{(3\div1)} = C_2H_3O_2Cl_3$
(empirical formula)

Chloral, $C_2HOCl_3 \rightarrow C_{(2\div1)}H_{(1\div1)}O_{(1\div1)}Cl_{(3\div1)} = C_2HOCl_3$ (empirical formula)

c. $2 \text{ mol C} \times \dfrac{12.01 \text{ g C}}{1 \text{ mol C}} = 24.02 \text{ g of C}$

$3 \text{ mol H} \times \dfrac{1.008 \text{ g H}}{1 \text{ mol H}} = 3.024 \text{ g of H}$

$2 \text{ mol O} \times \dfrac{16.00 \text{ g O}}{1 \text{ mol O}} = 32.00 \text{ g of O}$

$3 \text{ mol Cl} \times \dfrac{35.45 \text{ g Cl}}{1 \text{ mol Cl}} = 106.4 \text{ g of Cl}$

∴ Molar mass of $C_2H_3O_2Cl_3 = \overline{165.4 \text{ g}}$

Mass % Cl $= \dfrac{106.4 \text{ g Cl}}{165.4 \text{ g } C_2H_3O_2Cl_3} \times 100\% = 64.33\%$ Cl (4 SFs)

CI.19 a.

b. The molecular formula of acetone is C_3H_6O.
Molar mass of C_3H_6O
$= 3 \text{ mol of C} + 6 \text{ mol of H} + 1 \text{ mol of O} = 3(12.01 \text{ g}) + 6(1.008 \text{ g}) + 16.00 \text{ g} = 58.08 \text{ g}$

c. $C_3H_6O(l) + 4O_2(g) \xrightarrow{\Delta} 3CO_2(g) + 3H_2O(g) + 1790 \text{ kJ}$

d. Combustion is an exothermic reaction.

e. $2.58 \text{ g } C_3H_6O \times \dfrac{1 \text{ mol } C_3H_6O}{58.08 \text{ g } C_3H_6O} \times \dfrac{1790 \text{ kJ}}{1 \text{ mol } C_3H_6O} = 79.5 \text{ kJ}$ (3 SFs)

f. $15.0 \text{ mL } C_3H_6O \times \dfrac{0.786 \text{ g } C_3H_6O}{1 \text{ mL } C_3H_6O} \times \dfrac{1 \text{ mol } C_3H_6O}{58.08 \text{ g } C_3H_6O} \times \dfrac{4 \text{ mol } O_2}{1 \text{ mol } C_3H_6O} \times \dfrac{32.00 \text{ g } O_2}{1 \text{ mol } O_2}$
$= 26.0 \text{ g of } O_2$ (3 SFs)

11

Gases

After Whitney's asthma attack, she and her parents learned to identify and treat her asthma. Because her asthma is mild and intermittent, they follow a management plan to control the symptoms. Whitney participates in physical activity and exercise to maintain lung function and to prevent recurrent attacks. If her symptoms worsen, she uses a bronchodilator to help her breathing. If a child has a lung capacity of 2.0 L, a body temperature of 37 °C, and a lung pressure of 756 mmHg, what would the volume be at STP?

Credit: Adam Gault/Alamy

CHAPTER READINESS*

⚛ **Key Math Skill**

- Solving Equations (1.4)

🔬 **Core Chemistry Skills**

- Using Significant Figures in Calculations (2.3)

- Writing Conversion Factors from Equalities (2.5)
- Using Conversion Factors (2.6)
- Using Molar Mass as a Conversion Factor (7.3)
- Using Mole–Mole Factors (9.2)
- Converting Grams to Grams (9.3)

*These Key Math Skills and Core Chemistry Skills from previous chapters are listed here for your review as you proceed to the new material in this chapter.

LOOKING AHEAD

11.1 Properties of Gases

Learning Goal: Describe the kinetic molecular theory of gases and the units of measurement used for gases.

- In a gas, small particles (atoms or molecules) are far apart and moving randomly with high velocities.
- The attractive forces between the particles of a gas are usually very small.
- The actual volume occupied by gas molecules is extremely small compared with the volume that the gas occupies.

Gases

- Gas particles are in constant motion, moving rapidly in straight paths.
- The average kinetic energy of gas molecules is proportional to the Kelvin temperature.
- A gas is described by the physical properties of pressure (P), volume (V), temperature (T), and amount (n).
- A gas exerts pressure, which is the force that gas particles exert on the walls of a container.
- Units of gas pressure include torr (Torr), millimeters of mercury (mmHg), pascals (Pa), kilopascals (kPa), $lb/in.^2$ (psi), and atmospheres (atm).

♦ **Learning Exercise 11.1A**

Answer *true* (T) or *false* (F) for each of the following:

a. _____ Gases are composed of small particles.

b. _____ Gas molecules are usually close together.

c. _____ Gas molecules move rapidly because they are strongly attracted.

d. _____ Distances between gas molecules are large.

e. _____ Gas molecules travel in straight lines until they collide.

f. _____ Pressure is the force of gas particles striking the walls of a container.

g. _____ Kinetic energy decreases with increasing temperature.

h. _____ 450 K is a measure of temperature.

i. _____ 3.56 g is a measure of pressure.

Answers   **a.** T   **b.** F   **c.** F   **d.** T   **e.** T
             f. T   **g.** F   **h.** T   **i.** F

Units for Measuring Pressure

Unit	Abbreviation	Unit Equivalent to 1 atm
atmosphere	atm	1 atm (exact)
millimeters of Hg	mmHg	760 mmHg (exact)
torr	Torr	760 Torr (exact)
inches of Hg	inHg	29.9 inHg
pounds per square inch	$lb/in.^2$ (psi)	$14.7\ lb/in.^2$
pascal	Pa	101 325 Pa
kilopascal	kPa	101.325 kPa

♦ **Learning Exercise 11.1B**

Complete the following:

a. 1.50 atm = _____ Torr

b. 550 mmHg = _____ atm

c. 0.725 atm = _____ kPa

d. 1520 Torr = _____ atm

e. 30.5 psi = _____ mmHg

f. 87.6 kPa = _____ Torr

g. During the weather report on TV, the pressure is given as 29.4 inHg. What is this pressure in millimeters of mercury and in atmospheres?

Answers	**a.** 1140 Torr	**b.** 0.72 atm	**c.** 73.5 kPa	**d.** 2.00 atm
	e. 1580 mmHg	**f.** 657 Torr	**g.** 747 mmHg, 0.983 atm	

11.2 Pressure and Volume (Boyle's Law)

Learning Goal: Use the pressure–volume relationship (Boyle's law) to calculate the unknown pressure or volume when the temperature and amount of gas are constant.

- According to Boyle's law, pressure increases if volume decreases and pressure decreases if volume increases.
- If two properties change in opposite directions, the properties have an inverse relationship.
- The volume (V) of a gas changes inversely with the pressure (P) of the gas when T and n are held constant:

$$P_1 V_1 = P_2 V_2$$ The subscripts 1 and 2 represent initial and final conditions.

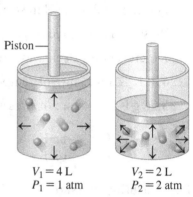

Piston—

$V_1 = 4\ L$
$P_1 = 1\ atm$

$V_2 = 2\ L$
$P_2 = 2\ atm$

Guide to Using the Gas Laws	
STEP 1	State the given and needed quantities.
STEP 2	Rearrange the gas law equation to solve for the unknown quantity.
STEP 3	Substitute values into the gas law equation and calculate.

Key Terms for Sections 11.1 and 11.2

Match the following key terms with the correct description:

    a. kinetic molecular theory   **b.** inverse relationship   **c.** Boyle's law
    d. pressure   **e.** atmospheric pressure

1. _____ the pressure exerted by the atmosphere

2. _____ a force exerted by gas particles when they collide with the sides of a container

3. _____ a relationship in which two properties change in opposite directions

4. _____ the volume of a gas varies inversely with the pressure of a gas, if there is no change in temperature and amount of gas

5. _____ the model that explains the behavior of gas particles

Answers	**1.** e	**2.** d	**3.** b	**4.** c	**5.** a

♦ **Learning Exercise 11.2**

Solve each of the following gas problems using Boyle's law:

a. A sample of 4.0 L of He has a pressure of 800. mmHg.
What is the pressure, in millimeters of mercury, when the volume is reduced to 1.0 L, if there is no change in temperature and amount of gas?

b. The maximum volume of air that can fill a person's lungs is 6250 mL at a pressure of 728 mmHg. Inhalation occurs as the pressure in the lungs drops to 721 mmHg with no change in temperature and amount of gas. To what volume, in milliliters, do the lungs expand, if there is no change in temperature and amount of gas?

c. A sample of O_2 at a pressure of 5.00 atm has a volume of 3.00 L. If the gas pressure is changed to 760. mmHg, what volume, in liters, will the O_2 occupy, if there is no change in temperature and amount of gas?

d. A sample of 250. mL of N_2 is initially at a pressure of 2.50 atm. If the pressure changes to 825 mmHg, what is the volume, in milliliters, if there is no change in temperature and amount of gas?

Answers   **a.** 3200 mmHg   **b.** 6310 mL   **c.** 15.0 L   **d.** 576 mL

11.3   Temperature and Volume (Charles's Law)

Learning Goal: Use the temperature–volume relationship (Charles's law) to calculate the unknown temperature or volume when the pressure and amount of gas are constant.

- According to Charles's law, at constant pressure and quantity, the volume of a gas increases when the temperature of the gas increases, and the volume decreases when the temperature decreases.

- In a direct relationship, two properties increase or decrease together.

- The volume (V) of a gas is directly related to its Kelvin temperature (T) when there is no change in the pressure or moles of the gas:

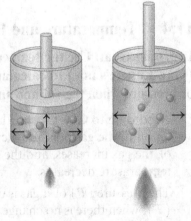

$T_1 = 200$ K
$V_1 = 1$ L

$T_2 = 400$ K
$V_2 = 2$ L

$$\frac{V_1}{T_1} = \frac{V_2}{T_2}$$ The subscripts 1 and 2 represent initial and final conditions.

♦ **Learning Exercise 11.3**

Solve each of the following problems using Charles's law:

a. A balloon has a volume of 2.5 L at a temperature of 0 °C. What is the volume, in liters, of the balloon when the temperature rises to 120 °C, if there is no change in pressure and amount of gas?

b. A balloon is filled with He to a volume of 6600 L at a temperature of 223 °C. To what temperature, in degrees Celsius, must the gas be cooled to decrease the volume to 4800 L, if there is no change in pressure and amount of gas?

c. A sample of 750. mL of Ne is heated from 120. °C to 350. °C. What is the volume, in milliliters, if there is no change in pressure and amount of gas?

d. What is the temperature, in degrees Celsius, of 350. mL of O_2 at 22 °C when its volume increases to 0.800 L, if there is no change in pressure and amount of gas?

Answers　　**a.** 3.6 L　　**b.** 88 °C　　**c.** 1190 mL　　**d.** 401 °C

11.4　Temperature and Pressure (Gay-Lussac's Law)

Learning Goal: Use the temperature–pressure relationship (Gay-Lussac's law) to calculate the unknown temperature or pressure when the volume and amount of gas are constant.

- According to Gay-Lussac's law, at constant volume and amount, the gas pressure increases when the temperature of the gas increases, and the pressure decreases when the temperature decreases.

- The pressure (P) of a gas is directly related to its Kelvin temperature (T), when there is no change in the volume or moles of the gas:

$$\frac{P_1}{T_1} = \frac{P_2}{T_2}$$　The subscripts 1 and 2 represent initial and final conditions.

$T_1 = 200$ K　　$T_2 = 400$ K
$P_1 = 1$ atm　　$P_2 = 2$ atm

♦ **Learning Exercise 11.4**

Solve each of the following problems using Gay-Lussac's law:

a. A sample of He has a pressure of 860. mmHg at a temperature of 225 K. What is the pressure, in millimeters of mercury, when the He reaches a temperature of 675 K, if there is no change in volume or amount of gas?

b. A balloon contains a gas with a pressure of 580. mmHg and a temperature of 227 °C. What is the pressure, in millimeters of mercury, of the gas when the temperature drops to 27 °C, if there is no change in volume or amount of gas?

c. A spray can contains a gas with a pressure of 3.0 atm at a temperature of 17 °C. What is the pressure, in atmospheres, inside the container if the temperature rises to 110 °C, if there is no change in volume or amount of gas?

d. A gas has a pressure of 1200. mmHg at 300. °C. What will the temperature, in degrees Celsius, be when the pressure falls to 1.10 atm, if there is no change in volume or amount of gas?

Answers   **a.** 2580 mmHg   **b.** 348 mmHg   **c.** 4.0 atm   **d.** 126 °C

11.5   The Combined Gas Law

Learning Goal: Use the combined gas law to calculate the unknown pressure, volume, or temperature of a gas when changes in two of these properties are given and the amount of gas is constant.

• When the amount of a gas is constant, the gas laws can be combined into a relationship of pressure (P), volume (V), and temperature (T):

$$\frac{P_1 V_1}{T_1} = \frac{P_2 V_2}{T_2}$$ The subscripts 1 and 2 represent initial and final conditions.

Key Terms for Sections 11.3 to 11.5

Match the following key terms with the correct description:

   a. vapor pressure   **b.** direct relationship   **c.** Charles's law
   d. combined gas law   **e.** Gay-Lussac's law

1. _____ a gas law that states that the volume of a gas changes directly with a change in Kelvin temperature when pressure and amount (moles) of the gas do not change

2. _____ the pressure exerted by the particles of vapor above a liquid

3. _____ a relationship in which two properties change in the same direction

4. ____ a gas law stating that the pressure of a gas changes directly with a change in Kelvin temperature when the number of moles of a gas and its volume remain constant

5. ____ a relationship that combines several gas laws relating pressure, volume, and temperature when the amount of gas remains constant

Answers   **1.** c   **2.** a   **3.** b   **4.** e   **5.** d

♦ Learning Exercise 11.5

Solve each of the following using the combined gas law:

a. A 4.0-L sample of N_2 has a pressure of 1200 mmHg at 220 K. What is the pressure, in millimeters of mercury, of the sample when the volume increases to 20. L at 440 K, if there is no change in the amount of gas?

b. A 25.0-mL bubble forms at the ocean depths when the pressure is 10.0 atm and the temperature is 5.0 °C. What is the volume, in milliliters, of that bubble at the ocean surface when the pressure is 760. mmHg and the temperature is 25 °C, if there is no change in the amount of gas?

c. A 35.0-mL sample of Ar has a pressure of 1.0 atm and a temperature of 15 °C. What is the volume, in milliliters, if the pressure goes to 2.0 atm and the temperature to 45 °C, if there is no change in the amount of gas?

d. A weather balloon is launched from the Earth's surface with a volume of 315 L, a temperature of 12 °C, and pressure of 0.930 atm. What is the volume, in liters, of the balloon in the upper atmosphere when the pressure is 116 mmHg and the temperature is −35 °C, if there is no change in the amount of gas?

e. A 10.0-L sample of gas is emitted from a volcano with a pressure of 1.20 atm and a temperature of 150.°C. What is the volume, in liters, of the gas when its pressure is 0.900 atm and the temperature is −40.°C, if there is no change in the amount of gas?

Answers   **a.** 480 mmHg   **b.** 268 mL   **c.** 19 mL   **d.** 1.60×10^3 L   **e.** 7.34 L

11.6 Volume and Moles (Avogadro's Law)

Learning Goal: Use Avogadro's law to calculate the unknown amount or volume of a gas when the pressure and temperature are constant.

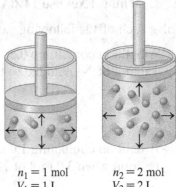

$n_1 = 1$ mol $n_2 = 2$ mol
$V_1 = 1$ L $V_2 = 2$ L

- If the number of moles of gas increases, the volume increases; if the number of moles of gas decreases, the volume decreases.

- Avogadro's law states that equal volumes of gases at the same temperature and pressure contain the same number of moles. The volume (V) of a gas is directly related to the number of moles of the gas when the pressure and temperature of the gas do not change:

$$\frac{V_1}{n_1} = \frac{V_2}{n_2}$$ The subscripts 1 and 2 represent initial and final conditions.

- At STP conditions, standard temperature ($0\,^\circ C$) and pressure (1 atm), 1.00 mol of a gas occupies a volume of 22.4 L, the *molar volume*.

Guide to Using Molar Volume	
STEP 1	State the given and needed quantities.
STEP 2	Write a plan to calculate the needed quantity.
STEP 3	Write the equalities and conversion factors including 22.4 L/mol at STP.
STEP 4	Set up the problem with factors to cancel units.

Example: How many liters would 2.00 mol of N_2 occupy at STP?

Solution: **STEP 1 State the given and needed quantities.**

Analyze the Problem	Given	Need	Connect
	2.00 mol of N_2 at STP	liters of N_2 at STP	molar volume (STP)

STEP 2 Write a plan to calculate the needed quantity.

moles of N_2 $\xrightarrow{\text{Molar volume}}$ liters of N_2 (STP)

STEP 3 Write the equalities and conversion factors including 22.4 L/mol at STP.

$$1 \text{ mol of } N_2 = 22.4 \text{ L of } N_2 \text{ (STP)}$$

$$\frac{22.4 \text{ L } N_2 \text{ (STP)}}{1 \text{ mol } N_2} \quad \text{and} \quad \frac{1 \text{ mol } N_2}{22.4 \text{ L } N_2 \text{ (STP)}}$$

STEP 4 Set up the problem with factors to cancel units.

$$2.00 \text{ mol } N_2 \times \frac{22.4 \text{ L } N_2 \text{ (STP)}}{1 \text{ mol } N_2} = 44.8 \text{ L of } N_2 \text{ (STP)}$$

♦ **Learning Exercise 11.6A**

Solve each of the following gas law problems using Avogadro's law:

a. A balloon containing 0.50 mol of He has a volume of 4.0 L. What is the volume, in liters, when 1.0 mol of N_2 is added to the balloon, if there is no change in pressure and temperature?

b. A balloon containing 1.0 mol of O_2 has a volume of 15 L. What is the volume, in liters, of the balloon when 3.0 mol of He is added, if there is no change in pressure and temperature?

Answers   **a.** 12 L   **b.** 60. L

♦ **Learning Exercise 11.6B**

Solve each of the following gas law problems using molar volume:

a. What is the volume, in liters, occupied by 18. g of N_2 at STP?

b. What is the mass, in grams, of 4.48 L of O_2 at STP?

Answers   **a** 14.8 L   **b.** 6.40 g

11.7 The Ideal Gas Law

Learning Goal: Use the ideal gas law equation to calculate the unknown *P*, *V*, *T*, or *n* of a gas when given three of the four values in the ideal gas law equation. Calculate the molar mass of a gas.

*   The ideal gas law $PV = nRT$ gives the relationship between the four properties used in the measurement of a gas: pressure, volume, moles, and temperature. When any three properties are given, the fourth can be calculated.
*   *R* is the universal gas constant: 0.0821 L · atm/mol · K or 62.4 L · mmHg/mol · K.
*   The ideal gas law can be used to calculate the mass or volume of a gas in a chemical reaction.
*   The molar mass of a gas can be calculated using the ideal gas law.

	$\dfrac{0.0821\ \text{L} \cdot \text{atm}}{\text{mol} \cdot \text{K}}$	$\dfrac{62.4\ \text{L} \cdot \text{mmHg}}{\text{mol} \cdot \text{K}}$
Ideal gas constant (*R*)		
Pressure (*P*)	atm	mmHg
Volume (*V*)	L	L
Amount (*n*)	mol	mol
Temperature (*T*)	K	K

	Guide to Using the Ideal Gas Law
STEP 1	State the given and needed quantities.
STEP 2	Rearrange the ideal gas law equation to solve for the needed quantity.
STEP 3	Substitute the gas data into the equation and calculate the needed quantity.

Study Note

Identify the three known properties for the ideal gas law, and arrange the equation to solve for the unknown.

Example: Solve the ideal gas law for P.

Solution: $PV = nRT$ $\qquad \dfrac{P\cancel{V}}{\cancel{V}} = \dfrac{nRT}{V}$ $\qquad P = \dfrac{nRT}{V}$

♦ **Learning Exercise 11.7A**

Use the ideal gas law to solve each of the following:

CORE CHEMISTRY SKILL

Using the Ideal Gas Law

a. What volume, in liters, is occupied by 0.250 mol of N_2 at 0 °C and 1.50 atm?

b. What is the temperature, in degrees Celsius, of 0.500 mol of He that occupies a volume of 15.0 L at a pressure of 1210 mmHg?

c. What is the pressure, in atmospheres, of 1.50 mol of Ne in a 5.00-L steel container at a temperature of 125 °C?

d. What is the pressure, in atmospheres, of 8.0 g of O_2 that has a volume of 245 mL at a temperature of 22 °C?

e. A single-patient hyperbaric chamber has a volume of 650. L. At a temperature of 22 °C, how many grams of O_2 are needed to give a pressure of 1.80 atm?

Answers   **a.** 3.74 L   **b.** 309 °C   **c.** 9.80 atm   **d.** 25 atm   **e.** 1550 g of O_2

Guide to Calculating the Molar Mass of a Gas	
STEP 1	State the given and needed quantities.
STEP 2	Rearrange the ideal gas law equation to solve for the number of moles.
STEP 3	Obtain the molar mass by dividing the given number of grams by the number of moles.

♦ **Learning Exercise 11.7B**

Use the ideal gas law to calculate the molar mass of each of the following gases:

a. A sample of gas has a mass of 0.650 g and a volume of 560. mL at STP. What is the molar mass of the gas?

b. A sample of gas with a mass of 0.412 g has a volume of 273 mL and a pressure of 746 mmHg at 25 °C. What is the molar mass of the gas?

Answers   **a.** 26.0 g/mol   **b.** 37.6 g/mol

11.8 Gas Laws and Chemical Reactions

Learning Goal: Calculate the mass or volume of a gas that reacts or forms in a chemical reaction.

- The ideal gas law or molar volume at STP is used to calculate the number of moles (and therefore, mass using the molar mass) or volume of a gas in a chemical reaction.
- Mole–mole factors are then used to calculate moles of other substances in a reaction.

Guide to Using the Ideal Gas Law for Reactions	
STEP 1	State the given and needed quantities.
STEP 2	Write a plan to convert the given quantity to the needed moles.
STEP 3	Write the equalities and conversion factors for molar mass and mole–mole factors.
STEP 4	Set up the problem to calculate moles of needed quantity.
STEP 5	Convert the moles of needed quantity to mass or volume using the molar mass or the ideal gas law equation.

♦ **Learning Exercise 11.8**

Use gas laws to calculate the quantity of a reactant or product in each of the following chemical reactions:

a. How many liters of hydrogen gas at STP are produced when 12.5 g of magnesium reacts?

$$Mg(s) + 2HCl(aq) \longrightarrow H_2(g) + MgCl_2(aq)$$

b. How many grams of KNO_3 must decompose to produce 35.8 L of O_2 at 28 °C and 745 mmHg?

$$2KNO_3(s) \longrightarrow 2KNO_2(s) + O_2(g)$$

c. At a temperature of 325 °C and a pressure of 1.20 atm, how many liters of CO_2 can be produced when 50.0 g of propane (C_3H_8) undergoes combustion?

$$C_3H_8(g) + 5O_2(g) \xrightarrow{\Delta} 3CO_2(g) + 4H_2O(g)$$

Answers   **a.** 11.5 L of H_2 **b.** 287 g of KNO_3 **c.** 139 L of CO_2

11.9 Partial Pressures (Dalton's Law)

Learning Goal: Use Dalton's law of partial pressures to calculate the total pressure of a mixture of gases.

• In a mixture of two or more gases, the total pressure is the sum of the partial pressures (subscripts 1, 2, 3 . . . represent the partial pressures of the individual gases).

$$P_{total} = P_1 + P_2 + P_3 + \cdots$$

• The partial pressure of a gas in a mixture is the pressure it would exert if it were the only gas in the container.

Key Terms for Sections 11.6 to 11.9

Match the following key terms with the correct description:

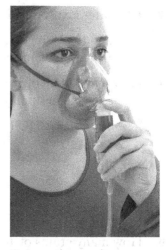

a. molar volume   **b.** STP   **c.** Dalton's law
d. partial pressure   **e.** Avogadro's law   **f.** ideal gas law

1. ____ the gas law stating that the total pressure exerted by a mixture of gases in a container is the sum of the pressures that each gas would exert alone

2. ____ the pressure exerted by a single gas in a gas mixture

3. ____ the volume occupied by 1 mol of a gas at STP

4. ____ the law that combines the four measured properties of a gas in the equation $PV = nRT$

5. ____ the gas law stating that the volume of a gas is directly related to the number of moles of gas when pressure and temperature do not change

6. ____ the standard conditions of 0 °C (273 K) and 1 atm, used for the comparison of gases

Oxygen therapy increases the amount of oxygen available to the tissues of the body.

Credit: Levent Konuk / Shutterstock

Answers   **1.** c   **2.** d   **3.** a   **4.** f   **5.** e   **6.** b

Guide to Calculating Partial Pressure	
STEP 1	Write the equation for the sum of the partial pressures.
STEP 2	Rearrange the equation to solve for the unknown pressure.
STEP 3	Substitute known pressures into the equation and calculate the unknown partial pressure.

◆ Learning Exercise 11.9A

Use Dalton's law to solve the following problems:

🔬 **CORE CHEMISTRY SKILL**

Calculating Partial Pressure

a. What is the total pressure, in millimeters of mercury, of a gas sample that contains O_2 at 0.500 atm, N_2 at 132 Torr, and He at 224 mmHg?

b. What is the total pressure, in atmospheres, of a gas sample that contains He at 285 mmHg and O_2 at 1.20 atm?

c. A gas sample containing N_2 and O_2 has a pressure of 1500. mmHg. If the partial pressure of the N_2 is 0.900 atm, what is the partial pressure, in millimeters of mercury, of the O_2 in the mixture?

Answers   **a.** 736 mmHg   **b.** 1.58 atm   **c.** 816 mmHg

♦ **Learning Exercise 11.9B**

Fill in the blanks by writing Increases or Decreases for a gas in a closed container:

	Pressure	Volume	Moles	Temperature
a.	_____	Increases	Constant	Constant
b.	Increases	Constant	_____	Constant
c.	Constant	Decreases	_____	Constant
d.	Increases	Constant	Constant	_____
e.	Constant	_____	Constant	Decreases
f.	_____	Constant	Increases	Constant

Answers   **a.** Decreases   **b.** Increases   **c.** Decreases   **d.** Increases   **e.** Decreases   **f.** Increases

Checklist for Chapter 11

You are ready to take the Practice Test for Chapter 11. Be sure you have accomplished the following learning goals for this chapter. If not, review the section listed at the end of the goal. Then apply your new skills and understanding to the Practice Test.

After studying Chapter 11, I can successfully:

_____ Describe the kinetic molecular theory of gases. (11.1)

_____ Change the units of pressure from one to another. (11.1)

_____ Use the pressure–volume relationship (Boyle's law) to calculate the unknown pressure or volume of a fixed amount of gas at constant temperature. (11.2)

_____ Use the temperature–volume relationship (Charles's law) to calculate the unknown temperature or volume of a fixed amount of gas at a constant pressure. (11.3)

_____ Use the temperature–pressure relationship (Gay-Lussac's law) to calculate the unknown temperature or pressure of a fixed amount of gas at a constant volume. (11.4)

_____ Use the combined gas law to calculate the unknown pressure, volume, or temperature of a fixed amount of gas when changes in two of these properties are given. (11.5)

_____ Describe the relationship between the amount of a gas and its volume (Avogadro's law) at constant P and T, and use this relationship in calculations. (11.6)

_____ Use the ideal gas law to solve for pressure, volume, temperature, or amount of a gas. (11.7)

_____ Use the gas laws to calculate the quantity of a reactant or product in a chemical reaction. (11.8)

_____ Calculate the total pressure of a gas mixture from the partial pressures. (11.9)

Practice Test for Chapter 11

The chapter sections to review are shown in parentheses at the end of each question.

For questions 1 through 5, use the kinetic molecular theory of gases to answer true (T) or false (F): (11.1)

1. ____ A gas does not have its own volume or shape.

2. ____ The molecules of a gas are moving extremely fast.

3. ____ The collisions of gas molecules with the walls of their container create pressure.

4. ____ Gas molecules are close together and move in straight lines.

5. ____ The attractive forces between molecules are very small.

6. What is the pressure, in atmospheres, of a gas with a pressure of 1200 mmHg? (11.1)
   A. 0.63 atm   B. 0.79 atm   C. 1.2 atm
   D. 1.6 atm   E. 2.0 atm

7. The relationship that the volume of a gas is inversely related to its pressure at constant temperature is known as (11.2)
   A. Boyle's law   B. Charles's law   C. Gay-Lussac's law
   D. Dalton's law   E. Avogadro's law

8. A 6.00-L sample of O_2 has a pressure of 660. mmHg. When the volume is reduced to 2.00 L with no change in temperature and amount of gas, it will have a pressure of (11.2)
   A. 1980 mmHg   B. 1320 mmHg   C. 330. mmHg
   D. 220. mmHg   E. 110. mmHg

9. If the temperature and amount of a gas does not change, but its volume doubles, its pressure will (11.2)
   A. double
   B. triple
   C. decrease to one-half the original pressure
   D. decrease to one-fourth the original pressure
   E. not change

10. If the temperature of a gas is increased, (11.3, 11.4)
    A. the pressure will decrease
    B. the volume will increase
    C. the volume will decrease
    D. the number of molecules will increase
    E. none of these

11. When a gas is heated in a closed metal container, the (11.4)
    A. pressure increases
    B. pressure decreases
    C. volume increases
    D. volume decreases
    E. number of molecules increases

12. A sample of N_2 at 110 K has a pressure of 1.0 atm. When the temperature is increased to 360 K with no change in volume and amount of gas, what is the pressure? (11.4)
    A. 0.50 atm   B. 1.0 atm   C. 1.5 atm
    D. 3.3 atm   E. 4.0 atm

13. A gas sample with a volume of 4.0 L has a pressure of 750 mmHg and a temperature of 77 °C. What is its volume at 277 °C and 250 mmHg, if there is no change in the amount of gas? (11.5)
    A. 7.6 L   B. 19 L   C. 2.1 L
    D. 0.00056 L   E. 3.3 L

14. A sample of 2.00 mol of gas initially at STP is converted to a volume of 5.0 L and a temperature of 27 °C. What is the pressure, in atmospheres, if there is no change in the amount of gas? (11.5, 11.6)

  A. 0.12 atm   **B.** 5.5 atm   **C.** 7.5 atm

  D. 9.8 atm   **E.** 10. atm

15. The conditions for standard temperature and pressure (STP) are (11.6)

  A. 0 K, 1 atm   **B.** 0 °C, 10 atm   **C.** 25 °C, 1 atm

  D. 273 K, 1 atm   **E.** 273 K, 0.5 atm

16. The volume occupied by 1.50 mol of CH_4 at STP is (11.6)

  A. 44.10 L   **B.** 33.6 L   **C.** 22.4 L

  D. 11.2 L   **E.** 5.60 L

17. How many grams of O_2 are present in 44.1 L of O_2 at STP? (11.6)

  A. 10.0 g   **B.** 16.0 g   **C.** 32.0 g

  D. 410.0 g   **E.** 63.0 g

18. The pressure of a gas will increase when (11.6)

  A. the volume increases

  B. the temperature decreases

  C. more molecules of gas are added

  D. molecules of gas are removed

  E. none of these

19. If two gases have the same volume, temperature, and pressure, they also have the same (11.6, 11.7)

  A. density   **B.** number of molecules   **C.** molar mass

  D. speed   **E.** size molecules

20. A gas sample with a mass of 0.330 g has a volume of 0.224 L at STP. The gas has a molar mass of (11.6, 11.7)

  A. 19.6 g/mol   **B.** 22.4 g/mol   **C.** 33.0 g/mol

  D. 44.8 g/mol   **E.** 50.9 g/mol

21. A sample of O_2 with a temperature of 56 °C contains 0.210 mol of gas and has a volume of 6.50 L. What is the pressure, in millimeters of mercury? (11.7)

  A. 113 mmHg   **B.** 0.151 mmHg   **C.** 663 mmHg

  D. 0.873 mmHg   **E.** 700. mmHg

22. What is the volume, in liters, of 0.50 mol of N_2 at 25 °C and 2.0 atm? (11.7)

  A. 0.51 L   **B.** 1.0 L   **C.** 4.2 L

  D. 6.1 L   **E.** 24 L

23. 2.50 g of $KClO_3$ decomposes by the equation: $2KClO_3(s) \longrightarrow 2KCl(s) + 3O_2(g)$. What volume of O_2 is produced at 25 °C and a pressure of 750. mmHg? (11.8)

  A. 0.00998 L   **B.** 0.759 L   **C.** 0.0636 L

  D. 0.506 L   **E.** 1.32 L

24. A gas mixture contains He with a partial pressure of 0.100 atm, O_2 with a partial pressure of 445 mmHg, and N_2 with a partial pressure of 235 mmHg. What is the total pressure, in atmospheres, for the gas mixture? (11.9)

  A. 0.995 atm   **B.** 1.39 atm   **C.** 1.69 atm

  D. 2.00 atm   **E.** 10.0 atm

25. A mixture of O_2 and N_2 has a total pressure of 1040 mmHg. If the O_2 has a partial pressure of 510 mmHg, what is the partial pressure of the N_2? (11.9)

  A. 240 mmHg   **B.** 530 mmHg   **C.** 770 mmHg

  D. 1040 mmHg   **E.** 1350 mmHg

26. 3.00 mol of He in a steel container has a pressure of 12.0 atm. What is the pressure after 4.00 mol of Ne is added, if there is no change in temperature? (11.9)

A. 5.14 atm	**B.** 16.0 atm	**C.** 28.0 atm
D. 32.0 atm	**E.** 45.0 atm	

Answers to the Practice Test

1. T	**2.** T	**3.** T	**4.** F	**5.** T
6. D	**7.** A	**8.** A	**9.** C	**10.** B
11. A	**12.** D	**13.** B	**14.** D	**15.** D
16. B	**17.** E	**18.** C	**19.** B	**20.** C
21. C	**22.** D	**23.** B	**24.** A	**25.** B
26. C				

Selected Answers and Solutions to Text Problems

11.1 a. At a higher temperature, gas particles have greater kinetic energy, which makes them move faster.
  b. Because there are great distances between the particles of a gas, they can be pushed closer together and still remain a gas.
  c. Gas particles are very far apart, meaning that the mass of a gas in a certain volume is very small, resulting in a low density.

11.3 a. The temperature of a gas can be expressed in kelvins.
  b. The volume of a gas can be expressed in milliliters.
  c. The amount of a gas can be expressed in grams.
  d. Pressure can be expressed in millimeters of mercury (mmHg).

11.5 Statements **a**, **d**, and **e** describe the pressure of a gas.

11.7 a. $2.00 \text{ atm} \times \dfrac{760 \text{ Torr}}{1 \text{ atm}} = 1520 \text{ Torr (3 SFs)}$

  b. $2.00 \text{ atm} \times \dfrac{14.7 \text{ lb/in.}^2}{1 \text{ atm}} = 29.4 \text{ lb/in.}^2 \text{ (3 SFs)}$

  c. $2.00 \text{ atm} \times \dfrac{760 \text{ mmHg}}{1 \text{ atm}} = 1520 \text{ mmHg (3 SFs)}$

  d. $2.00 \text{ atm} \times \dfrac{101.325 \text{ kPa}}{1 \text{ atm}} = 203 \text{ kPa (3 SFs)}$

11.9 As the scuba diver ascends to the surface, external pressure decreases. If the air in the lungs, which is at a higher pressure, were not exhaled, its volume would expand and severely damage the lungs. The pressure of the gas in the lungs must adjust to changes in the external pressure.

11.11 a. The pressure is greater in cylinder A. According to Boyle's law, a decrease in volume pushes the gas particles closer together, which will cause an increase in the pressure.
  b. From Boyle's law, we know that pressure is inversely related to volume. The mmHg units must be converted to atm for unit cancellation in the calculation, and because the pressure increases, volume must decrease.
  According to Boyle's law, $P_1 V_1 = P_2 V_2$, then

$$V_2 = V_1 \times \frac{P_1}{P_2} = 220 \text{ mL} \times \frac{650 \text{ mmHg}}{1.2 \text{ atm}} \times \frac{1 \text{ atm}}{760 \text{ mmHg}} = 160 \text{ mL (2 SFs)}$$

11.13 a. The pressure of the gas *increases* to two times the original pressure when the volume is halved.
  b. The pressure *decreases* to one-third the original pressure when the volume expands to three times its initial value.
  c. The pressure *increases* to 10 times the original pressure when the volume decreases to one-tenth of the initial volume.

11.15 From Boyle's law, we know that pressure is inversely related to volume (for example, pressure increases when volume decreases).
  a. Volume increases; pressure must decrease.

$$P_2 = P_1 \times \frac{V_1}{V_2} = 655 \text{ mmHg} \times \frac{10.0 \text{ L}}{20.0 \text{ L}} = 328 \text{ mmHg (3 SFs)}$$

b. Volume decreases; pressure must increase.

$$P_2 = P_1 \times \frac{V_1}{V_2} = 655 \text{ mmHg} \times \frac{10.0 \cancel{L}}{2.50 \cancel{L}} = 2620 \text{ mmHg (3 SFs)}$$

c. The mL units must be converted to L for unit cancellation in the calculation, and because the volume increases, pressure must decrease.

$$P_2 = P_1 \times \frac{V_1}{V_2} = 655 \text{ mmHg} \times \frac{10.0 \cancel{L}}{13\,800 \cancel{mL}} \times \frac{1000 \cancel{mL}}{1 \cancel{L}} = 475 \text{ mmHg (3 SFs)}$$

d. The mL units must be converted to L for unit cancellation in the calculation, and because the volume decreases, pressure must increase.

$$P_2 = P_1 \times \frac{V_1}{V_2} = 655 \text{ mmHg} \times \frac{10.0 \cancel{L}}{1250 \cancel{mL}} \times \frac{1000 \cancel{mL}}{1 \cancel{L}} = 5240 \text{ mmHg (3 SFs)}$$

11.17 From Boyle's law, we know that pressure is inversely related to volume.

a. Pressure decreases; volume must increase.

$$V_2 = V_1 \times \frac{P_1}{P_2} = 50.0 \text{ L} \times \frac{760. \cancel{mmHg}}{725 \cancel{mmHg}} = 52.4 \text{ L (3 SFs)}$$

b. The mmHg units must be converted to atm for unit cancellation in the calculation, and because the pressure increases, volume must decrease.

$$P_1 = 760. \cancel{mmHg} \times \frac{1 \text{ atm}}{760 \cancel{mmHg}} = 1.00 \text{ atm}$$

$$V_2 = V_1 \times \frac{P_1}{P_2} = 50.0 \text{ L} \times \frac{1.00 \cancel{atm}}{2.0 \cancel{atm}} = 25 \text{ L (2 SFs)}$$

c. The mmHg units must be converted to atm for unit cancellation in the calculation, and because the pressure decreases, volume must increase.

$$P_1 = 760. \cancel{mmHg} \times \frac{1 \text{ atm}}{760 \cancel{mmHg}} = 1.00 \text{ atm}$$

$$V_2 = V_1 \times \frac{P_1}{P_2} = 50.0 \text{ L} \times \frac{1.00 \cancel{atm}}{0.500 \cancel{atm}} = 100. \text{ L (3 SFs)}$$

d. The mmHg units must be converted to torr for unit cancellation in the calculation, and because the pressure increases, volume must decrease.

$$P_1 = 760. \cancel{mmHg} \times \frac{760 \text{ Torr}}{760 \cancel{mmHg}} = 760. \text{ Torr}$$

$$V_2 = V_1 \times \frac{P_1}{P_2} = 50.0 \text{ L} \times \frac{760. \cancel{Torr}}{850 \cancel{Torr}} = 45 \text{ L (2 SFs)}$$

11.19 Volume increases; initial pressure must have been higher.

$$P_1 = P_2 \times \frac{V_2}{V_1} = 3.62 \text{ atm} \times \frac{9.73 \cancel{L}}{5.40 \cancel{L}} = 6.52 \text{ atm (3 SFs)}$$

11.21 Pressure decreases; volume must increase.

$$V_2 = V_1 \times \frac{P_1}{P_2} = 5.0 \text{ L} \times \frac{5.0 \cancel{atm}}{1.0 \cancel{atm}} = 25 \text{ L (2 SFs)}$$

11.23 a. *Inspiration* begins when the diaphragm flattens, causing the lungs to expand. The increased volume of the thoracic cavity reduces the pressure in the lungs such that air flows into the lungs.

b. *Expiration* occurs as the diaphragm relaxes, causing a decrease in the volume of the lungs. The pressure of the air in the lungs increases, and air flows out of the lungs.

c. *Inspiration* occurs when the pressure within the lungs is less than that of the atmosphere.

11.25 According to Charles's law, there is a direct relationship between Kelvin temperature and volume (for example, volume increases when temperature increases, if the pressure and amount of gas remain constant).

a. Diagram C shows an increased volume corresponding to an increase in temperature.

b. Diagram A shows a decreased volume corresponding to a decrease in temperature.

c. Diagram B shows no change in volume, which corresponds to no net change in temperature.

11.27 According to Charles's law, the volume of a gas is directly related to the Kelvin temperature. In all gas law computations, temperatures must be in kelvins. (Temperatures in °C are converted to K by the addition of 273.) The initial temperature for all cases here is $T_1 = 15\,°C + 273 = 288$ K.

a. Volume increases; temperature must have increased.

$$T_2 = T_1 \times \frac{V_2}{V_1} = 288\text{ K} \times \frac{5.00\text{ L}}{2.50\text{ L}} = 576\text{ K} \quad 576\text{ K} - 273 = 303\,°C \text{ (3 SFs)}$$

b. Volume decreases; temperature must have decreased.

$$T_2 = T_1 \times \frac{V_2}{V_1} = 288\text{ K} \times \frac{1250\text{ mL}}{2.50\text{ L}} \times \frac{1\text{ L}}{1000\text{ mL}} = 144\text{ K} \quad 144\text{ K} - 273 = -129\,°C \text{ (3 SFs)}$$

c. Volume increases; temperature must have increased.

$$T_2 = T_1 \times \frac{V_2}{V_1} = 288\text{ K} \times \frac{7.50\text{ L}}{2.50\text{ L}} = 864\text{ K} \quad 864\text{ K} - 273 = 591\,°C \text{ (3 SFs)}$$

d. Volume increases; temperature must have increased.

$$T_2 = T_1 \times \frac{V_2}{V_1} = 288\text{ K} \times \frac{3550\text{ mL}}{2.50\text{ L}} \times \frac{1\text{ L}}{1000\text{ mL}} = 409\text{ K} \quad 409\text{ K} - 273 = 136\,°C \text{ (3 SFs)}$$

11.29 According to Charles's law, the volume of a gas is directly related to the Kelvin temperature. In all gas law computations, temperatures must be in kelvins. (Temperatures in °C are converted to K by the addition of 273.) The initial temperature for all cases here is $T_1 = 75\,°C + 273 = 348$ K.

a. When temperature decreases, volume must also decrease.

$$T_2 = 55\,°C + 273 = 328\text{ K}$$

$$V_2 = V_1 \times \frac{T_2}{T_1} = 2500\text{ mL} \times \frac{328\text{ K}}{348\text{ K}} = 2400\text{ mL (2 SFs)}$$

b. When temperature increases, volume must also increase.

$$V_2 = V_1 \times \frac{T_2}{T_1} = 2500\text{ mL} \times \frac{680.\text{ K}}{348\text{ K}} = 4900\text{ mL (2 SFs)}$$

c. When temperature decreases, volume must also decrease.

$$T_2 = -25\,°C + 273 = 248\text{ K}$$

$$V_2 = V_1 \times \frac{T_2}{T_1} = 2500\text{ mL} \times \frac{248\text{ K}}{348\text{ K}} = 1800\text{ mL (2 SFs)}$$

d. When temperature decreases, volume must also decrease.

$$V_2 = V_1 \times \frac{T_2}{T_1} = 2500 \text{ mL} \times \frac{240. \, K}{348 \, K} = 1700 \text{ mL (2 SFs)}$$

11.31 Volume decreases; initial temperature must have been higher.

$$T_2 = 32\,°C + 273 = 305 \text{ K}$$

$$T_1 = T_2 \times \frac{V_1}{V_2} = 305 \text{ K} \times \frac{0.256 \, L}{0.198 \, L} = 394 \text{ K} \quad 394 \text{ K} - 273 = 121\,°C \text{ (3 SFs)}$$

11.33 According to Gay-Lussac's law, temperature is directly related to pressure. In all gas law computations, temperatures must be in kelvins. (Temperatures in °C are converted to K by the addition of 273.)

a. When temperature decreases, pressure must also decrease.

$$T_1 = 155\,°C + 273 = 428 \text{ K} \qquad T_2 = 0\,°C + 273 = 273 \text{ K}$$

$$P_2 = P_1 \times \frac{T_2}{T_1} = 1200 \text{ Torr} \times \frac{1 \text{ mmHg}}{1 \text{ Torr}} \times \frac{273 \, K}{428 \, K} = 770 \text{ mmHg (2 SFs)}$$

b. When temperature increases, pressure must also increase.

$$T_1 = 12\,°C + 273 = 285 \text{ K} \qquad T_2 = 35\,°C + 273 = 308 \text{ K}$$

$$P_2 = P_1 \times \frac{T_2}{T_1} = 1.40 \text{ atm} \times \frac{760 \text{ mmHg}}{1 \text{ atm}} \times \frac{308 \, K}{285 \, K} = 1150 \text{ mmHg (3 SFs)}$$

11.35 According to Gay-Lussac's law, pressure is directly related to temperature. In all gas law computations, temperatures must be in kelvins. (Temperatures in °C are converted to K by the addition of 273.)

a. Pressure decreases; temperature must have decreased.

$$T_1 = 25\,°C + 273 = 298 \text{ K}$$

$$T_2 = T_1 \times \frac{P_2}{P_1} = 298 \text{ K} \times \frac{620. \text{ mmHg}}{740. \text{ mmHg}} = 250. \text{ K} \quad 250. \text{ K} - 273 = -23\,°C \text{ (2 SFs)}$$

b. Pressure increases; temperature must have increased.

$$T_1 = -18\,°C + 273 = 255 \text{ K}$$

$$T_2 = T_1 \times \frac{P_2}{P_1} = 255 \text{ K} \times \frac{1250 \text{ Torr}}{0.950 \text{ atm}} \times \frac{1 \text{ atm}}{760 \text{ Torr}} = 441 \text{ K} \quad 441 \text{ K} - 273 = 168\,°C \text{ (3 SFs)}$$

11.37 Pressure decreases; initial temperature must have been higher.

$$T_2 = 22\,°C + 273 = 295 \text{ K}$$

$$T_1 = T_2 \times \frac{P_1}{P_2} = 295 \text{ K} \times \frac{766 \text{ mmHg}}{744 \text{ mmHg}} = 304 \text{ K} \quad 304 \text{ K} - 273 = 31\,°C \text{ (2 SFs)}$$

11.39 a. On the top of a mountain, water boils below 100 °C because the atmospheric (external) pressure is less than 1 atm. The boiling point is the temperature at which the vapor pressure of a liquid becomes equal to the external (in this case, atmospheric) pressure.

b. Because the pressure inside a pressure cooker is greater than 1 atm, water boils above 100 °C. The higher temperature of the boiling water allows food to cook more quickly.

11.41 According to Gay-Lussac's law, pressure is directly related to temperature. In all gas law computations, temperatures must be in kelvins. (Temperatures in °C are converted to K by the addition of 273.)

When temperature increases, pressure must also increase.

$T_1 = 5\,°C + 273 = 278\text{ K}$ $T_2 = 22\,°C + 273 = 295\text{ K}$

$$P_2 = P_1 \times \frac{T_2}{T_1} = 1.8\text{ atm} \times \frac{295\text{ K}}{278\text{ K}} = 1.9\text{ atm (2 SFs)}$$

11.43 $\dfrac{P_1 V_1}{T_1} = \dfrac{P_2 V_2}{T_2}$ $\dfrac{P_1 V_1}{T_1} \times \dfrac{T_2 T_1}{P_1 V_1} = \dfrac{P_2 V_2}{T_2} \times \dfrac{T_2 T_1}{P_1 V_1}$ $\therefore T_2 = T_1 \times \dfrac{P_2}{P_1} \times \dfrac{V_2}{V_1}$

11.45 $T_1 = 25\,°C + 273 = 298\text{ K};$ $V_1 = 6.50\text{ L};$ $P_1 = 845\text{ mmHg}$

   a. $T_2 = 325\text{ K};$ $V_2 = 1850\text{ mL} = 1.85\text{ L};$ $P_2 = ?\text{ atm}$

$$P_2 = P_1 \times \frac{V_1}{V_2} \times \frac{T_2}{T_1} = 845\text{ mmHg} \times \frac{6.50\text{ L}}{1.85\text{ L}} \times \frac{325\text{ K}}{298\text{ K}} \times \frac{1\text{ atm}}{760\text{ mmHg}} = 4.26\text{ atm (3 SFs)}$$

   b. $T_2 = 12\,°C + 273 = 285\text{ K};$ $V_2 = 2.25\text{ L};$ $P_2 = ?\text{ atm}$

$$P_2 = P_1 \times \frac{V_1}{V_2} \times \frac{T_2}{T_1} = 845\text{ mmHg} \times \frac{6.50\text{ L}}{2.25\text{ L}} \times \frac{285\text{ K}}{298\text{ K}} \times \frac{1\text{ atm}}{760\text{ mmHg}} = 3.07\text{ atm (3 SFs)}$$

   c. $T_2 = 47\,°C + 273 = 320.\text{ K};$ $V_2 = 12.8\text{ L};$ $P_2 = ?\text{ atm}$

$$P_2 = P_1 \times \frac{V_1}{V_2} \times \frac{T_2}{T_1} = 845\text{ mmHg} \times \frac{6.50\text{ L}}{12.8\text{ L}} \times \frac{320.\text{ K}}{298\text{ K}} \times \frac{1\text{ atm}}{760\text{ mmHg}} = 0.606\text{ atm (3 SFs)}$$

11.47 $T_1 = 212\,°C + 273 = 485\text{ K};$ $V_1 = 124\text{ mL};$ $P_1 = 1.80\text{ atm}$

   $T_2 = ?\,°C;$ $V_2 = 138\text{ mL};$ $P_2 = 0.800\text{ atm}$

$$T_2 = T_1 \times \frac{P_2}{P_1} \times \frac{V_2}{V_1} = 485\text{ K} \times \frac{0.800\text{ atm}}{1.80\text{ atm}} \times \frac{138\text{ mL}}{124\text{ mL}} = 240.\text{ K}$$

   $240.\text{ K} - 273 = -33\,°C\text{ (2 SFs)}$

11.49 The volume increases because the number of gas particles in the tire or basketball is increased.

11.51 According to Avogadro's law, the volume of a gas is directly related to the number of moles of gas.

   $n_1 = 1.50\text{ mol Ne};$ $V_1 = 8.00\text{ L}$

   a. $V_2 = V_1 \times \dfrac{n_2}{n_1} = 8.00\text{ L} \times \dfrac{\frac{1}{2}(1.50)\text{ mol Ne}}{1.50\text{ mol Ne}} = 4.00\text{ L (3 SFs)}$

   b. $n_2 = 1.50\text{ mol Ne} + 3.50\text{ mol Ne} = 5.00\text{ mol of Ne}$

     $V_2 = V_1 \times \dfrac{n_2}{n_1} = 8.00\text{ L} \times \dfrac{5.00\text{ mol Ne}}{1.50\text{ mol Ne}} = 26.7\text{ L (3 SFs)}$

   c. $25.0\text{ g Ne} \times \dfrac{1\text{ mol Ne}}{20.18\text{ g Ne}} = 1.24\text{ mol of Ne added}$

     $n_2 = 1.50\text{ mol Ne} + 1.24\text{ mol Ne} = 2.74\text{ mol of Ne}$

     $V_2 = V_1 \times \dfrac{n_2}{n_1} = 8.00\text{ L} \times \dfrac{2.74\text{ mol Ne}}{1.50\text{ mol Ne}} = 14.6\text{ L (3 SFs)}$

11.53 At STP, 1 mol of any gas occupies a volume of 22.4 L.

   a. $44.8\text{ L }O_2\text{ (STP)} \times \dfrac{1\text{ mol }O_2}{22.4\text{ L }O_2\text{ (STP)}} = 2.00\text{ mol of }O_2\text{ (3 SFs)}$

   b. $2.50\text{ mol }N_2 \times \dfrac{22.4\text{ L }N_2\text{ (STP)}}{1\text{ mol }N_2} = 56.0\text{ L at STP (3 SFs)}$

c. $50.0 \, \text{g Ar} \times \dfrac{1 \, \text{mol Ar}}{39.95 \, \text{g Ar}} \times \dfrac{22.4 \, \text{L Ar (STP)}}{1 \, \text{mol Ar}} = 28.0 \, \text{L at STP (3 SFs)}$

d. $1620 \, \text{mL H}_2 \text{ (STP)} \times \dfrac{1 \, \text{L H}_2}{1000 \, \text{mL H}_2} \times \dfrac{1 \, \text{mol H}_2}{22.4 \, \text{L H}_2 \text{ (STP)}} \times \dfrac{2.016 \, \text{g H}_2}{1 \, \text{mol H}_2}$

$= 0.146 \, \text{g of H}_2 \text{ (3 SFs)}$

11.55 Analyze the Problem: Using the Ideal Gas Law to Calculate Pressure

Property	*P*	*V*	*n*	*R*	*T*
Given		10.0 L	2.00 mol	$\dfrac{0.0821 \, \text{L} \cdot \text{atm}}{\text{mol} \cdot \text{K}}$	27 °C 27 °C + 273 = 300. K
Need	? atm				

$$P = \frac{nRT}{V} = \frac{(2.00 \, \text{mol})\left(\dfrac{0.0821 \, \text{L} \cdot \text{atm}}{\text{mol} \cdot \text{K}}\right)(300. \, \text{K})}{(10.0 \, \text{L})} = 4.93 \, \text{atm (3 SFs)}$$

11.57 Analyze the Problem: Using the Ideal Gas Law to Calculate Moles

Property	*P*	*V*	*n*	*R*	*T*
Given	845 mmHg	20.0 L		$\dfrac{62.4 \, \text{L} \cdot \text{mmHg}}{\text{mol} \cdot \text{K}}$	22 °C 22 °C + 273 = 295 K
Need			? mol (? g)		

$$n = \frac{PV}{RT} = \frac{(845 \, \text{mmHg})(20.0 \, \text{L})}{\left(\dfrac{62.4 \, \text{L} \cdot \text{mmHg}}{\text{mol} \cdot \text{K}}\right)(295 \, \text{K})} = 0.918 \, \text{mol of O}_2$$

$$0.918 \, \text{mol O}_2 \times \frac{32.00 \, \text{g O}_2}{1 \, \text{mol O}_2} = 29.4 \, \text{g of O}_2 \text{ (3 SFs)}$$

11.59 $n = 25.0 \, \text{g N}_2 \times \dfrac{1 \, \text{mol N}_2}{28.02 \, \text{g N}_2} = 0.892 \, \text{mol of N}_2$

Analyze the Problem: Using the Ideal Gas Law to Calculate Temperature

Property	*P*	*V*	*n*	*R*	*T*
Given	630. mmHg	50.0 L	25.0 g (0.892 mol)	$\dfrac{62.4 \, \text{L} \cdot \text{mmHg}}{\text{mol} \cdot \text{K}}$	
Need					? K (? °C)

$$T = \frac{PV}{nR} = \frac{(630. \, \text{mmHg})(50.0 \, \text{L})}{(0.892 \, \text{mol})\left(\dfrac{62.4 \, \text{L} \cdot \text{mmHg}}{\text{mol} \cdot \text{K}}\right)} = 566 \, \text{K} - 273 = 293 \, \text{°C (3 SFs)}$$

11.61 a. $n = 450 \text{ mL (STP)} \times \dfrac{1 \text{ L}}{1000 \text{ mL}} \times \dfrac{1 \text{ mol}}{22.4 \text{ L (STP)}} = 0.020 \text{ mol}$

Molar mass $= \dfrac{\text{mass}}{\text{moles}} = \dfrac{0.84 \text{ g}}{0.020 \text{ mol}} = 42 \text{ g/mol } (2 \text{ SFs})$

b. $n = 1.00 \text{ L (STP)} \times \dfrac{1 \text{ mol}}{22.4 \text{ L (STP)}} = 0.0446 \text{ mol}$

Molar mass $= \dfrac{\text{mass}}{\text{moles}} = \dfrac{1.28 \text{ g}}{0.0446 \text{ mol}} = 28.7 \text{ g/mol } (3 \text{ SFs})$

c. Analyze the Problem: Using the Ideal Gas Law to Calculate Moles

Property	P	V	n	R	T
Given	685 mmHg	1.00 L		$\dfrac{62.4 \text{ L} \cdot \text{mmHg}}{\text{mol} \cdot \text{K}}$	22 °C 22 °C + 273 = 295 K
Need			? mol (? molar mass)		

$n = \dfrac{PV}{RT} = \dfrac{(685 \text{ mmHg})(1.00 \text{ L})}{\left(\dfrac{62.4 \text{ L} \cdot \text{mmHg}}{\text{mol} \cdot \text{K}}\right)(295 \text{ K})} = 0.0372 \text{ mol}$

Molar mass $= \dfrac{\text{mass}}{\text{moles}} = \dfrac{1.48 \text{ g}}{0.0372 \text{ mol}} = 39.8 \text{ g/mol } (3 \text{ SFs})$

d. Analyze the Problem: Using the Ideal Gas Law to Calculate Moles

Property	P	V	n	R	T
Given	0.95 atm	2.30 L		$\dfrac{0.0821 \text{ L} \cdot \text{atm}}{\text{mol} \cdot \text{K}}$	24 °C 24 °C + 273 = 297 K
Need			? mol (? molar mass)		

$n = \dfrac{PV}{RT} = \dfrac{(0.95 \text{ atm})(2.30 \text{ L})}{\left(\dfrac{0.0821 \text{ L} \cdot \text{atm}}{\text{mol} \cdot \text{K}}\right)(297 \text{ K})} = 0.090 \text{ mol}$

Molar mass $= \dfrac{\text{mass}}{\text{moles}} = \dfrac{2.96 \text{ g}}{0.090 \text{ mol}} = 33 \text{ g/mol } (2 \text{ SFs})$

11.63 Analyze the Problem: Using the Ideal Gas Law to Calculate Moles

Property	P	V	n	R	T
Given	1.6 atm	640 L		$\dfrac{0.0821 \text{ L} \cdot \text{atm}}{\text{mol} \cdot \text{K}}$	24 °C 24 °C + 273 = 297 K
Need			? mol (? g)		

$n = \dfrac{PV}{RT} = \dfrac{(1.6 \text{ atm})(640 \text{ L})}{\left(\dfrac{0.0821 \text{ L} \cdot \text{atm}}{\text{mol} \cdot \text{K}}\right)(297 \text{ K})} = 42 \text{ mol of } O_2$

$42 \text{ mol } O_2 \times \dfrac{32.00 \text{ g } O_2}{1 \text{ mol } O_2} = 1300 \text{ g of } O_2 \ (2 \text{ SFs})$

11.65 a. $8.25 \text{ g Mg} \times \dfrac{1 \text{ mol Mg}}{24.31 \text{ g Mg}} \times \dfrac{1 \text{ mol H}_2}{1 \text{ mol Mg}} \times \dfrac{22.4 \text{ L (STP)}}{1 \text{ mol H}_2} = 7.60 \text{ L of H}_2$

released at STP (3 SFs)

b. Analyze the Problem: Using the Ideal Gas Law to Calculate Moles

Property	P	V	n	R	T
Given	735 mmHg	5.00 L		$\dfrac{62.4 \text{ L} \cdot \text{mmHg}}{\text{mol} \cdot \text{K}}$	18 °C 18 °C + 273 = 291 K
Need			? mol H$_2$		

$$n = \frac{PV}{RT} = \frac{(735 \text{ mmHg})(5.00 \text{ L})}{\left(\dfrac{62.4 \text{ L} \cdot \text{mmHg}}{\text{mol} \cdot \text{K}}\right)(291 \text{ K})} = 0.202 \text{ mol of H}_2$$

Analyze the Problem: Reactant and Product

	Reactant	Product
Mass	? g Mg	
Molar Mass	1 mol of Mg = 24.31 g of Mg	
Moles		0.202 mol H$_2$
Equation	$\text{Mg}(s) + 2\text{HCl}(aq) \longrightarrow \text{H}_2(g) + \text{MgCl}_2(aq)$	

$$0.202 \text{ mol H}_2 \times \frac{1 \text{ mol Mg}}{1 \text{ mol H}_2} \times \frac{24.31 \text{ g Mg}}{1 \text{ mol Mg}} = 4.91 \text{ g of Mg (3 SFs)}$$

11.67 Analyze the Problem: Reactant and Reactant

	Reactant 1	Reactant 2
Mass	55.2 g C$_4$H$_{10}$	
Molar Mass	1 mol of C$_4$H$_{10}$ = 58.12 g of C$_4$H$_{10}$	
Moles		? mol O$_2$
Equation	$2\text{C}_4\text{H}_{10}(g) + 13\text{O}_2(g) \xrightarrow{\Delta} 8\text{CO}_2(g) + 10\text{H}_2\text{O}(g)$	

$$55.2 \text{ g C}_4\text{H}_{10} \times \frac{1 \text{ mol C}_4\text{H}_{10}}{58.12 \text{ g C}_4\text{H}_{10}} \times \frac{13 \text{ mol O}_2}{2 \text{ mol C}_4\text{H}_{10}} = 6.17 \text{ mol of O}_2$$

Analyze the Problem: Using the Ideal Gas Law to Calculate Volume

Property	P	V	n	R	T
Given	0.850 atm		6.17 mol O$_2$	$\dfrac{0.0821 \text{ L} \cdot \text{atm}}{\text{mol} \cdot \text{K}}$	25 °C 25 °C + 273 = 298 K
Need		? L			

$$V = \frac{nRT}{P} = \frac{(6.17 \text{ mol})\left(\dfrac{0.0821 \text{ L} \cdot \text{atm}}{\text{mol} \cdot \text{K}}\right)(298 \text{ K})}{(0.850 \text{ atm})} = 178 \text{ L of O}_2 \text{ (3 SFs)}$$

11.69 Analyze the Problem: Reactant and Reactant

	Reactant 1	Reactant 2
Mass	5.4 g Al	
Molar Mass	1 mol of Al = 26.98 g of Al	
Moles		? mol O_2 (? L O_2)
Molar Volume		1 mol of O_2 = 22.4 L of O_2 (STP)
Equation	$4Al(s) + 3O_2(g) \xrightarrow{\Delta} 2Al_2O_3(s)$	

$$5.4 \text{ g Al} \times \frac{1 \text{ mol Al}}{26.98 \text{ g Al}} \times \frac{3 \text{ mol } O_2}{4 \text{ mol Al}} \times \frac{22.4 \text{ L } O_2 \text{ (STP)}}{1 \text{ mol } O_2} = 3.4 \text{ L of } O_2 \text{ at STP (2 SFs)}$$

11.71 To obtain the total pressure of a gas mixture, add up all of the partial pressures using the same pressure unit.

$$P_{total} = P_{nitrogen} + P_{oxygen} + P_{helium} = 425 \text{ Torr} + 115 \text{ Torr} + 225 \text{ Torr} = 765 \text{ Torr (3 SFs)}$$

11.73 $P_{total} = P_{oxygen} + P_{nitrogen} + P_{carbon \ dioxide} + P_{water \ vapor}$
$= 93 \text{ mmHg} + 565 \text{ mmHg} + 38 \text{ mmHg} + 47 \text{ mmHg} = 743 \text{ mmHg}$

$$743 \text{ mmHg} \times \frac{1.00 \text{ atm}}{760 \text{ mmHg}} = 0.978 \text{ atm (3 SFs)}$$

11.75 Because the total pressure of a gas mixture is the sum of the partial pressures using the same pressure unit, addition and subtraction are used to obtain the "missing" partial pressure.

$$P_{nitrogen} = P_{total} - (P_{oxygen} + P_{helium}) = 925 \text{ Torr} - (425 \text{ Torr} + 75 \text{ Torr}) = 425 \text{ Torr (3 SFs)}$$

11.77 a. $P_{O_2} = P_{total} - P_{H_2O} = 765 \text{ mmHg} - 22 \text{ mmHg} = 743 \text{ mmHg (3 SFs)}$

b. $V = 256 \text{ mL} \times \dfrac{1 \text{ L}}{1000 \text{ mL}} = 0.256 \text{ L}; \quad T = 24\,°C + 273 = 297 \text{ K}$

$$n = \frac{PV}{RT} = \frac{(743 \text{ mmHg})(0.256 \text{ L})}{\left(\dfrac{62.4 \text{ L} \cdot \text{mmHg}}{\text{mol} \cdot \text{K}}\right)(297 \text{ K})} = 0.0103 \text{ mol of } O_2 \text{ (3 SFs)}$$

11.79 Analyze the Problem: Using the Ideal Gas Law to Calculate Moles

Property	*P*	*V*	*n*	*R*	*T*
Given	745 mmHg	3.2 L		$\dfrac{62.4 \text{ L} \cdot \text{mmHg}}{\text{mol} \cdot \text{K}}$	37 °C 37 °C + 273 = 310 K
Need			? mol air		

$$n = \frac{PV}{RT} = \frac{(745 \text{ mmHg})(3.2 \text{ L})}{\left(\dfrac{62.4 \text{ L} \cdot \text{mmHg}}{\text{mol} \cdot \text{K}}\right)(310 \text{ K})} = 0.12 \text{ mol of air}$$

and since air is 21% O_2,

$$0.12 \text{ mol air} \times \frac{21 \text{ mol } O_2}{100 \text{ mol air}} = 0.025 \text{ mol of } O_2 \text{ (2 SFs)}$$

11.81 a. True. The flask containing helium has more moles of helium and thus more atoms of helium to collide with the walls of the container to give a higher pressure.
  b. True. The mass and volume of each are the same, meaning that the mass/volume ratio or density is the same in both flasks.

11.83 a. 2; The fewest number of gas particles will exert the lowest pressure.
  b. 1; The greatest number of gas particles will exert the highest pressure.

11.85 a. A; Volume decreases when temperature decreases (at constant P and n).
  b. C; Volume increases when pressure decreases (at constant n and T).
  c. A; Volume decreases when the number of moles of gas decreases (at constant T and P).
  d. B; Doubling the Kelvin temperature doubles the volume, but when half of the gas escapes, the volume decreases by half. These two opposing effects cancel each other, and there is no overall change in the volume (at constant P).
  e. C; Increasing the moles of gas causes an increase in the volume to keep T and P constant.

11.87 a. $P_{O_2} = 650.\ \text{mmHg total} \times \dfrac{21\ \text{mmHg O}_2}{100\ \text{mmHg total}} = 140\ \text{mmHg (2 SFs)}$

  b. $P_{\text{total}} = 100.\ \text{mmHg O}_2 \times \dfrac{100\ \text{mmHg total}}{21\ \text{mmHg O}_2} = 480\ \text{mmHg (2 SFs)}$

11.89 $T_1 = 22\,°\text{C} + 273 = 295\ \text{K};\qquad P_1 = 755\ \text{mmHg};\qquad V_1 = 31\,000\ \text{L}$
$T_2 = -8\,°\text{C} + 273 = 265\ \text{K};\qquad P_2 = 658\ \text{mmHg};\qquad V_2 = ?\ \text{L}$

$V_2 = V_1 \times \dfrac{P_1}{P_2} \times \dfrac{T_2}{T_1} = 31\,000\ \text{L} \times \dfrac{755\ \text{mmHg}}{658\ \text{mmHg}} \times \dfrac{265\ \text{K}}{295\ \text{K}} = 32\,000\ \text{L (2 SFs)}$

11.91 $T_1 = 15\,°\text{C} + 273 = 288\ \text{K};\qquad P_1 = 745\ \text{mmHg};\qquad V_1 = 4250\ \text{mL}$

$T_2 = ?\,°\text{C};\qquad P_2 = 1.20\ \text{atm} \times \dfrac{760\ \text{mmHg}}{1\ \text{atm}} = 912\ \text{mmHg};$

$V_2 = 2.50\ \text{L} \times \dfrac{1000\ \text{mL}}{1\ \text{L}} = 2.50 \times 10^3\ \text{mL}$

$T_2 = T_1 \times \dfrac{V_2}{V_1} \times \dfrac{P_2}{P_1} = 288\ \text{K} \times \dfrac{2.50 \times 10^3\ \text{mL}}{4250\ \text{mL}} \times \dfrac{912\ \text{mmHg}}{745\ \text{mmHg}} = 207\ \text{K} - 273$

$= -66\,°\text{C (2 SFs)}$

11.93 Analyze the Problem: Using the Ideal Gas Law to Calculate Moles

Property	P	V	n	R	T
Given	2500. mmHg	2.00 L		$\dfrac{62.4\ \text{L} \cdot \text{mmHg}}{\text{mol} \cdot \text{K}}$	18 °C 18 °C + 273 = 291 K
Need			? mol		

$n = \dfrac{PV}{RT} = \dfrac{(2500.\ \text{mmHg})(2.00\ \text{L})}{\left(\dfrac{62.4\ \text{L} \cdot \text{mmHg}}{\text{mol} \cdot \text{K}}\right)(291\ \text{K})} = 0.275\ \text{mol of CH}_4$

$0.275\ \text{mol CH}_4 \times \dfrac{16.04\ \text{g CH}_4}{1\ \text{mol CH}_4} = 4.41\ \text{g of CH}_4\ \text{(3 SFs)}$

11.95 $V = 941 \text{ mL} \times \dfrac{1 \text{ L}}{1000 \text{ mL}} = 0.941 \text{ L}; \quad P = 748 \text{ Torr} \times \dfrac{1 \text{ mmHg}}{1 \text{ Torr}} = 748 \text{ mmHg}$

Analyze the Problem: Using the Ideal Gas Law to Calculate Moles

Property	P	V	n	R	T
Given	748 Torr (748 mmHg)	941 mL (0.941 L)		$\dfrac{62.4 \text{ L} \cdot \text{mmHg}}{\text{mol} \cdot \text{K}}$	20.0 °C 20.0 °C + 273 = 293 K
Need			? mol (? molar mass)		

$n = \dfrac{PV}{RT} = \dfrac{(748 \text{ mmHg})(0.941 \text{ L})}{\left(\dfrac{62.4 \text{ L} \cdot \text{mmHg}}{\text{mol} \cdot \text{K}}\right)(293 \text{ K})} = 0.0385 \text{ mol}$

$\text{Molar mass} = \dfrac{\text{mass}}{\text{moles}} = \dfrac{1.62 \text{ g}}{0.0385 \text{ mol}} = 42.1 \text{ g/mol} \ (3 \text{ SFs})$

11.97 Analyze the Problem: Using the Ideal Gas Law to Calculate Moles

Property	P	V	n	R	T
Given	1.20 atm	35.0 L		$\dfrac{0.0821 \text{ L} \cdot \text{atm}}{\text{mol} \cdot \text{K}}$	5 °C 5 °C + 273 = 278 K
Need			? mol (? g)		

$n = \dfrac{PV}{RT} = \dfrac{(1.20 \text{ atm})(35.0 \text{ L})}{\left(\dfrac{0.0821 \text{ L} \cdot \text{atm}}{\text{mol} \cdot \text{K}}\right)(278 \text{ K})} = 1.84 \text{ mol of } CO_2$

$1.84 \text{ mol } CO_2 \times \dfrac{44.01 \text{ g } CO_2}{1 \text{ mol } CO_2} = 81.0 \text{ g of } CO_2 \ (3 \text{ SFs})$

11.99 $25.0 \text{ g } Zn \times \dfrac{1 \text{ mol } Zn}{65.41 \text{ g } Zn} \times \dfrac{1 \text{ mol } H_2}{1 \text{ mol } Zn} \times \dfrac{22.4 \text{ L (STP)}}{1 \text{ mol } H_2} = 8.56 \text{ L of } H_2 \text{ released at STP } (3 \text{ SFs})$

11.101 Analyze the Problem: Using the Ideal Gas Law to Calculate Moles

Property	P	V	n	R	T
Given	725 mmHg	5.00 L		$\dfrac{62.4 \text{ L} \cdot \text{mmHg}}{\text{mol} \cdot \text{K}}$	375 °C 375 °C + 273 = 648 K
Need			? mol H_2O		

$n = \dfrac{PV}{RT} = \dfrac{(725 \text{ mmHg})(5.00 \text{ L})}{\left(\dfrac{62.4 \text{ L} \cdot \text{mmHg}}{\text{mol} \cdot \text{K}}\right)(648 \text{ K})} = 0.0896 \text{ mol of } H_2O$

Analyze the Problem: Reactant and Product

	Reactant	Product
Mass		? g NH_3
Molar Mass		1 mol of NH_3 = 17.03 g of NH_3
Moles	0.0896 mol H_2O	
Equation	$4NO_2(g) + 6H_2O(g) \longrightarrow 7O_2(g) + 4NH_3(g)$	

$$0.0896 \text{ mol } H_2O \times \frac{4 \text{ mol } NH_3}{6 \text{ mol } H_2O} \times \frac{17.03 \text{ g } NH_3}{1 \text{ mol } NH_3} = 1.02 \text{ g of } NH_3 \text{ (3 SFs)}$$

11.103 The partial pressure of each gas is proportional to the number of particles of each type of gas that is present. Thus, a ratio of partial pressure to total pressure is equal to the ratio of moles of that gas to the total number of moles of gases that are present:

$$\frac{P_{helium}}{P_{total}} = \frac{n_{helium}}{n_{total}}$$

Solving the equation for the partial pressure of helium yields:

$$P_{helium} = P_{total} \times \frac{n_{helium}}{n_{total}} = 2400 \text{ Torr} \times \frac{2.0 \text{ mol}}{8.0 \text{ mol}} = 600 \text{ Torr } (6.0 \times 10^2 \text{ Torr}) \text{ (2 SFs)}$$

$$P_{oxygen} = P_{total} \times \frac{n_{oxygen}}{n_{total}} = 2400 \text{ Torr} \times \frac{6.0 \text{ mol}}{8.0 \text{ mol}} = 1800 \text{ Torr } (1.8 \times 10^3 \text{ Torr}) \text{ (2 SFs)}$$

11.105 Because the partial pressure of nitrogen is to be reported in torr, the atm and mmHg units (for oxygen and argon, respectively) must be converted to torr, as follows:

$$P_{oxygen} = 0.60 \text{ atm} \times \frac{760 \text{ Torr}}{1 \text{ atm}} = 460 \text{ Torr} \quad \text{and}$$

$$P_{argon} = 425 \text{ mmHg} \times \frac{1 \text{ Torr}}{1 \text{ mmHg}} = 425 \text{ Torr}$$

$$\therefore P_{nitrogen} = P_{total} - (P_{oxygen} + P_{argon})$$
$$= 1250 \text{ Torr} - (460 \text{ Torr} + 425 \text{ Torr}) = 370 \text{ Torr (2 SFs)}$$

11.107 a. $P_{hydrogen} = P_{total} - P_{water} = 755 \text{ mmHg} - 21 \text{ mmHg} = 734 \text{ mmHg (2 SFs)}$

b. $V = 415 \text{ mL} \times \dfrac{1 \text{ L}}{1000 \text{ mL}} = 0.415 \text{ L}$

Analyze the Problem: Using the Ideal Gas Law to Calculate Moles

Property	P	V	n	R	T
Given	734 mmHg	415 mL (0.415 L)		$\dfrac{62.4 \text{ L} \cdot \text{mmHg}}{\text{mol} \cdot \text{K}}$	23 °C 23 °C + 273 = 296 K
Need			? mol		

$$n = \frac{PV}{RT} = \frac{(734 \text{ mmHg})(0.415 \text{ L})}{\left(\dfrac{62.4 \text{ L} \cdot \text{mmHg}}{\text{mol} \cdot \text{K}}\right)(296 \text{ K})} = 0.0165 \text{ mol of } H_2 \text{ (3 SFs)}$$

c. Analyze the Problem: Reactant and Product

	Reactant	Product
Mass	? g Al	
Molar Mass	1 mol of Al = 26.98 g of Al	
Moles		0.0165 mol H_2
Equation	$2Al(s) + 3H_2SO_4(aq) \longrightarrow 3H_2(g) + Al_2(SO_4)_3(aq)$	

$$0.0165 \text{ mol } H_2 \times \frac{2 \text{ mol Al}}{3 \text{ mol } H_2} \times \frac{26.98 \text{ g Al}}{1 \text{ mol Al}} = 0.297 \text{ g of Al (3 SFs)}$$

11.109 a. Analyze the Problem: Using the Ideal Gas Law to Calculate Moles

Property	P	V	n	R	T
Given	760. mmHg	762 mL (0.762 L)		$\dfrac{62.4 \text{ L} \cdot \text{mmHg}}{\text{mol} \cdot \text{K}}$	0.0 °C 0.0 °C + 273 = 273 K
Need			? mol		

$$n = \frac{PV}{RT} = \frac{(760. \text{ mmHg})(0.762 \text{ L})}{\left(\dfrac{62.4 \text{ L} \cdot \text{mmHg}}{\text{mol} \cdot \text{K}}\right)(273 \text{ K})} = 0.0340 \text{ mol}$$

$$\text{Molar mass} = \frac{\text{mass}}{\text{moles}} = \frac{1.020 \text{ g}}{0.0340 \text{ mol}} = 30.0 \text{ g/mol}$$

b. $0.815 \text{ g C} \times \dfrac{1 \text{ mol C}}{12.01 \text{ g C}} = 0.0679 \text{ mol of C (smaller number of moles)}$

mass of H = mass of compound − mass of C

$\qquad = 1.020 \text{ g} - 0.815 \text{ g} = 0.205 \text{ g of H}$

$0.205 \text{ g H} \times \dfrac{1 \text{ mol H}}{1.008 \text{ g H}} = 0.203 \text{ mol of H}$

$\dfrac{0.0679 \text{ mol C}}{0.0679} = 1.00 \text{ mol of C} \qquad \dfrac{0.203 \text{ mol H}}{0.0679} = 2.99 \text{ mol of H}$

∴ Empirical formula = $C_{1.00}H_{2.99} \rightarrow CH_3$

Empirical formula mass of CH_3 = 12.01 + 3(1.008) = 15.03 g

$$\text{Small integer} = \frac{\text{molar mass}}{\text{empirical formula mass of } CH_3} = \frac{30.0 \text{ g}}{15.03 \text{ g}} = 2$$

∴ Molecular formula = $C_{(1 \times 2)}H_{(3 \times 2)} = C_2H_6$

11.111 Analyze the Problem: Reactant and Product

	Reactant	Product
Mass	881 g C_3H_8	
Molar Mass	1 mol of C_3H_8 = 44.09 g of C_3H_8	
Moles		? mol CO_2 (? L CO_2)
Molar Volume		1 mol of CO_2 = 22.4 L of CO_2 (STP)
Equation	$C_3H_8(g) + 5O_2(g) \xrightarrow{\Delta} 3CO_2(g) + 4H_2O(g)$	

$881 \text{ g } C_3H_8 \times \dfrac{1 \text{ mol } C_3H_8}{44.09 \text{ g } C_3H_8} \times \dfrac{3 \text{ mol } CO_2}{1 \text{ mol } C_3H_8} \times \dfrac{22.4 \text{ L } CO_2 \text{ (STP)}}{1 \text{ mol } CO_2}$

$= 1340 \text{ L } CO_2 \text{ at STP (3 SFs)}$

11.113 **Analyze the Problem: Using the Ideal Gas Law to Calculate Moles**

Property	P	V	n	R	T
Given	1.00 atm	7.50 L		$\dfrac{0.0821 \text{ L} \cdot \text{atm}}{\text{mol} \cdot \text{K}}$	37 °C 37 °C + 273 = 310 K
Need			? mol O_2		

$$n = \frac{PV}{RT} = \frac{(1.00 \text{ atm})(7.50 \text{ L})}{\left(\dfrac{0.0821 \text{ L} \cdot \text{atm}}{\text{mol} \cdot \text{K}}\right)(310 \text{ K})} = 0.295 \text{ mol of } O_2$$

$$0.295 \text{ mol } O_2 \times \frac{6 \text{ mol } H_2O}{6 \text{ mol } O_2} \times \frac{18.02 \text{ g } H_2O}{1 \text{ mol } H_2O} = 5.32 \text{ g of } H_2O \text{ (smaller amount of product)}$$

$$18.0 \text{ g } C_6H_{12}O_6 \times \frac{1 \text{ mol } C_6H_{12}O_6}{180.16 \text{ g } C_6H_{12}O_6} \times \frac{6 \text{ mol } H_2O}{1 \text{ mol } C_6H_{12}O_6} \times \frac{18.02 \text{ g } H_2O}{1 \text{ mol } H_2O} = 10.8 \text{ g of } H_2O$$

Thus, O_2 is the limiting reactant, and 5.32 g of H_2O (3 SFs) can be produced.

12
Solutions

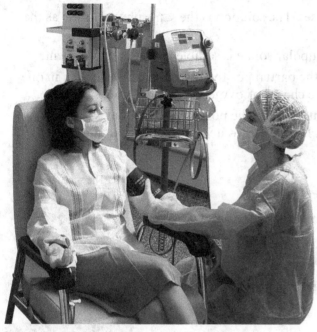

Michelle has hemodialysis three times a week for about 4 h each visit. The treatment removes harmful waste products and extra fluid. Her blood pressure is maintained because the electrolytes potassium and sodium are adjusted each time. Michelle measures the amount of fluid she takes in each day. Her diet is low in protein and includes fruits and vegetables that are low in potassium. She and her husband are now learning about performing hemodialysis at home. Soon she will have her own hemodialysis machine. A typical potassium level in blood serum is 0.091 g in 500. mL of serum. What is the molarity of the potassium ion in serum?

Credit: AJPhoto / Science Source

CHAPTER READINESS*

⚛ Key Math Skills

- Calculating Percentages (1.4)
- Solving Equations (1.4)
- Interpreting Graphs (1.4)

🐍 Core Chemistry Skills

- Writing Conversion Factors from Equalities (2.5)
- Using Conversion Factors (2.6)
- Writing Positive and Negative Ions (6.1)
- Using Molar Mass as a Conversion Factor (7.3)
- Using Mole–Mole Factors (9.2)
- Converting Grams to Grams (9.3)
- Identifying Intermolecular Forces (10.6)

*These Key Math Skills and Core Chemistry Skills from previous chapters are listed here for your review as you proceed to the new material in this chapter.

LOOKING AHEAD

12.1 Solutions

Learning Goal: Identify the solute and solvent in a solution; describe the formation of a solution.

- A solution is a homogeneous mixture that forms when a solute dissolves in a solvent. The solvent is the substance that is present in a greater amount.
- Solvents and solutes can be solids, liquids, or gases. The solution is the same physical state as the solvent.
- A polar solute is soluble in a polar solvent; a nonpolar solute is soluble in a nonpolar solvent.
- Water molecules form hydrogen bonds because the partial positive charge of the hydrogen in one water molecule is attracted to the partial negative charge of oxygen in another water molecule.
- An ionic solute dissolves in water, a polar solvent, because the polar water molecules attract and pull the positive and negative ions into solution. In solution, water molecules surround the ions in a process called hydration.

♦ **Learning Exercise 12.1A**

Water is polar, and hexane is nonpolar. In which solvent is each of the following solutes soluble?

a. bromine, Br_2, nonpolar _____   **b.** HCl, polar _____

c. cholesterol, nonpolar _____   **d.** vitamin D, nonpolar _____

e. vitamin C, polar _____

Answers   **a.** hexane   **b.** water   **c.** hexane
   d. hexane   **e.** water

Solute: The substance present in lesser amount

Salt

Water

Solvent: The substance present in greater amount

A solution consists of at least one solute dispersed in a solvent.
Credit: Pearson Education / Pearson Science

♦ **Learning Exercise 12.1B**

Indicate the solute and solvent in each of the following:

	Solute	Solvent
a. 10 g of KCl dissolved in 100 g of water	_____	_____
b. soda water: $CO_2(g)$ dissolved in water	_____	_____
c. an alloy composed of 80% Zn and 20% Cu	_____	_____
d. a mixture of O_2 (200 mmHg) and He (500 mmHg)	_____	_____
e. a solution of 40 mL of CCl_4 and 2 mL of Br_2	_____	_____

Answers   **a.** KCl; water   **b.** CO_2; water   **c.** Cu; Zn
   d. O_2; He   **e.** Br_2; CCl_4

12.2 Electrolytes and Nonelectrolytes

Learning Goal: Identify solutes as electrolytes or nonelectrolytes.

- Electrolytes conduct an electrical current because they produce ions in aqueous solutions.
- Strong electrolytes are completely dissociated into ions, whereas weak electrolytes are partially dissociated into ions.
- Nonelectrolytes do not form ions in solution but dissolve as molecules.

Key Terms for Sections 12.1 and 12.2

Match the following key terms with the correct description:

    a. solution   **b.** nonelectrolyte   **c.** solvent

    d. solute   **e.** strong electrolyte

1. _____ a substance that dissociates as molecules, not ions, when it dissolves in water

2. _____ the substance that comprises the lesser amount in a solution

3. _____ a solute that dissociates 100% into ions in solution

4. _____ the substance that comprises the greater amount in a solution

5. _____ a homogeneous mixture of at least two components called a solute and a solvent

Answers   **1.** b   **2.** d   **3.** e   **4.** c   **5.** a

♦ **Learning Exercise 12.2A**

Write an equation for the formation of an aqueous solution of each of the following strong electrolytes:

    a. LiCl

    b. $Mg(NO_3)_2$

    c. Na_3PO_4

    d. K_2SO_4

    e. $MgCl_2$

Answers

    a. $LiCl(s) \xrightarrow{H_2O} Li^+(aq) + Cl^-(aq)$

    b. $Mg(NO_3)_2(s) \xrightarrow{H_2O} Mg^{2+}(aq) + 2NO_3^-(aq)$

    c. $Na_3PO_4(s) \xrightarrow{H_2O} 3Na^+(aq) + PO_4^{3-}(aq)$

    d. $K_2SO_4(s) \xrightarrow{H_2O} 2K^+(aq) + SO_4^{2-}(aq)$

    e. $MgCl_2(s) \xrightarrow{H_2O} Mg^{2+}(aq) + 2Cl^-(aq)$

♦ **Learning Exercise 12.2B**

Indicate whether an aqueous solution of each of the following contains only ions, only molecules, or mostly molecules and a few ions. Write an equation for the formation of the solution.

a. glucose, $C_6H_{12}O_6$, a nonelectrolyte

b. NaOH, a strong electrolyte

c. KCl, a strong electrolyte

d. HF, a weak electrolyte

Answers

a. $C_6H_{12}O_6(s) \xrightarrow{H_2O} C_6H_{12}O_6(aq)$ only molecules

b. $NaOH(s) \xrightarrow{H_2O} Na^+(aq) + OH^-(aq)$ only ions

c. $KCl(s) \xrightarrow{H_2O} K^+(aq) + Cl^-(aq)$ only ions

d. $HF(aq) \rightleftharpoons H^+(aq) + F^-(aq)$ mostly molecules and a few ions

12.3 Solubility

Learning Goal: Define solubility; distinguish between an unsaturated and a saturated solution. Identify an ionic compound as soluble or insoluble.

- The amount of solute that dissolves depends on the nature of the solute and solvent.
- Solubility describes the maximum amount of a solute that dissolves in exactly 100. g of solvent at a given temperature.
- A saturated solution contains the maximum amount of dissolved solute at a certain temperature whereas an unsaturated solution contains less than this amount.
- An increase in temperature increases the solubility of most solids, but decreases the solubility of gases in water.
- Henry's law states that the solubility of gases in liquids is directly related to the pressure of that gas above the liquid.
- The solubility rules describe the kinds of ionic combinations that are soluble and insoluble in water.

CdS   PbI₂

If an ionic compound contains a combination of a cation and an anion such as Cd^{2+} and S^{2-} or Pb^{2+} and I^-, that ionic compound is insoluble.

Credit: Pearson Education / Pearson Science

♦ Learning Exercise 12.3A

Identify each of the following as a saturated solution (S) or an unsaturated solution (U):

a. _____ A sugar cube dissolves when added to a cup of coffee.

b. _____ A KCl crystal added to a KCl solution does not change in size.

c. _____ A layer of sugar forms in the bottom of a glass of iced tea.

d. _____ The rate of crystal formation is equal to the rate of dissolving.

e. _____ Upon heating, all the sugar in a solution dissolves.

Answers   a. U   b. S   c. S   d. S   e. U

♦ Learning Exercise 12.3B

Use the solubility data for $NaNO_3$ listed below to answer each of the following problems:

Temperature (°C)	Solubility (g of $NaNO_3$/100. g of H_2O)
40	100. g
60	120. g
80	150. g
100	200. g

a. How many grams of $NaNO_3$ will dissolve in 100. g water at 40 °C?

b. How many grams of $NaNO_3$ will dissolve in 300. g of water at 60 °C?

c. A solution is prepared using 200. g of water and 350. g of $NaNO_3$ at 80 °C. Will any solute remain undissolved? If so, how much?

d. Will 150. g of $NaNO_3$ all dissolve when added to 100. g of water at 100 °C?

Answers   a. 100. g   b. 360. g
c. 300. g of $NaNO_3$ will dissolve leaving 50. g of $NaNO_3$ that will not dissolve.
d. Yes, all 150. g of $NaNO_3$ will dissolve.

Solubility Rules for Ionic Compounds in Water

An ionic compound is soluble in water if it contains one of the following:

Positive Ions: Li^+, Na^+, K^+, Rb^+, Cs^+, NH_4^+

Negative Ions: NO_3^-, $C_2H_3O_2^-$

Cl^-, Br^-, I^- except when combined with Ag^+, Pb^{2+}, or Hg_2^{2+}

SO_4^{2-} except when combined with Ba^{2+}, Pb^{2+}, Ca^{2+}, Sr^{2+}, or Hg_2^{2+}

Ionic compounds that do not contain at least one of these ions are usually insoluble.

♦ **Learning Exercise 12.3C**

Predict whether the following ionic compounds are soluble (S) or insoluble (I) in water:

Using Solubility Rules

a. _____ NaCl   b. _____ AgNO$_3$ c. _____ PbCl$_2$

d. _____ Ag$_2$S   e. _____ BaSO$_4$ f. _____ Na$_2$CO$_3$

g. _____ K$_2$S   h. _____ MgCl$_2$ i. _____ BaS

Answers   **a.** S   **b.** S   **c.** I   **d.** I   **e.** I
              f. S   **g.** S   **h.** S   **i.** I

Guide to Writing an Equation for the Formation of an Insoluble Ionic Compound	
STEP 1	Write the ions of the reactants.
STEP 2	Write the combinations of ions and determine if any are insoluble.
STEP 3	Write the ionic equation including any solid.
STEP 4	Write the net ionic equation.

Example: Write the ionic and net ionic equations for the reaction that occurs when solutions of Sr(NO$_3$)$_2$ and Na$_2$SO$_4$ are mixed.

Solution: **STEP 1** **Write the ions of the reactants.**

$$Sr(NO_3)_2(aq) = Sr^{2+}(aq) + 2NO_3^-(aq) \qquad Na_2SO_4(aq) = 2Na^+ + SO_4^{2-}(aq)$$

STEP 2 **Write the combinations of ions and determine if any are insoluble.**

$$Na^+(aq) + NO_3^-(aq) \longrightarrow NaNO_3(aq) \qquad \text{a soluble ionic compound}$$
$$Sr^{2+}(aq) + SO_4^{2-}(aq) \longrightarrow SrSO_4(s) \qquad \text{an insoluble ionic compound}$$

STEP 3 **Write the ionic equation including any solid.** By breaking up all the ionic formulas into ions, we can write

$$Sr^{2+}(aq) + 2NO_3^-(aq) + 2Na^+(aq) + SO_4^{2-}(aq) \longrightarrow SrSO_4(s) + 2Na^+(aq) + 2NO_3^-(aq)$$

STEP 4 **Write the net ionic equation.** Remove the spectator ions, which appear on both sides of the ionic equation, to write the net ionic equation.

$$Sr^{2+}(aq) + SO_4^{2-}(aq) + \cancel{2Na^+(aq)} + \cancel{2NO_3^-(aq)} \longrightarrow SrSO_4(s) + \cancel{2Na^+(aq)} + \cancel{2NO_3^-(aq)}$$
$$Sr^{2+}(aq) + SO_4^{2-}(aq) \longrightarrow SrSO_4(s)$$

♦ **Learning Exercise 12.3D**

Predict whether an insoluble ionic compound forms in the following mixtures of soluble ionic compounds. If so, write the formula of the solid.

1. $NaCl(aq) + Pb(NO_3)_2(aq) \longrightarrow$ _____

2. $BaCl_2(aq) + Na_2SO_4(aq) \longrightarrow$ _____

3. $K_3PO_4(aq) + NaNO_3(aq) \longrightarrow$ _____

4. $Na_2S(aq) + AgNO_3(aq) \longrightarrow$ _____

Answers   **1.** Yes; $PbCl_2$ **2.** Yes; $BaSO_4$ **3.** None   **4.** Yes; Ag_2S

♦ **Learning Exercise 12.3E**

For the mixtures that form insoluble ionic compounds in Learning Exercise 12.3D, write the ionic equation and the net ionic equation.

a.

b.

c.

d.

Answers

a. $2Na^+(aq) + 2Cl^-(aq) + Pb^{2+}(aq) + 2NO_3^-(aq) \longrightarrow PbCl_2(s) + 2Na^+(aq) + 2NO_3^-(aq)$
$Pb^{2+}(aq) + 2Cl^-(aq) \longrightarrow PbCl_2(s)$

b. $Ba^{2+}(aq) + 2Cl^-(aq) + 2Na^+(aq) + SO_4^{2-}(aq) \longrightarrow BaSO_4(s) + 2Na^+(aq) + 2Cl^-(aq)$
$Ba^{2+}(aq) + SO_4^{2-}(aq) \longrightarrow BaSO_4(s)$

c. No insoluble ionic compound forms.

d. $2Na^+(aq) + S^{2-}(aq) + 2Ag^+(aq) + 2NO_3^-(aq) \longrightarrow Ag_2S(s) + 2Na^+(aq) + 2NO_3^-(aq)$
$2Ag^+(aq) + S^{2-}(aq) \longrightarrow Ag_2S(s)$

12.4   Solution Concentrations

Learning Goal: Calculate the concentration of a solute in a solution; use concentration as conversion factors to calculate the amount of solute or solution.

- The concentration of a solution is the relationship between the amount of solute, in grams, moles, or milliliters, and the amount of solution, in grams, milliliters, or liters.
- A mass percent (m/m) expresses the ratio of the mass of solute to the mass of solution multiplied by 100%. Mass units must be the same such as grams of solute/grams of solution × 100%.

$$\text{Mass percent (m/m)} = \frac{\text{mass of solute}}{\text{mass of solution}} \times 100\%$$

- A volume percent (v/v) expresses the ratio of the volume of solute to volume of solution multiplied by 100%. Volume units must be the same such as milliliters of solute/milliliters of solution × 100%.

$$\text{Volume percent (v/v)} = \frac{\text{volume of solute}}{\text{volume of solution}} \times 100\%$$

- A mass/volume percent (m/v) expresses the ratio of the mass of solute to the volume of solution multiplied by 100%. Mass units for the solute must be in grams and the volume units for the solution must be in milliliters.

$$\text{Mass/volume percent (m/v)} = \frac{\text{grams of solute}}{\text{milliliters of solution}} \times 100\%$$

- Molarity is a concentration term that indicates the number of moles of solute dissolved in 1 L (1000 mL) of solution.

$$\text{Molarity (M)} = \frac{\text{moles of solute}}{\text{liters of solution}}$$

- Concentrations can be used as conversion factors to relate the amount of solute and the volume or mass of the solution.

$$\frac{\text{amount of solute}}{\text{amount of solution}} \quad \text{and} \quad \frac{\text{amount of solution}}{\text{amount of solute}}$$

Key Terms for Sections 12.3 and 12.4

Match each of the following key terms with the correct description:

   a. mass percent (m/m)   **b.** concentration   **c.** solubility
   d. unsaturated   **e.** molarity

1. _____ the number of moles of solute in 1 L of solution

2. _____ the amount of solute dissolved in a certain amount of solution

3. _____ the maximum amount of solute that can dissolve in a solution at a certain temperature

4. _____ a solution with less than the maximum amount of solute that can dissolve

5. _____ the concentration of a solution in terms of mass of solute in 100. g of solution

Answers   **1.** e   **2.** b   **3.** c   **4.** d   **5.** a

Guide to Calculating Solution Concentration	
STEP 1	State the given and needed quantities.
STEP 2	Write the concentration expression.
STEP 3	Substitute solute and solution quantities into the expression and calculate.

Example: What is the mass percent (m/m) when 2.4 g of $NaHCO_3$ dissolves in water to make 120. g of a solution of $NaHCO_3$?

Solution: **STEP 1**   **State the given and needed quantities.**

Analyze the Problem	**Given**	**Need**	**Connect**
	2.4 g of $NaHCO_3$ solute, 120. g of $NaHCO_3$ solution	mass percent (m/m)	$\dfrac{\text{mass of solute}}{\text{mass of solution}} \times 100\%$

STEP 2   **Write the concentration expression.**

$$\text{Mass percent (m/m)} = \frac{\text{mass of solute}}{\text{mass of solution}} \times 100\%$$

STEP 3   **Substitute solute and solution quantities into the expression and calculate.**

$$\text{Mass percent (m/m)} = \frac{2.4 \text{ g } NaHCO_3}{120. \text{ g } NaHCO_3 \text{ solution}} \times 100\% = 2.0\% \text{ (m/m)}$$

 **CORE CHEMISTRY SKILL**

Calculating Concentration

♦ **Learning Exercise 12.4A**

Calculate the percent concentration for each of the following solutions:

a. mass percent (m/m) for 18.0 g of NaCl in 90.0 g of a solution of NaCl

b. mass/volume percent (m/v) for 4.0 g of KOH in 50.0 mL of a solution of KOH

c. mass/volume percent (m/v) for 5.0 g of KCl in 2.0 L of a solution of KCl

d. volume percent (v/v) for 18 mL of ethanol in 350 mL of a mouthwash solution

Answers   **a.** 20.0% (m/m)   **b.** 8.0% (m/v)   **c.** 0.25% (m/v)   **d.** 5.1% (v/v)

♦ **Learning Exercise 12.4B**

Calculate the molarity (M) of the following solutions:

   a. 1.45 mol of HCl in 0.250 L of a solution of HCl

   b. 10.0 mol of glucose ($C_6H_{12}O_6$) in 2.50 L of a glucose solution

   c. 80.0 g of NaOH in 1.60 L of a solution of NaOH

   d. 38.8 g of NaBr in 175 mL of a solution of NaBr

Answers　　　**a.** 5.80 M HCl solution　　　**b.** 4.00 M glucose solution
　　　　　　　　c. 1.25 M NaOH solution　　　**d.** 2.15 M NaBr solution

Guide to Using Concentration to Calculate Mass or Volume	
STEP 1	State the given and needed quantities.
STEP 2	Write a plan to calculate the mass or volume.
STEP 3	Write equalities and conversion factors.
STEP 4	Set up the problem to calculate the mass or volume.

Example: How many grams of KI are needed to prepare 225 g of a 4.0% (m/m) KI solution?

Solution: **STEP 1** **State the given and needed quantities.**

	Given	Need	Connect
Analyze the Problem	225 g of 4.0% (m/m) KI solution	grams of KI	mass percent factor $\dfrac{\text{g of solute}}{100.\ \text{g of solution}}$

   STEP 2 **Write a plan to calculate the mass.**

       grams of KI solution ⟶ grams of KI

   STEP 3 **Write equalities and conversion factors.**

$$4.0 \text{ g of KI} = 100.0 \text{ g of KI solution}$$

$$\frac{4.0 \text{ g KI}}{100.\ \text{g KI solution}} \quad \text{and} \quad \frac{100.\ \text{g KI solution}}{4.0 \text{ g KI}}$$

   STEP 4 **Set up the problem to calculate the mass.**

$$225 \text{ g KI solution} \times \frac{4.0 \text{ g KI}}{100.\ \text{g KI solution}} = 9.0 \text{ g of KI}$$

♦ Learning Exercise 12.4C

Calculate the number of grams of solute needed to prepare each
of the following solutions:

a. How many grams of glucose are needed to prepare 480. g of a 5.0% (m/m) glucose solution?

b. How many grams of lidocaine, a local anesthetic, are needed to prepare 50.0 g of
a 2.0% (m/m) lidocaine solution?

c. How many grams of KCl are needed to prepare 1250 g of a 4.00% (m/m) KCl solution?

d. How many grams of NaCl are needed to prepare 75.6 g of a 1.50% (m/m) NaCl solution?

Answers   **a.** 24 g   **b.** 1.0 g   **c.** 50.0 g   **d.** 1.13 g

♦ Learning Exercise 12.4D

Use percent concentration factors to calculate the mass of solution that contains the amount
of solute stated in each problem:

a. 8.75 g of NaCl from a 3.00% (m/m) NaCl solution

b. 24 g of glucose from a 5.0% (m/m) glucose solution

c. 10.5 g of KCl from a 2.5% (m/m) KCl solution

d. 12.0 g of NaOH from a 5.00% (m/m) NaOH solution

Answers   **a.** 292 g   **b.** 480 g   **c.** 420 g   **d.** 240. g

---

Study Note

Molarity is used as a conversion factor to convert between the amount of solute and the volume of solution.

Example: How many grams of NaOH are in 0.250 L of a 5.00 M NaOH solution?

Solution:

STEP 1   State the given and needed quantities.

Analyze the Problem	Given	Need	Connect
	0.250 L of a 5.00 M NaOH solution	grams of NaOH	molarity (mol/L), molar mass of NaOH

STEP 2   Write a plan to calculate the mass.

liters of solution $\longrightarrow$ moles of solution $\longrightarrow$ grams of solute

STEP 3   Write equalities and conversion factors.

1 L of solution = 5.00 mol of NaOH   1 mol of NaOH = 40.00 g of NaOH

$$\frac{5.00 \text{ mol NaOH}}{1 \text{ L solution}} \quad \text{and} \quad \frac{1 \text{ L solution}}{5.00 \text{ mol NaOH}} \qquad \frac{40.00 \text{ g NaOH}}{1 \text{ mol NaOH}} \quad \text{and} \quad \frac{1 \text{ mol NaOH}}{40.00 \text{ g NaOH}}$$

STEP 4   Set up the problem to calculate the mass.

$$0.250 \text{ L solution} \times \frac{5.00 \text{ mol NaOH}}{1 \text{ L solution}} \times \frac{40.00 \text{ g NaOH}}{1 \text{ mol NaOH}} = 50.0 \text{ g of NaOH}$$

◆ **Learning Exercise 12.4E**

Calculate the quantity of solute in each of the following solutions:

   a. How many moles of HCl are in 1.50 L of a 6.00 M HCl solution?

   b. How many moles of KOH are in 0.750 L of a 10.0 M KOH solution?

   c. How many grams of NaOH are needed to prepare 0.500 L of a 4.40 M NaOH solution?

   d. How many grams of NaCl are in 285 mL of a 1.75 M NaCl solution?

Answers   **a.** 9.00 mol of HCl   **b.** 7.50 mol of KOH
          c. 88.0 g of NaOH   **d.** 29.1 g of NaCl

♦ **Learning Exercise 12.4F**

Calculate the volume, in milliliters, needed of each solution to obtain the given quantity of solute:

a. 1.50 mol of $Mg(OH)_2$ from a 2.00 M $Mg(OH)_2$ solution

b. 0.150 mol of glucose from a 2.20 M glucose solution

c. 18.5 g of KI from a 3.00 M KI solution

d. 18.0 g of NaOH from a 6.00 M NaOH solution

Answers **a.** 750 mL **b.** 68.2 mL **c.** 37.1 mL **d.** 75.0 mL

12.5 Dilution of Solutions

Learning Goal: Describe the dilution of a solution; calculate the unknown concentration or volume when a solution is diluted.

- Dilution is the process of mixing a solution with solvent to obtain a solution with a lower concentration.
- For dilutions, use the expression $C_1V_1 = C_2V_2$ and solve for the unknown value.

	Guide to Calculating Dilution Quantities
STEP 1	Prepare a table of the concentrations and volumes of the solutions.
STEP 2	Rearrange the dilution expression to solve for the unknown quantity.
STEP 3	Substitute the known quantities into the dilution expression and calculate.

Example: What is the molarity when 150. mL of a 2.00 M NaCl solution is diluted to a volume of 400. mL?

Solution: **STEP 1** **Prepare a table of the concentrations and volumes of the solutions.**

	Given	Need	Connect
Analyze the Problem	C_1 = 2.00 M V_1 = 150. mL V_2 = 400. mL	C_2	$C_1V_1 = C_2V_2$ V increases, C decreases

STEP 2 **Rearrange the dilution expression to solve for the unknown quantity.**

$$C_1V_1 = C_2V_2 \qquad C_2 = C_1 \times \frac{V_1}{V_2}$$

STEP 3 **Substitute the known quantities into the dilution expression and calculate.**

$$C_2 = 2.00 \text{ M} \times \frac{150.\ \cancel{mL}}{400.\ \cancel{mL}} = 0.750 \text{ M}$$

♦ **Learning Exercise 12.5**

Solve each of the following dilution problems:

a. What is the molarity when 100. mL of a 5.0 M KCl solution is diluted with water to give a final volume of 200. mL?

b. What is the molarity of the diluted solution when 5.0 mL of a 1.5 M KCl solution is diluted to a total volume of 25 mL?

c. What is the molarity when 250 mL of an 8.0 M NaOH solution is diluted with 750 mL of water?

d. 160. mL of water is added to 40. mL of a 1.0 M NaCl solution. What is the molarity?

e. What volume of 6.0 M HCl is needed to prepare 300. mL of a 1.0 M HCl solution? How much water must be added?

Answers   **a.** 2.5 M   **b.** 0.30 M   **c.** 2.7 M   **d.** 0.20 M
e. V_1 = 50. mL; add 250. mL of water

12.6 Chemical Reactions in Solution

Learning Goal: Given the volume and concentration of a solution in a chemical reaction, calculate the amount of a reactant or product in the reaction.

- The volume and molarity of a solution are used to determine the moles or liters of a substance required or produced in a chemical reaction.
- The balanced equation can be used to convert moles of one substance to moles of another substance.
- Molar mass is used to convert moles to grams or grams to moles.
- The number of moles of a solute and the molarity of the solution are used to determine the volume of a solution in a chemical reaction.

Guide to Calculations Involving Solutions in Chemical Reactions	
STEP 1	State the given and needed quantities.
STEP 2	Write a plan to calculate the needed quantity or concentration.
STEP 3	Write equalities and conversion factors including mole–mole and concentration factors.
STEP 4	Set up the problem to calculate the needed quantity or concentration.

♦ **Learning Exercise 12.6A**

For the reaction,

$$2AgNO_3(aq) + H_2SO_4(aq) \longrightarrow Ag_2SO_4(s) + 2HNO_3(aq)$$

▶ **CORE CHEMISTRY SKILL**

Calculating the Quantity of a Reactant or Product for a Chemical Reaction in Solution

    a. How many milliliters of a 1.5 M $AgNO_3$ solution will react with 40.0 mL of a 1.0 M H_2SO_4 solution?

    b. How many grams of Ag_2SO_4 will be produced?

Answers

  a. 53 mL of $AgNO_3$ solution   **b.** 12 g of Ag_2SO_4

♦ **Learning Exercise 12.6B**

Calculate the number of milliliters of a 1.80 M KOH solution that react with 18.5 mL of a 2.20 M HNO_3 solution.

$$HNO_3(aq) + KOH(aq) \longrightarrow H_2O(l) + KNO_3(aq)$$

Answer

22.6 mL of KOH solution

♦ **Learning Exercise 12.6C**

How many liters of hydrogen gas can form at STP when 125 mL of a 3.50 M HCl solution reacts with excess aluminum?

$$2Al(s) + 6HCl(aq) \longrightarrow 3H_2(g) + 2AlCl_3(aq)$$

Answer

4.90 L of H_2

12.7 Molality and Freezing Point Lowering/Boiling Point Elevation

Learning Goal: Identify a mixture as a solution, a colloid, or a suspension. Using the molality, calculate the new freezing point and new boiling point for a solution.

- Colloids contain particles that do not settle out and pass through filters but not through semipermeable membranes.
- Suspensions are composed of large particles that settle out of solution.
- A solute added to water decreases the vapor pressure, decreases the freezing point, and increases the boiling point.
- The freezing point of a solution is lower than that of the solvent and boiling point is higher and depend only on the number of particles of solute in the solution.
- The freezing point and boiling point of a solution depend on the molality (*m*) of the solution.

$$\text{Molality } (m) = \frac{\text{moles of solute}}{\text{kilograms of solvent}}$$

- The freezing point and freezing point lowering (ΔT_f) of a solution can be calculated based on the molality of particles in the solution and the freezing point constant (K_f) for the solvent.

$$\Delta T_f = m \times K_f$$

- The boiling point and boiling point elevation (ΔT_b) of a solution can be calculated based on the molality of particles in the solution and the boiling point constant (K_b) for the solvent.

$$\Delta T_b = m \times K_b$$

♦ **Learning Exercise 12.7A**

Identify each of the following as characteristic of a solution, colloid, or suspension:

a. _____ the solute consists of single atoms, ions, or small molecules

b. _____ settles out with gravity

c. _____ retained by filters

d. _____ cannot diffuse through a cellular membrane

e. _____ contains large particles that are visible

Answers　　a. solution　　b. suspension　　c. suspension
　　　　　　　d. colloid, suspension　　e. suspension

Guide to Calculating Molality	
STEP 1	State the given and needed quantities.
STEP 2	Write the molality expression.
STEP 3	Substitute solute and solvent quantities into the expression and calculate.

♦ **Learning Exercise 12.7B**

Calculate the molality (m) of the following solutions:

a. 75.0 g of fructose, $C_6H_{12}O_6$, a nonelectrolyte, added to 225 g of water

b. 110. g of ethanol, C_2H_6O, a nonelectrolyte, dissolved in 850. g of water

c. 35.5 g of urea, CH_4N_2O, a nonelectrolyte, added to 465 g of water

Answers   **a.** $1.85\ m$ **b.** $2.81\ m$ **c.** $1.27\ m$

Guide to Calculating Freezing Point Lowering / Boiling Point Elevation	
STEP 1	State the given and needed quantities.
STEP 2	Determine the number of moles of solute particles and calculate the molality.
STEP 3	Calculate the temperature change and subtract from the freezing point or add to the boiling point.

Example: Calculate the freezing point of a solution containing 125 g of KCl in 500. g of water.

Solution: **STEP 1**   **State the given and needed quantities.**

Analyze the Problem	Given	Need	Connect
	125 g of KCl, 500. g of water	freezing point of KCl solution	molar mass, $\Delta T_f = m \times K_f$

    STEP 2   **Determine the number of moles of solute particles and calculate the molality.**

$$\text{moles of particles} = 125\ \text{g KCl} \times \frac{1\ \text{mol KCl}}{74.55\ \text{g KCl}} \times \frac{2\ \text{mol particles}}{1\ \text{mol KCl}} = 3.35\ \text{mol of particles (ions)}$$

$$\text{Molality}\ (m) = \frac{3.35\ \text{mol particles}}{0.500\ \text{kg water}} = 6.70\ m$$

    STEP 3   **Calculate the temperature change and subtract from the freezing point.**

$$\Delta T_f = 6.70\ m \times \frac{1.86\ °C}{m} = 12.5\ °C$$

$$T_{\text{solution}} = T_{\text{water}} - \Delta T_f$$

$$= 0.0\ °C - 12.5\ °C = -12.5\ °C$$

♦ Learning Exercise 12.7C

CORE CHEMISTRY SKILL
Calculating the Freezing Point/Boiling Point of a Solution

For the following solutes, each in 1000. g of water, calculate:

a. the molality of solute particles present in the solution

b. the freezing point of the solution

c. the boiling point of the solution

Solute	a. Molality of Solute Particles	b. Freezing Point	c. Boiling Point
1. 1.00 mol of fructose (nonelectrolyte)			
2. 1.50 mol of KCl (strong electrolyte)			
3. 1.25 mol of $Ca(NO_3)_2$ (strong electrolyte)			

Answers

Solute	a. Molality of Solute Particles	b. Freezing Point	c. Boiling Point
1. 1.00 mol of fructose (nonelectrolyte)	1.00 *m*	−1.86 °C	100.52 °C
2. 1.50 mol of KCl (strong electrolyte)	3.00 *m*	−5.58 °C	101.56 °C
3. 1.25 mol of $Ca(NO_3)_2$ (strong electrolyte)	3.75 *m*	−6.98 °C	101.95 °C

12.8 Properties of Solutions: Osmosis

Learning Goal: Describe how the number of particles in solution affects osmosis.

- In the process of osmosis, water (solvent) moves through a semipermeable membrane from the solution that has a lower solute concentration to a solution where the solute concentration is higher.
- Osmotic pressure is the pressure that prevents the flow of water into a more concentrated solution.
- Isotonic solutions have osmotic pressures equal to that of body fluids. A hypotonic solution has a lower osmotic pressure than body fluids; a hypertonic solution has a higher osmotic pressure.
- A red blood cell maintains its volume in an isotonic solution, but it swells (hemolysis) in a hypotonic solution and shrinks (crenation) in a hypertonic solution.
- In dialysis, water and small solute particles can pass through a dialyzing membrane, while larger particles such as blood cells cannot.

♦ **Learning Exercise 12.8A**

Fill in the blanks:

In osmosis, the direction of solvent flow is from the solution with (**a.**) [higher/lower] solute concentration to the solution with (**b.**) [higher/lower] solute concentration. A semipermeable membrane separates 5% (m/v) and 10% (m/v) sucrose solutions. The (**c.**) _____ % (m/v) solution has the greater osmotic pressure. Water will move from the (**d.**) _____ % (m/v) solution into the (**e.**) _____ % (m/v) solution. The compartment that contains the (**f.**) _____ % (m/v) solution increases in volume.

Answers   **a.** lower   **b.** higher   **c.** 10   **d.** 5   **e.** 10   **f.** 10

♦ **Learning Exercise 12.8B**

A semipermeable membrane separates a 2% starch solution from a 10% starch solution. Complete each of the following with 2% or 10%:

    a. Water will flow from the _____ starch solution to the _____ starch solution.

    b. The volume of the _____ starch solution will increase and the volume of the _____ starch solution will decrease.

Answers   **a.** 2%, 10%   **b.** 10%, 2%

♦ **Learning Exercise 12.8C**

Fill in the blanks:

A (**a.**) _____ % (m/v) NaCl solution and a (**b.**) _____ % (m/v) glucose solution are isotonic to the body fluids. A red blood cell placed in these solutions does not change in volume because these solutions are (**c.**) _____ tonic. When a red blood cell is placed in water, it undergoes (**d.**) _____ because water is (**e.**) _____ tonic. A 20% (m/v) glucose solution will cause a red blood cell to undergo (**f.**) _____ because the 20% (m/v) glucose solution is (**g.**) _____ tonic.

Answers   **a.** 0.9   **b.** 5   **c.** iso   **d.** hemolysis
             e. hypo   **f.** crenation   **g.** hyper

♦ **Learning Exercise 12.8D**

Indicate whether the following are

    a. hypotonic   **b.** hypertonic   **c.** isotonic

1. ___ 5% (m/v) glucose   **2.** ___ 3% (m/v) NaCl   **3.** ___ 2% (m/v) glucose

4. ___ water   **5.** ___ 0.9% (m/v) NaCl   **6.** ___ 10% (m/v) glucose

Answers   **1.** c   **2.** b   **3.** a   **4.** a   **5.** c   **6.** b

♦ **Learning Exercise 12.8E**

Indicate whether the following will cause a red blood cell to undergo

    a. crenation   **b.** hemolysis   **c.** no change

1. ____ 10% (m/v) NaCl   **2.** ____ 1% (m/v) glucose   **3.** ____ 5% (m/v) glucose

4. ____ 0.5% (m/v) NaCl   **5.** ____ 10% (m/v) glucose   **6.** ____ water

Answers   **1.** a   **2.** b   **3.** c   **4.** b   **5.** a   **6.** b

♦ **Learning Exercise 12.8F**

A dialysis bag contains starch, glucose, NaCl, protein, and urea.

    a. When the dialysis bag is placed in water, what components would you expect to dialyze through the bag? Why?

    b. Which components will stay inside the dialysis bag? Why?

Answers   **a.** Glucose, Na^+, Cl^- ions, and urea are solution particles. Solution particles will pass through semipermeable membranes.
    b. Starch and protein form colloidal particles; because of their size, colloidal particles are retained by semipermeable membranes.

Key Terms for Sections 12.5 to 12.8

Match each of the following key terms with the correct description:

    a. hypertonic   **b.** dilution   **c.** molality
    d. osmosis   **e.** semipermeable membrane   **f.** colloid

1. ____ the particles that pass through filters but are too large to pass through semipermeable membranes

2. ____ the number of moles of particles in exactly 1 kg of solvent

3. ____ a solution that has a higher osmotic pressure than the red blood cells of the body

4. ____ a process by which water is added to a solution and decreases its concentration

5. ____ a membrane that permits the passage of small particles while blocking larger particles

6. ____ the flow of solvent through a semipermeable membrane into a solution of higher solute concentration

Answers   **1.** f   **2.** c   **3.** a   **4.** b   **5.** e   **6.** d

Checklist for Chapter 12

You are ready to take the Practice Test for Chapter 12. Be sure you have accomplished the following learning goals for this chapter. If not, review the section listed at the end of the goal. Then apply your new skills and understanding to the Practice Test.

After studying Chapter 12, I can successfully:

_____ Identify the solute and solvent in a solution and describe the process of dissolving an ionic solute in water. (12.1)

_____ Identify the components in solutions of electrolytes and nonelectrolytes. (12.2)

_____ Identify a saturated and an unsaturated solution. (12.3)

_____ Identify an ionic compound as soluble or insoluble. (12.3)

_____ Write a chemical equation (or ionic or net ionic equation) to show the formation of an insoluble ionic compound. (12.3)

_____ Describe the effects of temperature and nature of a solute on its solubility in a solvent. (12.3)

_____ Determine the solubility of an ionic compound in water. (12.3)

_____ Calculate the percent concentration, m/m, v/v, and m/v, of a solute in a solution, and use percent concentration to calculate the amount of solute or solution. (12.4)

_____ Calculate the molarity of a solution. (12.4)

_____ Use molarity as a conversion factor to calculate the moles (or grams) of a solute or the volume of the solution. (12.4)

_____ Calculate the unknown concentration or volume when a solution is diluted. (12.5)

_____ Use the volume and molarity of a solution to calculate the grams or moles of a substance produced in a chemical reaction. (12.6)

_____ Identify a mixture as a solution, a colloid, or a suspension. (12.7)

_____ Calculate the molality of a solution. (12.7)

_____ Use the concentration of particles in a solution to calculate the freezing point or boiling point of the solution. (12.7)

_____ Identify a solution as isotonic, hypotonic, or hypertonic. (12.8)

_____ Explain the processes of osmosis and dialysis. (12.8)

Practice Test for Chapter 12

The chapter sections to review are shown in parentheses at the end of each question.

For questions 1 through 4, indicate if each of the following is more soluble in (W) water, a polar solvent, or (B) benzene, a nonpolar solvent: (12.1)

1. $I_2(s)$, nonpolar

2. $NaBr(s)$, polar

3. $KI(s)$, polar

4. C_6H_{12}, nonpolar

5. When dissolved in water, $Ca(NO_3)_2(s)$ dissociates into (12.2)
    A. $Ca^{2+}(aq) + (NO_3)_2{}^{2-}(aq)$
    B. $Ca^+(aq) + NO_3{}^-(aq)$
    C. $Ca^{2+}(aq) + 2NO_3{}^-(aq)$
    D. $Ca^{2+}(aq) + 2N^{5+}(aq) + 2O_3{}^{6-}(aq)$
    E. $CaNO_3{}^+(aq) + NO_3{}^-(aq)$

6. Ethanol (C_2H_6O) is a nonelectrolyte. When placed in water it (12.2)
   A. dissociates completely.   **B.** dissociates partially.
   C. does not dissociate.   **D.** makes the solution acidic.
   E. makes the solution basic.

7. The solubility of NH_4Cl is 46 g in 100. g of water at 40 °C. How much NH_4Cl can dissolve in 500. g of water at 40 °C? (12.3)
   A. 9.2 g   **B.** 46 g   **C.** 100 g   **D.** 184 g   **E.** 230 g

For questions 8 through 11, indicate if each of the following is soluble (S) *or insoluble* (I) *in water:* (12.3)

8. NaCl   **9.** AgCl   **10.** $BaSO_4$ **11.** FeO

12. Which of the following ionic compounds is soluble in water? (12.3)
   A. FeS   **B.** $BaCO_3$ **C.** K_2SO_4 **D.** PbS   **E.** MgO

13. The insoluble ionic compound that forms when a solution of NaCl mixes with a $Pb(NO_3)_2$ solution is (12.3)
   A. Na_2Pb **B.** $ClNO_3$ **C.** $NaNO_3$ **D.** $PbCl_2$ **E.** none

14. Which of the following ionic compounds is insoluble in water? (12.3)
   A. $CuCl_2$ **B.** $Pb(NO_3)_2$ **C.** K_2CO_3 **D.** $(NH_4)_2SO_4$ **E.** $CaCO_3$

15. A solution containing 1.20 g of sucrose in 50.0 g of solution has a percent (m/m) concentration of (12.4)
   A. 0.600%   **B.** 1.20%   **C.** 2.40%   **D.** 30.0%   **E.** 41.6%

16. The grams of lactose in 250 g of a 3.0% (m/m) lactose solution for infant formula are (12.4)
   A. 0.15 g   **B.** 1.2 g   **C.** 6.0 g   **D.** 7.5 g   **E.** 30 g

17. The mass of solution needed to obtain 0.40 g of glucose from a 5.0% (m/m) glucose solution is (12.4)
   A. 1.0 g   **B.** 2.0 g   **C.** 4.0 g   **D.** 5.0 g   **E.** 8.0 g

18. The amount of NaCl needed to prepare 50.0 g of a 4.00% (m/m) NaCl solution is (12.4)
   A. 20.0 g   **B.** 15.0 g   **C.** 10.0 g   **D.** 4.00 g   **E.** 2.00 g

19. A solution containing 6.0 g of NaCl in 1500 g of solution has a mass percent concentration of (12.4)
   A. 0.40% (m/m) **B.** 0.25% (m/m)   **C.** 4.0% (m/m)   **D.** 0.90% (m/m)   **E.** 2.5% (m/m)

20. The number of moles of KOH needed to prepare 2400 mL of a 2.0 M KOH solution is (12.4)
   A. 1.2 mol   **B.** 2.4 mol   **C.** 4.8 mol   **D.** 12 mol   **E.** 48 mol

21. The number of grams of NaOH needed to prepare 7.5 mL of a 5.0 M NaOH solution is (12.4)
   A. 1.5 g   **B.** 3.8 g   **C.** 6.7 g   **D.** 15 g   **E.** 38 g

For questions 22 through 24, consider a 20.0-g sample of a solution that contains 2.0 g of NaOH: (12.4)

22. The percent (m/m) concentration of the solution is
   A. 1.0%   **B.** 4.0%   **C.** 5.0%   **D.** 10.%   **E.** 20.%

23. The number of moles of NaOH in the sample is
   A. 0.050 mol   **B.** 0.40 mol   **C.** 1.0 mol   **D.** 2.5 mol   **E.** 4.0 mol

24. If the solution has a volume of 0.025 L, what is the molarity of the sample?
   A. 0.10 M   **B.** 0.5 M   **C.** 1.0 M   **D.** 1.5 M   **E.** 2.0 M

25. A 20.-mL sample of 5.0 M HCl solution is diluted with water to give 100. mL of solution. The final concentration of the HCl solution is (12.5)
   A. 10 M   **B.** 5.0 M   **C.** 2.0 M   **D.** 1.0 M   **E.** 0.50 M

26. Water is added to 200. mL of a 4.00 M KNO_3 solution to give 400. mL of solution. The final concentration of the diluted solution is (12.5)
    A. 1.00 M   **B.** 2.00 M   **C.** 4.00 M   **D.** 0.500 M   **E.** 0.100 M

27. 5.0 mL of a 2.0 M KOH solution is diluted with water to give 50.0 mL of solution. The final concentration of the KOH solution is (12.5)
    A. 1.5 M   **B.** 1.0 M   **C.** 20 M   **D.** 0.20 M   **E.** 0.50 M

28. What mass of Ag_2CO_3 is formed when 25.0 mL of a 0.111 M $AgNO_3$ solution reacts with excess Na_2CO_3? (12.6)
$$2AgNO_3(aq) + Na_2CO_3(aq) \longrightarrow Ag_2CO_3(s) + 2NaNO_3(aq)$$
    A. 276 g   **B.** 1.53 g   **C.** 0.765 g   **D.** 0.383 g   **E.** 0.192 g

For questions 29 through 33, indicate whether each statement describes a (12.7)
    A. solution   **B.** colloid   **C.** suspension

29. _____ can be separated by filtering

30. _____ can be separated by semipermeable membranes

31. _____ passes through semipermeable membranes

32. _____ contains single atoms, ions, or small molecules of solute

33. _____ settles out upon standing

34. If a solution contains 0.50 mol of $CaCl_2$ in 1000 g of water, what is the freezing point of the solution? (12.7)
    A. 0 °C   **B.** 2.8 °C   **C.** −2.8 °C   **D.** 0.93 °C   **E.** −0.93 °C

35. Two solutions that have identical osmotic pressures are (12.8)
    A. hypotonic   **B.** hypertonic   **C.** isotonic   **D.** isotopic   **E.** hyperactive

36. In osmosis, the net flow of water is (12.8)
    A. between solutions of equal concentrations
    B. from higher solute concentration to lower solute concentration
    C. from lower solute concentration to higher solute concentration
    D. from a colloid to a solution of equal concentration
    E. from lower solvent concentration to higher solvent concentration

37. A red blood cell undergoes hemolysis when placed in a solution that is (12.8)
    A. isotonic   **B.** hypotonic   **C.** hypertonic   **D.** colloidal   **E.** semitonic

38. A solution that has the same osmotic pressure as body fluids is (12.8)
    A. 0.1% (m/v) NaCl   **B.** 0.9% (m/v) NaCl   **C.** 5% (m/v) NaCl
    D. 10% (m/v) glucose   **E.** 15% (m/v) glucose

Answers to the Practice Test

1. B	2. W	3. W	4. B	5. C
6. C	7. E	8. S	9. I	10. I
11. I	12. C	13. D	14. E	15. C
16. D	17. E	18. E	19. A	20. C
21. A	22. D	23. A	24. E	25. D
26. B	27. D	28. D	29. C	30. B
31. A	32. A	33. C	34. C	35. C
36. C	37. B	38. B		

Selected Answers and Solutions to Text Problems

12.1 The component present in the smaller amount is the solute; the component present in the larger amount is the solvent.
  a. NaCl, solute; water, solvent
  b. water, solute; ethanol, solvent
  c. oxygen, solute; nitrogen, solvent

12.3 The K^+ and I^- ions at the surface of the solid are pulled into solution by the polar water molecules, where the hydration process surrounds separate ions with water molecules.

12.5 a. $CaCO_3$ (an ionic solute) would be soluble in water (a polar solvent).
  b. Retinol (a nonpolar solute) would be soluble in CCl_4 (a nonpolar solvent).
  c. Sucrose (a polar solute) would be soluble in water (a polar solvent).
  d. Cholesterol (a nonpolar solute) would be soluble in CCl_4 (a nonpolar solvent).

12.7 The strong electrolyte KF completely dissociates into K^+ and F^- ions when it dissolves in water. When the weak electrolyte HF dissolves in water, there are a few ions of H^+ and F^- present, but mostly dissolved HF molecules.

12.9 Strong electrolytes dissociate completely into ions.
  a. $KCl(s) \xrightarrow{H_2O} K^+(aq) + Cl^-(aq)$
  b. $CaCl_2(s) \xrightarrow{H_2O} Ca^{2+}(aq) + 2Cl^-(aq)$
  c. $K_3PO_4(s) \xrightarrow{H_2O} 3K^+(aq) + PO_4^{3-}(aq)$
  d. $Fe(NO_3)_3(s) \xrightarrow{H_2O} Fe^{3+}(aq) + 3NO_3^-(aq)$

12.11 a. $HC_2H_3O_2(l) \underset{H_2O}{\rightleftharpoons} H^+(aq) + C_2H_3O_2^-(aq)$ mostly molecules, a few ions
  An aqueous solution of a weak electrolyte like acetic acid will contain mostly $HC_2H_3O_2$ molecules, with a few H^+ ions and a few $C_2H_3O_2^-$ ions.
  b. $NaBr(s) \xrightarrow{H_2O} Na^+(aq) + Br^-(aq)$ ions only
  An aqueous solution of a strong electrolyte like NaBr will contain only the ions Na^+ and Br^-.
  c. $C_6H_{12}O_6(s) \xrightarrow{H_2O} C_6H_{12}O_6(aq)$ molecules only
  An aqueous solution of a nonelectrolyte like fructose will contain only $C_6H_{12}O_6$ molecules.

12.13 a. K_2SO_4 is a strong electrolyte because only ions are present in the K_2SO_4 solution.
  b. NH_3 is a weak electrolyte because only a few NH_4^+ and OH^- ions are present in the solution.
  c. $C_6H_{12}O_6$ is a nonelectrolyte because only $C_6H_{12}O_6$ molecules are present in the solution.

12.15 a. The solution must be saturated because no additional solute dissolves.
  b. The solution was unsaturated because the sugar cube dissolves completely.

12.17 a. At 20 °C, KCl has a solubility of 34 g of KCl in 100. g of H_2O. Because 25 g of KCl is less than the maximum amount that can dissolve in 100. g of H_2O at 20 °C, the KCl solution is unsaturated.
  b. At 20 °C, $NaNO_3$ has a solubility of 88 g of $NaNO_3$ in 100. g of H_2O. Using the solubility as a conversion factor, we can calculate the maximum amount of $NaNO_3$ that can dissolve in 25 g of H_2O:

$$25 \text{ g } \cancel{H_2O} \times \frac{88 \text{ g } NaNO_3}{100. \text{ g } \cancel{H_2O}} = 22 \text{ g of } NaNO_3 \text{ (2 SFs)}$$

Because 11 g of $NaNO_3$ is less than the maximum amount that can dissolve in 25 g of H_2O at 20 °C, the $NaNO_3$ solution is unsaturated.

c. At 20 °C, sugar has a solubility of 204 g of $C_{12}H_{22}O_{11}$ in 100. g of H_2O. Using the solubility as a conversion factor, we can calculate the maximum amount of sugar that can dissolve in 125 g of H_2O:

$$125 \text{ g } H_2O \times \frac{204 \text{ g sugar}}{100. \text{ g } H_2O} = 255 \text{ g of sugar (3 SFs)}$$

Because 400. g of $C_{12}H_{22}O_{11}$ exceeds the maximum amount that can dissolve in 125 g of H_2O at 20 °C, the sugar solution is saturated, and excess undissolved sugar will be present on the bottom of the container.

12.19 a. At 20 °C, KCl has a solubility of 34 g of KCl in 100. g of H_2O.
∴ 200. g of H_2O will dissolve:

$$200. \text{ g } H_2O \times \frac{34 \text{ g KCl}}{100. \text{ g } H_2O} = 68 \text{ g of KCl (2 SFs)}$$

At 20 °C, 68 g of KCl will remain in solution.
b. Since 80. g of KCl dissolves at 50 °C and 68 g remains in solution at 20 °C, the mass of solid KCl that crystallizes after cooling is
(80. g KCl − 68 g KCl =)12 g of KCl. (2 SFs)

12.21 a. In general, the solubility of solid solutes (like sugar) increases as temperature is increased.
b. The solubility of a gaseous solute (CO_2) is less at a higher temperature.
c. The solubility of a gaseous solute is less at a higher temperature, and the CO_2 pressure in the can is increased. When the can of warm soda is opened, more CO_2 is released, producing more spray.

12.23 a. Salts containing Li^+ ions are soluble.
b. Salts containing S^{2-} ions are usually insoluble.
c. Salts containing CO_3^{2-} ions are usually insoluble.
d. Salts containing K^+ ions are soluble.
e. Salts containing NO_3^- ions are soluble.

12.25 a. No solid forms; salts containing K^+ and Na^+ are soluble.
b. Solid silver sulfide (Ag_2S) forms:
$2AgNO_3(aq) + K_2S(aq) \longrightarrow Ag_2S(s) + 2KNO_3(aq)$
$2Ag^+(aq) + 2NO_3^-(aq) + 2K^+(aq) + S^{2-}(aq) \longrightarrow Ag_2S(s) + 2K^+(aq) + 2NO_3^-(aq)$
$2Ag^+(aq) + S^{2-}(aq) \longrightarrow Ag_2S(s)$ Net ionic equation
c. Solid calcium sulfate ($CaSO_4$) forms:
$CaCl_2(aq) + Na_2SO_4(aq) \longrightarrow CaSO_4(s) + 2NaCl(aq)$
$Ca^{2+}(aq) + 2Cl^-(aq) + 2Na^+(aq) + SO_4^{2-}(aq) \longrightarrow CaSO_4(s) + 2Na^+(aq) + 2Cl^-(aq)$
$Ca^{2+}(aq) + SO_4^{2-}(aq) \longrightarrow CaSO_4(s)$ Net ionic equation
d. Solid copper phosphate ($Cu_3(PO_4)_2$) forms:
$3CuCl_2(aq) + 2Li_3PO_4(aq) \longrightarrow Cu_3(PO_4)_2(s) + 6LiCl(aq)$
$3Cu^{2+}(aq) + 6Cl^-(aq) + 6Li^+(aq) + 2PO_4^{3-}(aq) \longrightarrow$
$\qquad\qquad\qquad Cu_3(PO_4)_2(s) + 6Li^+(aq) + 6Cl^-(aq)$
$3Cu^{2+}(aq) + 2PO_4^{3-}(aq) \longrightarrow Cu_3(PO_4)_2(s)$ Net ionic equation

12.27 A 5.00% (m/m) glucose solution can be made by adding 5.00 g of glucose to 95.00 g of water, while a 5.00% (m/v) glucose solution can be made by adding 5.00 g of glucose to enough water to make 100.0 mL of solution.

12.29 Mass percent (m/m) $= \dfrac{\text{mass of solute (g)}}{\text{mass of solution (g)}} \times 100\%$

a. mass of solution $= 25$ g KCl $+ 125$ g $H_2O = 150.$ g of solution

$$\dfrac{25 \text{ g KCl}}{150. \text{ g solution}} \times 100\% = 17\% \text{ (m/m) KCl solution (2 SFs)}$$

b. $\dfrac{12 \text{ g sucrose}}{225 \text{ g solution}} \times 100\% = 5.3\% \text{ (m/m) sucrose solution (2 SFs)}$

c. $\dfrac{8.0 \text{ g CaCl}_2}{80.0 \text{ g solution}} \times 100\% = 10.\% \text{ (m/m) CaCl}_2 \text{ solution (2 SFs)}$

12.31 Mass/volume percent (m/v) $= \dfrac{\text{mass of solute (g)}}{\text{volume of solution (mL)}} \times 100\%$

a. $\dfrac{75 \text{ g Na}_2SO_4}{250 \text{ mL solution}} \times 100\% = 30.\% \text{ (m/v) Na}_2SO_4 \text{ solution (2 SFs)}$

b. $\dfrac{39 \text{ g sucrose}}{355 \text{ mL solution}} \times 100\% = 11\% \text{ (m/v) sucrose solution (2 SFs)}$

12.33 a. $50. \text{ g solution} \times \dfrac{5.0 \text{ g KCl}}{100. \text{ g solution}} = 2.5 \text{ g of KCl (2 SFs)}$

b. $1250 \text{ mL solution} \times \dfrac{4.0 \text{ g NH}_4Cl}{100. \text{ mL solution}} = 50. \text{ g of NH}_4Cl \text{ (2 SFs)}$

c. $250. \text{ mL solution} \times \dfrac{10.0 \text{ mL acetic acid}}{100. \text{ mL solution}} = 25.0 \text{ mL of acetic acid (3 SFs)}$

12.35 $355 \text{ mL solution} \times \dfrac{22.5 \text{ mL alcohol}}{100. \text{ mL solution}} = 79.9 \text{ mL of alcohol (3 SFs)}$

12.37 a. $5.0 \text{ g LiNO}_3 \times \dfrac{100. \text{ g solution}}{25 \text{ g LiNO}_3} = 20. \text{ g of LiNO}_3 \text{ solution (2 SFs)}$

b. $40.0 \text{ g KOH} \times \dfrac{100. \text{ mL solution}}{10.0 \text{ g KOH}} = 400. \text{ mL of KOH solution (3 SFs)}$

c. $2.0 \text{ mL formic acid} \times \dfrac{100. \text{ mL solution}}{10.0 \text{ mL formic acid}} = 20. \text{ mL of formic acid solution (2 SFs)}$

12.39 Molarity (M) $= \dfrac{\text{moles of solute}}{\text{liters of solution}}$

a. $\dfrac{2.00 \text{ mol glucose}}{4.00 \text{ L solution}} = 0.500 \text{ M glucose solution (3 SFs)}$

b. $\dfrac{4.00 \text{ g KOH}}{2.00 \text{ L solution}} \times \dfrac{1 \text{ mol KOH}}{56.11 \text{ g KOH}} = 0.0356 \text{ M KOH solution (3 SFs)}$

c. $\dfrac{5.85 \text{ g NaCl}}{400. \text{ mL solution}} \times \dfrac{1 \text{ mol NaCl}}{58.44 \text{ g NaCl}} \times \dfrac{1000 \text{ mL solution}}{1 \text{ L solution}} = 0.250 \text{ M NaCl solution (3 SFs)}$

12.41 a. $2.00 \text{ L solution} \times \dfrac{1.50 \text{ mol NaOH}}{1 \text{ L solution}} \times \dfrac{40.00 \text{ g NaOH}}{1 \text{ mol NaOH}} = 120. \text{ g of NaOH (3 SFs)}$

b. $4.00 \text{ L solution} \times \dfrac{0.200 \text{ mol KCl}}{1 \text{ L solution}} \times \dfrac{74.55 \text{ g KCl}}{1 \text{ mol KCl}} = 59.6 \text{ g of KCl (3 SFs)}$

c. $25.0 \text{ mL solution} \times \dfrac{1 \text{ L solution}}{1000 \text{ mL solution}} \times \dfrac{6.00 \text{ mol HCl}}{1 \text{ L solution}} \times \dfrac{36.46 \text{ g HCl}}{1 \text{ mol HCl}}$

$= 5.47 \text{ g of HCl (3 SFs)}$

12.43 a. $3.00 \text{ mol KBr} \times \dfrac{1 \text{ L solution}}{2.00 \text{ mol KBr}} = 1.50 \text{ L of KBr solution (3 SFs)}$

b. $15.0 \text{ mol NaCl} \times \dfrac{1 \text{ L solution}}{1.50 \text{ mol NaCl}} = 10.0 \text{ L of NaCl solution (3 SFs)}$

c. $0.0500 \text{ mol Ca(NO}_3)_2 \times \dfrac{1 \text{ L solution}}{0.800 \text{ mol Ca(NO}_3)_2} \times \dfrac{1000 \text{ mL solution}}{1 \text{ L solution}}$

$= 62.5 \text{ mL of Ca(NO}_3)_2 \text{ solution (3 SFs)}$

12.45 a. $12.5 \text{ g Na}_2\text{CO}_3 \times \dfrac{1 \text{ mol Na}_2\text{CO}_3}{105.99 \text{ g Na}_2\text{CO}_3} \times \dfrac{1 \text{ L solution}}{0.120 \text{ mol Na}_2\text{CO}_3} \times \dfrac{1000 \text{ mL solution}}{1 \text{ L solution}}$

$= 983 \text{ mL of Na}_2\text{CO}_3 \text{ solution (3 SFs)}$

b. $0.850 \text{ mol NaNO}_3 \times \dfrac{1 \text{ L solution}}{0.500 \text{ mol NaNO}_3} \times \dfrac{1000 \text{ mL solution}}{1 \text{ L solution}}$

$= 1700 \text{ mL} \ (1.70 \times 10^3 \text{ mL}) \text{ of NaNO}_3 \text{ solution (3 SFs)}$

c. $30.0 \text{ g LiOH} \times \dfrac{1 \text{ mol LiOH}}{23.95 \text{ g LiOH}} \times \dfrac{1 \text{ L solution}}{2.70 \text{ mol LiOH}} \times \dfrac{1000 \text{ mL solution}}{1 \text{ L solution}}$

$= 464 \text{ mL of LiOH solution (3 SFs)}$

12.47 a. $1 \text{ L} \times \dfrac{100. \text{ mL solution}}{1 \text{ L}} \times \dfrac{20. \text{ g mannitol}}{100. \text{ mL solution}} = 20. \text{ g of mannitol (2 SFs)}$

b. $12 \text{ L} \times \dfrac{100. \text{ mL solution}}{1 \text{ L}} \times \dfrac{20. \text{ g mannitol}}{100. \text{ mL solution}} = 240 \text{ g of mannitol (2 SFs)}$

12.49 $100. \text{ g glucose} \times \dfrac{100. \text{ mL solution}}{5 \text{ g glucose}} \times \dfrac{1 \text{ L}}{1000 \text{ mL}} = 2 \text{ L of glucose solution (1 SF)}$

12.51 Adding water (solvent) to the soup increases the volume and dilutes the tomato soup concentration.

12.53 $C_1 V_1 = C_2 V_2$

a. $C_2 = C_1 \times \dfrac{V_1}{V_2} = 6.0 \text{ M} \times \dfrac{2.0 \text{ L}}{6.0 \text{ L}} = 2.0 \text{ M HCl solution (2 SFs)}$

b. $C_2 = C_1 \times \dfrac{V_1}{V_2} = 12 \text{ M} \times \dfrac{0.50 \text{ L}}{3.0 \text{ L}} = 2.0 \text{ M NaOH solution (2 SFs)}$

c. $C_2 = C_1 \times \dfrac{V_1}{V_2} = 25\% \times \dfrac{10.0 \text{ mL}}{100.0 \text{ mL}} = 2.5\% \text{ (m/v) KOH solution (2 SFs)}$

d. $C_2 = C_1 \times \dfrac{V_1}{V_2} = 15\% \times \dfrac{50.0 \text{ mL}}{250 \text{ mL}} = 3.0\% \text{ (m/v) H}_2\text{SO}_4 \text{ solution (2 SFs)}$

12.55 $C_1 V_1 = C_2 V_2$

    a. $V_2 = V_1 \times \dfrac{C_1}{C_2} = 20.0 \text{ mL} \times \dfrac{6.0 \text{ M}}{1.5 \text{ M}} = 80. \text{ mL of diluted HCl solution (2 SFs)}$

    b. $V_2 = V_1 \times \dfrac{C_1}{C_2} = 50.0 \text{ mL} \times \dfrac{10.0 \text{ \%}}{2.0 \text{ \%}} = 250 \text{ mL of diluted LiCl solution (2 SFs)}$

    c. $V_2 = V_1 \times \dfrac{C_1}{C_2} = 50.0 \text{ mL} \times \dfrac{6.00 \text{ M}}{0.500 \text{ M}} = 600. \text{ mL of diluted } H_3PO_4 \text{ solution (3 SFs)}$

    d. $V_2 = V_1 \times \dfrac{C_1}{C_2} = 75 \text{ mL} \times \dfrac{12 \text{ \%}}{5.0 \text{ \%}} = 180 \text{ mL of diluted glucose solution (2 SFs)}$

12.57 $C_1 V_1 = C_2 V_2$

    a. $V_1 = V_2 \times \dfrac{C_2}{C_1} = 255 \text{ mL} \times \dfrac{0.200 \text{ M}}{4.00 \text{ M}} = 12.8 \text{ mL of the } HNO_3 \text{ solution (3 SFs)}$

    b. $V_1 = V_2 \times \dfrac{C_2}{C_1} = 715 \text{ mL} \times \dfrac{0.100 \text{ M}}{6.00 \text{ M}} = 11.9 \text{ mL of the } MgCl_2 \text{ solution (3 SFs)}$

    c. $V_2 = 0.100 \text{ L} \times \dfrac{1000 \text{ mL}}{1 \text{ L}} = 100. \text{ mL}$

        $V_1 = V_2 \times \dfrac{C_2}{C_1} = 100. \text{ mL} \times \dfrac{0.150 \text{ M}}{8.00 \text{ M}} = 1.88 \text{ mL of the KCl solution (3 SFs)}$

12.59 $C_1 V_1 = C_2 V_2$

    $V_1 = V_2 \times \dfrac{C_2}{C_1} = 500. \text{ mL} \times \dfrac{5.0\%}{25\%} = 1.0 \times 10^2 \text{ mL of the glucose solution (2 SFs)}$

12.61 a. $50.0 \text{ mL solution} \times \dfrac{1 \text{ L solution}}{1000 \text{ mL solution}} \times \dfrac{1.50 \text{ mol KCl}}{1 \text{ L solution}} = 0.0750 \text{ mol of KCl}$

    $0.0750 \text{ mol KCl} \times \dfrac{1 \text{ mol } PbCl_2}{2 \text{ mol KCl}} \times \dfrac{278.1 \text{ g } PbCl_2}{1 \text{ mol } PbCl_2} = 10.4 \text{ g of } PbCl_2 \text{ (3 SFs)}$

    b. $50.0 \text{ mL solution} \times \dfrac{1 \text{ L solution}}{1000 \text{ mL solution}} \times \dfrac{1.50 \text{ mol KCl}}{1 \text{ L solution}} = 0.0750 \text{ mol of KCl}$

    $0.0750 \text{ mol KCl} \times \dfrac{1 \text{ mol } Pb(NO_3)_2}{2 \text{ mol KCl}} \times \dfrac{1 \text{ L solution}}{2.00 \text{ mol } Pb(NO_3)_2} \times \dfrac{1000 \text{ mL solution}}{1 \text{ L solution}}$

    $= 18.8 \text{ mL of } Pb(NO_3)_2 \text{ solution (3 SFs)}$

    c. $30.0 \text{ mL solution} \times \dfrac{1 \text{ L solution}}{1000 \text{ mL solution}} \times \dfrac{0.400 \text{ mol } Pb(NO_3)_2}{1 \text{ L solution}} \times \dfrac{2 \text{ mol KCl}}{1 \text{ mol } Pb(NO_3)_2}$

    $= 0.0240 \text{ mol of KCl}$

    $20.0 \text{ mL solution} \times \dfrac{1 \text{ L solution}}{1000 \text{ mL solution}} = 0.0200 \text{ L of solution}$

    $\text{Molarity (M)} = \dfrac{\text{moles of solute}}{\text{liters of solution}} = \dfrac{0.0240 \text{ mol KCl}}{0.0200 \text{ L solution}} = 1.20 \text{ M KCl solution (3 SFs)}$

12.63 a. $15.0 \text{ g Mg} \times \dfrac{1 \text{ mol Mg}}{24.31 \text{ g Mg}} \times \dfrac{2 \text{ mol HCl}}{1 \text{ mol Mg}} \times \dfrac{1 \text{ L solution}}{6.00 \text{ mol HCl}} \times \dfrac{1000 \text{ mL solution}}{1 \text{ L solution}}$

    $= 206 \text{ mL of HCl solution (3 SFs)}$

    b. $0.500 \text{ L solution} \times \dfrac{2.00 \text{ mol HCl}}{1 \text{ L solution}} \times \dfrac{1 \text{ mol } H_2}{2 \text{ mol HCl}} \times \dfrac{22.4 \text{ L } H_2 \text{ (STP)}}{1 \text{ mol } H_2}$

    $= 11.2 \text{ L of } H_2 \text{ at STP (3 SFs)}$

c. $n = \dfrac{PV}{RT} = \dfrac{(735 \text{ mmHg})(5.20 \text{ L})}{\left(\dfrac{62.4 \text{ L} \cdot \text{mmHg}}{\text{mol} \cdot \text{K}}\right)(298 \text{ K})} = 0.206 \text{ mol of } H_2 \text{ gas}$

$0.206 \text{ mol } H_2 \times \dfrac{2 \text{ mol HCl}}{1 \text{ mol } H_2} = 0.412 \text{ mol of HCl}$

$45.2 \text{ mL solution} \times \dfrac{1 \text{ L solution}}{1000 \text{ mL solution}} = 0.0452 \text{ L of solution}$

$\text{Molarity (M)} = \dfrac{\text{moles of solute}}{\text{liters of solution}} = \dfrac{0.412 \text{ mol HCl}}{0.0452 \text{ L solution}} = 9.12 \text{ M HCl solution (3 SFs)}$

12.65 a. $8.56 \text{ g Zn} \times \dfrac{1 \text{ mol Zn}}{65.41 \text{ g Zn}} \times \dfrac{2 \text{ mol HBr}}{1 \text{ mol Zn}} \times \dfrac{1 \text{ L solution}}{3.50 \text{ mol HBr}} \times \dfrac{1000 \text{ mL solution}}{1 \text{ L solution}}$

$= 74.8 \text{ mL of HBr solution (3 SFs)}$

b. $0.750 \text{ L solution} \times \dfrac{6.00 \text{ mol HBr}}{1 \text{ L solution}} \times \dfrac{1 \text{ mol } H_2}{2 \text{ mol HBr}} \times \dfrac{22.4 \text{ L } H_2 \text{ (STP)}}{1 \text{ mol } H_2}$

$= 50.4 \text{ L of } H_2 \text{ at STP (3 SFs)}$

c. $n = \dfrac{PV}{RT} = \dfrac{(725 \text{ mmHg})(0.620 \text{ L})}{\left(\dfrac{62.4 \text{ L} \cdot \text{mmHg}}{\text{mol} \cdot \text{K}}\right)(297 \text{ K})} = 0.0243 \text{ mol of } H_2 \text{ gas}$

$0.0243 \text{ mol } H_2 \times \dfrac{2 \text{ mol HBr}}{1 \text{ mol } H_2} = 0.0486 \text{ mol of HBr}$

$28.7 \text{ mL solution} \times \dfrac{1 \text{ L solution}}{1000 \text{ mL solution}} = 0.0287 \text{ L of solution}$

$\text{Molarity (M)} = \dfrac{\text{moles of solute}}{\text{liters of solution}} = \dfrac{0.0486 \text{ mol HBr}}{0.0287 \text{ L solution}} = 1.69 \text{ M HBr solution (3 SFs)}$

12.67 a. A solution cannot be separated by a semipermeable membrane.

b. A suspension settles out upon standing.

12.69 a. When 1.0 mol of glycerol (a nonelectrolyte) dissolves in water, it does not dissociate into ions and so will only produce 1.0 mol of particles. Similarly, 2.0 mol of ethylene glycol (also a nonelectrolyte) dissolves as molecules to produce only 2.0 mol of particles in water. The ethylene glycol solution has more particles in 1.0 kg of water and will thus have a lower freezing point.

b. When 0.50 mol of the strong electrolyte KCl dissolves in water, it will produce 1.0 mol of particles because each formula unit of KCl dissociates to give two particles, K^+ and Cl^-. When 0.50 mol of the strong electrolyte $MgCl_2$ dissolves in water, it will produce 1.5 mol of particles because each formula unit of $MgCl_2$ dissociates to give three particles, Mg^{2+} and $2Cl^-$. Thus, a solution of 0.50 mol of $MgCl_2$ in 1.0 kg of water will have the lower freezing point.

12.71 a. $325 \text{ g } CH_3OH \times \dfrac{1 \text{ mol } CH_3OH}{32.04 \text{ g } CH_3OH} = 10.1 \text{ mol of } CH_3OH$

$455 \text{ g } H_2O \times \dfrac{1 \text{ kg } H_2O}{1000 \text{ g } H_2O} = 0.455 \text{ kg of } H_2O$

$\text{Molality (m)} = \dfrac{\text{moles of solute}}{\text{kilograms of water}} = \dfrac{10.1 \text{ mol } CH_3OH}{0.455 \text{ kg } H_2O}$

$= 22.2 \text{ m } CH_3OH \text{ solution (3 SFs)}$

b. $640. \, \text{g} \, \overline{C_3H_8O_2} \times \dfrac{1 \text{ mol } C_3H_8O_2}{76.09 \text{ g } \overline{C_3H_8O_2}} = 8.41 \text{ mol of } C_3H_8O_2$

$\text{Molality } (m) = \dfrac{\text{moles of solute}}{\text{kilograms of water}} = \dfrac{8.41 \text{ mol } C_3H_8O_2}{1.22 \text{ kg } H_2O} = 6.89 \, m \, C_3H_8O_2 \text{ solution } (3 \text{ SFs})$

12.73 a. $22.2 \, m \, CH_3OH$ solution; freezing-point lowering: $\Delta T_f = m \times K_f$;

boiling-point elevation: $\Delta T_b = m \times K_b$

$\Delta T_f = m \times K_f = 22.2 \, \cancel{m} \times \dfrac{1.86 \, °C}{1 \, \cancel{m}} = 41.3 \, °C \, (3 \text{ SFs})$

$T_{\text{solution}} = T_{\text{water}} - \Delta T_f = 0.0 \, °C - 41.3 \, °C = -41.3 \, °C$

$\Delta T_b = m \times K_b = 22.2 \, \cancel{m} \times \dfrac{0.52 \, °C}{1 \, \cancel{m}} = 12 \, °C \, (2 \text{ SFs})$

$T_{\text{solution}} = T_{\text{water}} + \Delta T_b = 100. \, °C + 12 \, °C = 112 \, °C$

b. $6.89 \, m \, C_3H_8O_2$ solution; freezing-point lowering: $\Delta T_f = m \times K_f$;

boiling-point elevation: $\Delta T_b = m \times K_b$

$\Delta T_f = m \times K_f = 6.89 \, \cancel{m} \times \dfrac{1.86 \, °C}{1 \, \cancel{m}} = 12.8 \, °C \, (3 \text{ SFs})$

$T_{\text{solution}} = T_{\text{water}} - \Delta T_f = 0.0 \, °C - 12.8 \, °C = -12.8 \, °C$

$\Delta T_b = m \times K_b = 6.89 \, \cancel{m} \times \dfrac{0.52 \, °C}{1 \, \cancel{m}} = 3.6 \, °C \, (2 \text{ SFs})$

$T_{\text{solution}} = T_{\text{water}} + \Delta T_b = 100.0 \, °C + 3.6 \, °C = 103.6 \, °C$

12.75 a. The 10% (m/v) starch solution has the higher solute concentration, more solute particles, and therefore the higher osmotic pressure.

b. Initially, water will flow out of the 1% (m/v) starch solution into the more concentrated 10% (m/v) starch solution.

c. The volume of the 10% (m/v) starch solution will increase due to inflow of water.

12.77 Water will flow from a region of higher solvent concentration (which corresponds to a lower solute concentration) to a region of lower solvent concentration (which corresponds to a higher solute concentration).

a. B; The volume level will rise as water flows into compartment B, which contains the 10% (m/v) sucrose solution.

b. A; The volume level will rise as water flows into compartment A, which contains the 8% (m/v) albumin solution.

c. B; The volume level will rise as water flows into compartment B, which contains the 10% (m/v) starch solution.

12.79 A red blood cell has the same osmotic pressure as a 5% (m/v) glucose solution or a 0.9% (m/v) NaCl solution; these are isotonic solutions. Solutions with higher concentrations than these are hypertonic, and solutions with lower concentrations than these are hypotonic.

a. Distilled water is a hypotonic solution when compared with a red blood cell's contents.

b. A 1% (m/v) glucose solution is a hypotonic solution.

c. A 0.9% (m/v) NaCl solution is isotonic with a red blood cell's contents.

d. A 15% (m/v) glucose solution is a hypertonic solution.

12.81 Colloids cannot pass through the semipermeable dialysis membrane; water and solutions freely pass through semipermeable membranes.

a. Sodium and chloride ions will both pass through the membrane into the distilled water.

b. The amino acid alanine will pass through the semipermeable membrane into the distilled water; the colloid starch will not.

c. Sodium and chloride ions will both be present in the water surrounding the dialysis bag; the colloid starch will not.

d. Urea will diffuse through the dialysis bag into the surrounding water.

12.83 $0.075 \text{ g chlorpromazine} \times \dfrac{100. \text{ mL solution}}{2.5 \text{ g chlorpromazine}} = 3.0 \text{ mL of chlorpromazine solution (2 SFs)}$

12.85 $5.0 \text{ mL solution} \times \dfrac{10. \text{ g CaCl}_2}{100. \text{ mL solution}} = 0.50 \text{ g of CaCl}_2 \text{ (2 SFs)}$

12.87 a. 1; A solution will form because both the solute and the solvent are polar.

b. 2; Two layers will form because one component is nonpolar and the other is polar.

c. 1; A solution will form because both the solute and the solvent are nonpolar.

12.89 a. 3; A nonelectrolyte will show no dissociation.

b. 1; A weak electrolyte will show some dissociation, producing a few ions, but mostly remaining as molecules.

c. 2; A strong electrolyte will be completely dissociated into ions.

12.91 a. Beaker 3; Solid silver chloride (AgCl) will precipitate when the two solutions are mixed.

b. $NaCl(aq) + AgNO_3(aq) \longrightarrow AgCl(s) + NaNO_3(aq)$
$Na^+(aq) + Cl^-(aq) + Ag^+(aq) + NO_3^-(aq) \longrightarrow AgCl(s) + Na^+(aq) + NO_3^-(aq)$

c. $Ag^+(aq) + Cl^-(aq) \longrightarrow AgCl(s)$ Net ionic equation

12.93 A "brine" salt-water solution has a high concentration of Na^+ and Cl^- ions, which is hypertonic to the cucumber. The skin of the cucumber acts like a semipermeable membrane; therefore, water flows from the more dilute solution inside the cucumber into the more concentrated brine solution that surrounds it. The loss of water causes the cucumber to become a wrinkled pickle.

12.95 a. 2; Water will flow into the B (8% starch solution) side.

b. 1; Water will continue to flow equally in both directions; no change in volumes.

c. 3; Water will flow into the A (5% sucrose solution) side.

d. 2; Water will flow into the B (1% sucrose solution) side.

12.97 Because iodine is a nonpolar molecule, it will dissolve in hexane, a nonpolar solvent. Iodine does not dissolve in water because water is a polar solvent.

12.99 At 20 °C, KNO_3 has a solubility of 32 g of KNO_3 in 100. g of H_2O.

a. 200. g of H_2O will dissolve:

$$200. \text{ g H}_2\text{O} \times \dfrac{32 \text{ g KNO}_3}{100. \text{ g H}_2\text{O}} = 64 \text{ g of KNO}_3 \text{ (2 SFs)}$$

Because 32 g of KNO_3 is less than the maximum amount that can dissolve in 200. g of H_2O at 20 °C, the KNO_3 solution is unsaturated.

b. 50. g of H_2O will dissolve:

$$50. \text{ g H}_2\text{O} \times \dfrac{32 \text{ g KNO}_3}{100. \text{ g H}_2\text{O}} = 16 \text{ g of KNO}_3 \text{ (2 SFs)}$$

Because 19 g of KNO_3 exceeds the maximum amount that can dissolve in 50. g of H_2O at 20 °C, the KNO_3 solution is saturated, and excess undissolved KNO_3 will be present on the bottom of the container.

c. 150. g of H_2O will dissolve:

$$150. \text{ g } H_2O \times \frac{32 \text{ g KNO}_3}{100. \text{ g } H_2O} = 48 \text{ g of KNO}_3 \text{ (2 SFs)}$$

Because 68 g of KNO_3 exceeds the maximum amount that can dissolve in 150. g of H_2O at 20 °C, the KNO_3 solution is saturated, and excess undissolved KNO_3 will be present on the bottom of the container.

12.101 a. K^+ salts are soluble.

b. The SO_4^{2-} salt of Mg^{2+} is soluble.

c. Salts containing S^{2-} ions are usually insoluble.

d. Salts containing NO_3^- ions are soluble.

e. Salts containing OH^- ions are usually insoluble.

12.103 a. Solid silver chloride ($AgCl$) forms:

$AgNO_3(aq) + LiCl(aq) \longrightarrow AgCl(s) + LiNO_3(aq)$
$Ag^+(aq) + NO_3^-(aq) + Li^+(aq) + Cl^-(aq) \longrightarrow AgCl(s) + Li^+(aq) + NO_3^-(aq)$
$Ag^+(aq) + Cl^-(aq) \longrightarrow AgCl(s)$ Net ionic equation

b. none; no solid forms; salts containing K^+ and Na^+ are soluble.

c. Solid barium sulfate ($BaSO_4$) forms:

$Na_2SO_4(aq) + BaCl_2(aq) \longrightarrow BaSO_4(s) + 2NaCl(aq)$
$2Na^+(aq) + SO_4^{2-}(aq) + Ba^{2+}(aq) + 2Cl^-(aq) \longrightarrow BaSO_4(s) + 2Na^+(aq) + 2Cl^-(aq)$
$Ba^{2+}(aq) + SO_4^{2-}(aq) \longrightarrow BaSO_4(s)$ Net ionic equation

12.105 $80.0 \text{ g NaCl} \times \dfrac{100. \text{ g water}}{36.0 \text{ g NaCl}} = 222 \text{ g of water needed (3 SFs)}$

12.107 mass of solution = $15.5 \text{ g Na}_2SO_4 + 75.5 \text{ g H}_2O = 91.0 \text{ g of solution}$

$$\frac{15.5 \text{ g Na}_2SO_4}{91.0 \text{ g solution}} \times 100\% = 17.0\% \text{ (m/m) Na}_2SO_4 \text{ solution (3 SFs)}$$

12.109 $4.5 \text{ mL propyl alcohol} \times \dfrac{100. \text{ mL solution}}{12 \text{ mL propyl alcohol}} = 38 \text{ mL of propyl alcohol solution (2 SFs)}$

12.111 $86.0 \text{ g KOH} \times \dfrac{100. \text{ mL solution}}{12 \text{ g KOH}} \times \dfrac{1 \text{ L solution}}{1000 \text{ mL solution}} = 0.72 \text{ L of KOH solution (2 SFs)}$

12.113 $\dfrac{8.0 \text{ g NaOH}}{400. \text{ mL solution}} \times \dfrac{1 \text{ mol NaOH}}{40.00 \text{ g NaOH}} \times \dfrac{1000 \text{ mL solution}}{1 \text{ L solution}} = 0.50 \text{ M NaOH solution (2 SFs)}$

12.115 $15.2 \text{ g LiCl} \times \dfrac{1 \text{ mol LiCl}}{42.39 \text{ g LiCl}} \times \dfrac{1 \text{ L solution}}{1.75 \text{ mol LiCl}} \times \dfrac{1000 \text{ mL solution}}{1 \text{ L solution}}$

$= 205 \text{ mL of LiCl solution (3 SFs)}$

12.117 $60.0 \text{ g KNO}_3 \times \dfrac{1 \text{ mol KNO}_3}{101.11 \text{ g KNO}_3} \times \dfrac{1 \text{ L solution}}{2.50 \text{ mol KNO}_3} = 0.237 \text{ L of KNO}_3 \text{ solution (3 SFs)}$

12.119 a. $2.5 \text{ L solution} \times \dfrac{3.0 \text{ mol Al(NO}_3)_3}{1 \text{ L solution}} \times \dfrac{213.0 \text{ g Al(NO}_3)_3}{1 \text{ mol Al(NO}_3)_3}$

$= 1600 \text{ g of Al(NO}_3)_3 \text{ (2 SFs)}$

b. $75 \text{ mL solution} \times \dfrac{1 \text{ L solution}}{1000 \text{ mL solution}} \times \dfrac{0.50 \text{ mol } C_6H_{12}O_6}{1 \text{ L solution}} \times \dfrac{180.16 \text{ g } C_6H_{12}O_6}{1 \text{ mol } C_6H_{12}O_6}$

$= 6.8 \text{ g of } C_6H_{12}O_6 \ (2 \text{ SFs})$

c. $235 \text{ mL solution} \times \dfrac{1 \text{ L solution}}{1000 \text{ mL solution}} \times \dfrac{1.80 \text{ mol LiCl}}{1 \text{ L solution}} \times \dfrac{42.39 \text{ g LiCl}}{1 \text{ mol LiCl}}$

$= 17.9 \text{ g of LiCl} \ (3 \text{ SFs})$

12.121 $C_1V_1 = C_2V_2$

a. $C_2 = C_1 \times \dfrac{V_1}{V_2} = 0.200 \text{ M} \times \dfrac{25.0 \text{ mL}}{50.0 \text{ mL}} = 0.100 \text{ M NaBr solution} \ (3 \text{ SFs})$

b. $C_2 = C_1 \times \dfrac{V_1}{V_2} = 12.0\% \times \dfrac{15.0 \text{ mL}}{40.0 \text{ mL}} = 4.50\% \text{ (m/v) } K_2SO_4 \text{ solution} \ (3 \text{ SFs})$

c. $C_2 = C_1 \times \dfrac{V_1}{V_2} = 6.00 \text{ M} \times \dfrac{75.0 \text{ mL}}{255 \text{ mL}} = 1.76 \text{ M NaOH solution} \ (3 \text{ SFs})$

12.123 $C_1V_1 = C_2V_2$

a. $V_1 = V_2 \times \dfrac{C_2}{C_1} = 250 \text{ mL} \times \dfrac{3.0 \%}{10.0 \%} = 75 \text{ mL of the HCl solution} \ (2 \text{ SFs})$

b. $V_1 = V_2 \times \dfrac{C_2}{C_1} = 500. \text{ mL} \times \dfrac{0.90 \%}{5.0 \%} = 90. \text{ mL of the NaCl solution} \ (2 \text{ SFs})$

c. $V_1 = V_2 \times \dfrac{C_2}{C_1} = 350. \text{ mL} \times \dfrac{2.00 \text{ M}}{6.00 \text{ M}} = 117 \text{ mL of the NaOH solution} \ (3 \text{ SFs})$

12.125 $C_1V_1 = C_2V_2$

a. $V_2 = V_1 \times \dfrac{C_1}{C_2} = 25.0 \text{ mL} \times \dfrac{10.0 \%}{2.50 \%} = 100. \text{ mL of diluted HCl solution} \ (3 \text{ SFs})$

b. $V_2 = V_1 \times \dfrac{C_1}{C_2} = 25.0 \text{ mL} \times \dfrac{5.00 \text{ M}}{1.00 \text{ M}} = 125 \text{ mL of diluted HCl solution} \ (3 \text{ SFs})$

c. $V_2 = V_1 \times \dfrac{C_1}{C_2} = 25.0 \text{ mL} \times \dfrac{6.00 \text{ M}}{0.500 \text{ M}} = 300. \text{ mL of diluted HCl solution} \ (3 \text{ SFs})$

12.127 $60.0 \text{ mL solution} \times \dfrac{1 \text{ L solution}}{1000 \text{ mL solution}} \times \dfrac{2.00 \text{ mol Al(OH)}_3}{1 \text{ L solution}} \times \dfrac{3 \text{ mol HCl}}{1 \text{ mol Al(OH)}_3}$

$\times \dfrac{1 \text{ L solution}}{6.00 \text{ mol HCl}} \times \dfrac{1000 \text{ mL HCl solution}}{1 \text{ L solution}}$

$= 60.0 \text{ mL of HCl solution} \ (3 \text{ SFs})$

12.129 a. A solution of 0.580 mol of the nonelectrolyte lactose in 1.00 kg of water, which contains 0.580 mol of particles in 1.00 kg of water, is a 0.580-m solution and has a freezing point change of

$$\Delta T_f = m \times K_f = 0.580 \ m \times \dfrac{1.86 \ ^\circ\text{C}}{m} = 1.08 \ ^\circ\text{C}$$

The freezing point would be $T_{\text{solution}} = T_{\text{water}} - \Delta T_f$

$$= 0.00 \ ^\circ\text{C} - 1.08 \ ^\circ\text{C}$$

$$= -1.08 \ ^\circ\text{C}$$

b. $45.0 \text{ g KCl} \times \dfrac{1 \text{ mol KCl}}{74.55 \text{ g KCl}} = 0.604 \text{ mol of KCl (3 SFs)}$

A solution of 0.604 mol of the strong electrolyte KCl in 1.00 kg of water, which contains 1.208 mol of particles (0.604 mol of K^+ and 0.604 mol of Cl^-) in 1.00 kg of water, is a 1.21-m solution and has a freezing point change of

$$\Delta T_f = m \times K_f = 1.208 \text{ } m \times \dfrac{1.86 \text{ }°C}{m} = 2.25 \text{ }°C$$

The freezing point would be $T_{solution} = T_{water} - \Delta T_f$

$$= 0.00 \text{ }°C - 2.25 \text{ }°C$$
$$= -2.25 \text{ }°C$$

c. A solution of 1.5 mol of the strong electrolyte K_3PO_4 in 1.00 kg of water, which contains 6.0 mol of particles (4.5 mol of K^+ and 1.5 mol of PO_4^{3-}) in 1.00 kg of water, is a 6.0-m solution and has a freezing point change of

$$\Delta T_f = m \times K_f = 6.0 \text{ } m \times \dfrac{1.86 \text{ }°C}{m} = 11 \text{ }°C$$

The freezing point would be $T_{solution} = T_{water} - \Delta T_f$

$$= 0.0 \text{ }°C - 11 \text{ }°C$$
$$= -11 \text{ }°C$$

12.131 $450 \text{ mg Al(OH)}_3 \times \dfrac{1 \text{ g Al(OH)}_3}{1000 \text{ mg Al(OH)}_3} \times \dfrac{1 \text{ mol Al(OH)}_3}{78.00 \text{ g Al(OH)}_3} \times \dfrac{3 \text{ mol HCl}}{1 \text{ mol Al(OH)}_3} \times$

$\dfrac{1 \text{ L solution}}{0.20 \text{ mol HCl}} \times \dfrac{1000 \text{ mL solution}}{1 \text{ L solution}} = 87 \text{ mL of the HCl solution (2 SFs)}$

12.133 a. Solid silver sulfate (Ag_2SO_4) forms:

$2AgNO_3(aq) + Na_2SO_4(aq) \longrightarrow Ag_2SO_4(s) + 2NaNO_3(aq)$

$2Ag^+(aq) + 2NO_3^-(aq) + 2Na^+(aq) + SO_4^{2-}(aq) \longrightarrow$
$$Ag_2SO_4(s) + 2Na^+(aq) + 2NO_3^-(aq)$$

$2Ag^+(aq) + SO_4^{2-}(aq) \longrightarrow Ag_2SO_4(s)$ Net ionic equation

b. Solid lead(II) chloride ($PbCl_2$) forms:

$2KCl(aq) + Pb(NO_3)_2(aq) \longrightarrow PbCl_2(s) + 2KNO_3(aq)$

$2K^+(aq) + 2Cl^-(aq) + Pb^{2+}(aq) + 2NO_3^-(aq) \longrightarrow$
$$PbCl_2(s) + 2K^+(aq) + 2NO_3^-(aq)$$

$Pb^{2+}(aq) + 2Cl^-(aq) \longrightarrow PbCl_2(s)$ Net ionic equation

c. Solid calcium phosphate ($Ca_3(PO_4)_2$) forms:

$3CaCl_2(aq) + 2(NH_4)_3PO_4(aq) \longrightarrow Ca_3(PO_4)_2(s) + 6NH_4Cl(aq)$

$3Ca^{2+}(aq) + 6Cl^-(aq) + 6NH_4^+(aq) + 2PO_4^{3-}(aq) \longrightarrow$
$$Ca_3(PO_4)_2(s) + 6NH_4^+(aq) + 6Cl^-(aq)$$

$3Ca^{2+}(aq) + 2PO_4^{3-}(aq) \longrightarrow Ca_3(PO_4)_2(s)$ Net ionic equation

d. Solid barium sulfate ($BaSO_4$) forms:

$K_2SO_4(aq) + BaCl_2(aq) \longrightarrow BaSO_4(s) + 2KCl(aq)$

$2K^+(aq) + SO_4^{2-}(aq) + Ba^{2+}(aq) + 2Cl^-(aq) \longrightarrow$
$$BaSO_4(s) + 2K^+(aq) + 2Cl^-(aq)$$

$Ba^{2+}(aq) + SO_4^{2-}(aq) \longrightarrow BaSO_4(s)$ Net ionic equation

12.135 a. mass of NaCl = 25.50 g − 24.10 g = 1.40 g of NaCl

mass of solution = 36.15 g − 24.10 g = 12.05 g of solution

$$\text{Mass percent (m/m)} = \frac{1.40 \text{ g NaCl}}{12.05 \text{ g solution}} \times 100\% = 11.6\% \text{ (m/m) NaCl solution (3 SFs)}$$

b. $\text{Molarity (M)} = \dfrac{1.40 \text{ g NaCl}}{10.0 \text{ mL solution}} \times \dfrac{1 \text{ mol NaCl}}{58.44 \text{ g NaCl}} \times \dfrac{1000 \text{ mL solution}}{1 \text{ L solution}}$

= 2.40 M NaCl solution (3 SFs)

c. $C_1 V_1 = C_2 V_2$ $C_2 = C_1 \times \dfrac{V_1}{V_2} = 2.40 \text{ M} \times \dfrac{10.0 \text{ mL}}{60.0 \text{ mL}} = 0.400 \text{ M NaCl solution (3 SFs)}$

12.137 At 18 °C, KF has a solubility of 92 g of KF in 100. g of H_2O.

a. 25 g of H_2O will dissolve:

$$25 \text{ g } H_2O \times \frac{92 \text{ g KF}}{100. \text{ g } H_2O} = 23 \text{ g of KF (2 SFs)}$$

Because 35 g of KF exceeds the maximum amount that can dissolve in 25 g of H_2O at 18 °C, the KF solution is saturated, and excess undissolved KF will be present on the bottom of the container.

b. 50. g of H_2O will dissolve:

$$50. \text{ g } H_2O \times \frac{92 \text{ g KF}}{100. \text{ g } H_2O} = 46 \text{ g of KF (2 SFs)}$$

Because 42 g of KF is less than the maximum amount that can dissolve in 50. g of H_2O at 18 °C, the solution is unsaturated.

c. 150. g of H_2O will dissolve:

$$150. \text{ g } H_2O \times \frac{92 \text{ g KF}}{100. \text{ g } H_2O} = 140 \text{ g of KF (2 SFs)}$$

Because 145 g of KF exceeds the maximum amount that can dissolve in 150. g of H_2O at 18 °C, the KF solution is saturated, and excess undissolved KF will be present on the bottom of the container.

12.139 mass of solution: 70.0 g HNO_3 + 130.0 g H_2O = 200.0 g of solution

a. $\dfrac{70.0 \text{ g } HNO_3}{200.0 \text{ g solution}} \times 100\% = 35.0\% \text{ (m/m) } HNO_3 \text{ solution (3 SFs)}$

b. $200.0 \text{ g solution} \times \dfrac{1 \text{ mL solution}}{1.21 \text{ g solution}} = 165 \text{ mL of solution (3 SFs)}$

c. $\dfrac{70.0 \text{ g } HNO_3}{165 \text{ mL solution}} \times 100\% = 42.4\% \text{ (m/v) } HNO_3 \text{ solution (3 SFs)}$

d. $\dfrac{70.0 \text{ g } HNO_3}{165 \text{ mL solution}} \times \dfrac{1 \text{ mol } HNO_3}{63.02 \text{ g } HNO_3} \times \dfrac{1000 \text{ mL solution}}{1 \text{ L solution}}$

= 6.73 M HNO_3 solution (3 SFs)

12.141 Analyze the Problem: Using the Ideal Gas Law to Calculate Moles

Property	P	V	n	R	T
Given	745 mmHg	4.20 L		$\dfrac{62.4 \text{ L} \cdot \text{mmHg}}{\text{mol} \cdot \text{K}}$	35 °C 35 °C + 273 = 308 K
Need			? mol		

$$n = \frac{PV}{RT} = \frac{(745 \text{ mmHg})(4.20 \text{ L})}{\left(\dfrac{62.4 \text{ L} \cdot \text{mmHg}}{\text{mol} \cdot \text{K}}\right)(308 \text{ K})} = 0.163 \text{ mol of } H_2 \text{ gas}$$

$$0.163 \text{ mol } H_2 \times \frac{2 \text{ mol HCl}}{1 \text{ mol } H_2} = 0.326 \text{ mol of HCl}$$

$$355 \text{ mL solution} \times \frac{1 \text{ L solution}}{1000 \text{ mL solution}} = 0.355 \text{ L of solution}$$

$$\text{Molarity (M)} = \frac{\text{moles of solute}}{\text{liters of solution}} = \frac{0.326 \text{ mol HCl}}{0.355 \text{ L solution}} = 0.918 \text{ M HCl solution (3 SFs)}$$

12.143 a. When the strong electrolyte NaCl dissolves in water, each formula unit of NaCl dissociates to give two particles, Na^+ and Cl^-. Since freezing-point lowering depends on the number of particles in solution, only 0.60 mol of NaCl would be needed to have the same effect in 1 kg of water as 1.2 mol of the nonelectrolyte ethylene glycol.

b. When the strong electrolyte K_3PO_4 dissolves in water, each formula unit of K_3PO_4 dissociates to give four particles, three K^+ and one PO_4^{3-}. Since freezing-point lowering depends on the number of particles in solution, only 0.30 mol of K_3PO_4 would be needed to have the same effect in 1 kg of water as 1.2 mol of the nonelectrolyte ethylene glycol.

12.145 a. The boiling point of the NaCl solution is given as 101.04 °C. Since the boiling point for pure water is 100.0 °C, the ΔT_b is 101.04 °C − 100.0 °C = 1.04 °C.

Rearranging $\Delta T_b = m \times K_b$ to solve for m:

$$m = \frac{\Delta T_b}{K_b} = \frac{1.04 \text{ °C}}{0.52 \text{ °C}/m} = 2.0 \text{ } m \text{ NaCl solution (2 SFs)}$$

b. $\Delta T_f = m \times K_f = 2.0 \text{ } m \times \dfrac{1.86 \text{ °C}}{1 \text{ } m} = 3.7 \text{ °C (2 SFs)}$

$$T_{\text{solution}} = T_{\text{water}} - \Delta T_f = 0.0 \text{ °C} - 3.7 \text{ °C} = -3.7 \text{ °C}$$

Reaction Rates and Chemical Equilibrium

Credit: Dante Fenolio / Science Source

Peter, a chemical oceanographer, has worked for six months obtaining samples to test the levels of CO_2 dissolved in the seawater. The oceans absorb more than a quarter of the carbon dioxide that humans put into the air. From the data he collects, Peter hopes to determine the impact of dissolved CO_2 on conditions in the ocean. Equilibrium is established between the dissolved CO_2 and water, which combine to form H_2CO_3, carbonic acid.

$$CO_2(aq) + H_2O(l) \rightleftharpoons H_2CO_3(aq)$$

Carbonic acid, a weak acid, dissociates to yield hydrogen carbonate and hydrogen ion; the hydrogen carbonate dissociates to yield carbonate and hydrogen ion.

$$H_2CO_3(aq) \rightleftharpoons HCO_3^-(aq) + H^+(aq)$$

$$HCO_3^-(aq) \rightleftharpoons CO_3^{2-}(aq) + H^+(aq)$$

According to solubility rules, carbonate forms insoluble ionic compounds by reacting with certain cations in seawater. For example, Ca^{2+} and carbonate ions form calcium carbonate, which is the main constituent of clam shells, pearls, and coral reefs. Write the equilibrium constant expression for the formation of carbonic acid from carbon dioxide and water.

CHAPTER READINESS*

💠 Key Math Skills

- Solving Equations (1.4)
- Converting between Standard Numbers and Scientific Notation (1.5)

🔑 Core Chemistry Skills

- Using Significant Figures in Calculations (2.3)
- Balancing a Chemical Equation (8.2)
- Calculating Concentration (12.4)

*These Key Math Skills and Core Chemistry Skills from previous chapters are listed here for your review as you proceed to the new material in this chapter.

LOOKING AHEAD

13.1 Rates of Reactions

Learning Goal: Describe how temperature, concentration, and catalysts affect the rate of a reaction.

- The rate of a reaction is the speed at which reactants are consumed or products are formed.
- Collision theory is used to explain reaction rates and the factors that affect them.
- The rate of a reaction depends on temperature, concentration, and the presence of a catalyst.
- At higher temperatures, reaction rates increase because reactants move faster, collide more often, and produce more collisions with the required energy of activation. The opposite effects are observed at lower temperatures.
- Increasing the concentrations of reactants often increases the rate of a reaction because collisions occur more often.
- The reaction rate is increased by the addition of a catalyst since a catalyst provides an alternate reaction pathway with a lower energy of activation.

♦ **Learning Exercise 13.1A**

Indicate the effect of each of the following on the rate of a chemical reaction as:
      increase (I)   decrease (D)   no effect (N)

a. _____ adding a catalyst

b. _____ running the reaction at a lower temperature

c. _____ doubling the concentrations of the reactants

d. _____ removing a catalyst

e. _____ running the experiment under the same conditions in a different laboratory

f. _____ increasing the temperature

g. _____ using lower concentrations of reactants

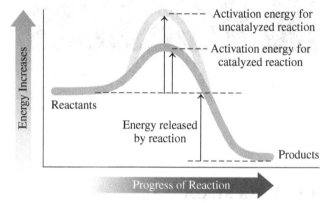

A catalyst lowers the activation energy.

Answers   **a.** I   **b.** D   **c.** I   **d.** D
                 e. N   **f.** I   **g.** D

♦ **Learning Exercise 13.1B**

For the following reaction, $NO_2(g) + CO(g) \longrightarrow NO(g) + CO_2(g)$, indicate the effect of each of the following on the reaction rate as:
      increase (I)   decrease (D)   no effect (N)

a. _____ adding CO

b. _____ running the experiment under the same conditions two days later

c. _____ removing NO_2

d. _____ adding a catalyst

e. _____ adding NO_2

Answers   **a.** I   **b.** N   **c.** D   **d.** I   **e.** I

13.2 Chemical Equilibrium

Learning Goal: Use the concept of reversible reactions to explain chemical equilibrium.

- A reversible reaction proceeds in both the forward and reverse directions.
- Chemical equilibrium is achieved when the rate of the forward reaction becomes equal to the rate of the reverse reaction.
- In a system at equilibrium, there is no change in the concentrations of reactants and products.
- At equilibrium, the concentrations of reactants are typically different from the concentrations of products.

♦ **Learning Exercise 13.2**

Indicate whether each of the following indicates a system at equilibrium (E) or not (NE):

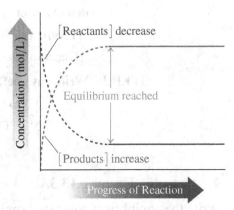

At equilibrium, the concentrations of the reactants and products are constant.

a. _____ The rate of the forward reaction is faster than the rate of the reverse reaction.

b. _____ There is no change in the concentrations of reactants and products.

c. _____ The rate of the forward reaction is equal to the rate of the reverse reaction.

d. _____ The concentrations of reactants are decreasing.

e. _____ The concentrations of products are increasing.

Answers   **a.** NE   **b.** E   **c.** E   **d.** NE   **e.** NE

13.3 Equilibrium Constants

Learning Goal: Calculate the equilibrium constant for a reversible reaction given the concentrations of reactants and products at equilibrium.

- The equilibrium constant expression for a system at equilibrium is the ratio of the concentrations of the products to the concentrations of the reactants with the concentration of each substance raised to an exponent that is equal to its coefficient in the balanced chemical equation.
- For the general equation, $aA + bB \rightleftharpoons cC + dD$, the equilibrium constant expression is written as:

$$K_c = \frac{[\text{Products}]}{[\text{Reactants}]} = \frac{[C]^c[D]^d}{[A]^a[B]^b}$$

- The square brackets around each substance indicate that the concentrations are given in mol/L.

Guide to Writing the Equilibrium Constant Expression	
STEP 1	Write the balanced chemical equation.
STEP 2	Write the concentrations of the products as the numerator and the reactants as the denominator.
STEP 3	Write any coefficient in the equation as an exponent.

Example: Write the equilibrium constant expression for the following reaction:

$$N_2(g) + O_2(g) \rightleftharpoons 2NO(g)$$

	Given	Need	Connect
Analyze the Problem	equation	equilibrium constant expression	$\dfrac{[products]}{[reactants]}$

Solution: **STEP 1** **Write the balanced chemical equation.**

$$N_2(g) + O_2(g) \rightleftharpoons 2NO(g)$$

STEP 2 **Write the concentrations of the products as the numerator and the reactants as the denominator.**

$$\frac{[\text{Products}]}{[\text{Reactants}]} \longrightarrow \frac{[NO]}{[N_2][O_2]}$$

STEP 3 **Write any coefficient in the equation as an exponent.**

$$K_c = \frac{[NO]^2}{[N_2][O_2]}$$

♦ **Learning Exercise 13.3A**

Write the equilibrium constant expression for each of the following reactions:

 **CORE CHEMISTRY SKILL**

Writing the Equilibrium Constant Expression

a. $2SO_3(g) \rightleftharpoons 2SO_2(g) + O_2(g)$

b. $2NO(g) + Br_2(g) \rightleftharpoons 2NOBr(g)$

c. $N_2(g) + 3H_2(g) \rightleftharpoons 2NH_3(g)$

d. $2NO_2(g) \rightleftharpoons N_2O_4(g)$

Answers

a. $K_c = \dfrac{[SO_2]^2[O_2]}{[SO_3]^2}$

b. $K_c = \dfrac{[NOBr]^2}{[NO]^2[Br_2]}$

c. $K_c = \dfrac{[NH_3]^2}{[N_2][H_2]^3}$

d. $K_c = \dfrac{[N_2O_4]}{[NO_2]^2}$

Guide to Calculating the K_c Value	
STEP 1	State the given and needed quantities.
STEP 2	Write the equilibrium constant expression, K_c.
STEP 3	Substitute equilibrium (molar) concentrations and calculate K_c.

♦ Learning Exercise 13.3B

CORE CHEMISTRY SKILL
Calculating an Equilibrium Constant

Calculate the numerical value of K_c using the following equilibrium concentrations:

a. $H_2(g) + I_2(g) \rightleftharpoons 2HI(g)$ $[H_2] = 0.28$ M $[I_2] = 0.28$ M $[HI] = 2.0$ M

b. $2NO_2(g) \rightleftharpoons N_2(g) + 2O_2(g)$ $[NO_2] = 0.60$ M $[N_2] = 0.010$ M $[O_2] = 0.020$ M

c. $N_2(g) + 3H_2(g) \rightleftharpoons 2NH_3(g)$ $[N_2] = 0.50$ M $[H_2] = 0.20$ M $[NH_3] = 0.80$ M

Answers

a. $K_c = \dfrac{[HI]^2}{[H_2][I_2]} = \dfrac{[2.0]^2}{[0.28][0.28]} = 51$

b. $K_c = \dfrac{[N_2][O_2]^2}{[NO_2]^2} = \dfrac{[0.010][0.020]^2}{[0.60]^2} = 1.1 \times 10^{-5}$

c. $K_c = \dfrac{[NH_3]^2}{[N_2][H_2]^3} = \dfrac{[0.80]^2}{[0.50][0.20]^3} = 1.6 \times 10^2$

♦ Learning Exercise 13.3C

Write the equilibrium constant expression for each of the following heterogeneous systems at equilibrium:

a. $H_2(g) + S(l) \rightleftharpoons H_2S(g)$

b. $H_2O(g) + C(s) \rightleftharpoons H_2(g) + CO(g)$

c. $2PbS(s) + 3O_2(g) \rightleftharpoons 2PbO(s) + 2SO_2(g)$

d. $SiH_4(g) + 2O_2(g) \rightleftharpoons SiO_2(s) + 2H_2O(g)$

Answers   **a.** $K_c = \dfrac{[H_2S]}{[H_2]}$ **b.** $K_c = \dfrac{[H_2][CO]}{[H_2O]}$

c. $K_c = \dfrac{[SO_2]^2}{[O_2]^3}$ **d.** $K_c = \dfrac{[H_2O]^2}{[SiH_4][O_2]^2}$

13.4   Using Equilibrium Constants

Learning Goal: Use an equilibrium constant to predict the extent of reaction and to calculate equilibrium concentrations.

- A large K_c indicates that a reaction at equilibrium has more products than reactants; a small K_c indicates that a reaction at equilibrium has more reactants than products.
- The concentration of one component in an equilibrium mixture is calculated from the K_c and the concentrations of all the other components.

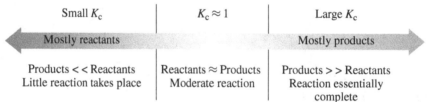

| Small K_c | $K_c \approx 1$ | Large K_c |

Mostly reactants   Mostly products

| Products << Reactants | Reactants ≈ Products | Products >> Reactants |
| Little reaction takes place | Moderate reaction | Reaction essentially complete |

The values of equilibrium constants can be less than 1, about equal to 1, or greater than 1.

◆ Learning Exercise 13.4A

Consider the reaction $2NOBr(g) \rightleftharpoons 2NO(g) + Br_2(g)$.

a. Write the equilibrium constant expression for the reaction.

b. If the K_c for the reaction is 2×10^3, does the equilibrium mixture contain mostly reactants, mostly products, or both reactants and products? Explain.

Answers

a. $K_c = \dfrac{[NO]^2[Br_2]}{[NOBr]^2}$

b. A large K_c ($>>1$) means that the equilibrium mixture contains mostly products.

♦ **Learning Exercise 13.4B**

Consider the reaction $2HI(g) \rightleftharpoons H_2(g) + I_2(g)$

a. Write the equilibrium constant expression for the reaction.

b. If the K_c for the reaction is 1.6×10^{-2}, does the equilibrium mixture contain mostly reactants, mostly products, or both reactants and products? Explain.

Answers

a. $K_c = \dfrac{[H_2][I_2]}{[HI]^2}$

b. A small K_c ($<<1$) means that the equilibrium mixture contains mostly reactants.

Guide to Using the Equilibrium Constant	
STEP 1	State the given and needed quantities.
STEP 2	Write the equilibrium constant expression, K_c, and solve for the needed concentration.
STEP 3	Substitute the equilibrium (molar) concentrations and calculate the needed concentration.

♦ **Learning Exercise 13.4C**

CORE CHEMISTRY SKILL

Write the equilibrium constant expression for the reaction and calculate the molar concentration of the indicated component for each of the following equilibrium systems:

Calculating Equilibrium Concentrations

a. $PCl_5(g) \rightleftharpoons PCl_3(g) + Cl_2(g)$ $K_c = 1.2 \times 10^{-2}$

$[PCl_5] = 2.50\ M$ $[PCl_3] = 0.50\ M$ $[Cl_2] = ?$

b. $CO(g) + H_2O(g) \rightleftharpoons CO_2(g) + H_2(g)$ $K_c = 1.6$

$[CO] = 1.0\ M$ $[H_2O] = 0.80\ M$ $[CO_2] = ?$ $[H_2] = 1.2\ M$

Answers

a. $K_c = \dfrac{[PCl_3][Cl_2]}{[PCl_5]}$ $[Cl_2] = 6.0 \times 10^{-2}\ M$

b. $K_c = \dfrac{[CO_2][H_2]}{[CO][H_2O]}$ $[CO_2] = 1.1\ M$

13.5 Changing Equilibrium Conditions: Le Châtelier's Principle

Learning Goal: Use Le Châtelier's principle to describe the changes made in equilibrium concentrations when reaction conditions change.

⚗ **CORE CHEMISTRY SKILL**

Using Le Châtelier's Principle

- Le Châtelier's principle states that a change in the concentration of a reactant or product, the temperature, or the volume of the container will cause the reaction to shift in the direction that relieves the stress.
- The addition of a catalyst does not affect the equilibrium position of a reaction.

♦ **Learning Exercise 13.5A**

For each of the following changes at equilibrium, indicate whether the equilibrium shifts in the direction of products (P), reactants (R), or does not change (N):

$$N_2(g) + O_2(g) + 180 \text{ kJ} \rightleftharpoons 2NO(g)$$

1. _____ adding more $O_2(g)$

2. _____ removing some $N_2(g)$

3. _____ removing some $NO(g)$

4. _____ increasing the temperature

5. _____ reducing the volume of the container

6. _____ increasing the volume of the container

Answers **1.** P **2.** R **3.** P **4.** P **5.** N **6.** N

♦ **Learning Exercise 13.5B**

For each of the following changes at equilibrium, indicate whether the equilibrium shifts in the direction of products (P), reactants (R), or does not change (N):

$$2NOBr(g) \rightleftharpoons 2NO(g) + Br_2(g) + 340 \text{ kJ}$$

1. _____ adding more $NO(g)$

2. _____ removing some $Br_2(g)$

3. _____ removing some $NOBr(g)$

4. _____ adding heat

5. _____ decreasing the temperature

6. _____ increasing the volume of the container

Answers **1.** R **2.** P **3.** R **4.** R **5.** P **6.** P

♦ **Learning Exercise 13.5C**

Use the following diagram of a gaseous reaction at equilibrium (the white atoms are nitrogen atoms and the black atoms are oxygen atoms) to answer questions **a** to **f**:

a. Write the balanced chemical equation for the reaction.

b. Write the equilibrium constant expression for the reaction.

Answer **c** to **f** with shifts in the direction of the products (P), shifts in the direction of the reactants (R), or no change (N).

c. What is the effect on the equilibrium when more NO_3 is added? _____

d. What is the effect on the equilibrium when more NO_2 is added? _____

e. What is the effect on the equilibrium when some NO is removed? _____

f. What is the effect on the equilibrium when the volume of the container is decreased? _____

Answers

a. $2NO_2(g) \rightleftarrows NO_3(g) + NO(g)$

b. $K_c = \dfrac{[NO_3][NO]}{[NO_2]^2}$

c. R   **d.** P   **e.** P   **f.** N

13.6   Equilibrium in Saturated Solutions

Learning Goal: Write the solubility product expression for a slightly soluble ionic compound and calculate K_{sp}; use K_{sp} to determine the solubility.

- The solubility of a slightly soluble ionic compound in an aqueous solution is represented by an equilibrium expression called the solubility product constant (K_{sp}).
- As in other heterogeneous equilibria, the concentration of the solid is constant and not included in the K_{sp} expression. For example, the K_{sp} of AgBr is: $K_{sp} = [Ag^+][Br^-]$

Key Terms for Sections 13.1 to 13.6

Match each of the following key terms with the correct description:

a. activation energy	**b.** equilibrium	**c.** Le Châtelier's principle
d. catalyst	**e.** equilibrium constant expression	**f.** collision theory
g. heterogeneous equilibrium	**h.** solubility product constant	

1. _____ a substance that lowers the activation energy and increases the rate of reaction

2. _____ equilibrium components are present in at least two different states

3. _____ states that a stress placed on a reaction at equilibrium causes the equilibrium to shift in a direction that relieves the stress

4. _____ describes the equilibrium of a slightly soluble ionic compound in an aqueous solution

5. _____ the ratio of the concentrations of products to reactants with each component raised to an exponent equal to its coefficient

6. _____ the energy that is required in a collision to break the bonds in the reactants

7. _____ a reaction requires that reactants collide to form products

8. _____ the condition in which the rate of the forward reaction is equal to the rate of the reverse reaction

Answers   **1.** d   **2.** g   **3.** c   **4.** h   **5.** e   **6.** a   **7.** f   **8.** b

♦ Learning Exercise 13.6A

CORE CHEMISTRY SKILL
Writing the Solubility Product Expression

For each of the following slightly soluble ionic compounds, write the equilibrium equation for dissociation and the solubility product expression:

a. MgF_2

b. NiS

c. Ag_2CO_3

d. $Zn(OH)_2$

Answers

a. $MgF_2(s) \rightleftarrows Mg^{2+}(aq) + 2F^-(aq)$ $\qquad K_{sp} = [Mg^{2+}][F^-]^2$
b. $NiS(s) \rightleftarrows Ni^{2+}(aq) + S^{2-}(aq)$ $\qquad K_{sp} = [Ni^{2+}][S^{2-}]$
c. $Ag_2CO_3(s) \rightleftarrows 2Ag^+(aq) + CO_3^{2-}(aq)$ $\qquad K_{sp} = [Ag^+]^2[CO_3^{2-}]$
d. $Zn(OH)_2(s) \rightleftarrows Zn^{2+}(aq) + 2OH^-(aq)$ $\qquad K_{sp} = [Zn^{2+}][OH^-]^2$

Guide to Calculating K_{sp}	
STEP 1	State the given and needed quantities.
STEP 2	Write the equilibrium equation for the dissociation of the slightly soluble ionic compound.
STEP 3	Write the solubility product expression K_{sp}.
STEP 4	Substitute the molar concentration of each ion into the K_{sp} expression and calculate.

♦ Learning Exercise 13.6B

CORE CHEMISTRY SKILL
Calculating a Solubility Product Constant

Calculate the numerical value of K_{sp} for each of the following slightly soluble ionic compounds given the equilibrium concentration of the ions:

a. CdS with $[Cd^{2+}] = 3 \times 10^{-14}$ M and $[S^{2-}] = 3 \times 10^{-14}$ M

b. $AgIO_3$ with $[Ag^+] = 1.7 \times 10^{-4}$ M and $[IO_3^-] = 1.7 \times 10^{-4}$ M

c. SrF_2 with $[Sr^{2+}] = 8.5 \times 10^{-4}$ M and $[F^-] = 1.7 \times 10^{-3}$ M

d. Ag_2SO_3 with $[Ag^+] = 3.2 \times 10^{-5}$ M and $[SO_3^{2-}] = 1.6 \times 10^{-5}$ M

Answers   **a.** $K_{sp} = 9 \times 10^{-28}$ **b.** $K_{sp} = 2.9 \times 10^{-8}$
　　　　　　　c. $K_{sp} = 2.5 \times 10^{-9}$ **d.** $K_{sp} = 1.6 \times 10^{-14}$

Guide to Calculating Molar Solubility from K_{sp}	
STEP 1	State the given and needed quantities.
STEP 2	Write the equilibrium equation for the dissociation of the slightly soluble ionic compound.
STEP 3	Write the solubility product expression, K_{sp}, using S.
STEP 4	Calculate the molar solubility, S.

♦ **Learning Exercise 13.6C**

▶ **CORE CHEMISTRY SKILL**

Calculate the molar solubility, S, of each of the following slightly soluble ionic compounds.

Calculating the Molar Solubility

a. $CaSO_4$ $K_{sp} = 9 \times 10^{-6}$

b. $ZnCO_3$ $K_{sp} = 1.4 \times 10^{-11}$

c. FeS $K_{sp} = 8 \times 10^{-19}$

d. CuI $K_{sp} = 1.3 \times 10^{-12}$

Answers   **a.** $S = 3 \times 10^{-3}$ M   **b.** $S = 3.7 \times 10^{-6}$ M
　　　　　　　c. $S = 9 \times 10^{-10}$ M   **d.** $S = 1.1 \times 10^{-6}$ M

Checklist for Chapter 13

You are ready to take the Practice Test for Chapter 13. Be sure you have accomplished the following learning goals for this chapter. If not, review the section listed at the end of the goal. Then apply your new skills and understanding to the Practice Test.

After studying Chapter 13, I can successfully:

_____ Describe the factors that increase or decrease the rate of a reaction. (13.1)

_____ Write the forward and reverse reactions of a reversible reaction. (13.2)

_____ Explain how equilibrium occurs when the rate of a forward reaction is equal to the rate of a reverse reaction. (13.2)

_____ Write the equilibrium constant expression for a reaction system at equilibrium. (13.3)

_____ Calculate the equilibrium constant from the equilibrium concentrations. (13.3)

_____ Write the equilibrium constant expression for a heterogeneous reaction system at equilibrium. (13.3)

_____ Use the equilibrium constant to determine whether an equilibrium mixture contains mostly reactants, mostly products, or about the same amounts of reactants and products. (13.4)

_____ Use the equilibrium constant expression to determine the equilibrium concentration of a component in the reaction. (13.4)

_____ Use Le Châtelier's principle to describe the shift in a system at equilibrium when stress is applied to the system. (13.5)

_____ Calculate the solubility product for a saturated solution of a slightly soluble ionic compound. (13.6)

_____ Use the solubility product expression to calculate the molar solubility of a slightly soluble ionic compound. (13.6)

Practice Test for Chapter 13

The chapter sections to review are shown in parentheses at the end of each question.

1. The number of effective molecular collisions increases when (13.1)
   A. more reactants are added.
   B. products are removed.
   C. the energy of collision is below the energy of activation.
   D. the reaction temperature is lowered.
   E. the reacting molecules have an incorrect orientation upon impact.

2. The energy of activation is lowered when (13.1)
   A. more reactants are added.
   B. products are removed.
   C. a catalyst is used.
   D. the reaction temperature is lowered.
   E. the reaction temperature is raised.

3. Food deteriorates more slowly in a refrigerator because (13.1)
   A. more reactants are added.
   B. products are removed.
   C. the energy of activation is higher.
   D. fewer collisions have the energy of activation.
   E. collisions have the wrong orientation upon impact.

4. A reaction reaches equilibrium when (13.2)
   A. the rate of the forward reaction is faster than the rate of the reverse reaction.
   B. the rate of the reverse reaction is faster than the rate of the forward reaction.
   C. the concentrations of reactants and products are changing.
   D. fewer collisions have the energy of activation.
   E. the rate of the forward reaction is equal to the rate of the reverse reaction.

5. The equilibrium constant expression for the following reaction is (13.3)
$$2NOCl(g) \rightleftarrows 2NO(g) + Cl_2(g)$$

A. $\dfrac{[NO][Cl_2]}{[NOCl]}$
B. $\dfrac{[NOCl_2]^2}{[NO]^2[Cl_2]}$
C. $\dfrac{[NOCl_2]}{[NO][Cl_2]}$

D. $\dfrac{[NO^2][Cl_2]}{[NOCl]}$
E. $\dfrac{[NO]^2[Cl_2]}{[NOCl]^2}$

6. The equilibrium constant expression for the following reaction is (13.3)
$$CO(g) + 2H_2(g) \rightleftarrows CH_3OH(g)$$

A. $[CO][2H_2]$
B. $\dfrac{[CO][H_2]}{[CH_3OH]}$
C. $\dfrac{[CH_3OH]}{[CO][H_2]^2}$

D. $\dfrac{1}{[CH_3OH]}$
E. $\dfrac{[CO][H_2]^2}{[CH_3OH]}$

7. The equilibrium constant expression for the following reaction is (13.3)
$$C_3H_8(g) + 5O_2(g) \rightleftarrows 3CO_2(g) + 4H_2O(g)$$

A. $\dfrac{[CO_2][H_2O]}{[C_3H_8][O_2]}$
B. $\dfrac{[C_3H_8][O_2]^5}{[CO_2]^3[H_2O]^4}$
C. $\dfrac{[CO_2]^3[H_2O]^4}{[C_3H_8][O_2]^5}$

D. $[CO_2]^3[H_2O]^4$
E. $\dfrac{[CO_2]^3}{[C_3H_8][O_2]^5}$

8. The equilibrium constant expression for the following heterogeneous reaction is (13.3)
$$MgO(s) \rightleftarrows CO_2(g) + MgCO_3(s)$$

A. $[MgO]$
B. $\dfrac{[CO_2][MgCO_3]}{[MgO]}$
C. $\dfrac{[MgO]}{[CO_2][MgCO_3]}$

D. $\dfrac{1}{[CO_2]}$
E. $[CO_2]$

9. The equilibrium constant expression for the following heterogeneous reaction is (13.3)
$$2PbS(s) + 3O_2(g) \rightleftarrows 2PbO(s) + 2SO_2(g)$$

A. $\dfrac{[PbO][SO_2]}{[PbS][O_2]}$
B. $\dfrac{[PbO]^2[SO_2]^2}{[PbS]^2[O_2]^3}$
C. $\dfrac{[SO_2]^2}{[O_2]^3}$

D. $\dfrac{[SO_2]}{[O_2]}$
E. $\dfrac{[O_2]^3}{[SO_2]^2}$

10. The equilibrium equation that has the following equilibrium constant expression is (13.3)

$$\frac{[H_2S]^2}{[H_2]^2[S_2]}$$

A. $H_2S(g) \rightleftharpoons H_2(g) + S_2(g)$
B. $2H_2S(g) \rightleftharpoons H_2(g) + S_2(g)$
C. $2H_2S(g) \rightleftharpoons 2H_2(g)$
D. $2H_2(g) + S_2(g) \rightleftharpoons 2H_2S(g)$
E. $2H_2(g) \rightleftharpoons 2H_2S(g)$

11. Use the given equilibrium concentrations to calculate the numerical value of the equilibrium constant for (13.3)
$$COBr_2(g) \rightleftharpoons CO(g) + Br_2(g)$$
$[COBr_2] = 0.93 \text{ M}$ $[CO] = 0.42 \text{ M}$ $[Br_2] = 0.42 \text{ M}$
A. 0.19   **B.** 0.39   **C.** 0.42
D. 2.2   **E.** 5.3

12. Use the given equilibrium concentrations to calculate the numerical value of the equilibrium constant for (13.3)
$$2NO(g) + O_2(g) \rightleftharpoons 2NO_2(g)$$
$[NO] = 2.7 \text{ M}$ $[O_2] = 1.0 \text{ M}$ $[NO_2] = 3.0 \text{ M}$
A. 0.81   **B.** 1.2   **C.** 1.1
D. 8.1   **E.** 9.0

13. Use the given equilibrium concentrations to calculate $[PCl_5]$ for the decomposition of PCl_5 that has a $K_c = 0.050$. (13.4)
$$PCl_5(g) \rightleftharpoons PCl_3(g) + Cl_2(g)$$
$[PCl_3] = 0.20 \text{ M}$ $[Cl_2] = 0.20 \text{ M}$
A. 0.01 M   **B.** 0.050 M   **C.** 0.20 M
D. 0.40 M   **E.** 0.80 M

14. The reaction that has a much greater concentration of products than reactants at equilibrium has a K_c value of (13.4)
A. 1.6×10^{-15} **B.** 2×10^{-11} **C.** 1.2×10^{-5}
D. 3×10^{-3} **E.** 1.4×10^{5}

15. The reaction that has a much greater concentration of reactants than products at equilibrium has a K_c value of (13.4)
A. 1.1×10^{-11} **B.** 2×10^{-2} **C.** 1.2×10^{2}
D. 2×10^{4} **E.** 1.3×10^{12}

16. The reaction that has about the same concentration of reactants and products at equilibrium has a K_c value of (13.4)
A. 1.4×10^{-12} **B.** 2×10^{-8} **C.** 1.2
D. 3×10^{2} **E.** 1.3×10^{7}

For questions 17 through 21, indicate how each of the following affects the equilibrium of the reaction shown:
$$PCl_5(g) + heat \rightleftharpoons PCl_3(g) + Cl_2(g) \quad (13.5)$$

A. shifts in the direction of the products   **B.** shifts in the direction of the reactants   **C.** does not change

17. adding more $Cl_2(g)$ **18.** cooling the reaction   **19.** removing some $PCl_3(g)$

20. adding more $PCl_5(g)$ **21.** removing some $PCl_5(g)$

For questions 22 through 26, indicate whether the equilibrium (13.5)

A. shifts in the direction of the products **B.** shifts in the direction of the reactants **C.** does not change

$$2NO(g) + O_2(g) \rightleftharpoons 2NO_2(g) + \text{heat}$$

22. adding more $NO(g)$ **23.** increasing temperature   **24.** adding a catalyst

25. adding more $O_2(g)$ **26.** removing $NO_2(g)$

27. The solubility product expression for the slightly soluble ionic compound $Ca_3(PO_4)_2$ is equal to (13.6)

    A. $[Ca^{2+}][PO_4^{3-}]$ **B.** $[Ca^{2+}]_3[PO_4^{3-}]_2$ **C.** $[Ca^{2+}]^3[PO_4^{3-}]^2$

    D. $[Ca^{2+}]^2[PO_4^{3-}]^3$ **E.** $\dfrac{[Ca^{2+}][PO_4^{3-}]}{[Ca_3(PO_4^{3-})_2]}$

28. The K_{sp} of CuI when a saturated solution has $[Cu^+] = 1 \times 10^{-6}$ M and $[I^-] = 1 \times 10^{-6}$ M is equal to (13.6)

    A. 1×10^{-12} **B.** 1×10^{-6} **C.** 1

    D. 2×10^{-36} **E.** 2×10^{-12}

29. What is the molar solubility, S, of $SrCO_3$ if it has a K_{sp} of 5.6×10^{-10}? (13.6)

    A. 2.8×10^{-10} M   **B.** 2.8×10^{-5} M   **C.** 5.6×10^{-10} M

    D. 2.3×10^{-20} M   **E.** 2.4×10^{-5} M

Answers to the Practice Test

1. A	**2.** C	**3.** D	**4.** E	**5.** E	**6.** C
7. C	**8.** E	**9.** C	**10.** D	**11.** A	**12.** B
13. E	**14.** E	**15.** A	**16.** C	**17.** B	**18.** B
19. A	**20.** A	**21.** B	**22.** A	**23.** B	**24.** C
25. A	**26.** A	**27.** C	**28.** A	**29.** E	

Selected Answers and Solutions to Text Problems

13.1 **a.** The rate of a reaction indicates how fast the products form or how fast the reactants are used up.

    b. At room temperature, more of the reactants will have the energy necessary to proceed to products (the activation energy) than at the lower temperature of the refrigerator, so the rate of formation of bread mold will be faster.

13.3 Adding $Br_2(g)$ molecules increases the concentration of reactants, which increases the number of collisions that take place between the reactants.

13.5 **a.** Adding more reactant increases the number of collisions that take place between the reactants, which increases the reaction rate.

    b. Increasing the temperature increases the kinetic energy of the reactant molecules, which increases the number of collisions and makes more collisions effective. The rate of reaction will be increased.

    c. Adding a catalyst lowers the energy of activation, which increases the reaction rate.

    d. Removing a reactant decreases the number of collisions that take place between the reactants, which decreases the reaction rate.

13.7 A reversible reaction is one in which a forward reaction converts reactants to products, while a reverse reaction converts products to reactants.

13.9 **a.** When the rate of the forward reaction is faster than the rate of the reverse reaction, the process is not at equilibrium.

    b. When the concentrations of the reactants and the products do not change, the process is at equilibrium.

    c. When the rate of either the forward or reverse reaction does not change, the process is at equilibrium.

13.11 The reaction has reached equilibrium because the number of reactants and products does not change; the concentrations of A and B have not changed between 3 h and 4 h.

13.13 In the expression for K_c, the products are divided by the reactants, with each concentration raised to a power equal to its coefficient in the balanced chemical equation:

    a. $K_c = \dfrac{[CS_2][H_2]^4}{[CH_4][H_2S]^2}$
        b. $K_c = \dfrac{[N_2][O_2]}{[NO]^2}$

    c. $K_c = \dfrac{[CS_2][O_2]^4}{[SO_3]^2[CO_2]}$
        d. $K_c = \dfrac{[H_2]^3[CO]}{[CH_4][H_2O]}$

13.15 $K_c = \dfrac{[XY]^2}{[X_2][Y_2]} = \dfrac{[6]^2}{[1][1]} = 36$

13.17 $K_c = \dfrac{[NO_2]^2}{[N_2O_4]} = \dfrac{[0.21]^2}{[0.030]} = 1.5\ (2\ \text{SFs})$

13.19 $K_c = \dfrac{[CH_4][H_2O]}{[CO][H_2]^3} = \dfrac{[1.8][2.0]}{[0.51][0.30]^3} = 260\ (2\ \text{SFs})$

13.21 **a.** only one state (gas) is present; homogeneous equilibrium

    b. solid and gaseous states present; heterogeneous equilibrium

    c. only one state (gas) is present; homogeneous equilibrium

    d. gaseous, liquid, and solid states present; heterogeneous equilibrium

13.23 a. $K_c = \dfrac{[O_2]^3}{[O_3]^2}$ **b.** $K_c = [CO_2][H_2O]$

c. $K_c = \dfrac{[C_6H_{12}]}{[C_6H_6][H_2]^3}$ **d.** $K_c = \dfrac{[H_2]^2}{[HCl]^4}$

13.25 $K_c = \dfrac{[CO_2]}{[CO]} = \dfrac{[0.052]}{[0.20]} = 0.26$ (2 SFs)

13.27 Calculate the value of K_c for each diagram and compare to the given value ($K_c = 4$).

for **A**: $K_c = \dfrac{[XY]^2}{[X_2][Y_2]} = \dfrac{[6]^2}{[1][1]} = 36$

for **B**: $K_c = \dfrac{[XY]^2}{[X_2][Y_2]} = \dfrac{[4]^2}{[2][2]} = 4$

for **C**: $K_c = \dfrac{[XY]^2}{[X_2][Y_2]} = \dfrac{[2]^2}{[3][3]} = 0.44$

Therefore, diagram **B** represents the equilibrium mixture.

13.29 a. A large K_c value indicates that the equilibrium mixture contains mostly products.
b. A large K_c value indicates that the equilibrium mixture contains mostly products.
c. A small K_c value indicates that the equilibrium mixture contains mostly reactants.

13.31 $K_c = \dfrac{[HI]^2}{[H_2][I_2]} = 54$

Rearrange the K_c expression to solve for $[H_2]$ and substitute in known values.

$[H_2] = \dfrac{[HI]^2}{K_c[I_2]} = \dfrac{[0.030]^2}{54[0.015]} = 1.1 \times 10^{-3}$ M (2 SFs)

13.33 $K_c = \dfrac{[NO]^2[Br_2]}{[NOBr]^2} = 2.0$

Rearrange the K_c expression to solve for $[NOBr]$ and substitute in known values.

$[NOBr]^2 = \dfrac{[NO]^2[Br_2]}{K_c} = \dfrac{[2.0]^2[1.0]}{2.0} = 2.0$

Take the square root of both sides of the equation.

$[NOBr] = \sqrt{2.0} = 1.4$ M (2 SFs)

13.35 a. Adding more reactant shifts equilibrium in the direction of the product.
b. Adding more product shifts equilibrium in the direction of the reactants.
c. Increasing the temperature of an endothermic reaction shifts equilibrium in the direction of the product.
d. Increasing the volume of the container shifts the equilibrium in the direction of the reactants, which has more moles of gas.
e. No shift in equilibrium occurs when a catalyst is added.

13.37 a. Adding more reactant shifts equilibrium in the direction of the product.
b. Increasing the temperature of an endothermic reaction shifts equilibrium in the direction of the product to remove heat.

    c. Removing product shifts equilibrium in the direction of the product.

    d. No shift in equilibrium occurs when a catalyst is added.

    e. Removing reactant shifts equilibrium in the direction of the reactants.

13.39 a. When an athlete first arrives at high altitude, the oxygen concentration is decreased.

    b. Removing reactant shifts equilibrium in the direction of the reactants.

13.41 a. $MgCO_3(s) \rightleftharpoons Mg^{2+}(aq) + CO_3^{2-}(aq)$; $K_{sp} = [Mg^{2+}][CO_3^{2-}]$

    b. $CaF_2(s) \rightleftharpoons Ca^{2+}(aq) + 2F^-(aq)$; $K_{sp} = [Ca^{2+}][F^-]^2$

    c. $Ag_3PO_4(s) \rightleftharpoons 3Ag^+(aq) + PO_4^{3-}(aq)$; $K_{sp} = [Ag^+]^3[PO_4^{3-}]$

13.43 $BaSO_4(s) \rightleftharpoons Ba^{2+}(aq) + SO_4^{2-}(aq)$;

    $K_{sp} = [Ba^{2+}][SO_4^{2-}] = [1 \times 10^{-5}][1 \times 10^{-5}] = 1 \times 10^{-10}$ (1 SF)

13.45 $Ag_2CO_3(s) \rightleftharpoons 2Ag^+(aq) + CO_3^{2-}(aq)$;

    $K_{sp} = [Ag^+]^2[CO_3^{2-}] = [2.6 \times 10^{-4}]^2[1.3 \times 10^{-4}] = 8.8 \times 10^{-12}$ (2 SFs)

13.47 $CuI(s) \rightleftharpoons Cu^+(aq) + I^-(aq)$; $K_{sp} = [Cu^+][I^-]$

Substitute S for the molarity of each ion into the K_{sp} expression.

$K_{sp} = S \times S = S^2 = 1 \times 10^{-12}$

Calculate the molar solubility (S) by taking the square root of both sides of the equation.

$S = \sqrt{1 \times 10^{-12}} = 1 \times 10^{-6}$ M (1 SF)

13.49 a. $CaCO_3(s) \rightleftharpoons Ca^{2+}(aq) + CO_3^{2-}(aq)$

    b. $K_{sp} = [Ca^{2+}][CO_3^{2-}]$

13.51 a. $K_c = \dfrac{[CO_2][H_2O]^2}{[CH_4][O_2]^2}$

    b. $K_c = \dfrac{[N_2]^2[H_2O]^6}{[NH_3]^4[O_2]^3}$

    c. $K_c = \dfrac{[CH_4]}{[H_2]^2}$

13.53 At equilibrium, the diagram shows mostly reactants and a few products, so the equilibrium constant K_c for the reaction would have a small value.

13.55 T_2 is lower than T_1. This would cause the exothermic reaction shown to shift in the direction of the products to add heat; more product is seen in the T_2 diagram.

13.57 a. Increasing the temperature of an exothermic reaction shifts equilibrium in the direction of the reactants.

    b. Decreasing volume favors the side of the reaction with fewer moles of gas, so there is a shift in the direction of the products.

    c. Adding a catalyst does not shift equilibrium.

    d. Adding more reactant shifts equilibrium in the direction of the products.

13.59 a. A large K_c value indicates that the equilibrium mixture contains mostly products.
  b. A K_c value close to 1 indicates that the equilibrium mixture contains both reactants and products.
  c. A small K_c value indicates that the equilibrium mixture contains mostly reactants.
  d. A K_c value close to 1 indicates that the equilibrium mixture contains both reactants and products.

13.61 a. $K_c = \dfrac{[N_2][H_2]^3}{[NH_3]^2}$

  b. $K_c = \dfrac{[3.0][0.50]^3}{[0.20]^2} = 9.4 \ (2 \ SFs)$

13.63 $K_c = \dfrac{[N_2O_4]}{[NO_2]^2} = 5.0$

Rearrange the K_c expression to solve for $[N_2O_4]$ and substitute in known values.

$[N_2O_4] = K_c[NO_2]^2 = 5.0[0.50]^2 = 1.3 \ M \ (2 \ SFs)$

13.65 a. When the reactant $[O_2]$ increases, the rate of the forward reaction increases to shift the equilibrium in the direction of the product.
  b. When the product $[O_2]$ increases, the rate of the reverse reaction increases to shift the equilibrium in the direction of the reactant.
  c. When the reactant $[O_2]$ increases, the rate of the forward reaction increases to shift the equilibrium in the direction of the product.
  d. When the product $[O_2]$ increases, the rate of the reverse reaction increases to shift the equilibrium in the direction of the reactants.

13.67 Decreasing the volume of an equilibrium mixture shifts the equilibrium in the direction of the side of the reaction that has the fewer number of moles of gas. No shift occurs when there are an equal number of moles of gas on both sides of the equation.
  a. With 3 mol of gas on the reactant side and 2 mol of gas on the product side, decreasing the volume will shift equilibrium in the direction of the product.
  b. With 2 mol of gas on the reactant side and 3 mol of gas on the product side, decreasing the volume will shift equilibrium in the direction of the reactant.
  c. With 6 mol of gas on reactant side and 0 mol of gas on the product side, decreasing the volume will shift equilibrium in the direction of the product.
  d. With 4 mol of gas on the reactant side and 5 mol of gas on the product side, decreasing the volume will shift equilibrium in the direction of the reactants.

13.69 $K_c = \dfrac{[CO][Cl_2]}{[COCl_2]} = 0.68$

Rearrange the K_c expression to solve for $[COCl_2]$ and substitute in known values.

$[COCl_2] = \dfrac{[CO][Cl_2]}{K_c} = \dfrac{[0.40][0.74]}{0.68} = 0.44 \ M \ (2 \ SFs)$

13.71 a. $CuCO_3(s) \rightleftharpoons Cu^{2+}(aq) + CO_3^{2-}(aq); K_{sp} = [Cu^{2+}][CO_3^{2-}]$
  b. $PbF_2(s) \rightleftharpoons Pb^{2+}(aq) + 2F^-(aq); K_{sp} = [Pb^{2+}][F^-]^2$
  c. $Fe(OH)_3(s) \rightleftharpoons Fe^{3+}(aq) + 3OH^-(aq); K_{sp} = [Fe^{3+}][OH^-]^3$

13.73 $FeS(s) \rightleftharpoons Fe^{2+}(aq) + S^{2-}(aq); K_{sp} = [Fe^{2+}][S^{2-}]$
$K_{sp} = [7.7 \times 10^{-10}][7.7 \times 10^{-10}] = 5.9 \times 10^{-19}$ (2 SFs)

13.75 $Mn(OH)_2(s) \rightleftharpoons Mn^{2+}(aq) + 2OH^-(aq); K_{sp} = [Mn^{2+}][OH^-]^2$
$K_{sp} = [3.7 \times 10^{-5}][7.4 \times 10^{-5}]^2 = 2.0 \times 10^{-13}$ (2 SFs)

13.77 $CdS(s) \rightleftharpoons Cd^{2+}(aq) + S^{2-}(aq); K_{sp} = [Cd^{2+}][S^{2-}]$
Substitute S for the molarity of each ion into the K_{sp} expression.
$K_{sp} = S \times S = 1.0 \times 10^{-24}$
Calculate the molar solubility (S) by taking the square root of both sides of the equation.
$S = \sqrt{1.0 \times 10^{-24}} = 1.0 \times 10^{-12}$ M (2 SFs)

13.79 a. $K_c = \dfrac{[NO]^2[Br_2]}{[NOBr]^2}$

b. When the concentrations are substituted into the expression, the result is 1.0, which is not equal to K_c (2.0). Therefore, the system is not at equilibrium.

$K_c = \dfrac{[NO]^2[Br_2]}{[NOBr]^2} = \dfrac{[1.0]^2[1.0]}{[1.0]^2} = 1.0$ (2 SFs)

c. Since the calculated value in part **b** is less than K_c, the rate of the forward reaction will initially increase.

d. When the system has reestablished equilibrium, the $[Br_2]$ and $[NO]$ will have increased, and the $[NOBr]$ will have decreased.

13.81 a. When more product molecules are added to an equilibrium mixture, the system shifts in the direction of the reactants. This will cause a decrease in the equilibrium concentration of the product CO.

b. When the temperature is increased for an endothermic reaction, the system shifts in the direction of the products to remove heat. This will cause an increase in the equilibrium concentration of the product CO.

c. Increasing the volume of the reaction container favors the side of the reaction with the greater number of moles of gas, so this system shifts in the direction of the products. This will cause an increase in the equilibrium concentration of the product CO.

d. Decreasing the volume of the reaction container favors the side of the reaction with the fewer moles of gas, so this system shifts in the direction of the reactants. This will cause a decrease in the equilibrium concentration of the product CO.

e. When a catalyst is added, the rates of both forward and reverse reactions increase; the equilibrium position does not change. No change will be observed in the equilibrium concentration of the product CO.

13.83 Increasing the volume of an equilibrium mixture shifts the system toward the side of the reaction that has the greater number of moles of gas; decreasing the volume shifts the equilibrium toward the side of the reaction that has the fewer moles of gas.

a. Since the product side has the greater number of moles of gas, increasing the volume of the container will increase the yield of product.

b. Since the product side has the greater number of moles of gas, increasing the volume of the container will increase the yield of products.

c. Since the product side has the fewer moles of gas, decreasing the volume of the container will increase the yield of product.

13.85 $Mg(OH)_2(s) \rightleftharpoons Mg^{2+}(aq) + 2OH^-(aq); K_{sp} = [Mg^{2+}][OH^-]^2$

$$\frac{9.7 \times 10^{-3} \text{ g Mg(OH)}_2}{1 \text{ L solution}} \times \frac{1 \text{ mol Mg(OH)}_2}{58.33 \text{ g Mg(OH)}_2} = 1.7 \times 10^{-4} \text{ M Mg(OH)}_2 \text{ solution}$$

$$\frac{1.7 \times 10^{-4} \text{ mol Mg(OH)}_2}{1 \text{ L solution}} \times \frac{1 \text{ mol Mg}^{2+}}{1 \text{ mol Mg(OH)}_2} = 1.7 \times 10^{-4} \text{ M Mg}^{2+} = [Mg^{2+}]$$

$$\frac{1.7 \times 10^{-4} \text{ mol Mg(OH)}_2}{1 \text{ L solution}} \times \frac{2 \text{ mol OH}^-}{1 \text{ mol Mg(OH)}_2} = 3.4 \times 10^{-4} \text{ M OH}^- = [OH^-]$$

$$K_{sp} = [Mg^{2+}][OH^-]^2 = [1.7 \times 10^{-4}][3.4 \times 10^{-4}]^2 = 2.0 \times 10^{-11} \text{ (2 SFs)}$$

14
Acids and Bases

Credit: Lisa S./Shutterstock

When Larry was in the emergency room after an automobile accident, a blood sample was analyzed by Brianna, a clinical laboratory technician. The results showed that Larry's blood pH was 7.30 and the partial pressure of CO_2 gas was above the desired level. Blood pH is typically in the range of 7.35 to 7.45, and a value less than 7.35 indicates a state of acidosis. Larry had respiratory acidosis because an increase in the partial pressure of CO_2 gas in the bloodstream decreases the pH. In the emergency room, Larry was given an IV containing bicarbonate to increase his blood pH.

$$HCO_3^-(aq) + H^+(aq) \rightleftharpoons H_2CO_3(aq) \rightleftharpoons CO_2(g) + H_2O(l)$$

Soon Larry's blood pH and partial pressure of CO_2 gas returned to normal. What is the pH of a sample of blood that has $[H_3O^+] = 3.5 \times 10^{-8}$ M?

CHAPTER READINESS*

✿ Key Math Skills

- Solving Equations (1.4)
- Converting between Standard Numbers and Scientific Notation (1.5)

⚗ Core Chemistry Skills

- Writing Ionic Formulas (6.2)
- Balancing a Chemical Equation (8.2)
- Using Concentration as a Conversion Factor (12.4)
- Writing the Equilibrium Expression (13.3)
- Calculating Equilibrium Concentrations (13.4)
- Using Le Châtelier's Principle (13.5)

*These Key Math Skills and Core Chemistry Skills from previous chapters are listed here for your review as you proceed to the new material in this chapter.

LOOKING AHEAD

14.1 Acids and Bases

Learning Goal: Describe and name acids and bases.

- In water, an Arrhenius acid produces hydrogen ions, H^+, and an Arrhenius base produces hydroxide ions, OH^-.
- An acid with a simple nonmetal anion is named by placing the prefix *hydro* in front of the name of the anion, and changing its *ide* ending to *ic acid*.
- An acid with a polyatomic anion is named as an *ic acid* when its anion ends in *ate* and as an *ous acid* when its anion ends in *ite*.
- Arrhenius bases are named as metal or ammonium hydroxides.

♦ **Learning Exercise 14.1A**

Indicate if each of the following characteristics describes an A (acid) or a B (base):

a. _____ turns litmus red

b. _____ tastes sour

c. _____ contains more OH^- than H_3O^+

d. _____ neutralizes bases

e. _____ tastes bitter

f. _____ turns litmus blue

g. _____ contains more H_3O^+ than OH^-

h. _____ neutralizes acids

Answers   **a.** A   **b.** A   **c.** B   **d.** A   **e.** B   **f.** B   **g.** A   **h.** B

♦ **Learning Exercise 14.1B**

Fill in the blanks with the formula or name of an acid or base.

a. HCl _____

b. _____ sodium hydroxide

c. _____ sulfurous acid

d. _____ nitric acid

e. $Ca(OH)_2$ _____

f. H_2CO_3 _____

g. $Al(OH)_3$ _____

h. _____ potassium hydroxide

i. $HClO_4$ _____

j. H_3PO_3 _____

Answers   **a.** hydrochloric acid   **b.** NaOH   **c.** H_2SO_3
d. HNO_3 **e.** calcium hydroxide   **f.** carbonic acid
g. aluminum hydroxide   **h.** KOH   **i.** perchloric acid
j. phosphorous acid

14.2 Brønsted–Lowry Acids and Bases

Learning Goal: Identify conjugate acid–base pairs for Brønsted–Lowry acids and bases.

- According to the Brønsted–Lowry theory, acids are hydrogen ion, H^+, donors, and bases are hydrogen ion, H^+, acceptors.
- In solution, hydrogen ions, H^+, from acids bond to polar water molecules to form hydronium ions, H_3O^+.
- Conjugate acid–base pairs are molecules or ions linked by the loss and gain of one hydrogen ion, H^+.
- Every hydrogen ion transfer reaction involves two acid–base conjugate pairs.

Guide to Writing Conjugate Acid–Base Pairs	
STEP 1	Identify the reactant that loses H^+ as the acid.
STEP 2	Identify the reactant that gains H^+ as the base.
STEP 3	Write the conjugate acid–base pairs.

◆ **Learning Exercise 14.2A**

Complete the following:

Acid	Conjugate Base
a. H_2O	_____
b. HSO_4^-	_____
c. _____	Cl^-
d. _____	CO_3^{2-}
e. HNO_3	_____
f. NH_4^+	_____
g. _____	HS^-
h. _____	$H_2PO_4^-$

Conjugate acid–base pair

Conjugate acid–base pair

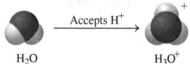

Answers   **a.** OH^- **b.** SO_4^{2-} **c.** HCl **d.** HCO_3^-
e. NO_3^- **f.** NH_3 **g.** H_2S **h.** H_3PO_4

◆ **Learning Exercise 14.2B**

Identify the conjugate acid–base pairs in each of the following reactions:

a. $NH_3(aq) + H_2O(l) \rightleftharpoons NH_4^+(aq) + OH^-(aq)$

b. $NH_4^+(aq) + SO_4^{2-}(aq) \rightleftharpoons NH_3(aq) + HSO_4^-(aq)$

c. $HCO_3^-(aq) + H_2O(l) \rightleftharpoons CO_3^{2-}(aq) + H_3O^+(aq)$

d. $HNO_3(aq) + OH^-(aq) \longrightarrow NO_3^-(aq) + H_2O(l)$

Answers   **a.** H_2O/OH^- and NH_4^+/NH_3
b. NH_4^+/NH_3 and HSO_4^-/SO_4^{2-}
c. HCO_3^-/CO_3^{2-} and H_3O^+/H_2O
d. HNO_3/NO_3^- and H_2O/OH^-

🦴 **CORE CHEMISTRY SKILL**

Identifying Conjugate Acid–Base Pairs

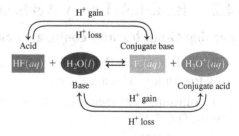

HF, an acid, loses one H^+ to form its conjugate base F^-. Water acts as a base by gaining one H^+ to form its conjugate acid H_3O^+.

♦ **Learning Exercise 14.2C**

Write the equilibrium equation with conjugate acid–base pairs starting with each of the following pairs of reactants:

a. HBr (acid) and CO_3^{2-} (base)

b. HSO_4^- (acid) and OH^- (base)

c. NH_4^+ (acid) and H_2O (base)

d. HCl (acid) and SO_4^{2-} (base)

Answers

a. $HBr(aq) + CO_3^{2-}(aq) \rightleftharpoons Br^-(aq) + HCO_3^-(aq)$

b. $HSO_4^-(aq) + OH^-(aq) \rightleftharpoons SO_4^{2-}(aq) + H_2O(l)$

c. $NH_4^+(aq) + H_2O(l) \rightleftharpoons NH_3(aq) + H_3O^+(aq)$

d. $HCl(aq) + SO_4^{2-}(aq) \rightleftharpoons Cl^-(aq) + HSO_4^-(aq)$

14.3 Strengths of Acids and Bases

Learning Goal: Write equations for the dissociation of strong and weak acids; identify the direction of reaction.

- Strong acids dissociate completely in water, and the H^+ is accepted by H_2O acting as a base.
- A weak acid dissociates slightly in water, producing only small amounts of H_3O^+.
- All hydroxides of Group 1A (1) and most hydroxides of Group 2A (2) are strong bases, which dissociate completely in water.
- In an aqueous ammonia solution, a weak base, NH_3, accepts only a small percentage of hydrogen ions to form the conjugate acid, NH_4^+.
- As the strength of an acid decreases, the strength of the conjugate base increases. By comparing relative strengths, the direction of an acid–base reaction can be predicted.

Study Note

Only six common acids are strong acids; other acids are weak acids.

HI	HBr
$HClO_4$	HCl
H_2SO_4	HNO_3

Example: Is H_2S a strong or a weak acid?

Solution: H_2S is a weak acid because it is not one of the six strong acids.

◆ **Learning Exercise 14.3A**

Identify each of the following as a strong acid, a weak acid, a strong base, or a weak base:

a. HNO_3 ⎯⎯⎯⎯

b. H_2CO_3 ⎯⎯⎯⎯

c. $H_2PO_4^-$ ⎯⎯⎯⎯

d. NH_3 ⎯⎯⎯⎯

e. LiOH ⎯⎯⎯⎯

f. H_3BO_3 ⎯⎯⎯⎯

g. $Ca(OH)_2$ ⎯⎯⎯⎯

h. H_2SO_4 ⎯⎯⎯⎯

Answers
a. strong acid **b.** weak acid **c.** weak acid
d. weak base **e.** strong base **f.** weak acid
g. strong base **h.** strong acid

Acetic acid $(HC_2H_3O_2)$ is a weak acid because it dissociates slightly in water, producing only a small percentage of H_3O^+.

Credit: Pearson Education/ Pearson Science

◆ **Learning Exercise 14.3B**

Using Table 14.3 in the text, identify the stronger acid in each of the following pairs of acids:

a. HCl or H_2CO_3 ⎯⎯⎯⎯

b. HNO_2 or HCN ⎯⎯⎯⎯

c. H_2S or HBr ⎯⎯⎯⎯

d. H_2SO_4 or HSO_4^- ⎯⎯⎯⎯

e. HF or H_3PO_4 ⎯⎯⎯⎯

Answers **a.** HCl **b.** HNO_2 **c.** HBr **d.** H_2SO_4 **e.** H_3PO_4

◆ **Learning Exercise 14.3C**

Using Table 14.3 in the text, indicate whether each of the following equilibrium mixtures contains mostly reactants or products:

a. $H_2SO_4(aq) + H_2O(l) \rightleftharpoons H_3O^+(aq) + HSO_4^-(aq)$ ⎯⎯⎯⎯⎯⎯

b. $I^-(aq) + H_3O^+(aq) \rightleftharpoons H_2O(l) + HI(aq)$ ⎯⎯⎯⎯⎯⎯

c. $NH_3(aq) + H_2O(l) \rightleftharpoons NH_4^+(aq) + OH^-(aq)$ ⎯⎯⎯⎯⎯⎯

d. $HCl(aq) + CO_3^{2-}(aq) \rightleftharpoons Cl^-(aq) + HCO_3^-(aq)$ ⎯⎯⎯⎯⎯⎯

Answers **a.** products **b.** reactants **c.** reactants **d.** products

14.4   Dissociation Constants for Acids and Bases

Learning Goal: Write the dissociation expression for a weak acid or weak base.

• For an acid dissociation expression K_a, the molar concentrations of the products are divided by the molar concentrations of the reactants. Because water is a pure liquid, its concentration is omitted.

$$HA(aq) + H_2O(l) \rightleftharpoons H_3O^+(aq) + A^-(aq) \qquad K_a = \frac{[H_3O^+][A^-]}{[HA]}$$

- The numerical value of the acid dissociation expression is the acid dissociation constant.
- Equilibrium mixtures of acids or bases with dissociation constants greater than one ($K > 1$) contain mostly products, whereas constants smaller than one ($K < 1$) contain mostly reactants.

Guide to Writing the Acid Dissociation Expression	
STEP 1	Write the balanced chemical equation.
STEP 2	Write the concentrations of the products as the numerator and the reactants as the denominator.

Example: Write the acid dissociation expression for hydrogen carbonate, H_2CO_3, a weak acid.

Solution: **STEP 1** Write the balanced chemical equation.

$$H_2CO_3(aq) + H_2O(l) \rightleftharpoons H_3O^+(aq) + HCO_3^-(aq)$$

STEP 2 Write the concentrations of the products as the numerator and the reactants as the denominator.

$$K_a = \frac{[H_3O^+][HCO_3^-]}{[H_2CO_3]}$$

◆ Learning Exercise 14.4A

Write the equation for the dissociation and the acid dissociation expression for each of the following weak acids:

a. HCN

b. HNO_2

c. H_2S (first dissociation only)

Answers

a. $HCN(aq) + H_2O(l) \rightleftharpoons H_3O^+(aq) + CN^-(aq)$ $\quad K_a = \dfrac{[H_3O^+][CN^-]}{[HCN]}$

b. $HNO_2(aq) + H_2O(l) \rightleftharpoons H_3O^+(aq) + NO_2^-(aq)$ $\quad K_a = \dfrac{[H_3O^+][NO_2^-]}{[HNO_2]}$

c. $H_2S(aq) + H_2O(l) \rightleftharpoons H_3O^+(aq) + HS^-(aq)$ $\quad K_a = \dfrac{[H_3O^+][HS^-]}{[H_2S]}$

♦ **Learning Exercise 14.4B**

For each of the following pairs of K_a values, indicate which one belongs to the weaker acid:

a. 5.2×10^{-5} or 3.8×10^{-3} _____   **b.** 3.0×10^{-8} or 1.6×10^{-10} _____

c. 4.5×10^{-2} or 7.2×10^{-6} _____

Answers   **a.** 5.2×10^{-5} **b.** 1.6×10^{-10} **c.** 7.2×10^{-6}

♦ **Learning Exercise 14.4C**

Indicate whether the equilibrium mixture contains mostly reactants or products for each of the following K_a values:

a. 5.2×10^{-5} _____   **b.** 3.0×10^{8} _____

c. 4.5×10^{-3} _____   **d.** 7.2×10^{15} _____

Answers   **a.** reactants   **b.** products   **c.** reactants   **d.** products

14.5   Dissociation of Water

Learning Goal: Use the water dissociation expression to calculate the $[H_3O^+]$ and $[OH^-]$ in an aqueous solution.

- In pure water, a few water molecules transfer hydrogen ions to other water molecules, producing small, but equal, amounts of each ion such that $[H_3O^+]$ and $[OH^-]$ each $= 1.0 \times 10^{-7}$ M at 25 °C.

$$H_2O(l) + H_2O(l) \rightleftharpoons H_3O^+(aq) + OH^-(aq)$$

- K_w, the water dissociation expression $[H_3O^+][OH^-] = [1.0 \times 10^{-7}][1.0 \times 10^{-7}]$ $= 1.0 \times 10^{-14}$ at 25 °C, applies to all aqueous solutions.

- In acidic solutions, the $[H_3O^+]$ is greater than the $[OH^-]$. In basic solutions, the $[OH^-]$ is greater than the $[H_3O^+]$.

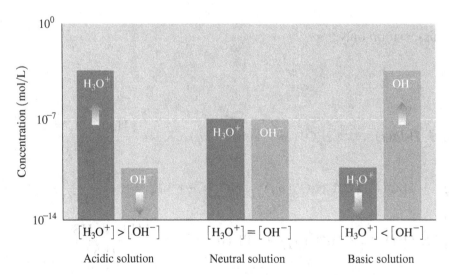

Guide to Calculating $[H_3O^+]$ and $[OH^-]$ in Aqueous Solutions	
STEP 1	State the given and needed quantities.
STEP 2	Write the K_w for water and solve for the unknown $[H_3O^+]$ or $[OH^-]$.
STEP 3	Substitute the known $[H_3O^+]$ or $[OH^-]$ into the equation and calculate.

Example: What is the $[H_3O^+]$ in an urine sample that has $[OH^-] = 2.0 \times 10^{-9}$ M?

Solution: **STEP 1 State the given and needed quantities.**

Analyze the Problem	Given	Need	Connect
	$[OH^-] = 2.0 \times 10^{-9}$ M	$[H_3O^+]$	$K_w = [H_3O^+][OH^-]$

STEP 2 Write the K_w for water and solve for the unknown $[H_3O^+]$.

$$[H_3O^+][OH^-] = 1.0 \times 10^{-14}$$

$$[H_3O^+] = \frac{1.0 \times 10^{-14}}{[OH^-]}$$

STEP 3 Substitute the known $[OH^-]$ into the equation and calculate.

$$[H_3O^+] = \frac{1.0 \times 10^{-14}}{[2.0 \times 10^{-9}]} = 5.0 \times 10^{-6} \text{ M}$$

A dipstick is used to measure the pH of a urine sample.

Credit: Alexander Gospodinov / Fotolia

♦ **Learning Exercise 14.5A**

Indicate whether each of the following solutions is acidic, basic, or neutral:

a. $[H_3O^+] = 2.5 \times 10^{-9}$ M _____

b. $[OH^-] = 1.6 \times 10^{-2}$ M _____

c. $[H_3O^+] = 7.9 \times 10^{-3}$ M _____

d. $[OH^-] = 2.9 \times 10^{-12}$ M _____

Answers   **a.** basic   **b.** basic   **c.** acidic   **d.** acidic

♦ **Learning Exercise 14.5B**

Use the K_w to calculate the $[H_3O^+]$ when the $[OH^-]$ has each of the following values:

🦴 **CORE CHEMISTRY SKILL**

Calculating $[H_3O^+]$ and $[OH^-]$ in Solutions

a. $[OH^-] = 1.0 \times 10^{-10}$ M $[H_3O^+] =$ _____

b. $[OH^-] = 2.0 \times 10^{-5}$ M $[H_3O^+] =$ _____

c. $[OH^-] = 4.5 \times 10^{-7}$ M $[H_3O^+] =$ _____

d. $[OH^-] = 8.0 \times 10^{-4}$ M $[H_3O^+] =$ _____

e. $[OH^-] = 5.5 \times 10^{-8}$ M $[H_3O^+] =$ _____

Answers   **a.** 1.0×10^{-4} M   **b.** 5.0×10^{-10} M   **c.** 2.2×10^{-8} M
               d. 1.3×10^{-11} M   **e.** 1.8×10^{-7} M

♦ **Learning Exercise 14.5C**

Use the K_w to calculate the $[OH^-]$ when the $[H_3O^+]$ has each of the following values:

a. $[H_3O^+] = 1.0 \times 10^{-3}$ M $[OH^-] = $ _____

b. $[H_3O^+] = 3.0 \times 10^{-10}$ M $[OH^-] = $ _____

c. $[H_3O^+] = 4.0 \times 10^{-6}$ M $[OH^-] = $ _____

d. $[H_3O^+] = 2.8 \times 10^{-13}$ M $[OH^-] = $ _____

e. $[H_3O^+] = 8.6 \times 10^{-7}$ M $[OH^-] = $ _____

Answers   **a.** 1.0×10^{-11} M   **b.** 3.3×10^{-5} M   **c.** 2.5×10^{-9} M
                   d. 3.6×10^{-2} M   **e.** 1.2×10^{-8} M

14.6 The pH Scale

Learning Goal: Calculate pH from $[H_3O^+]$; given the pH, calculate the $[H_3O^+]$ and $[OH^-]$ of a solution.

- The pH scale is a range of numbers from 0 to 14 related to the $[H_3O^+]$ of the solution.
- A neutral solution has a pH of 7.0. In acidic solutions, the pH is below 7.0, and in basic solutions, the pH is above 7.0.
- Mathematically, pH is the negative logarithm of the hydronium ion concentration: $pH = -\log[H_3O^+]$.
- The pOH scale is similar to the pH scale except that pOH is associated with the $[OH^-]$ of an aqueous solution.
- In any aqueous solution, the sum of the pH and pOH is equal to 14.00.

♦ **Learning Exercise 14.6A**

State whether the following pH values are acidic, basic, or neutral:

a. _____ blood plasma, pH = 7.4   **b.** _____ soft drink, pH = 2.8

c. _____ maple syrup, pH = 6.8   **d.** _____ beans, pH = 5.0

e. _____ tomatoes, pH = 4.2   **f.** _____ lemon juice, pH = 2.2

g. _____ saliva, pH = 7.0   **h.** _____ egg white, pH = 7.8

i. _____ lye, pH = 12.4   **j.** _____ strawberries, pH = 3.0

Answers   **a.** basic   **b.** acidic   **c.** acidic   **d.** acidic   **e.** acidic
                   f. acidic   **g.** neutral   **h.** basic   **i.** basic   **j.** acidic

Guide to Calculating pH of an Aqueous Solution	
STEP 1	State the given and needed quantities.
STEP 2	Enter the $[H_3O^+]$ into the pH equation and calculate.
STEP 3	Adjust the number of SFs on the *right* of the decimal point.

♦ Learning Exercise 14.6B

KEY MATH SKILL
Calculating pH from $[H_3O^+]$

Calculate the pH of each of the following solutions:

a. $[H_3O^+] = 1.0 \times 10^{-8}$ M ______ **b.** $[OH^-] = 1 \times 10^{-12}$ M ______

c. $[H_3O^+] = 1 \times 10^{-3}$ M ______ **d.** $[OH^-] = 1 \times 10^{-10}$ M ______

e. $[H_3O^+] = 3.4 \times 10^{-8}$ M ______ **f.** $[OH^-] = 7.8 \times 10^{-2}$ M ______

Answers **a.** 8.00 **b.** 2.0 **c.** 3.0 **d.** 4.0 **e.** 7.47 **f.** 12.89

♦ Learning Exercise 14.6C

Calculate the pOH of each of the following solutions:

a. $[OH^-] = 4.0 \times 10^{-6}$ M **b.** $[H_3O^+] = 5.0 \times 10^{-3}$ M

c. $[H_3O^+] = 7.5 \times 10^{-8}$ M **d.** $[OH^-] = 2.5 \times 10^{-10}$ M

Answers **a.** 5.40 **b.** 11.70 **c.** 6.88 **d.** 9.60

Guide to Calculating $[H_3O^+]$ from pH	
STEP 1	State the given and needed quantities.
STEP 2	Enter the pH value into the inverse log equation and calculate.
STEP 3	Adjust the SFs for the coefficient.

♦ Learning Exercise 14.6D

KEY MATH SKILL
Calculating $[H_3O^+]$ from pH

Complete the following table:

	$[H_3O^+]$	$[OH^-]$	pH
a.	______	1×10^{-12} M	______
b.	______	______	8.0
c.	5.0×10^{-11} M	______	______
d.	______	______	7.80
e.	______	______	4.25
f.	2.0×10^{-10} M	______	______

Answers

	$[H_3O^+]$	$[OH^-]$	pH
a.	1×10^{-2} M	1×10^{-12} M	2.0
b.	1×10^{-8} M	1×10^{-6} M	8.0
c.	5.0×10^{-11} M	2.0×10^{-4} M	10.30
d.	1.6×10^{-8} M	6.3×10^{-7} M	7.80
e.	5.6×10^{-5} M	1.8×10^{-10} M	4.25
f.	2.0×10^{-10} M	5.0×10^{-5} M	9.70

14.7   Reactions of Acids and Bases

Learning Goal: Write balanced equations for reactions of acids with metals, carbonates or bicarbonates, and bases.

- Acids react with many metals to yield hydrogen gas, H_2, and a salt.
- Acids react with carbonates and bicarbonates to yield CO_2, H_2O, and a salt.
- Acids neutralize bases in a reaction that produces water and a salt.
- The net ionic equation for any strong acid–strong base neutralization is $H^+(aq) + OH^-(aq) \longrightarrow H_2O(l)$.
- In a balanced neutralization equation, an equal number of moles of H^+ and OH^- must react.
- A titration is used to determine the volume or concentration of an acid or a base from the laboratory data.

◆ **Learning Exercise 14.7A**

Complete and balance each of the following reactions of acids:

CORE CHEMISTRY SKILL

Writing Equations for Reactions of Acids and Bases

a. ____ $Zn(s)$ + ____ $HCl(aq) \longrightarrow$ ____ + ____ $ZnCl_2(aq)$

b. ____ $HCl(aq)$ + ____ $Li_2CO_3(aq) \longrightarrow$ ____ + ____ + ____

c. ____ $HCl(aq)$ + ____ $NaHCO_3(aq) \longrightarrow$ ____ $CO_2(g)$ + ____ $H_2O(l)$ + ____

d. ____ $Al(s)$ + ____ $H_2SO_4(aq) \longrightarrow$ ____ + ____ $Al_2(SO_4)_3(aq)$

Answers
  a. $Zn(s) + 2HCl(aq) \longrightarrow H_2(g) + ZnCl_2(aq)$
  b. $2HCl(aq) + Li_2CO_3(aq) \longrightarrow CO_2(g) + H_2O(l) + 2LiCl(aq)$
  c. $HCl(aq) + NaHCO_3(aq) \longrightarrow CO_2(g) + H_2O(l) + NaCl(aq)$
  d. $2Al(s) + 3H_2SO_4(aq) \longrightarrow 3H_2(g) + Al_2(SO_4)_3(aq)$

Guide to Balancing an Equation for Neutralization	
STEP 1	Write the reactants and products.
STEP 2	Balance the H^+ in the acid with the OH^- in the base.
STEP 3	Balance the H_2O with the H^+ and the OH^-.
STEP 4	Write the formula of the salt from the remaining ions.

◆ **Learning Exercise 14.7B**

Complete and balance each of the following neutralization reactions:

a. ____ $H_2SO_4(aq)$ + ____ $NaOH(aq) \longrightarrow$ _____ + _____

b. ____ $HCl(aq)$ + ____ $Mg(OH)_2(s) \longrightarrow$ _____ + _____

c. ____ $HNO_3(aq)$ + ____ $Al(OH)_3(s) \longrightarrow$ _____ + _____

d. ____ $H_3PO_4(aq)$ + ____ $Ca(OH)_2(s) \longrightarrow$ _____ + _____

Answers
  a. $H_2SO_4(aq) + 2NaOH(aq) \longrightarrow 2H_2O(l) + Na_2SO_4(aq)$
  b. $2HCl(aq) + Mg(OH)_2(s) \longrightarrow 2H_2O(l) + MgCl_2(aq)$
  c. $3HNO_3(aq) + Al(OH)_3(s) \longrightarrow 3H_2O(l) + Al(NO_3)_3(aq)$
  d. $2H_3PO_4(aq) + 3Ca(OH)_2(s) \longrightarrow 6H_2O(l) + Ca_3(PO_4)_2(s)$

♦ **Learning Exercise 14.7C**

Complete and balance each of the following neutralization reactions:

a. _____ $H_3PO_4(aq)$ + _____ $KOH(aq) \longrightarrow$ _____ $H_2O(l)$ + _____

b. _____ + _____ $NaOH(aq) \longrightarrow$ _____ + _____ $Na_2CO_3(aq)$

c. _____ + _____ $\longrightarrow$ _____ + _____ $AlCl_3(aq)$

d. _____ + _____ $\longrightarrow$ _____ + _____ $Fe_2(SO_4)_3(aq)$

Answers
    a. $H_3PO_4(aq) + 3KOH(aq) \longrightarrow 3H_2O(l) + K_3PO_4(aq)$
    b. $H_2CO_3(aq) + 2NaOH(aq) \longrightarrow 2H_2O(l) + Na_2CO_3(aq)$
    c. $3HCl(aq) + Al(OH)_3(s) \longrightarrow 3H_2O(l) + AlCl_3(aq)$
    d. $3H_2SO_4(aq) + 2Fe(OH)_3(s) \longrightarrow 6H_2O(l) + Fe_2(SO_4)_3(aq)$

14.8   Acid–Base Titration

Learning Goal: Calculate the molarity or volume of an acid or base solution from titration information.

- In a laboratory procedure called titration, an acid or base sample is neutralized.
- A titration is used to determine the volume or concentration of an acid or a base from the laboratory data.

Guide to Calculations for an Acid–Base Titration	
STEP 1	State the given and needed quantities and concentrations.
STEP 2	Write a plan to calculate the molarity or volume.
STEP 3	State equalities and conversion factors, including concentrations.
STEP 4	Set up the problem to calculate the needed quantity.

Example: A 0.0157-L sample of a 0.165 M H_2SO_4 solution reacts completely with a 0.187 M NaOH solution. How many liters of the NaOH solution are needed?

$$H_2SO_4(aq) + 2NaOH(aq) \longrightarrow 2H_2O(l) + Na_2SO_4(aq)$$

Solution: **STEP 1**   **State the given and needed quantities and concentrations.**

	Given	Need	Connect
Analyze the Problem	0.0157 L of 0.165 M H_2SO_4 solution, 0.187 M NaOH solution	mL of NaOH solution	conversion factors (molarity, mole–mole)

    STEP 2   **Write a plan to calculate the volume.**

                Molarity of   Mole–mole   Molarity of
                H_2SO_4 factor   NaOH

    L of H_2SO_4 solution $\longrightarrow$ moles of H_2SO_4 $\longrightarrow$ moles of NaOH $\longrightarrow$ L of NaOH solution

STEP 3   State equalities and conversion factors, including concentrations.

$$1 \text{ L of } H_2SO_4 \text{ solution} = 0.165 \text{ mol of } H_2SO_4 \qquad 2 \text{ mol of NaOH} = 1 \text{ mol of } H_2SO_4$$

$$\frac{0.165 \text{ mol } H_2SO_4}{1 \text{ L } H_2SO_4 \text{ solution}} \text{ and } \frac{1 \text{ L } H_2SO_4 \text{ solution}}{0.165 \text{ mol } H_2SO_4} \qquad \frac{2 \text{ mol NaOH}}{1 \text{ mol } H_2SO_4} \text{ and } \frac{1 \text{ mol } H_2SO_4}{2 \text{ mol NaOH}}$$

$$1 \text{ L of NaOH solution} = 0.187 \text{ mol of NaOH}$$

$$\frac{0.187 \text{ mol NaOH}}{1 \text{ L NaOH solution}} \text{ and } \frac{1 \text{ L NaOH solution}}{0.187 \text{ mol NaOH}}$$

STEP 4   Set up the problem to calculate the needed quantity.

$$0.0157 \text{ L } \cancel{H_2SO_4 \text{ solution}} \times \frac{0.165 \text{ mol } \cancel{H_2SO_4}}{1 \text{ L } \cancel{H_2SO_4 \text{ solution}}} \times \frac{2 \text{ mol } \cancel{NaOH}}{1 \text{ mol } \cancel{H_2SO_4}} \times \frac{1 \text{ L NaOH solution}}{0.187 \text{ mol } \cancel{NaOH}}$$

$$= 0.0277 \text{ L of NaOH solution}$$

♦ **Learning Exercise 14.8**

Solve the following problems using the titration data given:

🔧 **CORE CHEMISTRY SKILL**

Calculating Molarity or Volume of an Acid or Base in a Titration

a. A 5.00-mL sample of HCl solution is placed in a flask. If 15.0 mL of a 0.200 M NaOH solution is required for neutralization, what is the molarity of the HCl solution?

$$HCl(aq) + NaOH(aq) \longrightarrow H_2O(l) + NaCl(aq)$$

b. How many milliliters of a 0.200 M KOH solution are required to neutralize completely 8.50 mL of a 0.500 M H_2SO_4 solution?

$$H_2SO_4(aq) + 2KOH(aq) \longrightarrow 2H_2O(l) + K_2SO_4(aq)$$

c. A 10.0-mL sample of H_3PO_4 solution is placed in a flask. If titration requires 42.0 mL of a 0.100 M NaOH solution for complete neutralization, what is the molarity of the H_3PO_4 solution?

$$H_3PO_4(aq) + 3NaOH(aq) \longrightarrow 3H_2O(l) + Na_3PO_4(aq)$$

d. A 24.6-mL sample of HCl solution reacts completely with 33.0 mL of a 0.222 M NaOH solution. What is the molarity of the HCl solution (see reaction in part **a**)?

Answers   **a.** 0.600 M HCl solution   **b.** 42.5 mL
              c. 0.140 M H_3PO_4 solution   **d.** 0.298 M HCl solution

14.9 Buffers

Learning Goal: Describe the role of buffers in maintaining the pH of a solution; calculate the pH of a buffer.

- A buffer solution resists a change in pH when small amounts of acid or base are added.
- A buffer contains either (1) a weak acid and its salt or (2) a weak base and its salt.
- The weak acid reacts with added OH^-, and the anion of the salt reacts with added H_3O^+, which maintains the pH of the solution.
- The weak base reacts with added H_3O^+, and the cation of the salt reacts with added OH^-, which maintains the pH of the solution.

Key Terms for Sections 14.1 to 14.9

Match each of the following key terms with the correct description:

    a. acid   **b.** base   **c.** pH   **d.** neutralization
    e. buffer   **f.** K_a **g.** dissociation

1. _____ a substance that forms OH^- in water and/or accepts H^+

2. _____ a reaction between an acid and a base to form water and a salt

3. _____ a substance that forms H^+ in water

4. _____ a mixture of a weak acid (or base) and its salt that maintains the pH of a solution

5. _____ a measure of the acidity of a solution

6. _____ the separation of a substance into ions in water

7. _____ the product of the molar concentrations of the ions from a weak acid divided by the molar concentration of the weak acid

Answers   **1.** b   **2.** d   **3.** a   **4.** e
                5. c   **6.** g   **7.** f

♦ **Learning Exercise 14.9A**

State whether each of the following represents a buffer system and explain why:

a. $HCl + NaCl$ **b.** K_2SO_4

c. H_2CO_3 **d.** $H_2CO_3 + NaHCO_3$

Answers   **a.** No. A strong acid and its salt do not make a buffer.
                b. No. A salt alone cannot act as a buffer.
                c. No. A weak acid alone cannot act as a buffer.
                d. Yes. A weak acid and its salt act as a buffer system.

♦ **Learning Exercise 14.9B**

A buffer system contains dihydrogen phosphate, $H_2PO_4^-$, and its conjugate base, hydrogen phosphate, HPO_4^{2-}.

$$H_2PO_4^-(aq) + H_2O(l) \rightleftharpoons H_3O^+(aq) + HPO_4^{2-}(aq)$$

a. The purpose of this buffer system is to _____
   1. maintain $H_2PO_4^-$ **2.** maintain HPO_4^{2-} **3.** maintain pH

b. The weak acid is needed to _____
   1. provide the conjugate base   **2.** neutralize added OH^- **3.** provide the conjugate acid

c. If H_3O^+ is added, it is neutralized by _____
   1. HPO_4^{2-} **2.** H_2O **3.** OH^-

d. When OH^- is added, the equilibrium shifts in the direction of _____
   1. the reactants   **2.** the products   **3.** does not change

Answers   **a.** 3   **b.** 2   **c.** 1   **d.** 2

Guide to Calculating pH of a Buffer	
STEP 1	State the given and needed quantities.
STEP 2	Write the K_a expression and rearrange for $[H_3O^+]$.
STEP 3	Substitute $[HA]$ and $[A^-]$ into the K_a expression.
STEP 4	Use $[H_3O^+]$ to calculate pH.

♦ **Learning Exercise 14.9C**

The K_a for acetic acid, $HC_2H_3O_2$, is 1.8×10^{-5}.

▷ **CORE CHEMISTRY SKILL**

Calculating the pH of a Buffer

a. What is the pH of a buffer that contains 1.0 M $HC_2H_3O_2$ and 1.0 M $NaC_2H_3O_2$?

b. What is the pH of a buffer that contains 1.0 M $HC_2H_3O_2$ and 0.10 M $NaC_2H_3O_2$?

c. What is the pH of a buffer that contains 0.10 M $HC_2H_3O_2$ and 1.0 M $NaC_2H_3O_2$?

Answers

a. $[H_3O^+] = 1.8 \times 10^{-5} \times \dfrac{[1.0]}{[1.0]} = 1.8 \times 10^{-5}$ M $pH = -\log[1.8 \times 10^{-5}] = 4.74$

b. $[H_3O^+] = 1.8 \times 10^{-5} \times \dfrac{[1.0]}{[0.1]} = 1.8 \times 10^{-4}$ M $pH = -\log[1.8 \times 10^{-4}] = 3.74$

c. $[H_3O^+] = 1.8 \times 10^{-5} \times \dfrac{[0.1]}{[1.0]} = 1.8 \times 10^{-6}$ M $pH = -\log[1.8 \times 10^{-6}] = 5.74$

Checklist for Chapter 14

You are ready to take the Practice Test for Chapter 14. Be sure you have accomplished the following learning goals for this chapter. If not, review the section listed at the end of the goal. Then apply your new skills and understanding to the Practice Test.

After studying Chapter 14, I can successfully:

_____ Describe the properties of Arrhenius acids and bases and write their names. (14.1)

_____ Describe the Brønsted–Lowry concept of acids and bases. (14.2)

_____ Write conjugate acid–base pairs an acid–base reaction. (14.2)

_____ Write equations for the dissociation of strong and weak acids and bases. (14.3)

_____ Write the expression for the dissociation of a weak acid or base. (14.4)

_____ Use K_a values to compare the strengths of acids. (14.4)

_____ Use the water dissociation expression to calculate $[H_3O^+]$ and $[OH^-]$. (14.5)

_____ Calculate pH from the $[H_3O^+]$ or $[OH^-]$ of a solution. (14.6)

_____ Write a balanced equation for the reactions of acids with metals, carbonates or bicarbonates, and bases. (14.7)

_____ Calculate the molarity or volume of an acid or base from titration information. (14.8)

_____ Describe the role of buffers in maintaining the pH of a solution. (14.9)

_____ Calculate the pH of a buffer solution. (14.9)

Practice Test for Chapter 14

The chapter sections to review are shown in parentheses at the end of each question.

1. An acid is a compound that when placed in water yields this characteristic ion. (14.1)
   A. H_3O^+ **B.** OH^- **C.** Na^+ **D.** Cl^- **E.** CO_3^{2-}

2. $MgCl_2$ would be classified as a(an) (14.1)
   A. acid   **B.** base   **C.** salt   **D.** buffer   **E.** nonelectrolyte

3. $Mg(OH)_2$ would be classified as a (14.1)
   A. weak acid   **B.** strong base   **C.** salt   **D.** buffer   **E.** nonelectrolyte

4. Which of the following would turn litmus blue? (14.1)
   A. HCl   **B.** NH_4Cl **C.** Na_2SO_4 **D.** KOH   **E.** $NaNO_3$

5. What is the name of $HClO_3$? (14.1)
   A. hypochlorous acid   **B.** chloric acid   **C.** chlorous acid
   D. chloric trioxide acid   **E.** perchloric acid

6. What is the name of NH_4OH? (14.1)
   A. ammonium oxide
   B. nitrogen tetrahydride hydroxide
   C. ammonium hydroxide
   D. perammonium hydroxide
   E. amine oxide hydride

7. Which of the following pairs is a conjugate acid–base pair? (14.2)
   A. HCl/HNO_3 **B.** HNO_2/NO_2^- **C.** $NaOH/KOH$ **D.** HSO_4^-/HCO_3^- **E.** H_2S/S^{2-}

8. The conjugate base of HSO_4^- is (14.2)
   A. SO_4^{2-} B. H_2SO_4 C. HS^- D. H_2S E. SO_3^{2-}

9. In which of the following reactions does H_2O act as an acid? (14.2)
   A. $H_3PO_4(aq) + H_2O(l) \longrightarrow H_3O^+(aq) + H_2PO_4^-(aq)$
   B. $H_2SO_4(aq) + H_2O(l) \longrightarrow H_3O^+(aq) + HSO_4^-(aq)$
   C. $H_2O(l) + HS^-(aq) \longrightarrow H_3O^+(aq) + S^{2-}(aq)$
   D. $NaOH(aq) + HCl(aq) \longrightarrow H_2O(l) + NaCl(aq)$
   E. $NH_3(g) + H_2O(l) \longrightarrow NH_4^+(aq) + OH^-(aq)$

10. Which of the following acids has the smallest K_a value? (14.3)
    A. HNO_3 B. H_2SO_4 C. HCl D. H_2CO_3 E. HBr

11. Using the following K_a values, identify the strongest acid in the group: (14.3)
    A. 7.5×10^{-3} B. 1.8×10^{-5} C. 4.5×10^8 D. 4.9×10^{-10} E. 3.2×10^4

12. Acetic acid is a weak acid because (14.3)
    A. it forms a dilute acid solution   B. it is isotonic
    C. it is slightly dissociated in water   D. it is a nonpolar molecule
    E. it can form a buffer

13. A weak base when added to water (14.3)
    A. makes the solution slightly basic   B. does not affect the pH
    C. dissociates completely   D. does not dissociate
    E. makes the solution slightly acidic

14. The weak acid formic acid dissociates in water.

$$HCHO_2(aq) + H_2O(l) \rightleftharpoons H_3O^+(aq) + CHO_2^-(aq)$$

The correctly written acid dissociation expression is: (14.4)

A. $K_a = \dfrac{[HCHO_2]}{[H_3O^+][CHO_2^-]}$ B. $K_a = \dfrac{[HCHO_2]}{[CHO_2^-]}$

C. $K_a = \dfrac{[H_3O^+][CHO_2^-]}{[HCHO_2][H_2O]}$ D. $K_a = \dfrac{[H_3O^+][CHO_2^-]}{[HCHO_2]}$

E. $K_a = \dfrac{[HCHO_2][H_2O]}{[H_3O^+][CHO_2^-]}$

15. In the K_w expression for pure H_2O, the $[H_3O^+]$ has the value (14.5)
    A. 1.0×10^{-7} M   B. 1.0×10^{-1} M   C. 1.0×10^{-14} M
    D. 1.0×10^{-6} M   E. 1.0×10^{12} M

For questions 16 and 17, consider a solution with $[H_3O^+] = 1 \times 10^{-11}$ M. (14.5)

16. The hydroxide ion concentration is
    A. 1×10^{-1} M   B. 1×10^{-3} M   C. 1×10^{-4} M   D. 1×10^{-7} M   E. 1×10^{-11} M

17. The solution is
    A. acidic   B. basic   C. neutral   D. a buffer   E. neutralized

For questions 18 and 19, consider a solution with $[OH^-] = 1 \times 10^{-5}$ M. (14.5)

18. The hydrogen ion concentration of the solution is
    A. 1×10^{-5} M   B. 1×10^{-7} M   C. 1×10^{-9} M   D. 1×10^{-10} M   E. 1×10^{-14} M

19. The solution is
    A. acidic   B. basic   C. neutral   D. a buffer   E. neutralized

20. What is the $[H_3O^+]$ of a solution that has a pH = 8.7? (14.6)
    A. 7×10^{-8} **B.** 2×10^{-9} **C.** 2×10^9 **D.** 8×10^{-7} **E.** 2×10^{-8}

21. Of the following pH values, which is the most acidic? (14.6)
    A. 8.0   **B.** 5.5   **C.** 1.5   **D.** 3.2   **E.** 9.0

22. Of the following pH values, which is the most basic? (14.6)
    A. 10.0   **B.** 4.0   **C.** 2.2   **D.** 11.5   **E.** 9.0

23. Which is an equation for neutralization of an acid and a base? (14.7)
    A. $CaCO_3(s) \longrightarrow CaO(s) + CO_2(g)$
    B. $Na_2SO_4(s) \longrightarrow 2Na^+(aq) + SO_4^{2-}(aq)$
    C. $H_2SO_4(aq) + 2NaOH(aq) \longrightarrow 2H_2O(l) + Na_2SO_4(aq)$
    D. $Na_2O(s) + SO_3(g) \longrightarrow Na_2SO_4(aq)$
    E. $H_2CO_3(aq) \longrightarrow CO_2(g) + H_2O(l)$

24. What is the molarity of a 10.0-mL sample of HCl solution that is neutralized by 15.0 mL of a 2.0 M NaOH solution? (14.8)
    A. 0.50 M HCl   **B.** 1.0 M HCl   **C.** 1.5 M HCl   **D.** 2.0 M HCl   **E.** 3.0 M HCl

25. In a titration, 6.0 mol of NaOH will completely neutralize _____ mol of H_2SO_4. (14.8)
    A. 1.0   **B.** 2.0   **C.** 3.0   **D.** 6.0   **E.** 11

26. What is the name given to components in the body that keep blood pH within its normal 7.35 to 7.45 range? (14.9)
    A. nutrients   **B.** buffers   **C.** metabolites   **D.** regufluids   **E.** neutralizers

27. A buffer system (14.9)
    A. maintains a pH of 7.0
    B. contains only a weak base
    C. contains only a salt
    D. contains a strong acid and its salt
    E. maintains the pH of a solution

28. Which of the following would act as a buffer system? (14.9)
    A. HCl
    B. Na_2CO_3
    C. $NaOH + NaNO_3$
    D. NH_4OH
    E. $KHCO_3 + H_2CO_3$

Answers to the Practice Test

1. A	**2.** C	**3.** B	**4.** D	**5.** B
6. C	**7.** B	**8.** A	**9.** E	**10.** D
11. C	**12.** C	**13.** A	**14.** D	**15.** A
16. B	**17.** B	**18.** C	**19.** B	**20.** B
21. C	**22.** D	**23.** C	**24.** E	**25.** C
26. B	**27.** E	**28.** E		

Selected Answers and Solutions to Text Problems

14.1 **a.** Acids taste sour.
  b. Acids neutralize bases.
  c. Acids produce H^+ ions in water.
  d. Barium hydroxide is the name of a base.
  e. Both acids and bases are electrolytes.

14.3 Acids containing a simple nonmetal anion use the prefix *hydro*, followed by the name of the anion with its *ide* ending changed to *ic acid*. When the anion is an oxygen-containing polyatomic ion, the *ate* ending of the polyatomic anion is replaced with *ic acid*. Acids with one oxygen less than the common *ic acid* name are named as *ous acids*. Bases are named as ionic compounds containing hydroxide anions.
  a. hydrochloric acid   **b.** calcium hydroxide   **c.** perchloric acid
  d. nitric acid   **e.** sulfurous acid   **f.** bromous acid

14.5 **a.** RbOH   **b.** HF   **c.** H_3PO_4
  d. LiOH   **e.** NH_4OH **f.** HIO_4

14.7 A Brønsted–Lowry acid donates a hydrogen ion (H^+), whereas a Brønsted–Lowry base accepts a hydrogen ion.
  a. HI is the acid (H^+ donor); H_2O is the base (H^+ acceptor).
  b. H_2O is the acid (H^+ donor); F^- is the base (H^+ acceptor).
  c. H_2S is the acid (H^+ donor); $CH_3-CH_2-NH_2$ is the base (H^+ acceptor).

14.9 To form the conjugate base, remove a hydrogen ion (H^+) from the acid.
  a. F^- **b.** OH^- **c.** HPO_3^{2-}
  d. SO_4^{2-} **e.** ClO_2^-

14.11 To form the conjugate acid, add a hydrogen ion (H^+) to the base.
  a. HCO_3^- **b.** H_3O^+ **c.** H_3PO_4
  d. HBr   **e.** $HClO_4$

14.13 The conjugate acid is an H^+ donor, and the conjugate base is an H^+ acceptor.
  a. In the reaction, the acid H_2CO_3 donates an H^+ to the base H_2O. The conjugate acid–base pairs are H_2CO_3/HCO_3^- and H_3O^+/H_2O.
  b. In the reaction, the acid NH_4^+ donates an H^+ to the base H_2O. The conjugate acid–base pairs are NH_4^+/NH_3 and H_3O^+/H_2O.
  c. In the reaction, the acid HCN donates an H^+ to the base NO_2^-. The conjugate acid–base pairs are HCN/CN^- and HNO_2/NO_2^-.
  d. In the reaction, the acid HF donates an H^+ to the base CHO_2^-. The conjugate acid–base pairs are HF/F^- and $HCHO_2/CHO_2^-$.

14.15 $NH_4^+(aq) + H_2O(l) \rightleftharpoons NH_3(aq) + H_3O^+(aq)$

14.17 A strong acid is a good H^+ donor, whereas its conjugate base is a poor H^+ acceptor.

14.19 Use Table 14.3 to answer (the stronger acid will be closer to the top of the table).
  a. HBr is the stronger acid.
  b. HSO_4^- is the stronger acid.
  c. H_2CO_3 is the stronger acid.

14.21 Use Table 14.3 to answer (the weaker acid will be closer to the bottom of the table).
a. HSO_4^- is the weaker acid.
b. HF is the weaker acid.
c. HCO_3^- is the weaker acid.

14.23 a. From Table 14.3, we see that H_2CO_3 is a weaker acid than H_3O^+ and that H_2O is a weaker base than HCO_3^-. Thus, the solution will contain mostly reactants at equilibrium.
b. From Table 14.3, we see that NH_4^+ is a weaker acid than H_3O^+ and that H_2O is a weaker base than NH_3. Thus, the solution will contain mostly reactants at equilibrium.
c. From Table 14.3, we see that NH_4^+ is a weaker acid than HNO_2 and that NO_2^- is a weaker base than NH_3. Thus, the solution will contain mostly products at equilibrium.

14.25 $NH_4^+(aq) + SO_4^{2-}(aq) \rightleftharpoons NH_3(aq) + HSO_4^-(aq)$

This equilibrium contains mostly reactants because NH_4^+ is a weaker acid than HSO_4^-, and SO_4^{2-} is a weaker base than NH_3.

14.27 a. True
b. False; a strong acid has a large value of K_a.
c. False; a strong acid has a weak conjugate base.
d. True
e. False; a strong acid is completely dissociated in aqueous solution.

14.29 The smaller the K_a value, the weaker the acid. The weaker acid has the stronger conjugate base.
a. H_2SO_3, which has a larger K_a value than HS^-, is the stronger acid.
b. The conjugate base HSO_3^- is formed by removing an H^+ from the acid H_2SO_3.
c. The stronger acid, H_2SO_3, has the weaker conjugate base, HSO_3^-.
d. The weaker acid, HS^-, has the stronger conjugate base, S^{2-}.
e. The stronger acid, H_2SO_3, dissociates more and produces more ions.

14.31 $H_3PO_4(aq) + H_2O(l) \rightleftharpoons H_3O^+(aq) + H_2PO_4^-(aq)$

The K_a expression is the ratio of the [products] divided by the [reactants] with [H_2O] considered constant and part of the K_a:

$$K_a = \frac{[H_3O^+][H_2PO_4^-]}{[H_3PO_4]} = 7.5 \times 10^{-3}$$

14.33 In pure water, a small fraction of the water molecules break apart to form H^+ and OH^-. The H^+ combines with H_2O to form H_3O^+. Every time an H^+ is formed, an OH^- is also formed. Therefore, the concentration of the two must be equal in pure water.

14.35 In an acidic solution, the [H_3O^+] is greater than the [OH^-], which means that at 25 °C, the [H_3O^+] is greater than 1.0×10^{-7} M and the [OH^-] is less than 1.0×10^{-7} M.

14.37 The value of $K_w = [H_3O^+][OH^-] = 1.0 \times 10^{-14}$ at 25 °C.

If [H_3O^+] needs to be calculated from [OH^-], then rearranging the K_w for [H_3O^+] gives

$$[H_3O^+] = \frac{1.0 \times 10^{-14}}{[OH^-]}.$$

If [OH^-] needs to be calculated from [H_3O^+], then rearranging the K_w for [OH^-] gives

$$[OH^-] = \frac{1.0 \times 10^{-14}}{[H_3O^+]}.$$

A neutral solution has $[OH^-] = [H_3O^+]$. If the $[OH^-] > [H_3O^+]$, the solution is basic; if the $[H_3O^+] > [OH^-]$, the solution is acidic.

a. $[OH^-] = \dfrac{1.0 \times 10^{-14}}{[H_3O^+]} = \dfrac{1.0 \times 10^{-14}}{[2.0 \times 10^{-5}]} = 5.0 \times 10^{-10}$ M; since $[H_3O^+] > [OH^-]$,

the solution is acidic.

b. $[OH^-] = \dfrac{1.0 \times 10^{-14}}{[H_3O^+]} = \dfrac{1.0 \times 10^{-14}}{[1.4 \times 10^{-9}]} = 7.1 \times 10^{-6}$ M; since $[OH^-] > [H_3O^+]$,

the solution is basic.

c. $[H_3O^+] = \dfrac{1.0 \times 10^{-14}}{[OH^-]} = \dfrac{1.0 \times 10^{-14}}{[8.0 \times 10^{-3}]} = 1.3 \times 10^{-12}$ M; since $[OH^-] > [H_3O^+]$,

the solution is basic.

d. $[H_3O^+] = \dfrac{1.0 \times 10^{-14}}{[OH^-]} = \dfrac{1.0 \times 10^{-14}}{[3.5 \times 10^{-10}]} = 2.9 \times 10^{-5}$ M; since $[H_3O^+] > [OH^-]$,

the solution is acidic.

14.39 The value of $K_w = [H_3O^+][OH^-] = 1.0 \times 10^{-14}$ at 25 °C.

When $[OH^-]$ is known, the $[H_3O^+]$ can be calculated by rearranging the K_w for $[H_3O^+]$:

$$[H_3O^+] = \dfrac{1.0 \times 10^{-14}}{[OH^-]}$$

a. $[H_3O^+] = \dfrac{1.0 \times 10^{-14}}{[OH^-]} = \dfrac{1.0 \times 10^{-14}}{[1.0 \times 10^{-9}]} = 1.0 \times 10^{-5}$ M (2 SFs)

b. $[H_3O^+] = \dfrac{1.0 \times 10^{-14}}{[OH^-]} = \dfrac{1.0 \times 10^{-14}}{[1.0 \times 10^{-6}]} = 1.0 \times 10^{-8}$ M (2 SFs)

c. $[H_3O^+] = \dfrac{1.0 \times 10^{-14}}{[OH^-]} = \dfrac{1.0 \times 10^{-14}}{[2.0 \times 10^{-5}]} = 5.0 \times 10^{-10}$ M (2 SFs)

d. $[H_3O^+] = \dfrac{1.0 \times 10^{-14}}{[OH^-]} = \dfrac{1.0 \times 10^{-14}}{[4.0 \times 10^{-13}]} = 2.5 \times 10^{-2}$ M (2 SFs)

14.41 The value of $K_w = [H_3O^+][OH^-] = 1.0 \times 10^{-14}$ at 25 °C.

When $[H_3O^+]$ is known, the $[OH^-]$ can be calculated by rearranging the K_w for $[OH^-]$:

$$[OH^-] = \dfrac{1.0 \times 10^{-14}}{[H_3O^+]}$$

a. $[OH^-] = \dfrac{1.0 \times 10^{-14}}{[H_3O^+]} = \dfrac{1.0 \times 10^{-14}}{[4.0 \times 10^{-2}]} = 2.5 \times 10^{-13}$ M (2 SFs)

b. $[OH^-] = \dfrac{1.0 \times 10^{-14}}{[H_3O^+]} = \dfrac{1.0 \times 10^{-14}}{[5.0 \times 10^{-6}]} = 2.0 \times 10^{-9}$ M (2 SFs)

c. $[OH^-] = \dfrac{1.0 \times 10^{-14}}{[H_3O^+]} = \dfrac{1.0 \times 10^{-14}}{[2.0 \times 10^{-4}]} = 5.0 \times 10^{-11}$ M (2 SFs)

d. $[OH^-] = \dfrac{1.0 \times 10^{-14}}{[H_3O^+]} = \dfrac{1.0 \times 10^{-14}}{[7.9 \times 10^{-9}]} = 1.3 \times 10^{-6}$ M (2 SFs)

14.43 In a neutral solution, the $[H_3O^+] = 1.0 \times 10^{-7}$ M at 25 °C.

$pH = -\log[H_3O^+] = -\log[1.0 \times 10^{-7}] = 7.00$. The pH value contains two *decimal places*, which represent the two significant figures in the coefficient 1.0.

14.45 An acidic solution has a pH less than 7.0 (or a pOH greater than 7.0). A basic solution has a pH greater than 7.0 (or a pOH less than 7.0). A neutral solution has a pH (or pOH) equal to 7.0.

a. basic (pH 7.38 > 7.0)   **b.** acidic (pH 2.8 < 7.0)

c. basic (pOH 2.8 < 7.0)   **d.** acidic (pH 5.52 < 7.0)

e. acidic (pH 4.2 < 7.0)   **f.** basic (pH 7.6 > 7.0)

14.47 Since pH is a logarithmic scale, an increase or decrease of 1 pH unit changes the $[H_3O^+]$ by a factor of 10. Thus, a pH of 3 ($[H_3O^+] = 10^{-3}$ M, or 0.001 M) is 10 times more acidic than a pH of 4 ($[H_3O^+] = 10^{-4}$ M, or 0.0001 M).

14.49 $pH = -\log[H_3O^+]$

Since the value of $K_w = [H_3O^+][OH^-] = 1.0 \times 10^{-14}$ at 25 °C, if $[H_3O^+]$ needs to be

calculated from $[OH^-]$, rearranging the K_w for $[H_3O^+]$ gives $[H_3O^+] = \dfrac{1.0 \times 10^{-14}}{[OH^-]}$.

a. $pH = -\log[H_3O^+] = -\log[1 \times 10^{-4}] = 4.0$ (1 SF on the right of the decimal point)

b. $pH = -\log[H_3O^+] = -\log[3 \times 10^{-9}] = 8.5$ (1 SF on the right of the decimal point)

c. $[H_3O^+] = \dfrac{1.0 \times 10^{-14}}{[1 \times 10^{-5}]} = 1 \times 10^{-9}$ M

$pH = -\log[1 \times 10^{-9}] = 9.0$ (1 SF on the right of the decimal point)

d. $[H_3O^+] = \dfrac{1.0 \times 10^{-14}}{[2.5 \times 10^{-11}]} = 4.0 \times 10^{-4}$ M

$pH = -\log[4.0 \times 10^{-4}] = 3.40$ (2 SFs on the right of the decimal point)

e. $pH = -\log[H_3O^+] = -\log[6.7 \times 10^{-8}] = 7.17$ (2 SFs on the right of the decimal point)

f. $[H_3O^+] = \dfrac{1.0 \times 10^{-14}}{[8.2 \times 10^{-4}]} = 1.2 \times 10^{-11}$ M

$pH = -\log[1.2 \times 10^{-11}] = 10.92$ (2 SFs on the right of the decimal point)

14.51 $[H_3O^+] = \dfrac{1.0 \times 10^{-14}}{[OH^-]}$; $[OH^-] = \dfrac{1.0 \times 10^{-14}}{[H_3O^+]}$; $pH = -\log[H_3O^+]$; $pOH = -\log[OH^-]$

$[H_3O^+] = 10^{-pH}$; $[OH^-] = 10^{-pOH}$

$[H_3O^+]$	$[OH^-]$	pH	pOH	Acidic, Basic, or Neutral?
1.0×10^{-8} M	1.0×10^{-6} M	8.00	6.00	Basic
3.2×10^{-4} M	3.1×10^{-11} M	3.49	10.51	Acidic
2.8×10^{-5} M	3.6×10^{-10} M	4.55	9.45	Acidic
1.0×10^{-12} M	1.0×10^{-2} M	12.00	2.00	Basic

$[H_3O^+]$ and $[OH^-]$ all have 2 SFs here, so all pH/pOH values have 2 decimal places on the right of the decimal point.

14.53 $[H_3O^+] = 10^{-pH} = 10^{-6.92} = 1.2 \times 10^{-7}$ M (2 SFs)

14.55 Acids react with active metals to form $H_2(g)$ and a salt of the metal. The reaction of acids with carbonates or bicarbonates yields CO_2, H_2O, and a salt. In a neutralization reaction, an acid and a base react to form H_2O and a salt.
    a. $ZnCO_3(s) + 2HBr(aq) \longrightarrow CO_2(g) + H_2O(l) + ZnBr_2(aq)$
    b. $Zn(s) + 2HCl(aq) \longrightarrow H_2(g) + ZnCl_2(aq)$
    c. $HCl(aq) + NaHCO_3(s) \longrightarrow CO_2(g) + H_2O(l) + NaCl(aq)$
    d. $H_2SO_4(aq) + Mg(OH)_2(s) \longrightarrow 2H_2O(l) + MgSO_4(aq)$

14.57 In balancing a neutralization equation, the number of H^+ and OH^- must be equalized by placing coefficients in front of the formulas for the acid and base.
    a. $2HCl(aq) + Mg(OH)_2(s) \longrightarrow 2H_2O(l) + MgCl_2(aq)$
    b. $H_3PO_4(aq) + 3LiOH(aq) \longrightarrow 3H_2O(l) + Li_3PO_4(aq)$

14.59 The products of a neutralization are water and a salt. In balancing a neutralization equation, the number of H^+ and OH^- must be equalized by placing coefficients in front of the formulas for the acid and base.
    a. $H_2SO_4(aq) + 2NaOH(aq) \longrightarrow 2H_2O(l) + Na_2SO_4(aq)$
    b. $3HCl(aq) + Fe(OH)_3(s) \longrightarrow 3H_2O(l) + FeCl_3(aq)$
    c. $H_2CO_3(aq) + Mg(OH)_2(s) \longrightarrow 2H_2O(l) + MgCO_3(s)$

14.61 To a measured volume of the formic acid solution, add a few drops of indicator. Place a solution of NaOH of known molarity in a buret. Add NaOH to the acid solution until one drop changes the color of the solution. Use the volume and molarity of the NaOH solution to calculate the moles of NaOH added to reach the endpoint. This equals the moles of formic acid in the sample. Then calculate the molarity of the formic acid solution from the calculated moles of formic acid and the known volume you started with.

14.63 In the titration equation, 1 mol of HCl reacts with 1 mol of NaOH.

$$28.6 \text{ mL NaOH solution} \times \frac{1 \text{ L solution}}{1000 \text{ mL solution}} \times \frac{0.145 \text{ mol NaOH}}{1 \text{ L solution}} \times \frac{1 \text{ mol HCl}}{1 \text{ mol NaOH}}$$

$$= 0.004 \ 15 \text{ mol of HCl}$$

$$5.00 \text{ mL HCl solution} \times \frac{1 \text{ L solution}}{1000 \text{ mL solution}} = 0.005 \ 00 \text{ L of HCl solution}$$

$$\text{Molarity (M) of HCl} = \frac{\text{moles of solute}}{\text{liters of solution}} = \frac{0.004 \ 15 \text{ mol HCl}}{0.005 \ 00 \text{ L solution}}$$

$$= 0.830 \text{ M HCl solution (3 SFs)}$$

14.65 In the titration equation, 1 mol of H_2SO_4 reacts with 2 mol of KOH.

$$38.2 \text{ mL KOH solution} \times \frac{1 \text{ L solution}}{1000 \text{ mL solution}} \times \frac{0.163 \text{ mol KOH}}{1 \text{ L solution}} \times \frac{1 \text{ mol H}_2\text{SO}_4}{2 \text{ mol KOH}}$$

$$= 0.003 \ 11 \text{ mol of H}_2\text{SO}_4$$

$$25.0 \text{ mL H}_2\text{SO}_4 \text{ solution} \times \frac{1 \text{ L solution}}{1000 \text{ mL solution}} = 0.0250 \text{ L of H}_2\text{SO}_4 \text{ solution}$$

$$\text{Molarity (M) of H}_2\text{SO}_4 = \frac{\text{moles of solute}}{\text{liters of solution}} = \frac{0.003 \ 11 \text{ mol H}_2\text{SO}_4}{0.0250 \text{ L solution}}$$

$$= 0.124 \text{ M H}_2\text{SO}_4 \text{ solution (3 SFs)}$$

14.67 In the titration equation, 1 mol of H_3PO_4 reacts with 3 mol of NaOH.

$$50.0 \text{ mL } H_3PO_4 \text{ solution} \times \frac{1 \text{ L } H_3PO_4 \text{ solution}}{1000 \text{ mL } H_3PO_4 \text{ solution}} \times \frac{0.0224 \text{ mol } H_3PO_4}{1 \text{ L } H_3PO_4 \text{ solution}} \times \frac{3 \text{ mol NaOH}}{1 \text{ mol } H_3PO_4}$$

$$\times \frac{1 \text{ L NaOH solution}}{0.204 \text{ mol NaOH}} \times \frac{1000 \text{ mL NaOH solution}}{1 \text{ L NaOH solution}} = 16.5 \text{ mL of NaOH solution (3 SFs)}$$

14.69 A buffer system contains a weak acid and a salt containing its conjugate base (or a weak base and a salt containing its conjugate acid).

  a. This is not a buffer system because it only contains the strong base NaOH and the neutral salt NaCl.

  b. This is a buffer system; it contains the weak acid H_2CO_3 and a salt containing its conjugate base HCO_3^-.

  c. This is a buffer system; it contains the weak acid HF and a salt containing its conjugate base F^-.

  d. This is not a buffer system because it only contains the neutral salts KCl and NaCl.

14.71 **a.** The purpose of this buffer system is to (3) maintain pH.

  b. The salt of the weak acid is needed to (1) provide the conjugate base and (2) neutralize added H_3O^+.

  c. If OH^- is added, it is neutralized by (3) H_3O^+.

  d. When H_3O^+ is added, the equilibrium shifts in the direction of the (1) reactants.

14.73 $HNO_2(aq) + H_2O(l) \rightleftharpoons NO_2^-(aq) + H_3O^+(aq)$

Rearrange the K_a for $[H_3O^+]$ and use it to calculate the pH.

$$[H_3O^+] = K_a \times \frac{[HNO_2]}{[NO_2^-]} = 4.5 \times 10^{-4} \times \frac{[0.10 \text{ M}]}{[0.10 \text{ M}]} = 4.5 \times 10^{-4} \text{ M}$$

$$pH = -\log[H_3O^+] = -\log[4.5 \times 10^{-4}] = 3.35 \text{ (2 SFs on the right of the decimal point)}$$

14.75 $HF(aq) + H_2O(l) \rightleftharpoons F^-(aq) + H_3O^+(aq)$

Rearrange the K_a for $[H_3O^+]$ and use it to calculate the pH.

$$[H_3O^+] = K_a \times \frac{[HF]}{[F^-]} = 3.5 \times 10^{-4} \times \frac{[0.10 \text{ M}]}{[0.10 \text{ M}]} = 3.5 \times 10^{-4} \text{ M}$$

$$pH = -\log[3.5 \times 10^{-4}] = 3.46 \text{ (2 SFs on the right of the decimal point)}$$

$$[H_3O^+] = K_a \times \frac{[HF]}{[F^-]} = 3.5 \times 10^{-4} \times \frac{[0.060 \text{ M}]}{[0.120 \text{ M}]} = 1.75 \times 10^{-4} \text{ M}$$

$$pH = -\log[1.75 \times 10^{-4}] = 3.76 \text{ (2 SFs on the right of the decimal point)}$$

∴ The solution with 0.10 M HF/0.10 M NaF is more acidic.

14.77 If you breathe fast, CO_2 is expelled and the equilibrium shifts to lower $[H_3O^+]$, which raises the pH.

14.79 If large amounts of HCO_3^- are lost, the equilibrium shifts to higher $[H_3O^+]$, which lowers the pH.

14.81 $pH = -\log[H_3O^+] = -\log[2.0 \times 10^{-4}] = 3.70$ (2 SFs on the right of the decimal point)

14.83 $[H_3O^+] = 10^{-pH} = 10^{-3.60} = 2.5 \times 10^{-4} \text{ M (2 SFs)}$

14.85 $CaCO_3(s) + 2HCl(aq) \longrightarrow CO_2(g) + H_2O(l) + CaCl_2(aq)$

14.87 From the neutralization equation in problem 14.85, 1 mol of $CaCO_3$ reacts with 2 mol of HCl.

$$100. \text{ mL HCl solution} \times \frac{1 \text{ L HCl solution}}{1000 \text{ mL HCl solution}} \times \frac{0.0400 \text{ mol HCl}}{1 \text{ L HCl solution}} \times \frac{1 \text{ mol } CaCO_3}{2 \text{ mol HCl}}$$

$$\times \frac{100.09 \text{ g } CaCO_3}{1 \text{ mol } CaCO_3} = 0.200 \text{ g of } CaCO_3 \text{ (3 SFs)}$$

14.89 a. This diagram represents a weak acid; only a few HX molecules dissociate into H_3O^+ and X^- ions.
  b. This diagram represents a strong acid; all of the HX molecules dissociate into H_3O^+ and X^- ions.

14.91 a. H_2SO_4 is an acid. **b.** RbOH is a base.
  c. $Ca(OH)_2$ is a base. **d.** HI is an acid.

14.93

Acid	Conjugate Base
H_2O	OH^-
HCN	CN^-
HNO_2	NO_2^-
H_3PO_4	$H_2PO_4^-$

14.95 a. Hyperventilation will lower the CO_2 concentration in the blood, which lowers the $[H_2CO_3]$, which decreases the $[H_3O^+]$ and increases the blood pH.
  b. Breathing into a paper bag will increase the CO_2 concentration in the blood, increase the $[H_2CO_3]$, increase $[H_3O^+]$, and lower the blood pH back toward the normal range.

14.97 An acidic solution has a pH less than 7.0. A neutral solution has a pH equal to 7.0. A basic solution has a pH greater than 7.0.
  a. acidic (pH 5.2 < 7.0) **b.** basic (pH 7.5 > 7.0)
  c. basic (pH 8.1 > 7.0) **d.** acidic (pH 2.5 < 7.0)

14.99 a. acid; bromous acid **b.** base; cesium hydroxide
  c. salt; magnesium nitrate **d.** acid; perchloric acid

14.101

Acid	Conjugate Base
HI	I^-
HCl	Cl^-
NH_4^+	NH_3
H_2S	HS^-

14.103 Use Table 14.3 to answer (the stronger acid will be closer to the top of the table).
  a. HF is the stronger acid.
  b. H_3O^+ is the stronger acid.
  c. HNO_2 is the stronger acid.
  d. HCO_3^- is the stronger acid.

14.105 $[H_3O^+] = \dfrac{1.0 \times 10^{-14}}{[OH^-]}$; $[OH^-] = \dfrac{1.0 \times 10^{-14}}{[H_3O^+]}$;

$pH = -\log[H_3O^+]$; $pOH = -\log[OH^-]$; $pH + pOH = 14.00$

a. $pH = -\log[H_3O^+] = -\log[2.0 \times 10^{-8}] = 7.70$ (2 SFs on the right of the decimal point)
since $pH + pOH = 14.00$, $pOH = 14.00 - pH = 14.00 - 7.70 = 6.30$

b. $pH = -\log[5.0 \times 10^{-2}] = 1.30$ (2 SFs on the right of the decimal point)
since $pH + pOH = 14.00$, $pOH = 14.00 - pH = 14.00 - 1.30 = 12.70$

c. $[H_3O^+] = \dfrac{1.0 \times 10^{-14}}{[3.5 \times 10^{-4}]} = 2.9 \times 10^{-11}$ M

$pH = -\log[2.9 \times 10^{-11}] = 10.54$ (2 SFs on the right of the decimal point)
$pOH = -\log[OH^-] = -\log[3.5 \times 10^{-4}] = 3.46$

d. $[H_3O^+] = \dfrac{1.0 \times 10^{-14}}{[0.0054]} = 1.9 \times 10^{-12}$ M

$pH = -\log[1.9 \times 10^{-12}] = 11.73$ (2 SFs on the right of the decimal point)
$pOH = -\log[OH^-] = -\log[0.0054] = 2.27$

14.107 a. basic ($pH > 7.0$)
b. acidic ($pH < 7.0$)
c. basic ($pH > 7.0$)
d. basic ($pH > 7.0$)

14.109 If the pH is given, the $[H_3O^+]$ can be found by using the relationship $[H_3O^+] = 10^{-pH}$.
The $[OH^-]$ can be found by rearranging $K_w = [H_3O^+][OH^-] = 1 \times 10^{-14}$.

a. $pH = 3.00$; $[H_3O^+] = 10^{-pH} = 10^{-3.00} = 1.0 \times 10^{-3}$ M (2 SFs)

$[OH^-] = \dfrac{1.0 \times 10^{-14}}{[H_3O^+]} = \dfrac{1.0 \times 10^{-14}}{[1.0 \times 10^{-3}]} = 1.0 \times 10^{-11}$ M (2 SFs)

b. $pH = 6.2$; $[H_3O^+] = 10^{-pH} = 10^{-6.2} = 6 \times 10^{-7}$ M (1 SF)

$[OH^-] = \dfrac{1.0 \times 10^{-14}}{[H_3O^+]} = \dfrac{1.0 \times 10^{-14}}{[6 \times 10^{-7}]} = 2 \times 10^{-8}$ M (1 SF)

c. $pH = 8.85$; $[H_3O^+] = 10^{-pH} = 10^{-8.85} = 1.4 \times 10^{-9}$ M (2 SFs)

$[OH^-] = \dfrac{1.0 \times 10^{-14}}{[H_3O^+]} = \dfrac{1.0 \times 10^{-14}}{[1.4 \times 10^{-9}]} = 7.1 \times 10^{-6}$ M (2 SFs)

d. $pH = 11.00$; $[H_3O^+] = 10^{-pH} = 10^{-11.00} = 1.0 \times 10^{-11}$ M (2 SFs)

$[OH^-] = \dfrac{1.0 \times 10^{-14}}{[H_3O^+]} = \dfrac{1.0 \times 10^{-14}}{[1.0 \times 10^{-11}]} = 1.0 \times 10^{-3}$ M (2 SFs)

14.111 a. Solution A, with a pH of 4.5, is more acidic.
b. In solution A, the $[H_3O^+] = 10^{-pH} = 10^{-4.5} = 3 \times 10^{-5}$ M (1 SF)

In solution B, the $[H_3O^+] = 10^{-pH} = 10^{-6.7} = 2 \times 10^{-7}$ M (1 SF)

c. In solution A, the $[OH^-] = \dfrac{1.0 \times 10^{-14}}{[H_3O^+]} = \dfrac{1.0 \times 10^{-14}}{[3 \times 10^{-5}]} = 3 \times 10^{-10}$ M (1 SF)

In solution B, the $[OH^-] = \dfrac{1.0 \times 10^{-14}}{[H_3O^+]} = \dfrac{1.0 \times 10^{-14}}{[2 \times 10^{-7}]} = 5 \times 10^{-8}$ M (1 SF)

14.113 $2.5 \text{ g HCl} \times \dfrac{1 \text{ mol HCl}}{36.46 \text{ g HCl}} \times \dfrac{1 \text{ mol H}_3\text{O}^+}{1 \text{ mol HCl}} = 0.069 \text{ mol of H}_3\text{O}^+ \text{ (2 SFs)}$

$[\text{H}_3\text{O}^+] = \dfrac{0.069 \text{ mol H}_3\text{O}^+}{0.425 \text{ L solution}} = 0.16 \text{ M (2 SFs)}$

$\text{pH} = -\log[\text{H}_3\text{O}^+] = -\log[0.16] = 0.80 \text{ (2 SFs on the right of the decimal point)}$

$\text{pOH} = 14.00 - \text{pH} = 14.00 - 0.80 = 13.20 \text{ (2 SFs on the right of the decimal point)}$

14.115 a. To form the conjugate base, remove a hydrogen ion (H^+) from the acid.

    1. HS^- **2.** H_2PO_4^-

  b. 1. $\dfrac{[\text{H}_3\text{O}^+][\text{HS}^-]}{[\text{H}_2\text{S}]}$ **2.** $\dfrac{[\text{H}_3\text{O}^+][\text{H}_2\text{PO}_4^-]}{[\text{H}_3\text{PO}_4]}$

  c. H_2S (see Table 14.3; the weaker acid will be closer to the bottom of the table)

14.117 a. $\text{NH}_4^+/\text{NH}_3$ and $\text{HNO}_3/\text{NO}_3^-$
    From Table 14.3, we see that NH_4^+ is a weaker acid than HNO_3 and that NO_3^- is a weaker base than NH_3. Thus, the equilibrium mixture will contain mostly products.
  b. $\text{H}_3\text{O}^+/\text{H}_2\text{O}$ and HBr/Br^-
    From Table 14.3, we see that H_3O^+ is a weaker acid than HBr and that Br^- is a weaker base than H_2O. Thus, the equilibrium mixture will contain mostly products.

14.119 a. $\text{ZnCO}_3(s) + \text{H}_2\text{SO}_4(aq) \longrightarrow \text{CO}_2(g) + \text{H}_2\text{O}(l) + \text{ZnSO}_4(aq)$
  b. $2\text{Al}(s) + 6\text{HNO}_3(aq) \longrightarrow 3\text{H}_2(g) + 2\text{Al}(\text{NO}_3)_3(aq)$

14.121 KOH (strong base) $\longrightarrow \text{K}^+(aq) + \text{OH}^-(aq)$ (100% dissociation)
$[\text{OH}^-] = 0.050 \text{ M} = 5.0 \times 10^{-2} \text{ M}$

  a. $[\text{H}_3\text{O}^+] = \dfrac{1.0 \times 10^{-14}}{[\text{OH}^-]} = \dfrac{1.0 \times 10^{-14}}{[5.0 \times 10^{-2}]} = 2.0 \times 10^{-13} \text{ M (2 SFs)}$

  b. $\text{pH} = -\log[\text{H}_3\text{O}^+] = -\log[2.0 \times 10^{-13}] = 12.70$ (2 SFs on the right of the decimal point)

  c. $\text{pOH} = 14.00 - \text{pH} = 14.00 - 12.70 = 1.30$ (2 SFs on the right of the decimal point)

  d. $\text{H}_2\text{SO}_4(aq) + 2\text{KOH}(aq) \longrightarrow 2\text{H}_2\text{O}(l) + \text{K}_2\text{SO}_4(aq)$
  e. In the titration equation, 1 mol of H_2SO_4 reacts with 2 mol of KOH.

$40.0 \text{ mL H}_2\text{SO}_4 \text{ solution} \times \dfrac{1 \text{ L H}_2\text{SO}_4 \text{ solution}}{1000 \text{ mL H}_2\text{SO}_4 \text{ solution}} \times \dfrac{0.035 \text{ mol H}_2\text{SO}_4}{1 \text{ L H}_2\text{SO}_4 \text{ solution}} \times \dfrac{2 \text{ mol KOH}}{1 \text{ mol H}_2\text{SO}_4}$

$\times \dfrac{1 \text{ L KOH solution}}{0.050 \text{ mol KOH}} \times \dfrac{1000 \text{ mL KOH solution}}{1 \text{ L KOH solution}} = 56 \text{ mL of KOH solution (2 SFs)}$

14.123 a. The $[\text{H}_3\text{O}^+]$ can be found by using the relationship $[\text{H}_3\text{O}^+] = 10^{-\text{pH}}$.

$[\text{H}_3\text{O}^+] = 10^{-\text{pH}} = 10^{-4.2} = 6 \times 10^{-5} \text{ M (1 SF)}$

$[\text{OH}^-] = \dfrac{1.0 \times 10^{-14}}{[\text{H}_3\text{O}^+]} = \dfrac{1.0 \times 10^{-14}}{[6 \times 10^{-5}]} = 2 \times 10^{-10} \text{ M (1 SF)}$

  b. $[\text{H}_3\text{O}^+] = 10^{-\text{pH}} = 10^{-6.5} = 3 \times 10^{-7} \text{ M (1 SF)}$

$[\text{OH}^-] = \dfrac{1.0 \times 10^{-14}}{[\text{H}_3\text{O}^+]} = \dfrac{1.0 \times 10^{-14}}{[3 \times 10^{-7}]} = 3 \times 10^{-8} \text{ M (1 SF)}$

c. In the titration equation, 1 mol of $CaCO_3$ reacts with 1 mol of H_2SO_4.

$$1.0 \text{ kL solution} \times \frac{1000 \text{ L solution}}{1 \text{ kL solution}} \times \frac{6 \times 10^{-5} \text{ mol } H_3O^+}{1 \text{ L solution}}$$

$$\times \frac{1 \text{ mol } H_2SO_4}{2 \text{ mol } H_3O^+} \times \frac{1 \text{ mol } CaCO_3}{1 \text{ mol } H_2SO_4} \times \frac{100.09 \text{ g } CaCO_3}{1 \text{ mol } CaCO_3} = 3 \text{ g of } CaCO_3 \text{ (1 SF)}$$

14.125 a. In the titration equation, 1 mol of HCl reacts with 1 mol of NaOH.

$$HCl(aq) + NaOH(aq) \longrightarrow H_2O(l) + NaCl(aq)$$

$$25.0 \text{ mL HCl solution} \times \frac{1 \text{ L HCl solution}}{1000 \text{ mL HCl solution}} \times \frac{0.288 \text{ mol HCl}}{1 \text{ L HCl solution}} \times \frac{1 \text{ mol NaOH}}{1 \text{ mol HCl}}$$

$$\times \frac{1 \text{ L NaOH solution}}{0.150 \text{ mol NaOH}} \times \frac{1000 \text{ mL NaOH solution}}{1 \text{ L NaOH solution}} = 48.0 \text{ mL of NaOH solution (3 SFs)}$$

b. In the titration equation, 1 mol of H_2SO_4 reacts with 2 mol of NaOH.

$$H_2SO_4(aq) + 2NaOH(aq) \longrightarrow 2H_2O(l) + Na_2SO_4(aq)$$

$$10.0 \text{ mL } H_2SO_4 \text{ solution} \times \frac{1 \text{ L } H_2SO_4 \text{ solution}}{1000 \text{ mL } H_2SO_4 \text{ solution}} \times \frac{0.560 \text{ mol } H_2SO_4}{1 \text{ L } H_2SO_4 \text{ solution}} \times \frac{2 \text{ mol NaOH}}{1 \text{ mol } H_2SO_4}$$

$$\times \frac{1 \text{ L NaOH solution}}{0.150 \text{ mol NaOH}} \times \frac{1000 \text{ mL NaOH solution}}{1 \text{ L NaOH solution}} = 74.7 \text{ mL of NaOH solution (3 SFs)}$$

14.127 In the titration equation, 1 mol of H_2SO_4 reacts with 2 mol of NaOH.

$$45.6 \text{ mL NaOH solution} \times \frac{1 \text{ L solution}}{1000 \text{ mL solution}} \times \frac{0.205 \text{ mol NaOH}}{1 \text{ L solution}} \times \frac{1 \text{ mol } H_2SO_4}{2 \text{ mol NaOH}}$$

$$= 0.004\ 67 \text{ mol of } H_2SO_4$$

$$20.0 \text{ mL } H_2SO_4 \text{ solution} \times \frac{1 \text{ L solution}}{1000 \text{ mL solution}} = 0.0200 \text{ L of } H_2SO_4 \text{ solution}$$

$$\text{Molarity (M) of } H_2SO_4 = \frac{\text{moles of solute}}{\text{liters of solution}} = \frac{0.004\ 67 \text{ mol } H_2SO_4}{0.0200 \text{ L solution}}$$

$$= 0.234 \text{ M } H_2SO_4 \text{ solution (3 SFs)}$$

14.129 This buffer solution is made from the weak acid H_3PO_4 and a salt containing its conjugate base $H_2PO_4^-$.
a. Add acid: $H_2PO_4^-(aq) + H_3O^+(aq) \longrightarrow H_3PO_4(aq) + H_2O(l)$
b. Add base: $H_3PO_4(aq) + OH^-(aq) \longrightarrow H_2PO_4^-(aq) + H_2O(l)$

c. $[H_3O^+] = K_a \times \dfrac{[H_3PO_4]}{[H_2PO_4^-]} = 7.5 \times 10^{-3} \times \dfrac{[0.50 \text{ M}]}{[0.20 \text{ M}]} = 1.9 \times 10^{-2} \text{ M}$

$pH = -\log[1.9 \times 10^{-2}] = 1.72$ (2 SFs on the right of the decimal point)

Selected Answers to Combining Ideas from Chapters 11 to 14

CI.21 a. CH_4

$$H : \overset{..}{\underset{..}{C}} : H \quad \text{or} \quad H - \overset{\displaystyle H}{\underset{\displaystyle H}{\overset{|}{\underset{|}{C}}}} - H$$

b. $7.0 \times 10^6 \text{ gal} \times \dfrac{4 \text{ qt}}{1 \text{ gal}} \times \dfrac{946.4 \text{ mL}}{1 \text{ qt}} \times \dfrac{0.45 \text{ g}}{1 \text{ mL}} \times \dfrac{1 \text{ kg}}{1000 \text{ g}} = 1.2 \times 10^7 \text{ kg of LNG (2 SFs)}$

c. $7.0 \times 10^6 \text{ gal} \times \dfrac{4 \text{ qt}}{1 \text{ gal}} \times \dfrac{946.4 \text{ mL}}{1 \text{ qt}} \times \dfrac{0.45 \text{ g}}{1 \text{ mL}} \times \dfrac{1 \text{ L } CH_4 \text{ (STP)}}{0.715 \text{ g } CH_4}$

$= 1.7 \times 10^{10} \text{ L of LNG (STP) (2 SFs)}$

d. $CH_4(g) + 2O_2(g) \xrightarrow{\Delta} CO_2(g) + 2H_2O(g) + 883 \text{ kJ}$

e. $7.0 \times 10^6 \text{ gal} \times \dfrac{4 \text{ qt}}{1 \text{ gal}} \times \dfrac{946.4 \text{ mL}}{1 \text{ qt}} \times \dfrac{0.45 \text{ g}}{1 \text{ mL}} \times \dfrac{1 \text{ mol } CH_4}{16.04 \text{ g}}$

$\times \dfrac{2 \text{ mol } O_2}{1 \text{ mol } CH_4} \times \dfrac{32.00 \text{ g } O_2}{1 \text{ mol } O_2} \times \dfrac{1 \text{ kg } O_2}{1000 \text{ g } O_2}$

$= 4.8 \times 10^7 \text{ kg of } O_2 \text{ (2 SFs)}$

f. $7.0 \times 10^6 \text{ gal} \times \dfrac{4 \text{ qt}}{1 \text{ gal}} \times \dfrac{946.4 \text{ mL}}{1 \text{ qt}} \times \dfrac{0.45 \text{ g}}{1 \text{ mL}} \times \dfrac{1 \text{ mol } CH_4}{16.04 \text{ g}} \times \dfrac{883 \text{ kJ}}{1 \text{ mol } CH_4}$

$= 6.6 \times 10^{11} \text{ kJ (2 SFs)}$

CI.23 a. $CS_2(g) + 3O_2(g) \xrightarrow{\Delta} CO_2(g) + 2SO_2(g)$

b. $25.0 \text{ g } CS_2 \times \dfrac{1 \text{ mol } CS_2}{76.15 \text{ g } CS_2} \times \dfrac{1 \text{ mol } CO_2}{1 \text{ mol } CS_2} \times \dfrac{44.01 \text{ g } CO_2}{1 \text{ mol } CO_2} = 14.4 \text{ g of } CO_2$

$30.0 \text{ g } O_2 \times \dfrac{1 \text{ mol } O_2}{32.00 \text{ g } O_2} \times \dfrac{1 \text{ mol } CO_2}{3 \text{ mol } O_2} \times \dfrac{44.01 \text{ g } CO_2}{1 \text{ mol } CO_2}$

$= 13.8 \text{ g of } CO_2 \text{ (smaller amount of product)}$

$\therefore O_2$ is the limiting reactant, and 13.8 g of CO_2 (3 SFs) can be produced.

c. $25.0 \text{ g } CS_2 \times \dfrac{1 \text{ mol } CS_2}{76.15 \text{ g } CS_2} = 0.328 \text{ mol of } CS_2 \text{ initially}$

$30.0 \text{ g } O_2 \times \dfrac{1 \text{ mol } O_2}{32.00 \text{ g } O_2} \times \dfrac{1 \text{ mol } CS_2}{3 \text{ mol } O_2} = 0.313 \text{ mol of } CS_2 \text{ consumed}$

$\therefore 0.328 \text{ mol} - 0.313 \text{ mol} = 0.015 \text{ mol of } CS_2 \text{ remaining} = n$

and $T = 125\,°C + 273 = 398 \text{ K}; \ V = 10.0 \text{ L}$

$$P_{CS_2} = \frac{nRT}{V} = \frac{(0.015 \text{ mol})\left(\dfrac{62.4 \text{ L} \cdot \text{mmHg}}{\text{mol} \cdot \text{K}}\right)(398 \text{ K})}{10.0 \text{ L}} = 37 \text{ mmHg (2 SFs)}$$

d. $30.0 \text{ g } O_2 \times \dfrac{1 \text{ mol } O_2}{32.00 \text{ g } O_2} \times \dfrac{1 \text{ mol } CO_2}{3 \text{ mol } O_2} = 0.313 \text{ mol of } CO_2$

$30.0 \text{ g } O_2 \times \dfrac{1 \text{ mol } O_2}{32.00 \text{ g } O_2} \times \dfrac{2 \text{ mol } SO_2}{3 \text{ mol } O_2} = 0.625 \text{ mol of } SO_2$

and 0.015 mol of CS_2 remaining (excess)

∴ Total moles of gas = 0.313 mol CO_2 + 0.625 mol SO_2 + 0.015 mol CS_2 = 0.953 mol

$$P_{\text{total}} = \frac{nRT}{V} = \frac{(0.953 \text{ mol})\left(\dfrac{62.4 \text{ L} \cdot \text{mmHg}}{\text{mol} \cdot \text{K}}\right)(398 \text{ K})}{10.0 \text{ L}} = 2370 \text{ mmHg (3 SFs)}$$

CI.25 a. $[H_2] = \dfrac{2.02 \text{ g } H_2}{10.0 \text{ L}} \times \dfrac{1 \text{ mol } H_2}{2.016 \text{ g } H_2} = 0.100 \text{ M}$

$[S_2] = \dfrac{10.3 \text{ g } S_2}{10.0 \text{ L}} \times \dfrac{1 \text{ mol } S_2}{64.14 \text{ g } S_2} = 0.0161 \text{ M}$

$[H_2S] = \dfrac{68.2 \text{ g } H_2S}{10.0 \text{ L}} \times \dfrac{1 \text{ mol } H_2S}{34.09 \text{ g } H_2S} = 0.200 \text{ M}$

$K_c = \dfrac{[H_2S]^2}{[H_2]^2[S_2]} = \dfrac{[0.200]^2}{[0.100]^2[0.0161]} = 248 \text{ (3 SFs)}$

b. If more H_2 (a reactant) is added, the equilibrium will shift in the direction of the products.

c. If the volume decreases from 10.0 L to 5.00 L (at constant temperature), the equilibrium will shift in the direction of the products (fewer moles of gas).

d. $[H_2] = \dfrac{0.300 \text{ mol } H_2}{5.00 \text{ L}} = 0.0600 \text{ M}$

$[H_2S] = \dfrac{2.50 \text{ mol } H_2S}{5.00 \text{ L}} = 0.500 \text{ M}$

$K_c = \dfrac{[H_2S]^2}{[H_2]^2[S_2]} = 248$

Rearrange the expression to solve for $[S_2]$ and substitute in known values.

$[S_2] = \dfrac{[H_2S]^2}{[H_2]^2 K_c} = \dfrac{[0.500]^2}{[0.0600]^2(248)} = 0.280 \text{ M (3 SFs)}$

CI.27 a. $2M(s) + 6HCl(aq) \longrightarrow 3H_2(g) + 2MCl_3(aq)$

b. $34.8 \text{ mL solution} \times \dfrac{1 \text{ L solution}}{1000 \text{ mL solution}} \times \dfrac{0.520 \text{ mol HCl}}{1 \text{ L solution}} \times \dfrac{3 \text{ mol } H_2}{6 \text{ mol HCl}}$

$= 0.009 \; 05 \text{ mol of } H_2 \text{ (3 SFs)}$

$T = 24\,°C + 273 = 297 \text{ K}$

Rearrange the ideal gas law $PV = nRT$ to solve for V,

$$V = \frac{nRT}{P} = \frac{(0.009 \; 05 \text{ mol})\left(\dfrac{62.4 \text{ L} \cdot \text{mmHg}}{\text{mol} \cdot \text{K}}\right)(297 \text{ K})}{(720. \text{ mmHg})} \times \frac{1000 \text{ mL}}{1 \text{ L}} = 233 \text{ mL of } H_2 \text{ (3 SFs)}$$

c. $34.8 \text{ mL solution} \times \dfrac{1 \text{ L solution}}{1000 \text{ mL solution}} \times \dfrac{0.520 \text{ mol HCl}}{1 \text{ L solution}} \times \dfrac{2 \text{ mol M}}{6 \text{ mol HCl}}$

$= 6.03 \times 10^{-3} \text{ mol of M (3 SFs)}$

d. $\dfrac{0.420 \text{ g M}}{6.03 \times 10^{-3} \text{ mol M}} = 69.7 \text{ g/mol of M (3 SFs)}; \therefore \text{ metal M is gallium}$

e. $2\text{Ga}(s) + 6\text{HCl}(aq) \longrightarrow 3\text{H}_2(g) + 2\text{GaCl}_3(aq)$

CI.29 a. $\dfrac{8.57 \text{ g } \cancel{\text{KOH}}}{850. \text{ mL } \cancel{\text{solution}}} \times \dfrac{1 \text{ mol KOH}}{56.11 \text{ g } \cancel{\text{KOH}}} \times \dfrac{1000 \text{ mL } \cancel{\text{solution}}}{1 \text{ L solution}} = 0.180 \text{ M KOH solution (3 SFs)}$

b. $[\text{H}_3\text{O}^+] = \dfrac{K_\text{w}}{[\text{OH}^-]} = \dfrac{1.0 \times 10^{-14}}{[0.180]} = 5.56 \times 10^{-14} \text{ M}$

$\text{pH} = -\log[\text{H}_3\text{O}^+] = -\log[5.56 \times 10^{-14}] = 13.255 \text{ (3 SFs on the right of the decimal point)}$

c. $2\text{KOH}(aq) + \text{H}_2\text{SO}_4(aq) \longrightarrow 2\text{H}_2\text{O}(l) + \text{K}_2\text{SO}_4(aq)$

d. $10.0 \text{ mL } \cancel{\text{KOH solution}} \times \dfrac{1 \text{ L } \cancel{\text{KOH solution}}}{1000 \text{ mL } \cancel{\text{KOH solution}}} \times \dfrac{0.180 \text{ mol } \cancel{\text{KOH}}}{1 \text{ L } \cancel{\text{KOH solution}}} \times \dfrac{1 \text{ mol } \cancel{\text{H}_2\text{SO}_4}}{2 \text{ mol } \cancel{\text{KOH}}}$

$\times \dfrac{1 \text{ L } \cancel{\text{H}_2\text{SO}_4 \text{ solution}}}{0.250 \text{ mol } \cancel{\text{H}_2\text{SO}_4}} \times \dfrac{1000 \text{ mL H}_2\text{SO}_4 \text{ solution}}{1 \text{ L } \cancel{\text{H}_2\text{SO}_4 \text{ solution}}}$

$= 3.60 \text{ mL of H}_2\text{SO}_4 \text{ solution (3 SFs)}$

CI.31 a. $[\text{H}_2\text{CO}_3] = \dfrac{0.0149 \text{ g } \cancel{\text{H}_2\text{CO}_3}}{200.0 \text{ mL } \cancel{\text{solution}}} \times \dfrac{1 \text{ mol H}_2\text{CO}_3}{62.03 \text{ g } \cancel{\text{H}_2\text{CO}_3}} \times \dfrac{1000 \text{ mL } \cancel{\text{solution}}}{1 \text{ L solution}}$

$= 0.001\,20 \text{ M (3 SFs)}$

b. $[\text{HCO}_3^-] = \dfrac{0.403 \text{ g } \cancel{\text{NaHCO}_3}}{200.0 \text{ mL } \cancel{\text{solution}}} \times \dfrac{1 \text{ mol } \cancel{\text{NaHCO}_3}}{84.01 \text{ g } \cancel{\text{NaHCO}_3}} \times \dfrac{1 \text{ mol HCO}_3^-}{1 \text{ mol } \cancel{\text{NaHCO}_3}} \times \dfrac{1000 \text{ mL } \cancel{\text{solution}}}{1 \text{ L solution}}$

$= 0.0240 \text{ M (3 SFs)}$

c. $[\text{H}_3\text{O}^+] = K_\text{a} \times \dfrac{[\text{H}_2\text{CO}_3]}{[\text{HCO}_3^-]} = 7.9 \times 10^{-7} \times \dfrac{[0.001\,20]}{[0.0240]} = 4.0 \times 10^{-8} \text{ M (2 SFs)}$

d. $\text{pH} = -\log[4.0 \times 10^{-8}] = 7.40 \text{ (2 SFs on the right of the decimal point)}$

e. $\text{HCO}_3^-(aq) + \text{H}_3\text{O}^+(aq) \longrightarrow \text{H}_2\text{CO}_3(aq) + \text{H}_2\text{O}(l)$

f. $\text{H}_2\text{CO}_3(aq) + \text{OH}^-(aq) \longrightarrow \text{HCO}_3^-(aq) + \text{H}_2\text{O}(l)$

Oxidation and Reduction

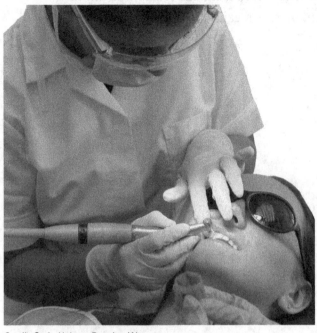

Credit: Craig Holmes Premium/Alamy

Kimberly has been delighted with the teeth whitening process and how her teeth look. The hydrogen peroxide gel caused a chemical reaction in the enamel coating of her teeth, oxidizing the brown and yellow stains on the teeth. Kimberly now avoids food or drinks, such as dark tea and coffee, that can stain her teeth. She now uses a teeth-whitening toothpaste once a week and will return to her dental hygienist in a year for a touch-up treatment. If hydrogen peroxide forms oxygen gas and water when it reacts, what is the balanced equation?

CHAPTER READINESS*

✥ Key Math Skills

- Using Positive and Negative Numbers in Calculations (1.4)
- Solving Equations (1.4)

⚗ Core Chemistry Skills

- Writing Positive and Negative Ions (6.1)
- Writing Ionic Formulas (6.2)
- Balancing a Chemical Equation (8.2)
- Identifying Oxidized and Reduced Substances (8.4)

*These Key Math Skills and Core Chemistry Skills from previous chapters are listed here for your review as you proceed to the new material in this chapter.

LOOKING AHEAD

15.1 Oxidation and Reduction
15.2 Balancing Oxidation–Reduction Equations Using Half-Reactions

15.3 Electrical Energy from Oxidation–Reduction Reactions
15.4 Oxidation–Reduction Reactions That Require Electrical Energy

15.1 Oxidation and Reduction

Learning Goal: Identify a reaction as an oxidation or a reduction. Assign and use oxidation numbers to identify elements that are oxidized or reduced.

- In an oxidation–reduction reaction, electrons are transferred from one reactant to another.
- In an oxidation, electrons are lost. In a reduction, there is a gain of electrons.

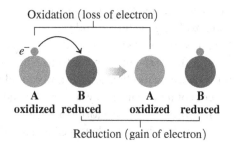

- A reducing agent (oxidized) provides the electrons, and an oxidizing agent (reduced) accepts electrons.
- An oxidation number is assigned to atoms to determine which substance is oxidized and which is reduced. These numbers do not necessarily represent actual charges.
- Oxidation numbers are assigned based on a set of rules.
- An increase in oxidation number indicates an oxidation; a decrease in oxidation number indicates a reduction.

Study Note

Oxidation **is** the **l**oss of electrons (OIL). **R**eduction **is** the **g**ain of electrons (RIG).

A balanced oxidation–reduction equation contains an oxidation of one reactant and a reduction of the other reactant. $Cu(s) + 2Ag^+(aq) \longrightarrow Cu^{2+}(aq) + 2Ag(s)$

    Oxidation: $Cu(s) \longrightarrow Cu^{2+}(aq) + 2\,e^-$
    Reduction: $2Ag^+(aq) + 2\,e^- \longrightarrow 2Ag(s)$

In a balanced oxidation–reduction equation, the loss of electrons must be equal to the gain of electrons.

♦ **Learning Exercise 15.1A**

Identify each of the following as oxidation (O) or reduction (R):

1. _____ loss of two electrons   **2.** _____ $Zn(s) \longrightarrow Zn^{2+}(aq) + 2\,e^-$

3. _____ $Cu^{2+}(aq) + e^- \longrightarrow Cu^+(aq)$ **4.** _____ gain of electrons

Answers   **1.** O   **2.** O   **3.** R   **4.** R

♦ **Learning Exercise 15.1B**

For each of the following oxidation–reduction reactions, identify the reactant that is oxidized and the reactant that is reduced:

a. $2Na(s) + Cl_2(g) \longrightarrow 2NaCl(aq)$

b. $Zn(s) + 2H^+(aq) \longrightarrow Zn^{2+}(aq) + H_2(g)$

c. $H_2S(aq) + I_2(s) \longrightarrow 2HI(aq) + S(s)$

d. $2Mg(s) + O_2(g) \longrightarrow 2MgO(s)$

e. $B_2O_3(s) + 3Mg(s) \longrightarrow 3MgO(s) + 2B(s)$

Answers
- **a.** Na loses electrons and is oxidized; Cl_2 gains electrons and is reduced.
- **b.** Zn loses electrons and is oxidized; H^+ gains electrons and is reduced.
- **c.** S^{2-} (in H_2S) loses electrons and is oxidized; I_2 gains electrons and is reduced.
- **d.** Mg loses electrons and is oxidized; O_2 gains electrons and is reduced.
- **e.** Mg loses electrons and is oxidized; B (in B_2O_3) gains electrons and is reduced.

Table 15.1 Rules for Assigning Oxidation Numbers

1. The sum of the oxidation numbers in a molecule is zero (0), or for a polyatomic ion is equal to its charge.
2. The oxidation number of an element (monatomic or diatomic) is zero (0).
3. The oxidation number of a monatomic ion is equal to its charge.
4. In compounds, the oxidation number of Group 1A (1) metals is $+1$, and that of Group 2A (2) metals is $+2$.
5. In compounds, the oxidation number of fluorine is -1. Other nonmetals in Group 7A (17) are -1 except when combined with oxygen or fluorine.
6. In compounds, the oxidation number of oxygen is -2 except in OF_2, where O is $+1$; in H_2O_2 and other peroxides, O is -1.
7. In compounds with nonmetals, the oxidation number of hydrogen is $+1$; in compounds with metals, the oxidation number of hydrogen is -1.

Chapter 15

♦ **Learning Exercise 15.1C**

⟩ **CORE CHEMISTRY SKILL**

Assign oxidation numbers to the elements in each of the following: Assigning Oxidation Numbers

a. PbO

b. Fe

c. NO_2

d. $CuCl_2$

e. H_2CO_3

f. HNO_3

g. K_3PO_3

h. $Cr_2O_7^{2-}$

Answers

a. PbO
$+2-2$

b. Fe
0

c. NO_2
$+4-2$
$N + 2(-2) = 0$
$N = +4$

d. $CuCl_2$
$+2-1$
$Cu + 2(-1) = 0$
$Cu = +2$

e. H_2CO_3
$+1+4-2$
$2(+1) + C + 3(-2) = 0$
$C = -2 + 6 = +4$

f. HNO_3
$+1+5-2$
$+1 + N + 3(-2) = 0$
$N = -1 + 6 = +5$

g. K_3PO_3
$+1+3-2$
$3(+1) + P + 3(-2) = 0$
$P = -3 + 6 = +3$

h. $Cr_2O_7^{2-}$
$+6-2$
$2Cr + 7(-2) = -2$
$2Cr = -2 + 14 = +12$
$Cr = +6$

Study Note

The oxidizing agent gains electrons from the oxidation. The reducing agent provides electrons for the reduction.

Guide to Using Oxidation Numbers	
STEP 1	Assign oxidation numbers to each element.
STEP 2	Identify the increase in oxidation number as oxidation; the decrease as reduction.

♦ **Learning Exercise 15.1D**

Complete each of the following statements about oxidation–reduction reactions by using the term *loses* or *gains*:

a. In reduction, an element _____ electrons.

b. The oxidizing agent _____ electrons.

c. The reducing agent _____ electrons.

d. The oxidation number increases when an element _____ electrons.

e. The oxidizing agent is the substance that _____ electrons.

f. The oxidation number decreases when an element _____ electrons.

g. In oxidation, an element _____ electrons.

Answers   **a.** gains   **b.** gains   **c.** loses   **d.** loses
            e. gains   **f.** gains   **g.** loses

Guide to Identifying Oxidizing and Reducing Agents	
STEP 1	Assign oxidation numbers to the components in the equation.
STEP 2	Identify the substance that is oxidized and the substance that is reduced.
STEP 3	Identify the oxidized substance as the reducing agent and the reduced substance as the oxidizing agent.

♦ **Learning Exercise 15.1E**

For each of the following, assign oxidation numbers and identify the reactant that is the reducing agent and the reactant that is the oxidizing agent:

CORE CHEMISTRY SKILL
Using Oxidation Numbers

a. $PbO(s) + CO(g) \longrightarrow Pb(s) + CO_2(g)$

reducing agent _____   oxidizing agent _____

b. $Fe_2O_3(s) + 3C(s) \longrightarrow 2Fe(s) + 3CO(g)$

CORE CHEMISTRY SKILL
Identifying Oxidizing and Reducing Agents

reducing agent _____   oxidizing agent _____

c. $Cu(s) + 4HNO_3(aq) \longrightarrow Cu(NO_3)_2(aq) + 2NO_2(g) + 2H_2O(l)$

reducing agent _____   oxidizing agent _____

Answers

a. $PbO(s) + CO(g) \longrightarrow Pb(s) + CO_2(g)$
       +2−2   +2−2   0   +4−2

reducing agent   CO   oxidizing agent   PbO

b. $Fe_2O_3(s) + 3C(s) \longrightarrow 2Fe(s) + 3CO(g)$
       +3−2   0   0   +2−2

reducing agent   C   oxidizing agent Fe_2O_3

c. $Cu(s) + 4HNO_3(aq) \longrightarrow Cu(NO_3)_2(aq) + 2NO_2(g) + 2H_2O(l)$
        0   +1+5−2   +2+5−2   +4−2   +1−2

reducing agent   Cu   oxidizing agent HNO_3

15.2 Balancing Oxidation–Reduction Equations Using Half-Reactions

Learning Goal: Balance oxidation–reduction equations using the half-reaction method.

- The half-reaction method of balancing oxidation–reduction equations is used to balance oxidation–reduction equations that are written in the ionic form.

- Using the half-reaction method for balancing redox equations, the elements are balanced, O is balanced by adding H_2O, and H is balanced by adding H^+. Then charge is balanced by adding electrons, and each half-reaction is multiplied by an integer to equalize electron loss and gain, and then the half-reactions are added together.

Guide to Balancing Redox Equations Using Half-Reactions	
STEP 1	Write two half-reactions for the equation.
STEP 2	For each half-reaction, balance the elements other than H and O. If the reaction takes place in acid or base, balance O by adding H_2O, and H by adding H^+.
STEP 3	Balance each half-reaction for charge by adding electrons.
STEP 4	Multiply each half-reaction by factors that equalize the loss and gain of electrons.
STEP 5	Add half-reactions and cancel electrons, and any identical ions or molecules. In base, add OH^- to neutralize H^+. Check the balance of atoms and charge.

Example: Balance the following oxidation–reduction equation in an acidic solution:

$$I^-(aq) + SO_4^{2-}(aq) \longrightarrow H_2S(g) + I_2(s)$$

Analyze the Problem	Given	Need	Connect
	Equation: $I^-(aq) + SO_4^{2-}(aq) \longrightarrow H_2S(g) + I_2(s)$	balanced equation	half-reactions, electron balance

Solution: **STEP 1** **Write two half-reactions for the equation.**

$$I^-(aq) \longrightarrow I_2(s)$$
$$SO_4^{2-}(aq) \longrightarrow H_2S(g)$$

STEP 2 **For each half-reaction, balance the elements other than H and O. If the reaction takes place in acid, balance O by adding H_2O, and H by adding H^+.**

$$2I^-(aq) \longrightarrow I_2(s)$$
$$10H^+(aq) + SO_4^{2-}(aq) \longrightarrow H_2S(g) + 4H_2O(l)$$

STEP 3 **Balance each half-reaction for charge by adding electrons.**

$$2I^-(aq) \longrightarrow I_2(s) + 2\,e^-$$
$$10H^+(aq) + SO_4^{2-}(aq) + 8\,e^- \longrightarrow H_2S(g) + 4H_2O(l)$$

STEP 4 **Multiply each half-reaction by factors that equalize the loss and gain of electrons.**

$$4 \times [2I^-(aq) \longrightarrow I_2(s) + 2\,e^-]$$
$$8I^-(aq) \longrightarrow 4I_2(s) + 8\,e^- \qquad \textit{8 electrons lost}$$
$$10H^+(aq) + SO_4^{2-}(aq) + 8\,e^- \longrightarrow H_2S(g) + 4H_2O(l) \qquad \textit{8 electrons gained}$$

STEP 5 Add half-reactions and cancel electrons, and any identical ions or molecules. Check the balance of atoms and charge.

$$10H^+(aq) + 8I^-(aq) + SO_4^{2-}(aq) + 8\cancel{e^-} \longrightarrow 4I_2(s) + H_2S(g) + 4H_2O(l) + 8\cancel{e^-}$$

Final balanced equation:

$$10H^+(aq) + 8I^-(aq) + SO_4^{2-}(aq) \longrightarrow 4I_2(s) + H_2S(g) + 4H_2O(l)$$

Check the balance of atoms and charge.

Reactants		Products
8I	=	8I
1S	=	1S
10H	=	10H
4O	=	4O
Charge: 0	=	0

♦ **Learning Exercise 15.2A**

Use half-reactions to balance the following oxidation–reduction equations in acidic solution:

▶ **CORE CHEMISTRY SKILL**

Using Half-Reactions to Balance Redox Equations

a. $NO_3^-(aq) + S(s) \longrightarrow NO_2(g) + H_2SO_4(aq)$

b. $MnO_4^-(aq) + Cl^-(aq) \longrightarrow Mn^{2+}(aq) + Cl_2(g)$

c. $Cr_2O_7^{2-}(aq) + Fe^{2+}(aq) \longrightarrow Cr^{3+}(aq) + Fe^{3+}(aq)$

d. $NO_2(g) + ClO^-(aq) \longrightarrow NO_3^-(aq) + Cl^-(aq)$

e. $BrO_3^-(aq) + MnO_2(s) \longrightarrow Br^-(aq) + MnO_4^-(aq)$

Answers

a. $6H^+(aq) + 6NO_3^-(aq) + S(s) \longrightarrow 6NO_2(g) + H_2SO_4(aq) + 2H_2O(l)$
b. $16H^+(aq) + 2MnO_4^-(aq) + 10Cl^-(aq) \longrightarrow 2Mn^{2+}(aq) + 5Cl_2(g) + 8H_2O(l)$
c. $14H^+(aq) + Cr_2O_7^{2-}(aq) + 6Fe^{2+}(aq) \longrightarrow 2Cr^{3+}(aq) + 6Fe^{3+}(aq) + 7H_2O(l)$
d. $2NO_2(g) + ClO^-(aq) + H_2O(l) \longrightarrow 2NO_3^-(aq) + Cl^-(aq) + 2H^+(aq)$
e. $BrO_3^-(aq) + 2MnO_2(s) + H_2O(l) \longrightarrow Br^-(aq) + 2MnO_4^-(aq) + 2H^+(aq)$

♦ **Learning Exercise 15.2B**

Use half-reactions to balance the following oxidation–reduction equations in basic solution:

a. $CO_2(g) + NH_2OH(aq) \longrightarrow CO(g) + N_2(g)$

b. $MnO_4^-(aq) + C_2O_4^{2-}(aq) \longrightarrow MnO_2(s) + CO_2(g)$

c. $Cr(OH)_3(s) + ClO_3^-(aq) \longrightarrow CrO_4^{2-}(aq) + Cl^-(aq)$

Answers

a. $CO_2(g) + 2NH_2OH(aq) \longrightarrow CO(g) + N_2(g) + 3H_2O(l)$

b. $2MnO_4^-(aq) + 3C_2O_4^{2-}(aq) + 4H_2O(l) \longrightarrow 2MnO_2(s) + 6CO_2(g) + 8OH^-(aq)$

c. $4OH^-(aq) + 2Cr(OH)_3(s) + ClO_3^-(aq) \longrightarrow 2CrO_4^{2-}(aq) + Cl^-(aq) + 5H_2O(l)$

♦ **Learning Exercise 15.2C**

Oxalic acid, $H_2C_2O_4$, present in plants such as spinach, reacts with permanganate, MnO_4^-.

$C_2O_4^{2-}(aq) + MnO_4^-(aq) \longrightarrow Mn^{2+}(aq) + CO_2(g)$

a. Balance the equation in acidic solution.

b. If 25.0 mL of a 0.0200 M MnO_4^- solution (from $KMnO_4$) is required to react with 5.00 mL of a $C_2O_4^{2-}$ solution, what is the molarity of the $C_2O_4^{2-}$ solution?

Answers

a. $16H^+(aq) + 5C_2O_4^{2-}(aq) + 2MnO_4^-(aq) \longrightarrow 2Mn^{2+}(aq) + 10CO_2(g) + 8H_2O(l)$

b. $0.0250 \text{ L solution} \times \dfrac{0.0200 \text{ mol MnO}_4^-}{1 \text{ L solution}} \times \dfrac{5 \text{ mol H}_2\text{C}_2\text{O}_4}{2 \text{ mol MnO}_4^-} = 0.001\ 25 \text{ mol of C}_2\text{O}_4^{2-}$

$$\text{Molarity (M)} = \dfrac{0.001\ 25 \text{ mol C}_2\text{O}_4^{2-}}{0.005\ 00 \text{ L solution}}$$

$$= 0.250 \text{ M C}_2\text{O}_4^{2-} \text{ solution}$$

♦ **Learning Exercise 15.2D**

Iodine reacts with a thiosulfate, $S_2O_3^{2-}$, solution.

$$I_2(s) + S_2O_3^{2-}(aq) \longrightarrow I^-(aq) + S_4O_6^{2-}(aq)$$

a. Balance the equation in acidic solution.

b. If 37.6 mL of a 0.250 M $S_2O_3^{2-}$ (from $Na_2S_2O_3$) solution is required to completely react with I_2, how many grams of I_2 are present in the sample?

Answers

a. $I_2(s) + 2S_2O_3^{2-}(aq) \longrightarrow 2I^-(aq) + S_4O_6^{2-}(aq)$

b. $0.0376 \text{ L solution} \times \dfrac{0.0250 \text{ mol S}_2\text{O}_3^{2-}}{1 \text{ L solution}} \times \dfrac{1 \text{ mol I}_2}{2 \text{ mol S}_2\text{O}_3^{2-}} \times \dfrac{253.8 \text{ g I}_2}{1 \text{ mol I}_2} = 1.19 \text{ g of I}_2$

15.3 Electrical Energy from Oxidation–Reduction Reactions

Learning Goal: Use the activity series to determine if an oxidation–reduction reaction is spontaneous. Write the half-reactions that occur in a voltaic cell and the cell notation.

- The activity series arranges elements in order of their ability to oxidize spontaneously when combined with the ions of any metal lower on the list.
- In an electrochemical cell, the half-reactions of an oxidation–reduction reaction are physically separated, so the electrons flow through an external circuit.
- Oxidation takes place at the anode; reduction takes place at the cathode.
- Electrons flow from the anode to the cathode.
- In a voltaic cell, the flow of electrons through an external circuit generates electrical energy.
- Batteries are examples of electrochemical cells that provide electrical energy.

Study Note

Suppose that a cell has a shorthand notation as follows:

$$Zn(s)\,|\,Zn^{2+}(aq)\,||\,Cu^{2+}(aq)\,|\,Cu(s)$$

The components of the oxidation half-cell (anode) are written on the left side in this shorthand notation, and the components of the reduction half-cell (cathode) are written on the right. A single vertical line separates the solid Zn anode from the ionic Zn^{2+} solution and the Cu^{2+} solution from the Cu cathode. Double vertical lines $||$ indicate the salt bridge, which separates the two half-cells.

◆ **Learning Exercise 15.3A**

Using the activity series in Table 15.3 in the text, predict the more active metal in each of the following pairs:

a. $Fe(s)$ or $Pb(s)$ **b.** $Ag(s)$ or $Na(s)$ **c.** $Zn(s)$ or $Ca(s)$

d. $Al(s)$ or $Pb(s)$ **e.** $Fe(s)$ or $Ni(s)$

Answers   **a.** $Fe(s)$ **b.** $Na(s)$ **c.** $Ca(s)$ **d.** $Al(s)$ **e.** $Fe(s)$

◆ **Learning Exercise 15.3B**

> ⊳ **CORE CHEMISTRY SKILL**
>
> Identifying Spontaneous Reactions

Using the activity series, predict whether each of the following reactions occurs spontaneously or not, and explain:

a. $Cu(s) + Sn^{2+}(aq) \longrightarrow Cu^{2+}(aq) + Sn(s)$

b. $Pb^{2+}(aq) + Mg(s) \longrightarrow Mg^{2+}(aq) + Pb(s)$

c. $2Ag(s) + Ni^{2+}(aq) \longrightarrow Ni(s) + 2Ag^{+}(aq)$

d. $Zn(s) + 2H^{+}(aq) \longrightarrow Zn^{2+}(aq) + H_2(g)$

e. $2Cr(s) + 3Ca^{2+}(aq) \longrightarrow 2Cr^{3+}(aq) + 3Ca(s)$

Answers

a. not spontaneous: $Cu(s)$ is less active than $Sn(s)$
b. spontaneous: $Mg(s)$ is more active than $Pb(s)$
c. not spontaneous: $Ag(s)$ is less active than $Ni(s)$
d. spontaneous: $Zn(s)$ is more active than $H_2(g)$
e. not spontaneous: $Cr(s)$ is less active than $Ca(s)$

♦ **Learning Exercise 15.3C**

For each of the following voltaic cells, write the shorthand cell notation, the half-reactions at the anode and cathode, and the balanced equation:

a. a Mg anode in a solution containing Mg^{2+}, and a Ag cathode in a solution containing Ag^+

b. an Al anode in a solution containing Al^{3+}, and a Cr cathode in a solution containing Cr^{3+}

c. an Fe anode in a solution containing Fe^{2+}, and a Ni cathode in a solution containing Ni^{2+}

Answers

a. $Mg(s)\,|\,Mg^{2+}(aq)\,||\,Ag^+(aq)\,|\,Ag(s)$

Anode: $\qquad\qquad Mg(s) \longrightarrow Mg^{2+}(aq) + 2\,e^-$

Cathode: $\underline{\qquad [Ag^+(aq) + e^- \longrightarrow Ag(s)] \times 2 \qquad}$

$\qquad\quad Mg(s) + 2Ag^+(aq) \longrightarrow Mg^{2+}(aq) + 2Ag(s)$

b. $Al(s)\,|\,Al^{3+}(aq)\,||\,Cr^{3+}(aq)\,|\,Cr(s)$

Anode: $\qquad\qquad Al(s) \longrightarrow Al^{3+}(aq) + 3\,e^-$

Cathode: $\underline{\quad Cr^{3+}(aq) + 3\,e^- \longrightarrow Cr(s) \qquad\qquad}$

$\qquad\quad Cr^{3+}(aq) + Al(s) \longrightarrow Cr(s) + Al^{3+}(aq)$

c. $Fe(s)\,|\,Fe^{2+}(aq)\,||\,Ni^{2+}(aq)\,|\,Ni(s)$

Anode: $\qquad\qquad Fe(s) \longrightarrow Fe^{2+}(aq) + 2\,e^-$

Cathode: $\underline{\quad Ni^{2+}(aq) + 2\,e^- \longrightarrow Ni(s) \qquad\qquad}$

$\qquad\quad Fe(s) + Ni^{2+}(aq) \longrightarrow Fe^{2+}(aq) + Ni(s)$

♦ **Learning Exercise 15.3D**

Write the half-reactions for the anode and cathode and the overall cell reaction from the following cell notations:

a. $Sr(s) | Sr^{2+}(aq) || Sn^{4+}(aq), Sn^{2+}(aq) | Pt(s)$

b. $Cr(s) | Cr^{3+}(aq) || Ni^{2+}(aq) | Ni(s)$

c. $Ag(s) | Ag^{+}(aq) || Au^{3+}(aq) | Au(s)$

Answers

a. Anode: $\qquad\qquad\qquad Sr(s) \longrightarrow Sr^{2+}(aq) + 2\,e^{-}$

Cathode: $\qquad \dfrac{Sn^{4+}(aq) + 2\,e^{-} \longrightarrow Sn^{2+}(aq)}{Sr(s) + Sn^{4+}(aq) \longrightarrow Sr^{2+}(aq) + Sn^{2+}(aq)}$

b. Anode: $\qquad\qquad\qquad [Cr(s) \longrightarrow Cr^{3+}(aq) + 3\,e^{-}] \times 2$

Cathode: $\qquad \dfrac{[Ni^{2+}(aq) + 2\,e^{-} \longrightarrow Ni(s)] \times 3}{2Cr(s) + 3Ni^{2+}(aq) \longrightarrow 2Cr^{2+}(aq) + 3Ni(s)}$

c. Anode: $\qquad\qquad\qquad [Ag(s) \longrightarrow Ag^{+}(aq) + e^{-}] \times 3$

Cathode: $\qquad \dfrac{Au^{3+}(aq) + 3\,e^{-} \longrightarrow Au(s)}{3Ag(s) + Au^{3+}(aq) \longrightarrow 3Ag^{+}(aq) + Au(s)}$

15.4 Oxidation–Reduction Reactions That Require Electrical Energy

Learning Goal: Describe the half-cell reactions and the overall reactions that occur in electrolysis.

- An oxidation–reduction reaction is not spontaneous when a less active metal is combined with the ions of a more active metal.
- In an electrolytic cell, a process called electrolysis uses electrical energy to drive a nonspontaneous reaction.
- Electrolytic cells are used in the electrolysis of molten salts and electroplating.

Key Terms for Sections 15.1 to 15.4

Match each of the following key terms with the correct description:

  a. cathode   **b.** oxidation   **c.** reduction

  d. oxidizing agent   **e.** reducing agent   **f.** voltaic cell

1. _____ the reactant that is reduced

2. _____ the gain of electrons by a substance

3. _____ a cell that uses an oxidation–reduction reaction to produce electrical energy

4. _____ the loss of electrons by a substance

5. _____ the reactant that is oxidized

6. _____ the electrode in an electrochemical cell where reduction takes place

Answers   **1.** d   **2.** c   **3.** f   **4.** b   **5.** e   **6.** a

♦ **Learning Exercise 15.4**

An electrolytic cell is used to electroplate nickel on the surface of an iron object. If the electrolytic cell uses a nickel bar as the anode, and the iron object as the cathode, indicate the following:

  a. What happens at the anode when electrical energy flows through the cell?

  b. What is the equation for the half-reaction that occurs at the anode?

  c. What happens at the cathode when electrical energy flows through the cell?

  d. What is the equation for the half-reaction that occurs at the cathode?

Answers

  a. At the anode, the nickel metal in the bar is oxidized to nickel(II) ions to make a Ni^{2+} solution.
  b. $Ni(s) \longrightarrow Ni^{2+}(aq) + 2\,e^-$
  c. At the cathode, the nickel(II) ions in the solution are reduced to nickel metal that plates on the surface of the iron object.
  d. $Ni^{2+}(aq) + 2\,e^- \longrightarrow Ni(s)$

Checklist for Chapter 15

You are ready to take the Practice Test for Chapter 15. Be sure you have accomplished the following learning goals for this chapter. If not, review the section listed at the end of the goal. Then apply your new skills and understanding to the Practice Test.

After studying Chapter 15, I can successfully:

_____ Identify what is oxidized and reduced in an oxidation–reduction reaction. (15.1)

_____ Assign oxidation numbers to the atoms in an equation. (15.1)

_____ Use oxidation numbers to determine the oxidized element, reduced element, oxidizing agent, and reducing agent. (15.1)

_____ Balance an oxidation–reduction equation using half-reactions. (15.2)

_____ Use the activity series for metals to determine whether an oxidation–reduction reaction is spontaneous. (15.3)

_____ Write the half-reactions for the anode and the cathode of a voltaic cell. (15.3)

_____ Write the shorthand cell notation for the half-reactions for the anode and the cathode of a voltaic cell. (15.3)

_____ Draw the cell diagram for a voltaic cell. (15.3)

_____ Write the half-cell reactions for an electrolytic cell. (15.4)

_____ Describe the process taking place in electroplating. (15.4)

Practice Test for Chapter 15

The chapter sections to review are shown in parentheses at the end of each question.

For questions 1 through 5, identify whether the element in each is oxidized (O) *or reduced* (R). (15.1)

1. _____ gains electrons

2. _____ an oxidizing agent

3. _____ loses electrons

4. _____ a reducing agent

5. _____ oxidation number increases

For questions 6 through 10, indicate the oxidation number of the element in each. (15.1)

A. +1   B. +2   C. +3   D. +4   E. +5

6. _____ N in HNO_2

7. _____ P in P_2O_5

8. _____ C in HCO_3^-

9. _____ Mn in MnO_2

10. _____ Ca in $CaSO_4$

For questions 11 through 15, indicate the oxidation number of the element in each. (15.1)

A. 0   B. −1   C. −2   D. −3   E. −4

11. _____ P in Na_3P

12. _____ N in N_2H_4

13. _____ C in CH_4

14. _____ Br in Br_2

15. _____ F in OF_2

For questions 16 through 20, use the following unbalanced equation: (15.1)

$$NO(g) + Br_2(g) \longrightarrow NOBr(g)$$

16. The oxidation numbers of N in the reactants and products are
   A. +1, +2   **B.** +2, +3   **C.** +3, +2   **D.** −2, −3   **E.** +2, −3

17. The oxidation numbers of Br in the reactants and products are
   A. 0, 0   **B.** −1, 0   **C.** 0, +1   **D.** −1, −1   **E.** 0, −1

18. The oxidation numbers of O in the reactants and products are
   A. 0, 0   **B.** −2, 0   **C.** +3, +2   **D.** −2, −2   **E.** 0, −2

19. The change in the oxidation number of N is
   A. a decrease of 1   **B.** a decrease of 2   **C.** an increase of 1
   D. an increase of 2   **E.** no change

20. The oxidizing agent is
   A. NO   **B.** Br_2 **C.** NOBr

For questions 21 through 23, consider the following half-reaction in acidic solution: (15.2)

$$ClO_2^- \longrightarrow Cl^-$$

21. What is the number of H_2O in the balanced half-reaction?
   A. 1   **B.** 2   **C.** 3   **D.** 4   **E.** 5

22. What is the number of H^+ in the balanced half-reaction?
   A. 1   **B.** 2   **C.** 3   **D.** 4   **E.** 5

23. How many electrons are in the balanced half-reaction?
   A. 1   **B.** 2   **C.** 3   **D.** 4   **E.** 5

For questions 24 through 26, consider the following half-reaction in acidic solution: (15.2)

$$Cr_2O_7^{2-} \longrightarrow Cr^{3+}$$

24. What is the number of H_2O in the balanced half-reaction?
   A. 2   **B.** 4   **C.** 5   **D.** 7   **E.** 14

25. What is the number of H^+ in the balanced half-reaction?
   A. 2   **B.** 6   **C.** 7   **D.** 12   **E.** 14

26. How many electrons are in the balanced half-reaction?
   A. 2   **B.** 4   **C.** 6   **D.** 10   **E.** 14

27. Which equation has the following shorthand cell notation? (15.3)
   $$Al(s) \,|\, Al^{3+}(aq) \,||\, Fe^{3+}(aq) \,|\, Fe(s)$$
   A. $Al(s) + Fe^{3+}(aq) \longrightarrow Al^{3+}(aq) + Fe(s)$
   B. $Al^{3+}(aq) + Fe(s) \longrightarrow Al(s) + Fe^{3+}(aq)$
   C. $Al^{3+}(aq) + Fe(s) \longrightarrow Al(s) + Fe^{3+}(aq)$
   D. $Al^{3+}(aq) + Fe^{3+}(aq) \longrightarrow Fe(s) + Al(s)$
   E. $Al(s) + Fe(s) \longrightarrow Fe^{3+}(aq) + Al^{3+}(aq)$

For questions 28 through 30, consider the following metals and ions in order from most to least active:

$$Mg(s) \longrightarrow Mg^{2+}(aq) + 2\,e^-$$
$$Cr(s) \longrightarrow Cr^{3+}(aq) + 3\,e^-$$
$$Fe(s) \longrightarrow Fe^{2+}(aq) + 2\,e^-$$
$$Pb(s) \longrightarrow Pb^{2+}(aq) + 2\,e^-$$
$$Ag(s) \longrightarrow Ag^+(aq) + e^-$$

28. Which of the following reactions is spontaneous? (15.3)

A. $3Pb(s) + 2Cr^{3+}(aq) \longrightarrow 3Pb^{2+}(aq) + 2Cr(s)$

B. $Mg(s) + 2Ag^+(aq) \longrightarrow Mg^{2+}(aq) + 2Ag(s)$

C. $Fe^{2+}(aq) + Pb(s) \longrightarrow Fe(s) + Pb^{2+}(aq)$

D. $2Ag(s) + Pb^{2+}(aq) \longrightarrow Pb(s) + 2Ag^+(aq)$

E. $3Ag(s) + Cr^{3+}(aq) \longrightarrow 3Ag^+(aq) + Cr(s)$

29. Which of the following reactions requires an external source of energy? (15.4)

A. $3Mg(s) + 2Cr^{3+}(aq) \longrightarrow 2Cr(s) + 3Mg^{2+}(aq)$

B. $Pb(s) + 2Ag^+(aq) \longrightarrow Pb^{2+}(aq) + 2Ag(s)$

C. $Mg(s) + Pb^{2+}(aq) \longrightarrow Mg^{2+}(aq) + Pb(s)$

D. $Mg(s) + 2Ag^+(aq) \longrightarrow Mg^{2+}(aq) + 2Ag(s)$

E. $3Pb(s) + 2Cr^{3+}(aq) \longrightarrow 3Pb^{2+}(aq) + 2Cr(s)$

30. In the electroplating of a bowl, an aluminum anode is placed in a solution of $Al(NO_3)_3$. The bowl to be aluminum-plated is the cathode. The anode and cathode are wired to a battery. The reaction at the anode is (15.4)

A. $Al^{3+}(aq) + 3\,e^- \longrightarrow Al(s)$

B. $Al^{3+}(aq) \longrightarrow Al(s) + 3\,e^-$

C. $Al(s) + 3\,e^- \longrightarrow Al^{3+}(aq)$

D. $Al(s) \longrightarrow Al^{3+}(aq) + 3\,e^-$

E. $Al(s) + Al^{3+}(aq) \longrightarrow 2Al(s)$

Answers to the Practice Test

1. R	**2.** R	**3.** O	**4.** O	**5.** O
6. C	**7.** E	**8.** D	**9.** D	**10.** B
11. D	**12.** C	**13.** E	**14.** A	**15.** B
16. B	**17.** E	**18.** D	**19.** C	**20.** B
21. B	**22.** D	**23.** D	**24.** D	**25.** E
26. C	**27.** A	**28.** B	**29.** E	**30.** D

Selected Answers and Solutions to Text Problems

15.1 Oxidation is the loss of electrons; reduction is the gain of electrons.
   a. Al^{3+} gains electrons to form Al; this is a reduction.
   b. Ca loses electrons to form Ca^{2+}; this is an oxidation.
   c. Fe^{3+} gains an electron to form Fe^{2+}; this is a reduction.

15.3 An oxidized substance has lost electrons; a reduced substance has gained electrons.
   a. Al loses electrons and is oxidized; O_2 gains electrons and is reduced.
   b. Zn loses electrons and is oxidized; H^+ gains electrons and is reduced.
   c. Br^- loses electrons and is oxidized; F_2 gains electrons and is reduced.

15.5 **a.** An element has an oxidation number of zero (Rule 2); Cu = 0.
   b. An element has an oxidation number of zero (Rule 2); F in F_2 = 0.
   c. The oxidation number of a monatomic ion is equal to its charge (Rule 3); Fe^{2+} = +2.
   d. The oxidation number of a monatomic ion is equal to its charge (Rule 3); Cl^- = −1.

15.7 **a.** In KCl, the oxidation number of K is +1 (Rule 4). The oxidation number of Cl is −1
   (Rule 5). Oxidation numbers: K = +1, Cl = −1
   K + Cl = 0
   (+1) + (−1) = 0
   b. In MnO_2, the oxidation number of O is −2 (Rule 6). Because MnO_2 is neutral, the
   oxidation number of Mn is calculated as +4. Oxidation numbers: Mn = +4, O = −2
   Mn + 2O = 0
   Mn + 2(−2) = 0
   ∴ Mn = +4
   c. In NO, the oxidation number of O is −2 (Rule 6). Because NO is neutral, the
   oxidation number of N is calculated as +2. Oxidation numbers: N = +2, O = −2
   N + O = 0
   N + (−2) = 0
   ∴ N = +2
   d. In Mn_2O_3, the oxidation number of O is −2 (Rule 6). Because Mn_2O_3 is neutral, the
   oxidation number of Mn is calculated as +3. Oxidation numbers: Mn = +3, O = −2
   2Mn + 3O = 0
   2Mn + 3(−2) = 0
   2Mn = +6
   ∴ Mn = +3

15.9 **a.** Li_3PO_4 is an ionic compound composed of the Li^+ and PO_4^{3-} ions. The oxidation
   number of the monatomic ion Li^+ is +1 (Rule 3). In the polyatomic ion PO_4^{3-}, the
   oxidation number of O is −2 (Rule 6). Because the sum of the oxidation numbers in
   PO_4^{3-} must equal −3, the oxidation number of P is calculated as +5. Oxidation numbers:
   Li = +1, P = +5, O = −2
   P + 4O = −3
   P + 4(−2) = −3
   ∴ P = +5
   b. In SO_3^{2-}, the oxidation number of O is −2 (Rule 6). Because the sum of the oxidation num-
   bers in the polyatomic ion SO_3^{2-} must equal −2 (Rule 1), the oxidation number of
   S is calculated as +4. Oxidation numbers: S = +4, O = −2
   S + 3O = −2
   S + 3(−2) = −2
   ∴ S = +4

    c. In Cr_2S_3, the oxidation number of S is -2. Because Cr_2S_3 is neutral, the oxidation number of Cr is calculated as $+3$. Oxidation numbers: $Cr = +3$, $S = -2$

$$2Cr + 3S = 0$$
$$2Cr + 3(-2) = 0$$
$$2Cr = +6$$
$$\therefore Cr = +3$$

    d. In NO_3^-, the oxidation number of O is -2 (Rule 6). Because the sum of the oxidation numbers in the polyatomic ion NO_3^- must equal -1 (Rule 1), the oxidation number of N is calculated as $+5$. Oxidation numbers: $N = +5$, $O = -2$

$$N + 3O = -1$$
$$N + 3(-2) = -1$$
$$\therefore N = +5$$

15.11 a. In HSO_4^-, the oxidation number of H is $+1$ (Rule 7) and the oxidation number of O is -2 (Rule 6). Because the sum of the oxidation numbers in the polyatomic ion HSO_4^- must equal -1 (Rule 1), the oxidation number of S is calculated as $+6$.

Oxidation numbers: $H = +1$, $S = +6$, $O = -2$

$$H + S + 4O = -1$$
$$(+1) + S + 4(-2) = -1$$
$$\therefore S = +6$$

    b. In H_3PO_3, the oxidation number of H is $+1$ (Rule 7) and the oxidation number of O is -2 (Rule 6). Because H_3PO_3 is neutral, the oxidation number of P is calculated as $+3$.

Oxidation numbers: $H = +1$, $P = +3$, $O = -2$

$$3H + P + 3O = 0$$
$$3(+1) + P + 3(-2) = 0$$
$$\therefore P = +3$$

    c. In $Cr_2O_7^{2-}$, the oxidation number of O is -2 (Rule 6). Because the sum of the oxidation numbers in the polyatomic ion $Cr_2O_7^{2-}$ must equal -2 (Rule 1), the oxidation number of Cr is calculated as $+6$. Oxidation numbers: $Cr = +6$, $O = -2$

$$2Cr + 7O = -2$$
$$2Cr + 7(-2) = -2$$
$$2Cr = +12$$
$$\therefore Cr = +6$$

    d. Na_2CO_3 is an ionic compound composed of the Na^+ and CO_3^{2-} ions. The oxidation number of the monatomic ion Na^+ is $+1$ (Rule 3). In the polyatomic ion CO_3^{2-}, the oxidation number of O is -2 (Rule 6). Because the sum of the oxidation numbers in CO_3^{2-} must equal -2, the oxidation number of C is calculated as $+4$.

Oxidation numbers: $Na = +1$, $C = +4$, $O = -2$

$$2Na + C + 3O = 0$$
$$2(+1) + C + 3(-2) = 0$$
$$\therefore C = +4$$

15.13 a. HNO_3 $H = +1$, $O = -2$

$$HNO_3 \rightarrow (+1) + N + 3(-2) = 0 \quad \therefore N = +5$$

    b. C_3H_6 $H = +1$

$$C_3H_6 \rightarrow 3C + 6(+1) = 0 \quad \therefore 3C = -6 \quad \text{and } C = -2$$

    c. K_3PO_4 $K = +1$, $O = -2$

$$K_3PO_4 \rightarrow 3(+1) + P + 4(-2) = 0 \quad \therefore P = +5$$

    d. CrO_4^{2-} $O = -2$

$$CrO_4^{2-} \rightarrow Cr + 4(-2) = -2 \quad \therefore Cr = +6$$

15.15 a. The substance that is oxidized is the reducing agent; it provides electrons for reduction.
  b. The substance that gains electrons is the oxidizing agent; it accepts the electrons lost in an oxidation, and so is reduced.

15.17 a. Li is oxidized; Li is the reducing agent. Cl (in Cl_2) is reduced; Cl_2 is the oxidizing agent.
  b. Br^- (in NaBr) is oxidized; NaBr is the reducing agent. Cl (in Cl_2) is reduced; Cl_2 is the oxidizing agent.
  c. Pb is oxidized; Pb is the reducing agent. O (in O_2) is reduced; O_2 is the oxidizing agent.
  d. Al is oxidized; Al is the reducing agent. Ag^+ is reduced; Ag^+ is the oxidizing agent.

15.19 Assign oxidation numbers and determine which one increases and which one decreases. The substance with an increase in oxidation number is oxidized and is also the reducing agent. The substance with a decrease in oxidation number is reduced and is also the oxidizing agent.

  a. $2NiS + 3O_2 \longrightarrow 2NiO + 2SO_2$

   +2−2 0 +2−2 +4−2

  Ni: $+2 \rightarrow +2$ No change
  S: $-2 \rightarrow +4$ Oxidation number increases (oxidation)
  O: $0 \rightarrow -2$ Oxidation number decreases (reduction)
  ∴ S^{2-} (in NiS) is oxidized; NiS is the reducing agent. O (in O_2) is reduced; O_2 is the oxidizing agent.

  b. $Sn^{2+} + 2Fe^{3+} \longrightarrow Sn^{4+} + 2Fe^{2+}$

   +2 +3 +4 +2

  Sn: $+2 \rightarrow +4$ Oxidation number increases (oxidation)
  Fe: $+3 \rightarrow +2$ Oxidation number decreases (reduction)
  ∴ Sn^{2+} is oxidized; Sn^{2+} is the reducing agent. Fe^{3+} is reduced; Fe^{3+} is the oxidizing agent.

  c. $CH_4 + 2O_2 \longrightarrow CO_2 + 2H_2O$

   −4+1 0 +4−2 +1−2

  C: $-4 \rightarrow +4$ Oxidation number increases (oxidation)
  H: $+1 \rightarrow +1$ No change
  O: $0 \rightarrow -2$ Oxidation number decreases (reduction)
  ∴ C (in CH_4) is oxidized; CH_4 is the reducing agent. O (in O_2) is reduced; O_2 is the oxidizing agent.

  d. $2Cr_2O_3 + 3Si \longrightarrow 4Cr + 3SiO_2$

   +3−2 0 0 +4−2

  Cr: $+3 \rightarrow 0$ Oxidation number decreases (reduction)
  O: $-2 \rightarrow -2$ No change
  Si: $0 \rightarrow +4$ Oxidation number increases (oxidation)
  ∴ Si is oxidized; Si is the reducing agent. Cr^{3+} (in Cr_2O_3) is reduced; Cr_2O_3 is the oxidizing agent.

15.21 a. $Sn^{2+}(aq) \longrightarrow Sn^{4+}(aq)$
  Balance charge with e^-: $\quad Sn^{2+}(aq) \longrightarrow Sn^{4+}(aq) + 2\,e^-$
  b. $Mn^{2+}(aq) \longrightarrow MnO_4^-(aq)$
  Balance O with H_2O: $\quad Mn^{2+}(aq) + 4H_2O(l) \longrightarrow MnO_4^-(aq)$
  Balance H with H^+: $\quad Mn^{2+}(aq) + 4H_2O(l) \longrightarrow MnO_4^-(aq) + 8H^+(aq)$
  Balance charge with e^-: $\quad Mn^{2+}(aq) + 4H_2O(l) \longrightarrow MnO_4^-(aq) + 8H^+(aq) + 5\,e^-$
  c. $NO_2^-(aq) \longrightarrow NO_3^-(aq)$
  Balance O with H_2O: $\quad NO_2^-(aq) + H_2O(l) \longrightarrow NO_3^-(aq)$
  Balance H with H^+: $\quad NO_2^-(aq) + H_2O(l) \longrightarrow NO_3^-(aq) + 2H^+(aq)$
  Balance charge with e^-: $\quad NO_2^-(aq) + H_2O(l) \longrightarrow NO_3^-(aq) + 2H^+(aq) + 2\,e^-$
  d. $ClO_3^-(aq) \longrightarrow ClO_2(aq)$
  Balance O with H_2O: $\quad ClO_3^-(aq) \longrightarrow ClO_2(aq) + H_2O(l)$
  Balance H with H^+: $\quad 2H^+(aq) + ClO_3^-(aq) \longrightarrow ClO_2(aq) + H_2O(l)$
  Balance charge with e^-: $\quad e^- + 2H^+(aq) + ClO_3^-(aq) \longrightarrow ClO_2(aq) + H_2O(l)$

15.23 Write half-reactions, multiply by small numbers to equalize electrons lost and gained, add together, and combine common species.

a. $Ag(s) \longrightarrow Ag^+(aq) + e^-$
$e^- + 2H^+(aq) + NO_3^-(aq) \longrightarrow NO_2(g) + H_2O(l)$

Overall: $2H^+(aq) + Ag(s) + NO_3^-(aq) \longrightarrow Ag^+(aq) + NO_2(g) + H_2O(l)$

b. $[3\,e^- + 4H^+(aq) + NO_3^-(aq) \longrightarrow NO(g) + 2H_2O(l)] \times 4$
$[2H_2O(l) + S(s) \longrightarrow SO_2(g) + 4H^+(aq) + 4\,e^-] \times 3$

Overall: $4H^+(aq) + 4NO_3^-(aq) + 3S(s) \longrightarrow 4NO(g) + 3SO_2(g) + 2H_2O(l)$

c. $2S_2O_3^{2-}(aq) \longrightarrow S_4O_6^{2-}(aq) + 2\,e^-$
$2\,e^- + Cu^{2+}(aq) \longrightarrow Cu(s)$

Overall: $2S_2O_3^{2-}(aq) + Cu^{2+}(aq) \longrightarrow S_4O_6^{2-}(aq) + Cu(s)$

15.25 Write half-reactions, multiply by small numbers to equalize electrons lost and gained, add together, and combine common species.

a. $2Fe(s) + 3H_2O(l) \longrightarrow Fe_2O_3(s) + 6H^+(aq) + 6\,e^-$
$6\,e^- + 10H^+(aq) + 2CrO_4^{2-}(aq) \longrightarrow Cr_2O_3(s) + 5H_2O(l)$

Overall (in acid): $4H^+(aq) + 2Fe(s) + 2CrO_4^{2-}(aq) \longrightarrow Fe_2O_3(s) + Cr_2O_3(s) + 2H_2O$
$\therefore$ in base: $\underbrace{4H^+(aq) + 4OH^-(aq)}_{4H_2O(l)} + 2Fe(s) + 2CrO_4^{2-}(aq) \longrightarrow$
$Fe_2O_3(s) + Cr_2O_3(s) + 2H_2O(l) + 4OH^-(aq)$

Overall (in base): $2Fe(s) + 2CrO_4^{2-}(aq) + 2H_2O(l) \longrightarrow Fe_2O_3(s) + Cr_2O_3(s) + 4OH^-(aq)$

b. $[CN^-(aq) + H_2O(l) \longrightarrow CNO^-(aq) + 2H^+(aq) + 2\,e^-] \times$
$[3\,e^- + 4H^+(aq) + MnO_4^-(aq) \longrightarrow MnO_2(s) + 2H_2O(l)] \times 2$

Overall (in acid): $2H^+(aq) + 3CN^-(aq) + 2MnO_4^-(aq) \longrightarrow$
$3CNO^-(aq) + 2MnO_2(s) + H_2O$
$\therefore$ in base: $\underbrace{2H^+(aq) + 2OH^-(aq)}_{2H_2O(l)} + 3CN^-(aq) + 2MnO_4^-(aq) \longrightarrow$
$3CNO^-(aq) + 2MnO_2(s) + H_2O(l) + 2OH^-(aq)$

Overall (in base): $3CN^-(aq) + 2MnO_4^-(aq) + H_2O(l) \longrightarrow$
$3CNO^-(aq) + 2MnO_2(s) + 2OH^-(aq)$

15.27 In the activity series, a metal oxidizes spontaneously when combined with the ions in solutions of any metal below it.

a. Since Au is below H_2 in the activity series, the reaction will not be spontaneous.
b. Since Fe is above Ni in the activity series, the reaction will be spontaneous.
c. Since Ag is below Cu in the activity series, the reaction will not be spontaneous.

15.29 Oxidation occurs at the anode; reduction occurs at the cathode. In the shorthand cell notation, the oxidation half-cell is written on the left side of the double vertical line, and the reduction half-cell is written on the right side.

a. Anode reaction: $Pb(s) \mid Pb^{2+}(aq) = Pb(s) \longrightarrow Pb^{2+}(aq) + 2\,e^-$
Cathode reaction: $Cu^{2+}(aq) \mid Cu(s) = Cu^{2+}(aq) + 2\,e^- \longrightarrow Cu(s)$
Overall cell reaction: $Pb(s) + Cu^{2+}(aq) \longrightarrow Pb^{2+}(aq) + Cu(s)$

b. Anode reaction: $Cr(s) \mid Cr^{2+}(aq) = [Cr(s) \longrightarrow Cr^{2+}(aq) + 2\,e^-] \times 1$
Cathode reaction: $Ag^+(aq) \mid Ag(s) = [Ag^+(aq) + e^- \longrightarrow Ag(s)] \times 2$
Overall cell reaction: $Cr(s) + 2Ag^+(aq) \longrightarrow Cr^{2+}(aq) + 2Ag(s)$

15.31 a. The anode is a Cd metal electrode in a Cd^{2+} solution. The anode reaction is
$$Cd(s) \longrightarrow Cd^{2+}(aq) + 2\,e^-$$
The cathode is a Sn metal electrode in a Sn^{2+} solution. The cathode reaction is
$$Sn^{2+}(aq) + 2\,e^- \longrightarrow Sn(s)$$
The shorthand notation for this cell is
$$Cd(s)\,|\,Cd^{2+}(aq)\,\|\,Sn^{2+}(aq)\,|\,Sn(s)$$
b. The anode is a Zn metal electrode in a Zn^{2+} solution. The anode reaction is
$$Zn(s) \longrightarrow Zn^{2+}(aq) + 2\,e^-$$
The cathode is a C(graphite) electrode, where Cl_2 gas is reduced to Cl^-.
The cathode reaction is
$$Cl_2(g) + 2\,e^- \longrightarrow 2Cl^-(aq)$$
The shorthand notation for this cell is
$$Zn(s)\,|\,Zn^{2+}(aq)\,\|\,Cl_2(g)\,|\,Cl^-(aq)\,|\,C\ (\text{electrode})$$

15.33 a. The $Cd(s)$ has lost electrons, which makes the half-reaction an oxidation.
b. Cd metal is oxidized.
c. Because this is oxidation, it takes place at the anode.

15.35 a. The $Zn(s)$ has lost electrons, which makes the half-reaction an oxidation.
b. Zn metal is oxidized.
c. Because this is oxidation, it takes place at the anode.

15.37 a. The half-reaction to plate tin is $Sn^{2+}(aq) + 2\,e^- \longrightarrow Sn(s)$.
b. The reduction of Sn^{2+} to Sn occurs at the cathode, which is the iron can.
c. The oxidation of Sn to Sn^{2+} occurs at the anode, which is the tin bar.

15.39 Since Fe is above Sn in the activity series, if the Fe is exposed to air and water, Fe will be oxidized and rust will form. To protect iron, Sn would have to be *more* active than Fe and it is not.

15.41 a. H_2 loses electrons and is oxidized; O_2 gains electrons and is reduced.
b. Since O_2 is reduced, it is the oxidizing agent.
c. Since H_2 is oxidized, it is the reducing agent.
d. $H_2(g) + O_2(g) \longrightarrow H_2O_2(aq)$

15.43 a. Electrons are lost in an oxidation.
b. An oxidizing agent undergoes reduction.
c. O_2 gains electrons to form OH^-; this is a reduction.
d. Br_2 gains electrons to form $2Br^-$; this is a reduction.
e. Sn^{2+} loses electrons to form Sn^{4+}; this is an oxidation.

15.45 a. VO_2 $\quad$ $O = -2$ $\quad\quad\quad\quad$ Calculate: $V + 2(-2) = 0$ $\quad\quad\quad\quad$ $\therefore V = +4$
b. Ag_2CrO_4 $\quad$ $Ag = +1, O = -2$ $\quad$ Calculate: $2(+1) + Cr + 4(-2) = 0$ $\quad\therefore Cr = +6$
c. $S_2O_8{}^{2-}$ $\quad$ $O = -2$ $\quad\quad\quad\quad$ Calculate: $2S + 8(-2) = -2 \therefore 2S = +14$ and $S = +7$
d. $FeSO_4$ $\quad$ $Fe = +2, O = -2$ $\quad$ Calculate: $(+2) + S + 4(-2) = 0$ $\quad\therefore S = +6$

15.47 Reactions **a** and **c** both involve loss and gain of electrons; **a** and **c** are oxidation–reduction reactions.
a. Yes. Ca: $0 \rightarrow +2$ $\quad\therefore$ oxidation $\quad\quad$ H: $+1 \rightarrow 0$ $\quad\therefore$ reduction
b. No. No change in oxidation numbers. Ca $= +2$, C $= +4$, O $= -2$
c. Yes. Al: $0 \rightarrow +3$ $\quad\therefore$ oxidation $\quad\quad$ O: $0 \rightarrow -2$ $\quad\therefore$ reduction

15.49 $2Cr_2O_3(s) + 3Si(s) \longrightarrow 4Cr(s) + 3SiO_2(s)$
$\quad\quad$ $+3\,-2$ $\quad\quad\quad$ 0 $\quad\quad\quad\quad$ 0 $\quad\quad$ $+4\,-2$
Cr: $+3 \rightarrow 0$ $\quad$ Oxidation number decreases (reduction)
Si: $0 \rightarrow +4$ $\quad$ Oxidation number increases (oxidation)

    a. Cr^{3+} (in Cr_2O_3) is reduced.
    b. Si is oxidized.
    c. Cr_2O_3 is the oxidizing agent.
    d. Si is the reducing agent.

15.51 a. Balance charge with e^-: $Zn(s)$ $\longrightarrow$ $Zn^{2+}(aq) + 2\,e^-$

    b. Balance O with H_2O: $SnO_2^{2-}(aq) + H_2O(l)$ $\longrightarrow$ $SnO_3^{2-}(aq)$
        Balance H with H^+: $SnO_2^{2-}(aq) + H_2O(l)$ $\longrightarrow$ $SnO_3^{2-}(aq) + 2H^+(aq)$
        Balance charge with e^-: $SnO_2^{2-}(aq) + H_2O(l)$ $\longrightarrow$ $SnO_3^{2-}(aq) + 2H^+(aq) + 2\,e^-$
    c. Balance O with H_2O: $SO_3^{2-}(aq) + H_2O(l)$ $\longrightarrow$ $SO_4^{2-}(aq)$
        Balance H with H^+: $SO_3^{2-}(aq) + H_2O(l)$ $\longrightarrow$ $SO_4^{2-}(aq) + 2H^+(aq)$
        Balance charge with e^-: $SO_3^{2-}(aq) + H_2O(l)$ $\longrightarrow$ $SO_4^{2-}(aq) + 2H^+(aq) + 2\,e^-$
    d. Balance O with H_2O: $NO_3^-(aq)$ $\longrightarrow$ $NO(g) + 2H_2O(l)$
        Balance H with H^+: $4H^+(aq) + NO_3^-(aq)$ $\longrightarrow$ $NO(g) + 2H_2O(l)$
        Balance charge with e^-: $3\,e^- + 4H^+(aq) + NO_3^-(aq)$ $\longrightarrow$ $NO(g) + 2H_2O(l)$

15.53 a. $Fe(s) \longrightarrow Fe^{2+}(aq) + 2\,e^-$
    b. $Ni^{2+}(aq) + 2\,e^- \longrightarrow Ni(s)$
    c. Fe is the anode.
    d. Ni is the cathode.
    e. The electrons flow from Fe to Ni.
    f. $Fe(s) + Ni^{2+}(aq) \longrightarrow Fe^{2+}(aq) + Ni(s)$
    g. $Fe(s)\,|\,Fe^{2+}(aq)\,||\,Ni^{2+}(aq)\,|\,Ni(s)$

15.55 Reactions **b**, **c**, and **d** all involve loss and gain of electrons; **b**, **c**, and **d** are oxidation–reduction reactions.
    a. No. No change in oxidation numbers. $Ag = +1$, $N = +5$, $Na = +1$, $Cl = -1$, $O = -2$
    b. Yes. Li: $0 \rightarrow +1$ ∴ oxidation   N: $0 \rightarrow -3$ ∴ reduction
    c. Yes. Ni: $0 \rightarrow +2$ ∴ oxidation   Pb: $+2 \rightarrow 0$ ∴ reduction
    d. Yes. K: $0 \rightarrow +1$ ∴ oxidation   H: $+1 \rightarrow 0$ ∴ reduction

15.57 a. Fe^{3+} gains an electron to form Fe^{2+}; this is a reduction.
    b. Fe^{2+} is loses an electron to form Fe^{3+}; this is an oxidation.

15.59 a. Co_2O_3 $O = -2$ Calculate: $2Co + 3(-2) = 0$ ∴ $2Co = +6$ and $Co = +3$
    b. $KMnO_4$ $K = +1$, $O = -2$ Calculate: $+1 + Mn + 4(-2) = 0$ ∴ $Mn = +7$
    c. $SbCl_5$ $Cl = -1$ Calculate: $Sb + 5(-1) = 0$ ∴ $Sb = +5$
    d. ClO_3^- $O = -2$ Calculate: $Cl + 3(-2) = -1$ ∴ $Cl = +5$
    e. PO_4^{3-} $O = -2$ Calculate: $P + 4(-2) = -3$ ∴ $P = +5$

15.61 a. $2FeCl_2(aq) + Cl_2(g) \longrightarrow 2FeCl_3(aq)$
       $+2-1$ 0 $+3-1$

       Fe: $+2 \rightarrow +3$ Oxidation number increases (oxidation)
       Cl: $0 \rightarrow -1$ Oxidation number decreases (reduction)
       Fe^{2+} (in $FeCl_2$) is oxidized; $FeCl_2$ is the reducing agent.
       Cl (in Cl_2) is reduced; Cl_2 is the oxidizing agent.
    b. $2H_2S(g) + 3O_2(g) \longrightarrow 2H_2O(l) + 2SO_2(g)$
       $+1-2$ 0 $+1-2$ $+4-2$

       H: $+1 \rightarrow +1$ No change
       S: $-2 \rightarrow +4$ Oxidation number increases (oxidation)
       O: $0 \rightarrow -2$ Oxidation number decreases (reduction)
       S (in H_2S) is oxidized; H_2S is the reducing agent.
       O (in O_2) is reduced; O_2 is the oxidizing agent.

c. $P_2O_5(s) + 5C(s) \longrightarrow 2P(s) + 5CO(g)$

    $+5-2$ 0 0 $+2-2$

P: $+5 \rightarrow 0$ Oxidation number decreases (reduction)

O: $-2 \rightarrow -2$ No change

C: $0 \rightarrow +2$ Oxidation number increases (oxidation)

C is oxidized; C is the reducing agent.

P (in P_2O_5) is reduced; P_2O_5 is the oxidizing agent.

15.63 Write half-reactions, multiply by small numbers to equalize electrons lost and gained, add together, and combine common species.

a. $Zn(s) \longrightarrow Zn^{2+}(aq) + 2\,e^-$

$[e^- + 2H^+(aq) + NO_3^-(aq) \longrightarrow NO_2(g) + H_2O(l)] \times 2$

Overall: $4H^+(aq) + Zn(s) + 2NO_3^-(aq) \longrightarrow Zn^{2+}(aq) + 2NO_2(g) + 2H_2O(l)$

b. $[5\,e^- + 8H^+(aq) + MnO_4^-(aq) \longrightarrow Mn^{2+}(aq) + 4H_2O(l)] \times 2$

$[SO_3^{2-}(aq) + H_2O(l) \longrightarrow SO_4^{2-}(aq) + 2H^+(aq) + 2\,e^-] \times 5$

Overall: $6H^+(aq) + 2MnO_4^-(aq) + 5SO_3^{2-}(aq) \longrightarrow 2Mn^{2+}(aq) + 5SO_4^{2-}(aq) + 3H_2O(l)$

c. $[2I^-(aq) \longrightarrow I_2(s) + 2\,e^-] \times 3$

$6\,e^- + 6H^+(aq) + ClO_3^-(aq) \longrightarrow Cl^-(aq) + 3H_2O(l)$

Overall: $6H^+(aq) + ClO_3^-(aq) + 6I^-(aq) \longrightarrow Cl^-(aq) + 3I_2(s) + 3H_2O(l)$

d. $[C_2O_4^{2-}(aq) \longrightarrow 2CO_2(g) + 2\,e^-] \times 3$

$6\,e^- + 14H^+(aq) + Cr_2O_7^{2-}(aq) \longrightarrow 2Cr^{3+}(aq) + 7H_2O(l)$

Overall: $14H^+(aq) + Cr_2O_7^{2-}(aq) + 3C_2O_4^{2-}(aq) \longrightarrow 2Cr^{3+}(aq) + 6CO_2(g) + 7H_2O(l)$

15.65 a. Since Cr is above Ni in the activity series, the reaction will be spontaneous.

b. Since Cu is below Zn in the activity series, the reaction will not be spontaneous.

c. Since Zn is above Pb in the activity series, the reaction will be spontaneous.

15.67 a. The anode is Mg.

b. The cathode is Ni.

c. The half-reaction at the anode is $Mg(s) \longrightarrow Mg^{2+}(aq) + 2\,e^-$.

d. The half-reaction at the cathode is $Ni^{2+}(aq) + 2\,e^- \longrightarrow Ni(s)$.

e. The overall reaction is $Mg(s) + Ni^{2+}(aq) \longrightarrow Mg^{2+}(aq) + Ni(s)$.

f. The shorthand cell notation is $Mg(s)\,|\,Mg^{2+}(aq)\,\|\,Ni^{2+}(aq)\,|\,Ni(s)$.

15.69 In the activity series, Fe is above Ni, Pb, and Ag, which means it is oxidized. However, it is below Ca and Al, which means it cannot be oxidized and therefore Fe will not reduce Ca^{2+} or Al^{3+}.

a. $Ca^{2+}(aq)$ will not be reduced by an iron strip.

b. $Ag^+(aq)$ will be reduced by an iron strip.

c. $Ni^{2+}(aq)$ will be reduced by an iron strip.

d. $Al^{3+}(aq)$ will not be reduced by an iron strip.

e. $Pb^{2+}(aq)$ will be reduced by an iron strip.

15.71 a. $Pb(s) + SO_4^{2-}(aq) \longrightarrow PbSO_4(s) + 2\,e^-$
   b. $Pb(s)$ loses electrons; it is oxidized.
   c. The oxidation half-reaction takes place at the anode.

15.73

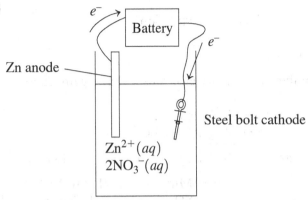

   a. The anode is a bar of zinc.
   b. The cathode is the steel bolt.
   c. The half-reaction at the anode is $Zn(s) \longrightarrow Zn^{2+}(aq) + 2\,e^-$
   d. The half-reaction at the cathode is $Zn^{2+}(aq) + 2\,e^- \longrightarrow Zn(s)$
   e. The purpose of the zinc coating is to prevent rusting of the bolt by H_2O and O_2.

15.75 a. Br_2 $\therefore Br = 0$
   b. $HBrO_2$ $H = +1, O = -2$ Calculate: $+1 + Br + 2(-2) = 0$ $\therefore Br = +3$
   c. BrO_3^- $O = -2$ Calculate: $Br + 3(-2) = -1$ $\therefore Br = +5$
   d. $NaBrO_4$ $Na = +1, O = -2$ Calculate: $+1 + Br + 4(-2) = 0$ $\therefore Br = +7$

15.77 $6\,e^- + 14H^+(aq) + Cr_2O_7^{2-}(aq)$ $\longrightarrow 2Cr^{3+}(aq) + 7H_2O(l)$
    $[NO_2^-(aq) + H_2O(l)$ $\longrightarrow NO_3^-(aq) + 2H^+(aq) + 2\,e^-]\times 3$

Overall: $8H^+(aq) + Cr_2O_7^{2-}(aq) + 3NO_2^-(aq) \longrightarrow 2Cr^{3+}(aq) + 3NO_3^-(aq) + 4H_2O(l)$

15.79 a. $[Ag(s)$ $\longrightarrow Ag^+(aq) + e^-]\times 3$
     $3\,e^- + 4H^+(aq) + NO_3^-(aq)$ $\longrightarrow NO(g) + 2H_2O(l)$

Overall: $4H^+(aq) + 3Ag(s) + NO_3^-(aq) \longrightarrow 3Ag^+(aq) + NO(g) + 2H_2O(l)$

   b. $15.0\ \text{g Ag} \times \dfrac{1\ \text{mol Ag}}{107.9\ \text{g Ag}} \times \dfrac{1\ \text{mol NO}}{3\ \text{mol Ag}} \times \dfrac{22.4\ \text{L NO (STP)}}{1\ \text{mol NO}} = 1.04\ \text{L of NO}(g)$ at STP (3 SFs)

15.81 a. $[CuS(s) + 4H_2O(l)$ $\longrightarrow CuSO_4(aq) + 8H^+(aq) + 8\,e^-]\times 3$
     $[3\,e^- + 3H^+(aq) + HNO_3(aq) \longrightarrow NO(g) + 2H_2O(l)]\times 8$

Overall: $3CuS(s) + 8HNO_3(aq) \longrightarrow 3CuSO_4(aq) + 8NO(g) + 4H_2O(l)$

   b. $24.8\ \text{g CuS} \times \dfrac{1\ \text{mol CuS}}{95.62\ \text{g CuS}} \times \dfrac{8\ \text{mol HNO}_3}{3\ \text{mol CuS}} \times \dfrac{1000\ \text{mL solution}}{16.0\ \text{mol HNO}_3}$
     $= 43.2\ \text{mL of HNO}_3$ solution (3 SFs)

15.83 $Br^-(aq) + 3H_2O(l)$ $\longrightarrow BrO_3^-(aq) + 6H^+(aq) + 6\,e^-$
    $[3\,e^- + 4H^+(aq) + MnO_4^-(aq)$ $\longrightarrow MnO_2(s) + 2H_2O(l)]\times 2$

Overall (in acid): $2H^+(aq) + Br^-(aq) + 2MnO_4^-(aq) \longrightarrow BrO_3^-(aq) + 2MnO_2(s) + H_2O(l)$
$\therefore$ in base: $\underbrace{2H^+(aq) + 2OH^-(aq)}_{2H_2O(l)} + Br^-(aq) + 2MnO_4^-(aq) \longrightarrow BrO_3^-(aq) + 2MnO_2(s)$
                                                       $+ H_2O(l) + 2OH^-(aq)$

Overall (in base): $H_2O(l) + Br^-(aq) + 2MnO_4^-(aq) \longrightarrow$
                                         $BrO_3^-(aq) + 2MnO_2(s) + 2OH^-(aq)$

15.85 a. Since Zn is below Ca in the activity series, the reaction will not be spontaneous.
   b. Since Al is above Sn in the activity series, the reaction will be spontaneous.
   c. Since Mg is above H_2 in the activity series, the reaction will be spontaneous.

15.87

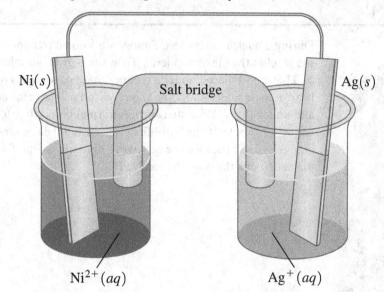

Ni(*s*) Salt bridge Ag(*s*)

$Ni^{2+}(aq)$ $Ag^{+}(aq)$

a. Ni(*s*) is the anode.
b. Ag(*s*) is the cathode.
c. The half-reaction at the anode is $Ni(s) \longrightarrow Ni^{2+}(aq) + 2\,e^{-}$.
d. The half-reaction at the cathode is $Ag^{+}(aq) + e^{-} \longrightarrow Ag(s)$.
e. The overall cell reaction is $Ni(s) + 2Ag^{+}(aq) \longrightarrow Ni^{2+}(aq) + 2Ag(s)$.

16
Nuclear Chemistry

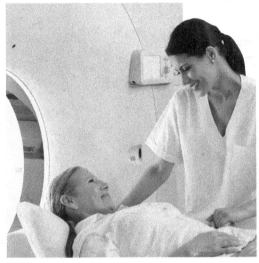

During a nuclear stress test, Simone walks on a treadmill until she reaches the maximum level. Then she is given an injection of Tl-201 and placed under a scanner that takes images of her heart muscle under stress. The comparison of her scans at rest and under stress shows that she has normal blood flow to her heart muscle. If the half-life of Tl-201 is 3.0 days, how many days will pass for the activity to decrease to one-fourth the activity of the dose she received?

Credit: Tyler Olson / Shutterstock

CHAPTER READINESS*

⚛ Key Math Skills

- Using Positive and Negative Numbers in Calculations (1.4)
- Solving Equations (1.4)
- Interpreting Graphs (1.4)

🧪 Core Chemistry Skills

- Using Conversion Factors (2.6)
- Counting Protons and Neutrons (4.4)
- Writing Atomic Symbols for Isotopes (4.5)

*These Key Math Skills and Core Chemistry Skills from previous chapters are listed here for your review as you proceed to the new material in this chapter.

LOOKING AHEAD

16.1 Natural Radioactivity
16.2 Nuclear Reactions
16.3 Radiation Measurement

16.4 Half-Life of a Radioisotope
16.5 Medical Applications Using Radioactivity

16.6 Nuclear Fission and Fusion

16.1 Natural Radioactivity

Learning Goal: Describe alpha, beta, positron, and gamma radiation.

- Radioactive isotopes have unstable nuclei that break down (decay), spontaneously emitting alpha (α), beta (β), positron (β^+), or gamma (γ) radiation.
- An alpha particle is the same as a helium nucleus; it contains two protons and two neutrons.
- A beta particle is a high-energy electron, and a positron is a high-energy positive particle. A gamma ray is high-energy radiation.
- Because radiation can damage cells in the body, proper protection must be used: shielding, time limitation, and distance.

Study Note

It is important to learn the symbols for the radiation particles in order to describe the different types of radiation:

$_2^4\text{He}$ or α	$_{-1}^{0}e$ or β	$_{+1}^{0}e$ or β^+	$_0^0\gamma$ or γ	$_1^1\text{H}$ or p	$_0^1n$ or n
alpha particle	beta particle	positron	gamma ray	proton	neutron

♦ **Learning Exercise 16.1A**

Match the terms in column A with the descriptions in column B.

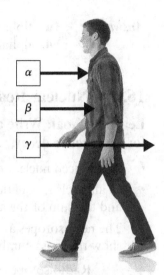

A		**B**
1. _____ $_8^{18}\text{O}$		**a.** symbol for a beta particle
2. _____ γ		**b.** symbol for an alpha particle
3. _____ β^+		**c.** an atom that emits radiation
4. _____ radioisotope		**d.** symbol for a positron
5. _____ $_2^4\text{He}$		**e.** symbol for an isotope of oxygen
6. _____ β		**f.** symbol for gamma radiation

Answers   **1.** e   **2.** f   **3.** d   **4.** c   **5.** b   **6.** a

Different types of radiation penetrate the body to different depths.
Credit: 4x6 / Getty Images

♦ **Learning Exercise 16.1B**

Discuss some things you can do to minimize the amount of radiation received if you work with a radioactive substance. Describe how each method helps to limit the amount of radiation you would receive.

Answer

Three ways to minimize exposure to radiation are (1) use shielding, (2) keep time short in the radioactive area, and (3) keep as much distance as possible from the radioactive materials. Shielding such as clothing and gloves stop alpha and beta particles from reaching your skin, whereas lead or concrete will absorb gamma rays. Limiting the time spent near radioactive samples reduces exposure time. Increasing the distance from a radioactive source reduces the intensity of radiation. Wearing a film badge will monitor the amount of radiation you receive.

♦ **Learning Exercise 16.1C**

What type(s) of radiation (alpha, beta, and/or gamma) would each of the following shielding materials protect you from?

a. heavy clothing _____   **b.** skin _____

c. paper _____   **d.** concrete _____

e. lead wall _____

Answers   **a.** alpha, beta   **b.** alpha   **c.** alpha
               d. alpha, beta, gamma   **e.** alpha, beta, gamma

16.2   Nuclear Reactions

Learning Goal: Write a balanced nuclear equation showing mass numbers and atomic numbers for radioactive decay.

- A balanced nuclear equation represents the changes in the nuclei of radioisotopes.
- In all nuclear equations, the sum of the mass numbers for the reactants and the products is equal and the sum of the atomic numbers for the reactants and the products is equal.
- The new isotopes and the type of radiation emitted can be determined from the symbols that show the mass numbers and atomic numbers of the isotopes in the nuclear reaction.

> Radioisotope $\longrightarrow$ new nucleus + radiation
>
> *Total of the mass numbers is equal on both sides of the equation.*
>
> $^{11}_{6}C \longrightarrow \, ^{7}_{4}Be + \, ^{4}_{2}He$
>
> *Total of the atomic numbers is equal on both sides of the equation.*

- A new radioactive isotope is produced when a nonradioactive isotope is bombarded by a small particle such as alpha particles, protons, neutrons, or small nuclei.

$^{10}_{5}B$	+	$^{4}_{2}He$	$\longrightarrow$	$^{13}_{7}N$	+	$^{1}_{0}n$
Stable nucleus		*Bombarding particle (α)*		*New nucleus*		*Neutron emitted*

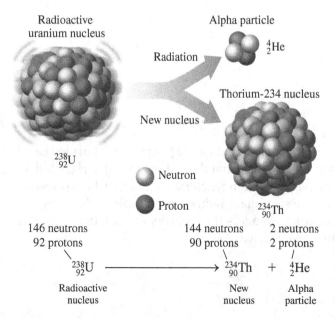

Radioactive uranium nucleus

Alpha particle

Radiation

$^{4}_{2}He$

Thorium-234 nucleus

New nucleus

$^{238}_{92}U$

○ Neutron

● Proton

$^{234}_{90}Th$

146 neutrons   144 neutrons   2 neutrons
92 protons   90 protons   2 protons

$^{238}_{92}U \longrightarrow \, ^{234}_{90}Th + \, ^{4}_{2}He$

Radioactive nucleus   New nucleus   Alpha particle

In the nuclear equation for alpha decay, the mass number of the new nucleus decreases by 4 and its atomic number decreases by 2.

Guide to Completing a Nuclear Equation	
STEP 1	Write the incomplete nuclear equation.
STEP 2	Determine the missing mass number.
STEP 3	Determine the missing atomic number.
STEP 4	Determine the symbol of the new nucleus.
STEP 5	Complete the nuclear equation.

♦ **Learning Exercise 16.2A**

Complete each of the following nuclear equations:

CORE CHEMISTRY SKILL

Writing Nuclear Equations

a. $^{66}_{29}Cu \longrightarrow {}^{66}_{30}Zn + ?$ _____

b. $^{127}_{53}I \longrightarrow ? + {}^{1}_{0}n$ _____

c. $^{238}_{92}U \longrightarrow ? + {}^{4}_{2}He$ _____

d. $^{24}_{11}Na \longrightarrow ? + {}^{0}_{-1}e$ _____

e. $? \longrightarrow {}^{30}_{14}Si + {}^{0}_{-1}e$ _____

Answers   a. $^{0}_{-1}e$ b. $^{126}_{53}I$ c. $^{234}_{90}Th$ d. $^{24}_{12}Mg$ e. $^{30}_{13}Al$

♦ **Learning Exercise 16.2B**

Complete each of the following bombardment reactions:

a. $? + {}^{40}_{20}Ca \longrightarrow {}^{40}_{19}K + {}^{1}_{1}H$ _____

b. $^{1}_{0}n + {}^{27}_{13}Al \longrightarrow {}^{24}_{11}Na + ?$ _____

c. $^{1}_{0}n + {}^{10}_{5}B \longrightarrow ? + {}^{4}_{2}He$ _____

d. $? + {}^{23}_{11}Na \longrightarrow {}^{23}_{12}Mg + {}^{1}_{0}n$ _____

e. $^{1}_{1}H + {}^{197}_{79}Au \longrightarrow ? + {}^{1}_{0}n$ _____

Answers   a. $^{1}_{0}n$ b. $^{4}_{2}He$ c. $^{7}_{3}Li$ d. $^{1}_{1}H$ e. $^{197}_{80}Hg$

16.3 Radiation Measurement

Learning Goal: Describe the detection and measurement of radiation.

- A Geiger counter may be used to detect radiation. When radiation passes through the gas in the counter tube, some atoms of gas are ionized, producing an electrical current.
- The activity of a radioactive sample measures the number of nuclear transformations per second. The curie (Ci) is equal to 3.7×10^{10} disintegrations/s. The SI unit is the becquerel (Bq), which is equal to 1 disintegration/s.
- The radiation dose absorbed by a gram of body tissue is measured in the unit rad (radiation absorbed dose) or the SI unit gray (Gy), which is equal to 100 rad.
- The biological damage to the body caused by different types of radiation is measured in the unit rem (radiation equivalent) or the SI unit sievert (Sv), which is equal to 100 rem.

♦ **Learning Exercise 16.3A**

Match each type of measurement unit with the radiation process measured.

    a. curie   **b.** becquerel   **c.** rad   **d.** gray   **e.** rem

1. _____ an activity of 1 disintegration/s

2. _____ the amount of radiation absorbed by 1 g of material

3. _____ an activity of 3.7×10^{10} disintegrations/s

4. _____ a unit measuring the biological damage caused by different kinds of radiation

5. _____ a unit that measures the absorbed dose of radiation, which is equal to 100 rad

Answers   **1.** b   **2.** c   **3.** a   **4.** e   **5.** d

♦ **Learning Exercise 16.3B**

    a. A sample of Ir-192 used in brachytherapy to treat breast and prostate cancers has an activity of 35 μCi. What is its activity in becquerels?

    b. If an absorbed dose is 15 mrem, what is the absorbed dose in sieverts?

Answers   **a.** 1.3×10^{6} Bq   **b.** 1.5×10^{-4} Sv

16.4   Half-Life of a Radioisotope

Learning Goal: Given the half-life of a radioisotope, calculate the amount of radioisotope remaining after one or more half-lives.

- The half-life of a radioactive sample is the time required for one-half of the sample to decay (emit radiation).
- Most radioisotopes used in medicine, such as Tc-99m and I-131, have short half-lives. By comparison, many naturally occurring radioisotopes, such as C-14, Ra-226, and U-238, have long half-lives. For example, potassium-42 has a half-life of 12 h, whereas potassium-40 takes 1.3×10^{9} yr for one-half of the radioactive sample to decay.

Guide to Using Half-Lives	
STEP 1	State the given and needed quantities.
STEP 2	Write a plan to calculate the unknown quantity.
STEP 3	Write the half-life equality and conversion factors.
STEP 4	Set up the problem to calculate the needed quantity.

Example: Chromium-51, which is used to measure blood volume, has a half-life of 28 days. How many micrograms of a 16-μg sample of chromium-51 will remain active after 84 days?

Solution: **STEP 1   State the given and needed quantities.**

	Given	Need	Connect
Analyze the Problem	16 μg of Cr-51, 84 days elapsed, half-life = 28 days	micrograms of Cr-51 remaining	number of half-lives

STEP 2   Write a plan to calculate the unknown quantity.

84 days $\xrightarrow{\textit{half-life}}$ number of half-lives

16 micrograms of Cr-51 $\xrightarrow{\textit{Number of half-lives}}$ micrograms of Cr-51 remaining

STEP 3   Write the half-life equality and conversion factors.

1 half-life of Cr-51 = 28 days

$$\frac{28 \text{ days}}{1 \text{ half-life}} \quad \text{and} \quad \frac{1 \text{ half-life}}{28 \text{ days}}$$

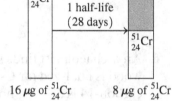

STEP 4   Set up the problem to calculate the needed quantity.

$$\text{Number of half-lives} = 84 \text{ days} \times \frac{1 \text{ half-life}}{28 \text{ days}} = 3.0 \text{ half-lives}$$

Now, we can calculate the quantity of Cr-51 that remains active.

16 μg $\xrightarrow{\text{1 half-life}}$ 8.0 μg $\xrightarrow{\text{2 half-lives}}$ 4.0 μg $\xrightarrow{\text{3 half-lives}}$ 2.0 μg

---

♦ **Learning Exercise 16.4A**

An 80.-mg sample of iodine-125 is used to treat prostate cancer. If iodine-125 has a half-life of 60. days, how many milligrams are radioactive

CORE CHEMISTRY SKILL

Using Half-Lives

a. after one half-life?

b. after two half-lives?

c. after 240 days?

Answers   **a.** 40. mg   **b.** 20. mg   **c.** 5.0 mg

♦ **Learning Exercise 16.4B**

a. $^{99m}_{43}$Tc, which is used to image the heart, brain, and lungs, has a half-life of 6.0 h. If a technician picked up a 16-mg sample at 8 A.M., how many milligrams of the radioactive sample will remain active at 8 P.M. that same day?

b. Phosphorus-32, which is used to treat leukemia, has a half-life of 14.3 days. How many micrograms of a 240-μg sample will be radioactive after 71.5 days?

c. Iodine-131, which is used for the treatment of thyroid cancer, has a half-life of 8.0 days. How many days will it take for 160 mg of I-131 to decay to 10. mg?

d. An archaeologist finds some pieces of a wooden boat at an ancient site. When a sample of the wood is analyzed for C-14, one-eighth of the original amount of C-14 remains. If the half-life of carbon-14 is 5730 yr, how many years ago was the boat made?

Answers   **a.** 4.0 mg **b.** 7.5 μg **c.** 32 days **d.** 17 200 yr

16.5   Medical Applications Using Radioactivity

Learning Goal: Describe the use of radioisotopes in medicine.

- Nuclear medicine uses radioactive isotopes that go to specific sites in the body.
- For diagnostic work, radioisotopes are used that emit gamma rays and produce nonradioactive products.
- By detecting the radiation emitted by medical radioisotopes, evaluations can be made about the location and extent of an injury, disease, tumor, blood flow, or level of function of a particular organ.

♦ **Learning Exercise 16.5**

Write the atomic symbol for each of the following radioactive isotopes:

a. _____ iodine-131, used to study thyroid gland activity

b. _____ phosphorus-32, used to locate and treat brain tumors

c. _____ sodium-24, used to determine blood flow and to locate a blood clot or embolism

d. _____ nitrogen-13, used in positron emission tomography

Answers   **a.** $^{131}_{53}$I **b.** $^{32}_{15}$P **c.** $^{24}_{11}$Na **d.** $^{13}_{7}$N

16.6   Nuclear Fission and Fusion

Learning Goal: Describe the processes of nuclear fission and fusion.

- In fission, a large nucleus breaks apart into smaller nuclei, releasing one or more types of radiation and a great amount of energy.
- A chain reaction is a fission reaction that will continue once initiated.
- In fusion, small nuclei combine to form a larger nucleus, which releases great amounts of energy.
- Nuclear fission is currently used to produce electrical power while nuclear fusion applications are still in the experimental stage.

Key Terms for Sections 16.1 to 16.6

Match each of the following key terms with the correct description:

    a. radiation   **b.** half-life   **c.** curie   **d.** nuclear fission
    e. alpha particle   **f.** positron   **g.** rem

1. _____ a particle identical to a helium nucleus produced in a radioactive nucleus

2. _____ the time required for one-half of a radioactive sample to undergo radioactive decay

3. _____ a unit of radiation measurement equal to 3.7×10^{10} disintegrations per second

4. _____ a process in which large nuclei split into smaller nuclei with the release of energy

5. _____ energy or particles released by radioactive atoms

6. _____ a measure of biological damage caused by radiation

7. _____ a particle produced when a proton is transformed into a neutron

Answers   **1.** e   **2.** b   **3.** c   **4.** d   **5.** a   **6.** g   **7.** f

♦ **Learning Exercise 16.6A**

Discuss the nuclear processes of fission and fusion for the production of energy.

Answer

Nuclear fission is a splitting of the atom into smaller nuclei accompanied by the release of large amounts of energy and radiation. In the process of *nuclear fusion*, two or more nuclei combine to form a heavier nucleus and release a large amount of energy. However, fusion requires a considerable amount of energy to initiate the process. Because the extremely high temperatures are difficult to maintain, fusion technology is still in the experimental stage.

♦ **Learning Exercise 16.6B**

Balance each of the following nuclear equations by writing the atomic symbol for the missing particle and state if each is a fission or fusion reaction:

a. $^{235}_{92}U + ^{1}_{0}n \longrightarrow ^{143}_{54}Xe + ? + 3^{1}_{0}n$ _____

b. $^{2}_{1}H + ^{2}_{1}H \longrightarrow ? + ^{1}_{0}n$ _____

c. $^3_1H + ? \longrightarrow \, ^4_2He + \, ^1_0n$ _____

d. $^{235}_{92}U + \, ^1_0n \longrightarrow \, ^{91}_{36}Kr + ? + 3^1_0n$ _____

Answers   **a.** $^{90}_{38}Sr$, fission   **b.** ^3_2He, fusion   **c.** ^2_1H, fusion   **d.** $^{142}_{56}Ba$, fission

Checklist for Chapter 16

You are ready to take the Practice Test for Chapter 16. Be sure you have accomplished the following learning goals for this chapter. If not, review the section listed at the end of the goal. Then apply your new skills and understanding to the Practice Test.

After studying Chapter 16, I can successfully:

_____ Describe alpha, beta, positron, and gamma radiation. (16.1)

_____ Write a nuclear equation showing mass numbers and atomic numbers for radioactive decay. (16.2)

_____ Write a nuclear equation for the formation of a radioactive isotope. (16.2)

_____ Describe the detection and measurement of radiation. (16.3)

_____ Calculate the amount of radioisotope remaining after one or more half-lives. (16.4)

_____ Calculate the amount of time for a specific amount of a radioisotope to decay given its half-life. (16.4)

_____ Describe the uses of radioisotopes in medicine. (16.5)

_____ Describe the processes of nuclear fission and fusion. (16.6)

Practice Test for Chapter 16

The chapter sections to review are shown in parentheses at the end of each question.

1. The correctly written symbol for an isotope of sulfur is (16.1)
   A. $^{30}_{16}Su$ **B.** $^{14}_{30}Si$ **C.** $^{30}_{16}S$ **D.** $^{30}_{16}Si$ **E.** $^{16}_{30}S$

2. Alpha particles are composed of (16.1)
   A. protons   **B.** neutrons   **C.** electrons
   D. protons and electrons   **E.** protons and neutrons

3. Gamma radiation is a type of radiation that (16.1)
   A. originates in the electron shells
   B. is most dangerous
   C. is least dangerous
   D. is the heaviest
   E. travels the shortest distance

4. The charge on an alpha particle is (16.1)
   A. −1   **B.** +1   **C.** −2   **D.** +2   **E.** +4

5. Beta particles are (16.1)
   A. protons   **B.** neutrons   **C.** electrons
   D. protons and electrons   **E.** protons and neutrons

For questions 6 through 10, select from the following: (16.1)

A. $_{-1}^{0}X$ **B.** $_{2}^{4}X$ **C.** $_{1}^{1}X$ **D.** $_{0}^{1}X$ **E.** $_{0}^{0}X$

6. an alpha particle

7. a beta particle

8. a gamma ray

9. a proton

10. a neutron

11. Shielding from gamma rays is provided by (16.1)
    A. skin   **B.** paper   **C.** clothing   **D.** lead   **E.** air

12. The skin will provide shielding from (16.1)
    A. alpha particles   **B.** beta particles   **C.** gamma rays
    D. ultraviolet rays   **E.** X-rays

13. The radioisotope iodine-131 is used as a radioactive tracer for studying thyroid activity. The correctly written symbol for iodine-131 is (16.5)
    A. I   **B.** $_{131}$I   **C.** $_{53}^{131}$I   **D.** $_{131}^{53}$I   **E.** $_{53}^{78}$I

14. When an atom emits an alpha particle, its mass number will (16.2)
    A. increase by 1   **B.** increase by 2   **C.** increase by 4
    D. decrease by 4   **E.** not change

15. When a nucleus emits a positron, the atomic number of the new nucleus (16.2)
    A. increases by 1   **B.** increases by 2   **C.** decreases by 1
    D. decreases by 2   **E.** does not change

16. When a nucleus emits a gamma ray, the atomic number of the new nucleus (16.2)
    A. increases by 1   **B.** increases by 2   **C.** decreases by 1
    D. decreases by 2   **E.** does not change

For questions 17 through 20, select the particle that completes each of the following equations: (16.2)

A. neutron   **B.** alpha particle   **C.** beta particle   **D.** gamma ray

17. $_{50}^{126}Sn \longrightarrow _{51}^{126}Sb + ?$

18. $_{43}^{99m}Tc \longrightarrow _{43}^{99}Tc + ?$

19. $_{30}^{69}Zn \longrightarrow _{31}^{69}Ga + ?$

20. $_{62}^{149}Sm \longrightarrow _{60}^{145}Nd + ?$

21. What symbol completes the following reaction? (16.2)
$$_{0}^{1}n + _{7}^{14}N \longrightarrow ? + _{1}^{1}H$$
    A. $_{8}^{15}O$ **B.** $_{6}^{15}C$ **C.** $_{8}^{14}O$ **D.** $_{6}^{14}C$ **E.** $_{7}^{15}N$

22. To complete this nuclear equation, you need to write (16.2)
$$? + _{26}^{54}Fe \longrightarrow _{28}^{57}Ni + _{0}^{1}n$$
    A. an alpha particle   **B.** a beta particle   **C.** gamma
    D. neutron   **E.** proton

23. The name of the unit used to measure the number of disintegrations per second is (16.3)
    A. curie   **B.** rad   **C.** rem   **D.** gray   **E.** sievert

24. The rem and the sievert are units used to measure (16.3)
   A. activity of a radioactive sample
   B. biological damage of different types of radiation
   C. radiation absorbed
   D. background radiation
   E. half-life of a radioactive sample

25. Radiation can cause (16.3)
   A. nausea   B. a lower white cell count   C. fatigue
   D. hair loss   E. all of these

26. The time required for a radioisotope to decay is measured by its (16.4)
   A. half-life   B. protons   C. activity   D. fusion   E. radioisotope

27. Oxygen-15, used in imaging brain function and blood flow, has a half-life of 2 min. How many half-lives have occurred in the 10 min it takes to prepare the sample? (16.4)
   A. 2   B. 3   C. 4   D. 5   E. 6

28. Iodine-131, used to treat Graves' disease as well as thyroid and prostate cancers, has a half-life of 8.0 days. How many days will it take for a 160-mg sample to decay to 10 mg? (16.4)
   A. 8.0 days   B. 16 days   C. 32 days   D. 40. days   E. 48 days

29. Phosphorus-32, used to treat leukemia and pancreatic cancer, has a half-life of 14.3 days. After 28.6 days, how many milligrams of a 120-mg sample will still be radioactive? (16.4)
   A. 240 mg   B. 120 mg   C. 60. mg   D. 30. mg   E. 15 mg

30. Radioisotopes used in medical diagnosis (16.5)
   A. have short half-lives   B. emit only gamma rays   C. locate in specific organs
   D. produce nonradioactive nuclei   E. all of these

31. The "splitting" of a large nucleus to form smaller particles accompanied by a release of energy is called (16.6)
   A. radioisotope   B. nuclear fission   C. nuclear fusion   D. bombardment   E. half-life

32. The process of combining small nuclei to form larger nuclei is (16.6)
   A. radioisotope   B. nuclear fission   C. nuclear fusion   D. bombardment   E. half-life

33. A fusion reaction (16.6)
   A. occurs in the Sun
   B. forms larger nuclei from smaller nuclei
   C. requires extremely high temperatures
   D. releases a large amount of energy
   E. all of these

Answers to the Practice Test

1. C	**2.** E	**3.** B	**4.** D	**5.** C
6. B	**7.** A	**8.** E	**9.** C	**10.** D
11. D	**12.** A	**13.** C	**14.** D	**15.** C
16. E	**17.** C	**18.** D	**19.** C	**20.** B
21. D	**22.** A	**23.** A	**24.** B	**25.** E
26. A	**27.** D	**28.** C	**29.** D	**30.** E
31. B	**32.** C	**33.** E		

Selected Answers and Solutions to Text Problems

16.1 a. ^4_2He is the symbol for an alpha particle.
  b. $^0_{+1}e$ is the symbol for a positron.
  c. $^0_0\gamma$ is the symbol for gamma radiation.

16.3 a. Since the element is potassium, the element symbol is K and the atomic number is 19. The potassium isotopes with mass number 39, 40, and 41 will have the atomic symbol $^{39}_{19}\text{K}$, $^{40}_{19}\text{K}$, and $^{41}_{19}\text{K}$, respectively.
  b. Each isotope has 19 protons and 19 electrons, but they differ in the number of neutrons present. Potassium-39 has $39 - 19 = 20$ neutrons, potassium-40 has $40 - 19 = 21$ neutrons, and potassium-41 has $41 - 19 = 22$ neutrons.

16.5 a. beta particle ($^0_{-1}e, \beta$)
  b. alpha particle ($^4_2\text{He}, \alpha$)
  c. neutron ($^1_0n, n$)
  d. argon-38 ($^{38}_{18}\text{Ar}$)
  e. carbon-14 ($^{14}_6\text{C}$)

16.7 a. Since the element is copper, the element symbol is Cu and the atomic number is 29. The copper isotope with mass number 64 will have the atomic symbol $^{64}_{29}\text{Cu}$.
  b. Since the element is selenium, the element symbol is Se and the atomic number is 34. The selenium isotope with mass number 75 will have the atomic symbol $^{75}_{34}\text{Se}$.
  c. Since the element is sodium, the element symbol is Na and the atomic number is 11. The sodium isotope with mass number 24 will have the atomic symbol $^{24}_{11}\text{Na}$.
  d. Since the element is nitrogen, the element symbol is N and the atomic number is 7. The nitrogen isotope with mass number 15 will have the atomic symbol $^{15}_7\text{N}$.

16.9

Medical Use	Atomic Symbol	Mass Number	Number of Protons	Number of Neutrons
Heart imaging	$^{201}_{81}\text{Tl}$	201	81	120
Radiation therapy	$^{60}_{27}\text{Co}$	60	27	33
Abdominal scan	$^{67}_{31}\text{Ga}$	67	31	36
Hyperthyroidism	$^{131}_{53}\text{I}$	131	53	78
Leukemia treatment	$^{32}_{15}\text{P}$	32	15	17

16.11 a. 1. Alpha particles do not penetrate the skin.
  b. 3. Gamma radiation requires shielding protection that includes lead or thick concrete.
  c. 1. Alpha particles can be very harmful if ingested.

16.13 The mass number of the radioactive atom is reduced by four when an alpha particle (^4_2He) is emitted. The unknown product will have an atomic number that is two less than the atomic number of the radioactive atom.
  a. $^{208}_{84}\text{Po} \longrightarrow {}^{204}_{82}\text{Pb} + {}^4_2\text{He}$
  b. $^{232}_{90}\text{Th} \longrightarrow {}^{228}_{88}\text{Ra} + {}^4_2\text{He}$
  c. $^{251}_{102}\text{No} \longrightarrow {}^{247}_{100}\text{Fm} + {}^4_2\text{He}$
  d. $^{220}_{86}\text{Rn} \longrightarrow {}^{216}_{84}\text{Po} + {}^4_2\text{He}$

16.15 The mass number of the radioactive atom is not changed when a beta particle ($_{-1}^{0}e$) is emitted. The unknown product will have an atomic number that is one greater than the atomic number of the radioactive atom.

a. $_{11}^{25}Na \longrightarrow {}_{12}^{25}Mg + {}_{-1}^{0}e$

b. $_{8}^{20}O \longrightarrow {}_{9}^{20}F + {}_{-1}^{0}e$

c. $_{38}^{92}Sr \longrightarrow {}_{39}^{92}Y + {}_{-1}^{0}e$

d. $_{26}^{60}Fe \longrightarrow {}_{27}^{60}Co + {}_{-1}^{0}e$

16.17 The mass number of the radioactive atom is not changed when a positron ($_{+1}^{0}e$) is emitted. The unknown product will have an atomic number that is one less than the atomic number of the radioactive atom.

a. $_{14}^{26}Si \longrightarrow {}_{13}^{26}Al + {}_{+1}^{0}e$

b. $_{27}^{54}Co \longrightarrow {}_{26}^{54}Fe + {}_{+1}^{0}e$

c. $_{37}^{77}Rb \longrightarrow {}_{36}^{77}Kr + {}_{+1}^{0}e$

d. $_{45}^{93}Rh \longrightarrow {}_{44}^{93}Ru + {}_{+1}^{0}e$

16.19 Balance the mass numbers and the atomic numbers in each nuclear equation.

a. $_{13}^{28}Al \longrightarrow {}_{14}^{28}Si + {}_{-1}^{0}e \qquad ? = {}_{14}^{28}Si \qquad$ beta decay

b. $_{73}^{180m}Ta \longrightarrow {}_{73}^{180}Ta + {}_{0}^{0}\gamma \qquad ? = {}_{0}^{0}\gamma \qquad$ gamma emission

c. $_{29}^{66}Cu \longrightarrow {}_{30}^{66}Zn + {}_{-1}^{0}e \qquad ? = {}_{-1}^{0}e \qquad$ beta decay

d. $_{92}^{238}U \longrightarrow {}_{90}^{234}Th + {}_{2}^{4}He \qquad ? = {}_{92}^{238}U \qquad$ alpha decay

e. $_{80}^{188}Hg \longrightarrow {}_{79}^{188}Au + {}_{+1}^{0}e \qquad ? = {}_{79}^{188}Au \qquad$ positron emission

16.21 Balance the mass numbers and the atomic numbers in each nuclear equation.

a. $_{0}^{1}n + {}_{4}^{9}Be \longrightarrow {}_{4}^{10}Be \qquad\qquad ? = {}_{4}^{10}Be$

b. $_{0}^{1}n + {}_{52}^{131}Te \longrightarrow {}_{53}^{132}I + {}_{-1}^{0}e \qquad ? = {}_{53}^{132}I$

c. $_{0}^{1}n + {}_{13}^{27}Al \longrightarrow {}_{11}^{24}Na + {}_{2}^{4}He \qquad ? = {}_{13}^{27}Al$

d. $_{2}^{4}He + {}_{7}^{14}N \longrightarrow {}_{8}^{17}O + {}_{1}^{1}H \qquad ? = {}_{8}^{17}O$

16.23 a. **2.** <u>Absorbed dose</u> can be measured in rad.

b. **3.** <u>Biological damage</u> can be measured in mrem.

c. **1.** <u>Activity</u> can be measured in mCi.

d. **2.** <u>Absorbed dose</u> can be measured in Gy.

16.25 $8 \; \cancel{mGy} \times \dfrac{1 \; \cancel{Gy}}{1000 \; \cancel{mGy}} \times \dfrac{100 \; rad}{1 \; \cancel{Gy}} = 0.8 \; rad \; (1 \; SF)$

Thus, a technician exposed to a 5-rad dose of radiation received more radiation than one exposed to 8 mGy (0.8 rad) of radiation.

16.27 a. $70.0 \; \cancel{kg \; body \; mass} \times \dfrac{4.20 \; \mu Ci}{1 \; \cancel{kg \; body \; mass}} = 294 \; \mu Ci \; (3 \; SFs)$

b. From 50 rad of gamma radiation (we use a biological damage factor of 1, so 1 rad of gamma radiation = 1 rem):

$50 \; \cancel{rad} \times \dfrac{1 \; Gy}{100 \; \cancel{rad}} = 0.5 \; Gy \; (1 \; SF)$

16.29 a. **2.** two half-lives:

$34 \; \cancel{days} \times \dfrac{1 \; half\text{-}life}{17 \; \cancel{days}} = 2.0 \; half\text{-}lives$

b. 1. one half-life:

$$20 \text{ min} \times \frac{1 \text{ half-life}}{20 \text{ min}} = 1 \text{ half-life}$$

c. 3. three half-lives:

$$21 \text{ h} \times \frac{1 \text{ half-life}}{7 \text{ h}} = 3 \text{ half-lives}$$

16.31 a. After one half-life, one-half of the sample would be radioactive:

$$80.0 \text{ mg of } {}^{99m}_{43}\text{Tc} \xrightarrow{1 \text{ half-life}} 40.0 \text{ mg of } {}^{99m}_{43}\text{Tc} \text{ (3 SFs)}$$

b. After two half-lives, one-fourth of the sample would still be radioactive:

$$80.0 \text{ mg of } {}^{99m}_{43}\text{Tc} \xrightarrow{1 \text{ half-life}} 40.0 \text{ mg of } {}^{99m}_{43}\text{Tc} \xrightarrow{2 \text{ half-lives}} 20.0 \text{ mg of } {}^{99m}_{43}\text{Tc} \text{ (3 SFs)}$$

c. $18 \text{ h} \times \dfrac{1 \text{ half-life}}{6.0 \text{ h}} = 3.0 \text{ half-lives}$

$$80.0 \text{ mg of } {}^{99m}_{43}\text{Tc} \xrightarrow{1 \text{ half-life}} 40.0 \text{ mg of } {}^{99m}_{43}\text{Tc} \xrightarrow{2 \text{ half-lives}}$$

$$20.0 \text{ mg of } {}^{99m}_{43}\text{Tc} \xrightarrow{3 \text{ half-lives}} 10.0 \text{ mg of } {}^{99m}_{43}\text{Tc} \text{ (3 SFs)}$$

d. $24 \text{ h} \times \dfrac{1 \text{ half-life}}{6.0 \text{ h}} = 4.0 \text{ half-lives}$

$$80.0 \text{ mg of } {}^{99m}_{43}\text{Tc} \xrightarrow{1 \text{ half-life}} 40.0 \text{ mg of } {}^{99m}_{43}\text{Tc} \xrightarrow{2 \text{ half-lives}} 20.0 \text{ mg of } {}^{99m}_{43}\text{Tc} \xrightarrow{3 \text{ half-lives}}$$

$$10.0 \text{ mg of } {}^{99m}_{43}\text{Tc} \xrightarrow{4 \text{ half-lives}} 5.00 \text{ mg of } {}^{99m}_{43}\text{Tc} \text{ (3 SFs)}$$

16.33 The radiation level in a radioactive sample is cut in half with each half-life; the half-life of Sr-85 is 65 days.

a. For the radiation level to drop to one-fourth of its original level, $\frac{1}{4} = \frac{1}{2} \times \frac{1}{2}$ or two half-lives

$$2 \text{ half-lives} \times \frac{65 \text{ days}}{1 \text{ half-life}} = 130 \text{ days} \text{ (2 SFs)}$$

b. For the radiation level to drop to one-eighth of its original level, $\frac{1}{8} = \frac{1}{2} \times \frac{1}{2} \times \frac{1}{2}$ or three half-lives

$$3 \text{ half-lives} \times \frac{65 \text{ days}}{1 \text{ half-life}} = 195 \text{ days} \text{ (2 SFs)}$$

16.35 a. Since the elements calcium and phosphorus are part of bone, any calcium or phosphorus atom, regardless of isotope, will be carried to and become part of the bony structures of the body. Once there, the radiation emitted by any radioactive isotope can be used to diagnose or treat bone diseases.

b. Strontium (Sr) acts much like calcium (Ca) because both are Group 2A (2) elements. The body will accumulate radioactive strontium in bones in the same way that it incorporates calcium. Radioactive strontium is harmful to children because the radiation it produces causes more damage in cells that are dividing rapidly.

16.37 $4.0 \text{ mL solution} \times \dfrac{45 \text{ } \mu\text{Ci}}{1 \text{ mL solution}} = 180 \text{ } \mu\text{Ci of selenium-75} \text{ (2 SFs)}$

16.39 Nuclear fission is the splitting of a large atom into smaller fragments with a simultaneous release of large amounts of energy.

16.41 ${}^{1}_{0}n + {}^{235}_{92}\text{U} \longrightarrow {}^{131}_{50}\text{Sn} + {}^{103}_{42}\text{Mo} + 2{}^{1}_{0}n + \text{energy} \quad ? = {}^{103}_{42}\text{Mo}$

16.43 a. Neutrons bombard a nucleus in the <u>fission</u> process.
  b. The nuclear process that occurs in the Sun is <u>fusion</u>.
  c. <u>Fission</u> is the process in which a large nucleus splits into smaller nuclei.
  d. <u>Fusion</u> is the process in which small nuclei combine to form larger nuclei.

16.45 a. $74 \text{ MBq} \times \dfrac{1 \times 10^6 \text{ Bq}}{1 \text{ MBq}} \times \dfrac{1 \text{ Ci}}{3.7 \times 10^{10} \text{ Bq}} = 2.0 \times 10^{-3} \text{ Ci (2 SFs)}$

  b. $74 \text{ MBq} \times \dfrac{1 \times 10^6 \text{ Bq}}{1 \text{ MBq}} \times \dfrac{1 \text{ Ci}}{3.7 \times 10^{10} \text{ Bq}} \times \dfrac{1 \times 10^3 \text{ mCi}}{1 \text{ Ci}} = 2.0 \text{ mCi (2 SFs)}$

16.47 For the activity to drop to one-eighth of its original level, $\frac{1}{8} = \frac{1}{2} \times \frac{1}{2} \times \frac{1}{2}$ or three half-lives

$3 \text{ half-lives} \times \dfrac{3.0 \text{ days}}{1 \text{ half-life}} = 9.0 \text{ days (2 SFs)}$

16.49

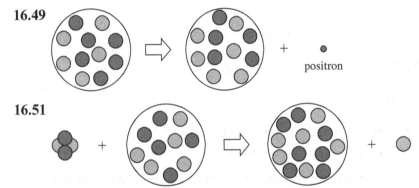

16.51

16.53 Half of a radioactive sample decays with each half-life:

$6.4 \ \mu\text{Ci of } ^{14}_{6}\text{C} \xrightarrow{1 \text{ half-life}} 3.2 \ \mu\text{Ci of } ^{14}_{6}\text{C} \xrightarrow{2 \text{ half-lives}} 1.6 \ \mu\text{Ci of } ^{14}_{6}\text{C} \xrightarrow{3 \text{ half-lives}}$
  $0.80 \ \mu\text{Ci of } ^{14}_{6}\text{C}$

  $\therefore$ The activity of carbon-14 drops to $0.80 \ \mu\text{Ci}$ in three half-lives or 3×5730 yr, which makes the age of the painting 17 200 yr (2 SFs).

16.55 a. $^{25}_{11}\text{Na}$ has 11 protons and $25 - 11 = 14$ neutrons.
  b. $^{61}_{28}\text{Ni}$ has 28 protons and $61 - 28 = 33$ neutrons.
  c. $^{84}_{37}\text{Rb}$ has 37 protons and $84 - 37 = 47$ neutrons.
  d. $^{110}_{47}\text{Ag}$ has 47 protons and $110 - 47 = 63$ neutrons.

16.57 a. gamma emission
  b. positron emission
  c. alpha decay

16.59 a. $^{225}_{90}\text{Th} \longrightarrow ^{221}_{88}\text{Ra} + ^{4}_{2}\text{He}$
  b. $^{210}_{83}\text{Bi} \longrightarrow ^{206}_{81}\text{Tl} + ^{4}_{2}\text{He}$
  c. $^{137}_{55}\text{Cs} \longrightarrow ^{137}_{56}\text{Ba} + ^{0}_{-1}e$
  d. $^{126}_{50}\text{Sn} \longrightarrow ^{126}_{51}\text{Sb} + ^{0}_{-1}e$
  e. $^{18}_{9}\text{F} \longrightarrow ^{18}_{8}\text{O} + ^{0}_{+1}e$

16.61 a. $^{4}_{2}\text{He} + ^{14}_{7}\text{N} \longrightarrow ^{17}_{8}\text{O} + ^{1}_{1}\text{H}$ $? = ^{17}_{8}\text{O}$
  b. $^{4}_{2}\text{He} + ^{27}_{13}\text{Al} \longrightarrow ^{30}_{14}\text{Si} + ^{1}_{1}\text{H}$ $? = ^{1}_{1}\text{H}$
  c. $^{1}_{0}n + ^{235}_{92}\text{U} \longrightarrow ^{90}_{38}\text{Sr} + 3^{1}_{0}n + ^{143}_{54}\text{Xe}$ $? = ^{143}_{54}\text{Xe}$
  d. $^{23m}_{12}\text{Mg} \longrightarrow ^{23}_{12}\text{Mg} + ^{0}_{0}\gamma$ $? = ^{23}_{12}\text{Mg}$

16.63 a. $^{16}_{8}O + ^{16}_{8}O \longrightarrow ^{28}_{14}Si + ^{4}_{2}He$

    b. $^{18}_{8}O + ^{249}_{98}Cf \longrightarrow ^{263}_{106}Sg + 4^{1}_{0}n$

    c. $^{222}_{86}Rn \longrightarrow ^{218}_{84}Po + ^{4}_{2}He$

    d. $^{80}_{38}Sr \longrightarrow ^{80}_{37}Rb + ^{0}_{+1}e$

16.65 First, calculate the number of half-lives that have elapsed:

$$24 \text{ h} \times \frac{1 \text{ half-life}}{6.0 \text{ h}} = 4.0 \text{ half-lives}$$

Now we can calculate the number of milligrams of technicium-99m that remain:

120 mg of $^{99m}_{43}Tc$ $\xrightarrow{\text{1 half-life}}$ 60. mg of $^{99m}_{43}Tc$ $\xrightarrow{\text{2 half-lives}}$ 30. mg of $^{99m}_{43}Tc$ $\xrightarrow{\text{3 half-lives}}$
15 mg of $^{99m}_{43}Tc$ $\xrightarrow{\text{4 half-lives}}$ 7.5 mg of $^{99m}_{43}Tc$ (2 SFs)

16.67 In the fission process, an atom splits into smaller nuclei with a simultaneous release of large amounts of energy. In fusion, two (or more) small nuclei combine (fuse) to form a larger nucleus, with a simultaneous release of large amounts of energy.

16.69 Fusion occurs naturally in the Sun and other stars.

16.71 $120 \text{ nCi} \times \dfrac{1 \times 10^{-9} \text{ Ci}}{1 \text{ nCi}} \times \dfrac{3.7 \times 10^{10} \text{ Bq}}{1 \text{ Ci}} = 4400 \text{ Bq} = 4.4 \times 10^{3} \text{ Bq (2 SFs)}$

16.73 Half of a radioactive sample decays with each half-life:

1.2 mg of $^{32}_{15}P$ $\xrightarrow{\text{1 half-life}}$ 0.60 mg of $^{32}_{15}P$ $\xrightarrow{\text{2 half-lives}}$ 0.30 mg of $^{32}_{15}P$

∴ Two half-lives must have elapsed during this time (28.6 days), yielding the half-life for phosphorus-32:

$$\frac{28.6 \text{ days}}{2 \text{ half-lives}} = 14.3 \text{ days/half-life (3 SFs)}$$

16.75 a. $^{47}_{20}Ca \longrightarrow ^{47}_{21}Sc + ^{0}_{-1}e$

    b. First, calculate the number of half-lives that have elapsed:

$$18 \text{ days} \times \frac{1 \text{ half-life}}{4.5 \text{ days}} = 4.0 \text{ half-lives}$$

Now we can calculate the number of milligrams of calcium-47 that remain:

16 mg of $^{47}_{20}Ca$ $\xrightarrow{\text{1 half-life}}$ 8.0 mg of $^{47}_{20}Ca$ $\xrightarrow{\text{2 half-lives}}$ 4.0 mg of $^{47}_{20}Ca$ $\xrightarrow{\text{3 half-lives}}$
2.0 mg of $^{47}_{20}Ca$ $\xrightarrow{\text{4 half-lives}}$ 1.0 mg of $^{47}_{20}Ca$

    c. Half of a radioactive sample decays with each half-life:

4.8 mg of $^{47}_{20}Ca$ $\xrightarrow{\text{1 half-life}}$ 2.4 mg of $^{47}_{20}Ca$ $\xrightarrow{\text{2 half-lives}}$ 1.2 mg of $^{47}_{20}Ca$

∴ Two half-lives have elapsed.

$$2 \text{ half-lives} \times \frac{4.5 \text{ days}}{1 \text{ half-life}} = 9.0 \text{ days (2 SFs)}$$

16.77 a. $^{180}_{80}Hg \longrightarrow ^{176}_{78}Pt + ^{4}_{2}He$

    b. $^{198}_{79}Au \longrightarrow ^{198}_{80}Hg + ^{0}_{-1}e$

    c. $^{82}_{37}Rb \longrightarrow ^{82}_{36}Kr + ^{0}_{+1}e$

16.79 $^{1}_{0}n + ^{238}_{92}U \longrightarrow ^{239}_{93}Np + ^{0}_{-1}e$

16.81 Half of a radioactive sample decays with each half-life:

$$64 \ \mu\text{Ci of } ^{201}_{81}\text{Tl} \xrightarrow{\text{1 half-life}} 32 \ \mu\text{Ci of } ^{201}_{81}\text{Tl} \xrightarrow{\text{2 half-lives}} 16 \ \mu\text{Ci of } ^{201}_{81}\text{Tl} \xrightarrow{\text{3 half-lives}}$$

$$8.0 \ \mu\text{Ci of } ^{201}_{81}\text{Tl} \xrightarrow{\text{4 half-lives}} 4.0 \ \mu\text{Ci of } ^{201}_{81}\text{Tl}$$

∴ Four half-lives must have elapsed during this time (12 days), yielding the half-life for thallium-201:

$$\frac{12 \text{ days}}{4 \text{ half-lives}} = 3.0 \text{ days/half-life (2 SFs)}$$

16.83 First, calculate the number of half-lives that have elapsed:

$$27 \ \cancel{\text{days}} \times \frac{1 \text{ half-life}}{4.5 \ \cancel{\text{days}}} = 6.0 \text{ half-lives}$$

Because the activity of a radioactive sample is cut in half with each half-life, the activity must have been double its present value before each half-life. For 6.0 half-lives, we need to double the value six times:

$$1.0 \ \mu\text{Ci of } ^{47}_{20}\text{Ca} \xleftarrow{\text{1 half-life}} 2.0 \ \mu\text{Ci of } ^{47}_{20}\text{Ca} \xleftarrow{\text{2 half-lives}} 4.0 \ \mu\text{Ci of } ^{47}_{20}\text{Ca} \xleftarrow{\text{3 half-lives}} 8.0 \ \mu\text{Ci of } ^{47}_{20}\text{Ca}$$

$$\xleftarrow{\text{4 half-lives}} 16 \ \mu\text{Ci of } ^{47}_{20}\text{Ca} \xleftarrow{\text{5 half-lives}} 32 \ \mu\text{Ci of } ^{47}_{20}\text{Ca} \xleftarrow{\text{6 half-lives}} 64 \ \mu\text{Ci of } ^{47}_{20}\text{Ca}$$

∴ The initial activity of the sample was 64 μCi (2 SFs).

16.85 $^{48}_{20}\text{Ca} + ^{244}_{94}\text{Pu} \longrightarrow ^{292}_{114}\text{Fl}$

16.87 First, calculate the number of half-lives that have elapsed since the technician was exposed:

$$36 \ \cancel{\text{h}} \times \frac{1 \text{ half-life}}{12 \ \cancel{\text{h}}} = 3.0 \text{ half-lives}$$

Because the activity of a radioactive sample is cut in half with each half-life, the activity must have been double its present value before each half-life. For 3.0 half-lives, we need to double the value three times:

$$2.0 \ \mu\text{Ci of } ^{42}_{19}\text{K} \xleftarrow{\text{1 half-life}} 4.0 \ \mu\text{Ci of } ^{42}_{19}\text{K} \xleftarrow{\text{2 half-lives}} 8.0 \ \mu\text{Ci of } ^{42}_{19}\text{K} \xleftarrow{\text{3 half-lives}} 16 \ \mu\text{Ci of } ^{42}_{19}\text{K}$$

∴ The activity of the sample when the technician was exposed was 16 μCi (2 SFs).

Selected Answers to Combining Ideas from Chapters 15 and 16

CI.33 a. $C_2O_4^{2-}(aq) \longrightarrow 2CO_2(g)$

Balance charge with e^-: $C_2O_4^{2-}(aq) \longrightarrow 2CO_2(g) + 2\,e^-$ balanced oxidation half-reaction

b. $MnO_4^-(aq) \longrightarrow Mn^{2+}(aq)$

Balance O with H_2O: $MnO_4^-(aq) \longrightarrow Mn^{2+}(aq) + 4H_2O(l)$

Balance H with H^+: $8H^+(aq) + MnO_4^-(aq) \longrightarrow Mn^{2+}(aq) + 4H_2O(l)$

Balance charge with e^-: $5\,e^- + 8H^+(aq) + MnO_4^-(aq) \longrightarrow Mn^{2+}(aq) + 4H_2O(l)$

                                         balanced reduction half-reaction

c. $2 \times [\,5\,e^- + 8H^+(aq) + MnO_4^-(aq) \longrightarrow Mn^{2+}(aq) + 4H_2O(l)\,]$

                    $5 \times [\,C_2O_4^{2-}(aq) \longrightarrow 2CO_2(g) + 2\,e^-\,]$

Overall: $16H^+(aq) + 2MnO_4^-(aq) + 5C_2O_4^{2-}(aq) \longrightarrow 2Mn^{2+}(aq) + 10CO_2(g) + 8H_2O(l)$

d. In the titration equation, 2 mol of MnO_4^- reacts with 5 mol of $C_2O_4^{2-}$.

$$0.758 \text{ g Na}_2\text{C}_2\text{O}_4 \times \frac{1 \text{ mol Na}_2\text{C}_2\text{O}_4}{134.00 \text{ g Na}_2\text{C}_2\text{O}_4} \times \frac{1 \text{ mol C}_2\text{O}_4^{2-}}{1 \text{ mol Na}_2\text{C}_2\text{O}_4} \times \frac{2 \text{ mol MnO}_4^-}{5 \text{ mol C}_2\text{O}_4^{2-}} \times \frac{1 \text{ mol KMnO}_4}{1 \text{ mol MnO}_4^-}$$

$= 0.002\ 26$ mol of $KMnO_4$

$$24.6 \text{ mL KMnO}_4 \text{ solution} \times \frac{1 \text{ L solution}}{1000 \text{ mL solution}} = 0.0246 \text{ L of KMnO}_4 \text{ solution}$$

$$\text{Molarity of KMnO}_4 = \frac{\text{moles of solute}}{\text{liters of solution}} = \frac{0.002\ 26 \text{ mol KMnO}_4}{0.0246 \text{ L solution}}$$

$= 0.0919$ M $KMnO_4$ solution (3 SFs)

CI.35 a. $0.121 \text{ g Mg} \times \dfrac{1 \text{ mol Mg}}{24.31 \text{ g Mg}} \times \dfrac{1 \text{ mol H}_2}{1 \text{ mol Mg}} = 0.004\ 98$ mol of H_2 (smaller amount of product)

$$50.0 \text{ mL solution} \times \frac{1 \text{ L}}{1000 \text{ mL}} \times \frac{1.00 \text{ mol HCl}}{1 \text{ L solution}} \times \frac{1 \text{ mol H}_2}{2 \text{ mol HCl}} = 0.0250 \text{ mol of H}_2$$

$\therefore$ Mg is the limiting reactant.

b. $T = 33\,°C + 273 = 306$ K; $P = 750.$ mmHg; $n = 0.004\ 98$ mol of H_2 (from **a**)

$$V = \frac{nRT}{P} = \frac{(0.004\ 98 \text{ mol})\left(\dfrac{62.4 \text{ L} \cdot \text{mmHg}}{\text{mol} \cdot \text{K}}\right)(306 \text{ K})}{750. \text{ mmHg}} \times \frac{1000 \text{ mL}}{1 \text{ L}}$$

$= 127$ mL of H_2 (3 SFs)

c. $50.0 \text{ mL solution} \times \dfrac{1.00 \text{ g}}{1 \text{ mL}} = 50.0$ g of solution

$\Delta T = T_{final} - T_{initial} = 33\,°C - 22\,°C = 11\,°C$

$$50.0 \text{ g solution} \times \frac{4.184 \text{ J}}{\text{g} \,°C} \times 11\,°C = 2.3 \times 10^3 \text{ J released (2 SFs)}$$

d. $\dfrac{-2.3 \times 10^3 \text{ J}}{0.121 \text{ g Mg}} = -1.9 \times 10^4$ J/g of Mg (2 SFs)

$$\frac{-2.3 \times 10^3 \text{ J}}{0.121 \text{ g Mg}} \times \frac{1 \text{ kJ}}{1000 \text{ J}} \times \frac{24.31 \text{ g Mg}}{1 \text{ mol Mg}} = -460 \text{ kJ/mol of Mg (2 SFs)}$$

CI.37 a.

Isotope	Number of Protons	Number of Neutrons	Number of Electrons
$^{27}_{14}Si$	14	13	14
$^{28}_{14}Si$	14	14	14
$^{29}_{14}Si$	14	15	14
$^{30}_{14}Si$	14	16	14
$^{31}_{14}Si$	14	17	14

b. Electron configuration of Si: $1s^2 2s^2 2p^6 3s^2 3p^2$
Abbreviated electron configuration: $[Ne]3s^2 3p^2$

c. $^{28}_{14}Si$ $27.977 \text{ amu} \times \dfrac{92.230}{100} = 25.803 \text{ amu}$

$^{29}_{14}Si$ $28.976 \text{ amu} \times \dfrac{4.683}{100} = 1.357 \text{ amu}$

$^{30}_{14}Si$ $29.974 \text{ amu} \times \dfrac{3.087}{100} = 0.9253 \text{ amu}$

$$\text{Atomic mass of Si} = 28.085 \text{ amu}$$

d. $^{27}_{14}Si \longrightarrow {}^{27}_{13}Al + {}^{0}_{+1}e$
$^{31}_{14}Si \longrightarrow {}^{31}_{15}P + {}^{0}_{-1}e$

e.

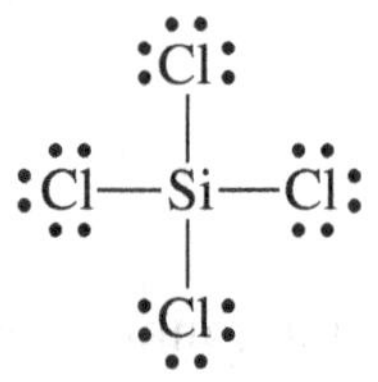

$$:\ddot{C}l\!-\!Si\!-\!\ddot{C}l:$$

The central atom Si has four electron groups bonded to four chlorine atoms; $SiCl_4$ has a tetrahedral shape.

f. Half of a radioactive sample decays with each half-life:

16 μCi of $^{31}_{14}Si$ $\xrightarrow{\text{1 half-life}}$ 8.0 μCi of $^{31}_{14}Si$ $\xrightarrow{\text{2 half-lives}}$ 4.0 μCi of $^{31}_{14}Si$ $\xrightarrow{\text{3 half-lives}}$ 2.0 μCi of $^{31}_{14}Si$
Therefore, 3 half-lives have passed.

$$3 \text{ half-lives} \times \dfrac{2.6 \text{ h}}{1 \text{ half-life}} = 7.8 \text{ h (2 SFs)}$$

CI.39 a. $^{238}_{92}U \longrightarrow {}^{234}_{90}Th + {}^{4}_{2}He$ $? = {}^{4}_{2}He$
b. $^{234}_{90}Th \longrightarrow {}^{234}_{91}Pa + {}^{0}_{-1}e$ $? = {}^{234}_{91}Pa$
c. $^{226}_{88}Ra \longrightarrow {}^{222}_{86}Rn + {}^{4}_{2}He$ $? = {}^{226}_{88}Ra$

Organic Chemistry

For 3 months after her house fire, Diane remained in the hospital's burn unit, where she continued to have treatment for burns, including removing damaged tissue, skin grafts, and preventing infection. For some of the large burn areas, some of Diane's own skin cells were removed and grown for three weeks to give a layer of skin that covered the burn areas. By covering the burn areas, the incidence of dehydration and infection were decreased. After 3 months, she was discharged.

The arson investigators determined that gasoline was the primary accelerant used to start the fire at Diane's house. They also determined there was kerosene and some diesel fuel. About 70% of kerosene consists of alkanes; about 30% of kerosene is aromatic hydrocarbons. Write a balanced chemical equation for the combustion of $C_{11}H_{24}$, which is a compound in kerosene.

Credit: Andrew Poertner / AP Images

CHAPTER READINESS*

❀ Core Chemistry Skills

- Balancing a Chemical Equation (8.2)
- Drawing Lewis Structures (10.1)
- Predicting Shape (10.3)

*These Core Chemistry Skills from previous chapters are listed here for your review as you proceed to the new material in this chapter.

LOOKING AHEAD

17.1 Alkanes

Learning Goal: Write the IUPAC names and draw the condensed or line-angle structural formulas for alkanes.

- Organic compounds are compounds of carbon and hydrogen that have covalent bonds, have low melting and boiling points, burn vigorously, are mostly nonpolar molecules, and are usually more soluble in nonpolar solvents than in water.

- Alkanes are hydrocarbons that have only single bonds, C—C and C—H.

- Each carbon in an alkane has four bonds arranged so that the bonded atoms are in the corners of a tetrahedron.

- In the IUPAC system, each substituent is numbered and listed alphabetically in front of the name of the longest chain.

- Structural isomers are compounds that have the same molecular formula but different arrangements of atoms.
- Carbon groups that are substituents are named as alkyl groups, by replacing the *ane* of the alkane name with *yl*. For example, CH_3— is methyl (from CH_4 methane), and CH_3—CH_2— is ethyl (from CH_3—CH_3 ethane).
- When a halogen atom, —F, —Cl, —Br, or —I, replaces a hydrogen atom in an alkane, it named as a substituent (*fluoro, chloro, bromo, iodo*).
- In an IUPAC (International Union of Pure and Applied Chemistry) name, the prefix indicates the number of carbon atoms, and the suffix describes the family of the compound. For example, in the name *propane*, the prefix *prop* indicates a chain of three carbon atoms and the ending *ane* indicates single bonds (alkane). The names of the first six alkanes follow:

Name	Number of Carbon Atoms	Condensed Structural Formula	Line-Angle Structural Formula
Methane	1	CH_4	
Ethane	2	CH_3—CH_3	
Propane	3	CH_3—CH_2—CH_3	
Butane	4	CH_3—CH_2—CH_2—CH_3	
Pentane	5	CH_3—CH_2—CH_2—CH_2—CH_3	
Hexane	6	CH_3—CH_2—CH_2—CH_2—CH_2—CH_3	

- Alkanes with seven or more carbon atoms in a chain are named using the prefix *hept* (7), *oct* (8), *non* (9), or *dec* (10) and the suffix *ane*, which indicates single bonds (alkane).
- In combustion, an alkane reacts rapidly with oxygen at a high temperature to produce carbon dioxide, water, and a great amount of heat.

♦ **Learning Exercise 17.1A**

Identify each of the following as characteristic of organic (O) or inorganic (I) compounds:

a. _____ have covalent bonds   **b.** _____ have low boiling points

c. _____ burn in air   **d.** _____ are soluble in water

e. _____ have high melting points   **f.** _____ have ionic bonds

g. _____ contain carbon and hydrogen **h.** _____ do not burn in air

Answers   **a.** O   **b.** O   **c.** O   **d.** I   **e.** I
             f. I   **g.** O   **h.** I

Vegetable oil, a mixture of organic compounds, is not soluble in water.

Credit: Pearson Education / Pearson Science

	Guide to Drawing Structural Formulas for Alkanes
STEP 1	Draw the carbon chain.
STEP 2	Draw the expanded structural formula by adding the hydrogen atoms using single bonds to each of the carbon atoms.
STEP 3	Draw the condensed structural formula by combining the H atoms with each C atom.
STEP 4	Draw the line-angle structural formula as a zigzag line in which the ends and corners represent C atoms.

◆ **Learning Exercise 17.1B**

🔺 **CORE CHEMISTRY SKILL**

Naming and Drawing Alkanes

Draw the condensed structural formula and give the name for the continuous-chain alkane with each of the following molecular formulas:

a. C_2H_6 _____

b. C_4H_{10} _____

c. C_5H_{12} _____

d. C_6H_{14} _____

Answers   **a.** CH_3—CH_3, ethane

   b. CH_3—CH_2—CH_2—CH_3, butane

   c. CH_3—CH_2—CH_2—CH_2—CH_3, pentane

   d. CH_3—CH_2—CH_2—CH_2—CH_2—CH_3, hexane

◆ **Learning Exercise 17.1C**

Draw the line-angle structural formula for each of the following:

a. butane   _____

b. pentane   _____

c. hexane   _____

d. octane   _____

Answers   **a.**    **b.**    **c.**

   d.

Guide to Naming Alkanes with Substituents	
STEP 1	Write the alkane name for the longest chain of carbon atoms.
STEP 2	Number the carbon atoms from the end nearer a substituent.
STEP 3	Give the location and name for each substituent (alphabetical order) as a prefix to the name of the main chain.

Example: Write the IUPAC name for the following compound:

$$CH_3-CH_2-\overset{\overset{\displaystyle CH_3}{|}}{CH}-CH_3$$

Analyze the Problem	**Given**	**Need**	**Connect**
	four-carbon chain, one methyl group	IUPAC name	position of substituents on the carbon chain

Solution:

STEP 1 **Write the alkane name for the longest chain of carbon atoms.** In this compound, the longest chain has four carbon atoms, which is named butane.

STEP 2 **Number the carbon atoms from the end nearer a substituent.** In this compound, the methyl group (substituent) is on carbon 2.

STEP 3 **Give the location and name for each substituent (alphabetical order) as a prefix to the name of the main chain.** The compound with a methyl group on carbon 2 of a four-carbon chain butane is named 2-methylbutane.

♦ **Learning Exercise 17.1D**

Write the IUPAC name for each of the following compounds:

a. 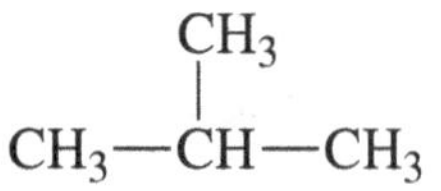 CH$_3$—CH—CH$_3$ with CH$_3$ _____

b. 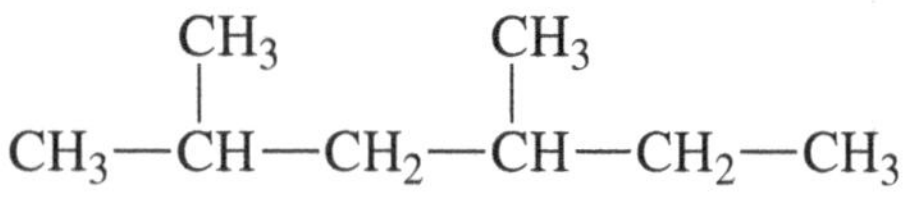 CH$_3$—CH—CH$_2$—CH—CH$_2$—CH$_3$ with CH$_3$, CH$_3$ _____

c. _____

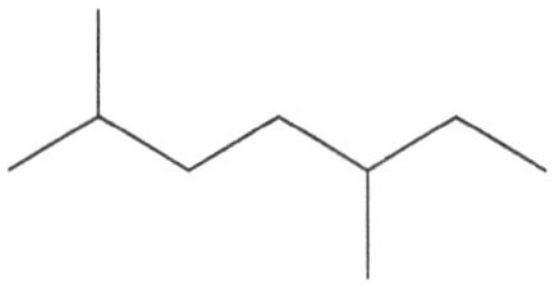

d. 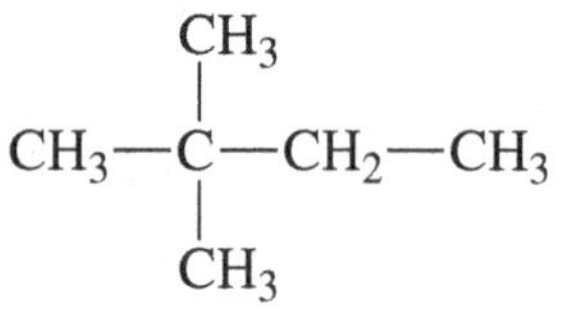 CH$_3$—C—CH$_2$—CH$_3$ with CH$_3$, CH$_3$ _____

Answers   **a.** methylpropane   **b.** 2,4-dimethylhexane   **c.** 2,5-dimethylheptane
d. 2,2-dimethylbutane

Guide to Drawing Structural Formulas for Alkanes with Substituents	
STEP 1	Draw the main chain of carbon atoms.
STEP 2	Number the chain and place the substituents on the carbons indicated by the numbers.
STEP 3	For the condensed structural formula, add the correct number of hydrogen atoms to give four bonds to each C atom.

♦ **Learning Exercise 17.1E**

Draw the condensed structural formulas for **a**, **b**, and **c** and the line-angle structural formulas for **d** and **e**.

a. methylbutane

b. 2-methylpentane

c. 4-ethyl-2-methylhexane

d. 2,2,4-trimethylheptane

e. 3-ethyloctane

Answers

a. 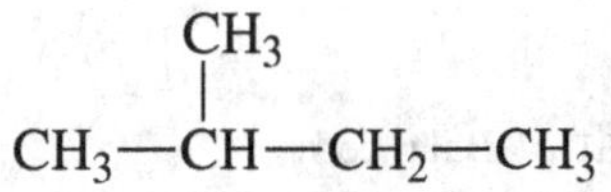 $CH_3-\overset{\overset{\displaystyle CH_3}{|}}{CH}-CH_2-CH_3$

b. 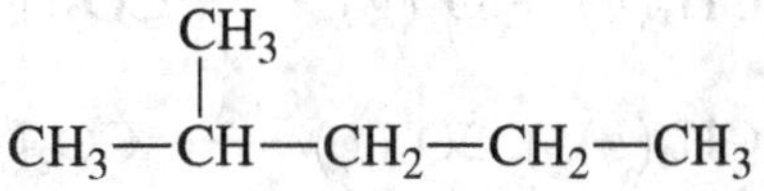 $CH_3-\overset{\overset{\displaystyle CH_3}{|}}{CH}-CH_2-CH_2-CH_3$

c. 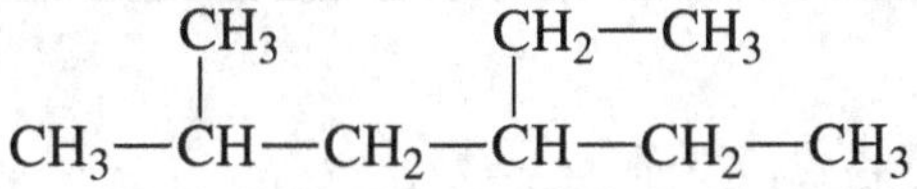 $CH_3-\overset{\overset{\displaystyle CH_3}{|}}{CH}-CH_2-\overset{\overset{\displaystyle CH_2-CH_3}{|}}{CH}-CH_2-CH_3$

d.

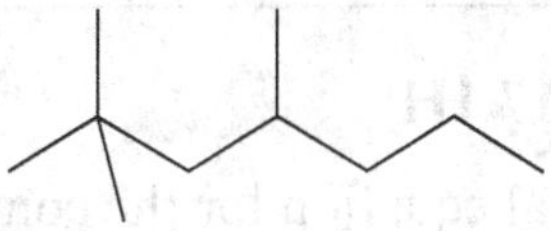

e.

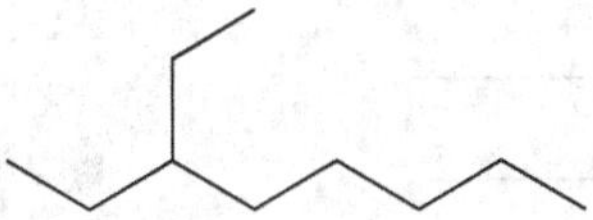

♦ **Learning Exercise 17.1F**

Write the IUPAC name for each of the following:

a. CH_3-CH_2-Br

b. 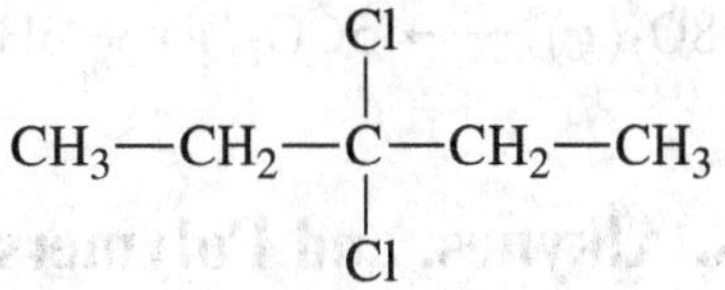 $CH_3-CH_2-\overset{\overset{\displaystyle Cl}{|}}{\underset{\underset{\displaystyle Cl}{|}}{C}}-CH_2-CH_3$

c.

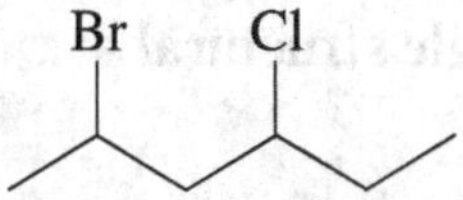

d. 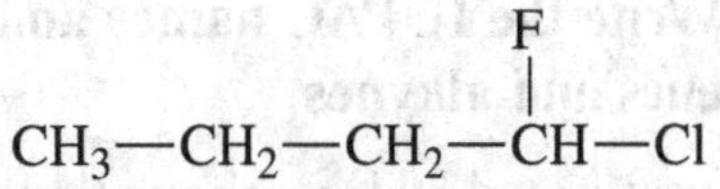 $CH_3-CH_2-CH_2-\overset{\overset{\displaystyle F}{|}}{CH}-Cl$

Answers
      a. bromoethane
      c. 2-bromo-4-chlorohexane

      b. 3,3-dichloropentane
      d. 1-chloro-1-fluorobutane

♦ **Learning Exercise 17.1G**

Draw the condensed structural formulas for **a**, **b**, and **c** and the line-angle structural formula for **d**.

a. chloroethane

b. bromomethane

c. 3-bromo-1-chloropentane

d. 1,1-dichlorohexane

Answers

a. CH_3-CH_2-Cl

b. CH_3-Br

c.
$$Cl-CH_2-CH_2-\overset{\overset{\displaystyle Br}{|}}{CH}-CH_2-CH_3$$

d.

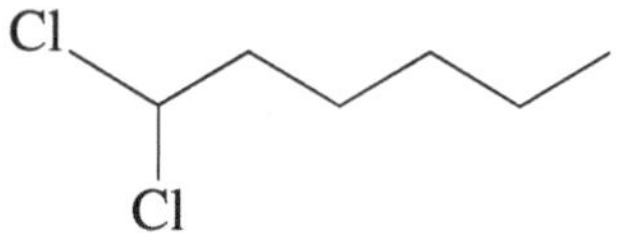

Study Note

Example: Write the balanced chemical equation for the combustion of methane.

Solution: Write the molecular formulas for the reactants: methane (CH_4) and oxygen (O_2). Write the products CO_2 and H_2O and balance the equation.

$$CH_4(g) + O_2(g) \xrightarrow{\Delta} CO_2(g) + H_2O(g) + \text{heat}$$

$$CH_4(g) + 2O_2(g) \xrightarrow{\Delta} CO_2(g) + 2H_2O(g) + \text{heat} \quad \text{Balanced}$$

♦ **Learning Exercise 17.1H**

Write a balanced chemical equation for the complete combustion of each of the following:

a. propane, gas   _____

b. hexane, liquid   _____

c. methylbutane, gas   _____

Answers

a. $C_3H_8(g) + 5O_2(g) \xrightarrow{\Delta} 3CO_2(g) + 4H_2O(g) + \text{heat}$

b. $2C_6H_{14}(l) + 19O_2(g) \xrightarrow{\Delta} 12CO_2(g) + 14H_2O(g) + \text{heat}$

c. $C_5H_{12}(g) + 8O_2(g) \xrightarrow{\Delta} 5CO_2(g) + 6H_2O(g) + \text{heat}$

17.2 Alkenes, Alkynes, and Polymers

Learning Goal: Write the IUPAC names and draw the condensed or line-angle structural formulas for alkenes and alkynes.

- Alkenes are unsaturated hydrocarbons that contain one or more carbon–carbon double bonds.
- In alkenes, each carbon in the double bond is connected to two other groups in addition to the other carbon of the double bond, all being in the same plane and having bond angles of 120°.
- Alkynes are unsaturated hydrocarbons that contain at least one carbon–carbon triple bond.

- The addition of small molecules to the double bond is a characteristic reaction of alkenes.
- Hydrogenation adds hydrogen atoms to the double bond of an alkene to yield an alkane.

$$H_2C{=}CH_2 + H_2 \xrightarrow{\text{Pt}} CH_3{-}CH_3$$

- Polymers are large molecules prepared from the bonding of many small units called *monomers*.
- Many synthetic polymers are made from small alkene monomers.
- The IUPAC names of alkenes are derived by changing the *ane* ending of the parent alkane to *ene*. For example, the IUPAC name of $H_2C{=}CH_2$ is ethene. It has a common name of ethylene. In alkenes, the longest carbon chain containing the double bond is numbered from the end nearer the double bond.

$$CH_3{-}CH{=}CH_2 \qquad H_2C{=}CH{-}CH_2{-}CH_3 \qquad CH_3{-}CH{=}CH{-}CH_3$$

Propene (propylene)   1-Butene   2-Butene

Ethene

- The IUPAC names of alkynes are derived by changing the *ane* ending of the parent alkane to *yne*. In alkynes, the longest carbon chain containing the triple bond is numbered from the end nearer the triple bond.

$$HC{\equiv}CH \qquad CH_3{-}C{\equiv}CH \qquad CH_3{-}CH_2{-}C{\equiv}CH$$

Ethyne   Propyne   1-Butyne

Ethyne

Ball-and-stick models show the double bond of an alkene and the triple bond of an alkyne.

◆ **Learning Exercise 17.2A**

Classify each of the following as an alkane, alkene, or alkyne:

a. _____ $CH_3{-}CH_2{-}CH_3$

b. _____

c. _____ $CH_3{-}C{\equiv}C{-}CH_3$

d. _____ $\underset{\qquad\qquad\quad CH_3}{CH_3{-}CH_2{-}CH{=}C{-}CH_2{-}CH_3}$

Answers   **a.** alkane   **b.** alkene   **c.** alkyne   **d.** alkene

Guide to Naming Alkenes and Alkynes	
STEP 1	Name the longest carbon chain that contains the double or triple bond.
STEP 2	Number the carbon chain from the end nearer the double or triple bond.
STEP 3	Give the location and name for each substituent (alphabetical order) as a prefix to the alkene or alkyne name.

◆ **Learning Exercise 17.2B**

Write the IUPAC name for each of the following alkenes:

a. $\underset{\qquad\quad CH_3}{CH_3{-}C{=}CH{-}CH_3}$

b. $\underset{\qquad\qquad\qquad Cl \qquad\quad CH_3}{H_2C{=}CH{-}CH{-}CH_2{-}CH{-}CH_3}$

_____   _____

$$\text{c. } CH_3-CH=\overset{\overset{\displaystyle CH_2-CH_3}{|}}{C}-CH_2-CH_3$$

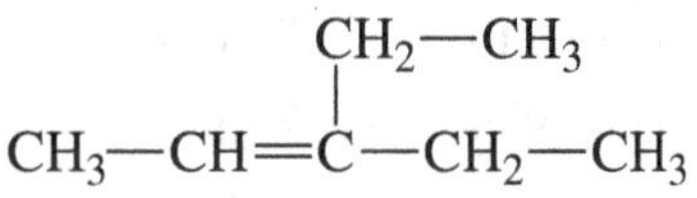

d. 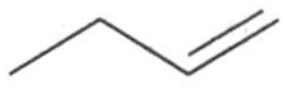 [line-angle structure]

_____

Answers   **a.** 2-methyl-2-butene   **b.** 3-chloro-5-methyl-1-hexene
             c. 3-ethyl-2-pentene   **d.** 1-butene

◆ Learning Exercise 17.2C

Write the IUPAC name for each of the following alkynes:

a. $HC\equiv CH$ **b.** $CH_3-C\equiv CH$

_____   _____

c. $CH_3-CH_2-C\equiv CH$ **d.** 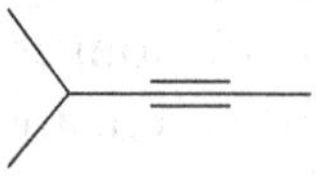 [line-angle structure]

_____   _____

Answers   **a.** ethyne   **b.** propyne
             c. 1-butyne   **d.** 4-methyl-2-pentyne

◆ Learning Exercise 17.2D

Draw the condensed structural formulas for **a** and **b** and the
line-angle structural formulas for **c** and **d**.

a. 2-pentyne   **b.** 2-chloro-2-butene

c. 3-bromo-2-methyl-2-pentene   **d.** 4-methyl-2-hexene

Answers

a. $CH_3-C\equiv C-CH_2-CH_3$ **b.** 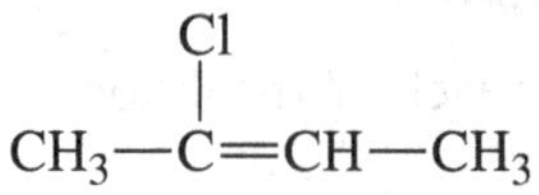 $CH_3-\overset{\overset{\displaystyle Cl}{|}}{C}=CH-CH_3$

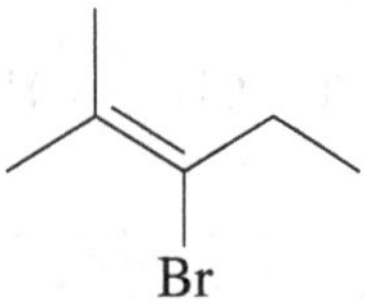

c. [line-angle structure with Br]   **d.** 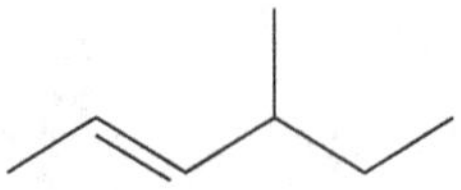 [line-angle structure]

♦ **Learning Exercise 17.2E**

Draw the condensed or line-angle structural formulas for the products of the following reactions:

a. $CH_3-CH_2-CH=CH_2 + H_2 \xrightarrow{Pt}$

b. $CH_3-CH=CH_2 + H_2 \xrightarrow{Pt}$

c. 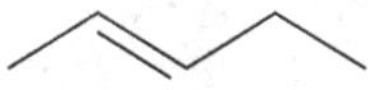 ╱╲╱ $+ H_2 \xrightarrow{Ni}$

Hydrogenation is used to convert unsaturated fats in vegetable oils to saturated fats that make a more solid product.
Credit: Pearson Education / Pearson Science

Answers **a.** $CH_3-CH_2-CH_2-CH_3$ **b.** $CH_3-CH_2-CH_3$

c. ╱╲╱╲

♦ **Learning Exercise 17.2F**

Draw the condensed structural formula for the starting monomer for each of the following polymers:

a. 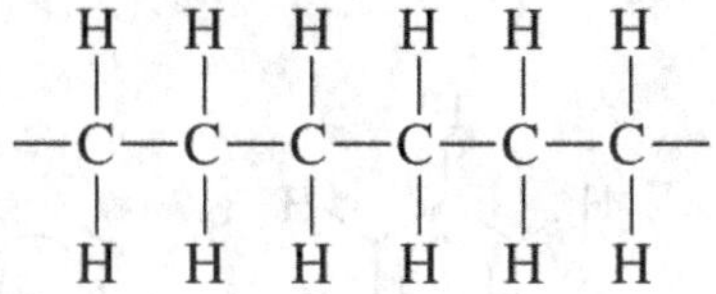
```
    H  H  H  H  H  H
    |  |  |  |  |  |
 —C—C—C—C—C—C—
    |  |  |  |  |  |
    H  H  H  H  H  H
```

b.
```
    H  CH₃ H  CH₃ H  CH₃
    |  |   |  |   |  |
 —C—C—C—C—C—C—
    |  |   |  |   |  |
    H  H   H  H   H  H
```

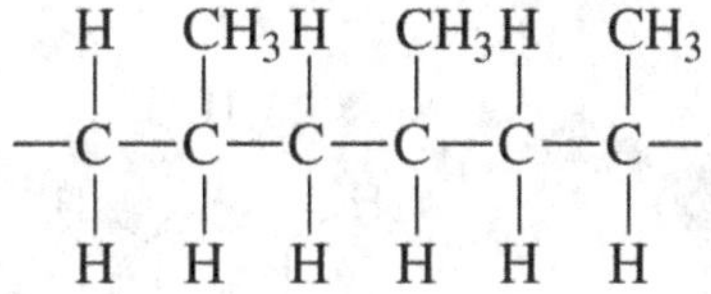

c.
```
    F  F  F  F  F  F
    |  |  |  |  |  |
 —C—C—C—C—C—C—
    |  |  |  |  |  |
    F  F  F  F  F  F
```

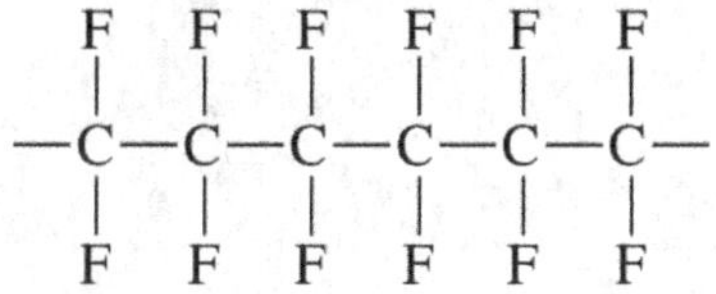

CORE CHEMISTRY SKILL

Writing Equations for Hydrogenation and Polymerization

Answers **a.** $H_2C=CH_2$ **b.** $H_2C=\overset{\overset{\displaystyle CH_3}{|}}{CH}$ **c.** $F_2C=CF_2$

♦ **Learning Exercise 17.2G**

Draw the structure of the polymer formed from the addition of three monomers of vinylacetate,

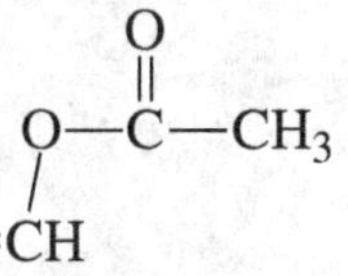

$$H_2C=CH-O-\overset{\overset{\displaystyle O}{\|}}{C}-CH_3$$

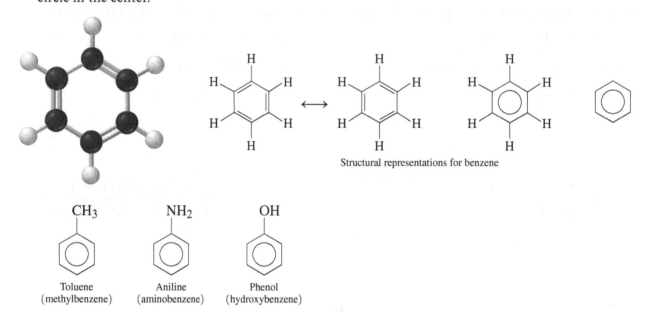

Answer

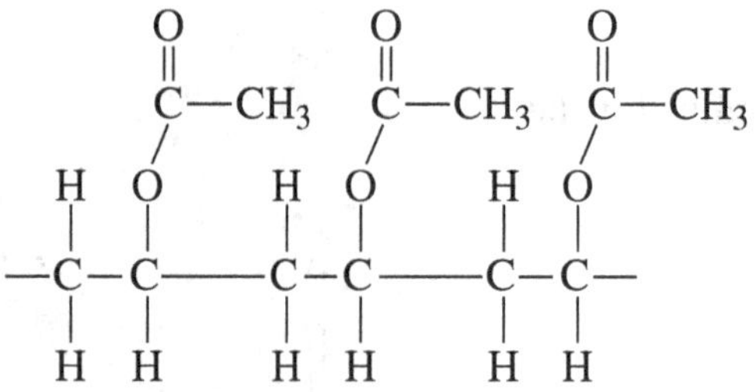

17.3   Aromatic Compounds

Learning Goal: Describe the bonding in benzene; name aromatic compounds, and draw their line-angle structural formulas.

- Most aromatic compounds contain benzene, a cyclic structure containing six CH units. The structure of benzene can be represented as a hexagon with alternating single and double bonds or with a circle in the center.

Structural representations for benzene

Toluene
(methylbenzene)

Aniline
(aminobenzene)

Phenol
(hydroxybenzene)

- When benzene has only one substituent, the ring is not numbered. The names of many aromatic compounds use the parent name benzene, although many common names were retained as IUPAC names, such as toluene, aniline, and phenol.

Key Terms for Sections 17.1 to 17.3

Match the following key terms with the correct description:

a. line-angle structural formula	**b.** alkene	**c.** alkane
d. hydrogenation	**e.** combustion	**f.** aromatic
g. condensed structural formula	**h.** substituent	**i.** alkyne

1. _____ groups of atoms such as an alkyl group or a halogen bonded to a carbon chain

2. _____ a type of formula for carbon compounds that indicates the carbon atoms as corners or ends of a zigzag line

3. _____ contains a cyclic structure with six CH units

4. _____ a hydrocarbon that contains only carbon–carbon and carbon–hydrogen single bonds

5. _____ the addition of H_2 to a carbon–carbon double bond

6. _____ the chemical reaction of an alkane and oxygen that yields CO_2, H_2O, and heat

7. _____ the type of formula that shows the arrangement of the carbon atoms grouped with their attached H atoms

8. _____ a compound that contains a carbon–carbon double bond

9. _____ a compound that contains a carbon–carbon triple bond

Answers   1. h   2. a   3. f   4. c   5. d
            6. e   7. g   8. b   9. i

Guide to Naming Aromatic Compounds	
STEP 1	Write the name for the aromatic compound.
STEP 2	If there are two substituents, number the aromatic ring.
STEP 3	Name a substituent as a prefix.

♦ **Learning Exercise 17.3**

Write the IUPAC name for each of the following:

Answers   a. bromobenzene   b. methylbenzene (toluene)
            c. 1,3-dichlorobenzene   d. 4-bromotoluene

17.4   Alcohols and Ethers

Learning Goal: Write the IUPAC and common names for alcohols and ethers; draw the condensed or line-angle structural formulas when given their names.

- An alcohol contains the hydroxyl group (—OH) attached to a carbon chain.
- In the IUPAC system, alcohols are named by replacing the *e* of the alkane name with *ol*. The location of the —OH group is given by numbering the carbon chain.

- Simple alcohols are generally named by their common names with the alkyl name preceding the term *alcohol.* For example, CH_3—OH is methyl alcohol, and CH_3—CH_2—OH is ethyl alcohol.

CH_3—OH CH_3—CH_2—OH CH_3—CH_2—CH_2—OH
Methanol   Ethanol   1-Propanol
(methyl alcohol)   (ethyl alcohol)   (propyl alcohol)

- When a hydroxyl group is attached to a benzene ring, the compound is a phenol.
- In ethers, an oxygen atom is connected by single bonds to two alkyl or aromatic groups.
- In the common names of ethers, the alkyl groups are listed alphabetically followed by the name *ether.*
- Thiols are similar to alcohols, except they have an —SH group in place of the —OH group.

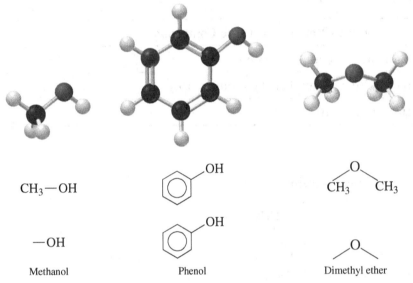

CH_3—OH

—OH

Methanol

OH

OH

Phenol

O
CH_3 CH_3

O

Dimethyl ether

Guide to Naming Alcohols	
STEP 1	Name the longest carbon chain attached to the —OH group by replacing the *e* in the corresponding alkane name with *ol*. Name an aromatic alcohol as a *phenol*.
STEP 2	Number the chain from the end nearer the —OH group.
STEP 3	Give the location and name for each substituent relative to the —OH group.

♦ **Learning Exercise 17.4A**

Write the IUPAC and common name (if any) for each of the following compounds:

a. CH_3—CH_2—CH_2—OH

b.
$$\overset{\displaystyle OH}{CH_3-CH-CH_2-CH_2-CH_3}$$

c. (structure with OH)

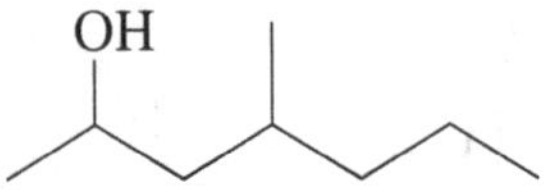

d. (phenol structure with OH and Cl)

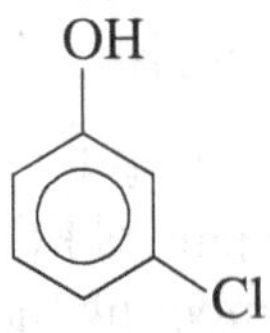

Answers   **a.** 1-propanol (propyl alcohol)   **b.** 2-pentanol
                  c. 4-methyl-2-heptanol   **d.** 3-chlorophenol

♦ **Learning Exercise 17.4B**

Draw the condensed structural formulas for **a** and **b** and the line-angle structural formulas for **c** and **d**.

  a. 2-butanol
  b. 2-chloro-1-propanol

  c. 2-bromoethanol
  d. 2-chlorophenol

Answers

  a. $CH_3-\overset{\displaystyle OH}{\underset{\displaystyle |}{CH}}-CH_2-CH_3$ **b.** $CH_3-\overset{\displaystyle Cl}{\underset{\displaystyle |}{CH}}-CH_2-OH$

  c. 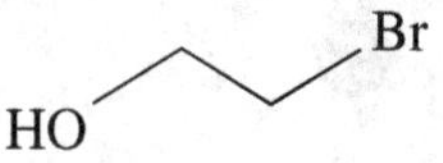 HO ⟍⟋⟍ Br   **d.** 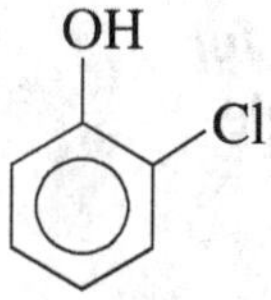 $\overset{\displaystyle OH}{}$ ⬡ Cl

♦ **Learning Exercise 17.4C**

Write the common name for each of the following:

  a. CH_3-O-CH_3 **b.** ⟍⟋O⟍⟋

  c. $CH_3-CH_2-CH_2-CH_2-O-CH_3$ **d.** $CH_3-O-CH_2-CH_3$

Answers   **a.** dimethyl ether   **b.** diethyl ether
             c. butyl methyl ether   **d.** ethyl methyl ether

♦ **Learning Exercise 17.4D**

Draw the condensed structural formulas for **a** and **b** and the line-angle structural formulas for **c** and **d**.

a. ethyl propyl ether

b. isopropyl methyl ether

c. dipropyl ether

d. ethyl pentyl ether

Answers

a. $CH_3-CH_2-O-CH_2-CH_2-CH_3$

b. 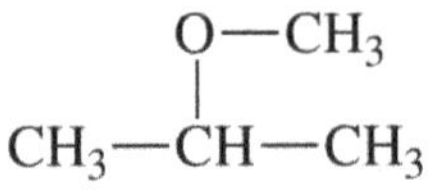 $CH_3-\overset{\overset{\displaystyle O-CH_3}{\displaystyle |}}{C}H-CH_3$

c. ∧∨∧–O–∨∧

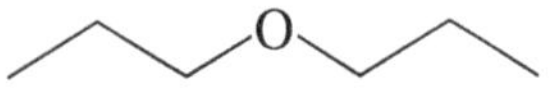

d.  ∧∨∧∨∧–O–∨

17.5   Aldehydes and Ketones

Learning Goal: Write the IUPAC and common names for aldehydes and ketones; draw the condensed or line-angle structural formulas when given their names.

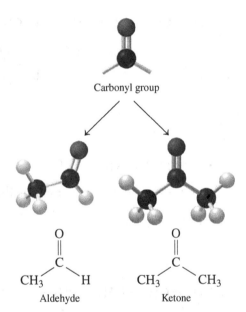

Carbonyl group

- The carbonyl group contains a carbon–oxygen double bond ($C=O$).
- In an aldehyde, the carbonyl group appears at the end of a carbon chain and is attached to at least one hydrogen atom.
- In a ketone, the carbonyl group is attached to two carbon groups.
- In the IUPAC system, aldehydes and ketones are named by replacing the *e* in the longest chain containing the carbonyl group with *al* for aldehydes and *one* for ketones. The location of the carbonyl group in a ketone is given if there are more than four carbon atoms in the chain.

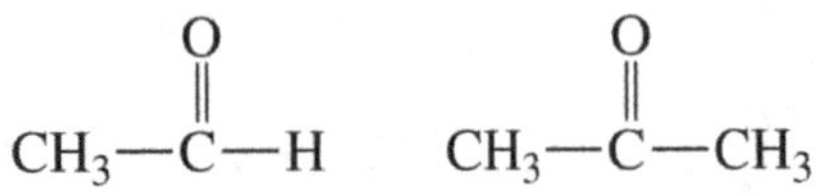

$$CH_3-\overset{\overset{\displaystyle O}{\displaystyle ||}}{C}-H \qquad CH_3-\overset{\overset{\displaystyle O}{\displaystyle ||}}{C}-CH_3$$

Ethanal
(acetaldehyde)

Propanone
(dimethyl ketone)

$$\overset{\overset{\displaystyle O}{\displaystyle ||}}{\underset{CH_3}{C}}\diagdown H \qquad \overset{\overset{\displaystyle O}{\displaystyle ||}}{\underset{CH_3}{C}}\diagdown CH_3$$

Aldehyde   Ketone

The carbonyl group is found in aldehydes and ketones.

♦ **Learning Exercise 17.5A**

Classify each of the compounds below as one of the following:

    a. alcohol   **b.** aldehyde   **c.** ketone   **d.** ether

1. _____ $CH_3-CH_2-CH_2-\overset{\overset{\textstyle O}{\|}}{C}-H$ 2. _____

3. 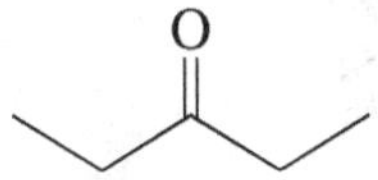 _____   4. _____ $CH_3-CH_2-O-CH_3$

5. _____ $CH_3-\overset{\overset{\textstyle O}{\|}}{C}-CH_2-CH_3$ 6. _____ $CH_3-\overset{\overset{\textstyle O}{\|}}{C}-H$

Answers   **1.** b   **2.** a   **3.** c
               4. d   **5.** c   **6.** b

Guide to Naming Aldehydes	
STEP 1	Name the longest carbon chain by replacing the *e* in the alkane name with *al*.
STEP 2	Name and number any substituents by counting the carbonyl group as carbon 1.

♦ **Learning Exercise 17.5B**

Write the IUPAC and common name, if any, for each of the following aldehydes:

CORE CHEMISTRY SKILL

Naming Aldehydes and Ketones

a. $CH_3-\overset{\overset{\textstyle O}{\|}}{C}-H$ **b.**

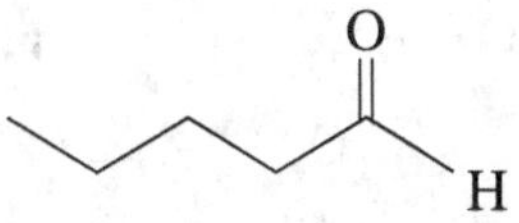

c. $CH_3-CH_2-\overset{\overset{\textstyle CH_3}{|}}{C}H-CH_2-CH_2-\overset{\overset{\textstyle O}{\|}}{C}-H$ **d.**

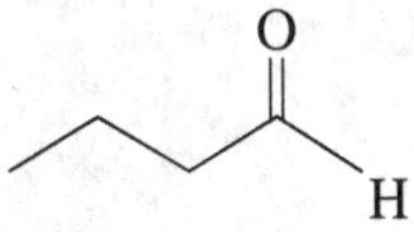

Answers   **a.** ethanal (acetaldehyde)   **b.** pentanal
               c. 4-methylhexanal   **d.** butanal (butyraldehyde)

Guide to Naming Ketones	
STEP 1	Name the longest carbon chain by replacing the *e* in the alkane name with *one*.
STEP 2	Number the carbon chain from the end nearer the carbonyl group and indicate its location.
STEP 3	Name and number any substituents on the carbon chain.

◆ **Learning Exercise 17.5C**

Write the IUPAC name and common name, if any, for each of the following ketones:

a. CH₃—C(=O)—CH₃

b. (line-angle structure of a ketone)

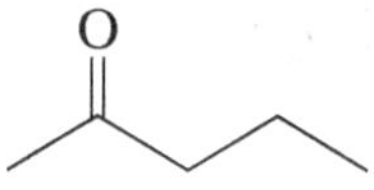

c. CH₃—CH₂—C(=O)—CH₂—CH₃

d. 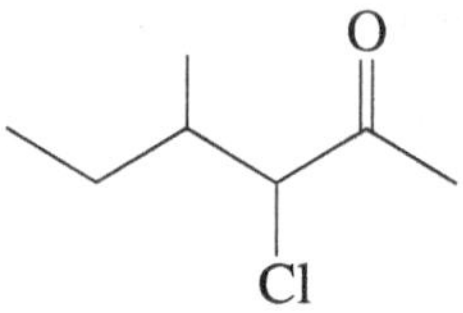 (line-angle structure with Cl)

Answers   a. propanone (dimethyl ketone, acetone)   b. 2-pentanone (methyl propyl ketone)
c. 3-pentanone (diethyl ketone)   d. 3-chloro-4-methyl-2-hexanone

◆ **Learning Exercise 17.5D**

Draw the condensed structural formulas for **a** to **d** and the line-angle structural formulas for **e** and **f**.

a. ethanal   b. 2-methylbutanal   c. 2-chloropropanal

d. ethyl methyl ketone   e. 3-hexanone   f. benzaldehyde

Answers   a. CH₃—C(=O)—H   b. CH₃—CH₂—CH(CH₃)—C(=O)—H   c. CH₃—CH(Cl)—C(=O)—H

d. CH₃—CH₂—C(=O)—CH₃   e. 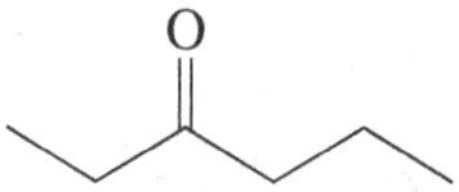 (line-angle structure)   f. 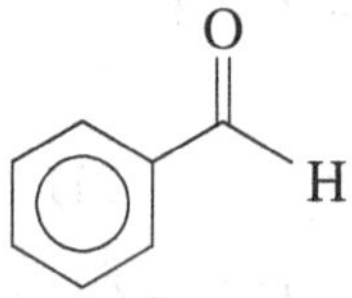 (benzaldehyde structure)

17.6   Carboxylic Acids and Esters

Learning Goal: Write the IUPAC and common names for carboxylic acids and esters; draw the condensed or line-angle structural formulas when given their names.

- The IUPAC names of carboxylic acids replace the *e* of the corresponding alkane with *oic acid*. Simple acids usually are

The sour taste of vinegar is due to ethanoic acid (acetic acid).

Credit: Pearson Education / Pearson Science

named by the common system using the prefixes *form* (1C), *acet* (2C), *propion* (3C), or *butyr* (4C), followed by *ic acid*.

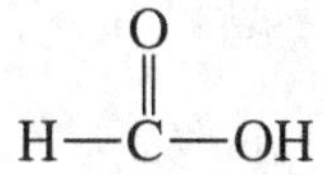

$$H-\overset{\displaystyle O}{\overset{\|}{C}}-OH \qquad CH_3-\overset{\displaystyle O}{\overset{\|}{C}}-OH \qquad CH_3-CH_2-CH_2-\overset{\displaystyle O}{\overset{\|}{C}}-OH$$

Methanoic acid 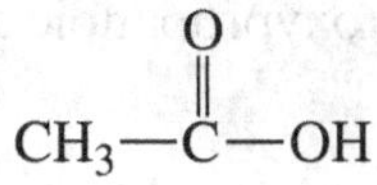   Ethanoic acid   Butanoic acid   Pentanoic acid
(formic acid)   (acetic acid)   (butyric acid)

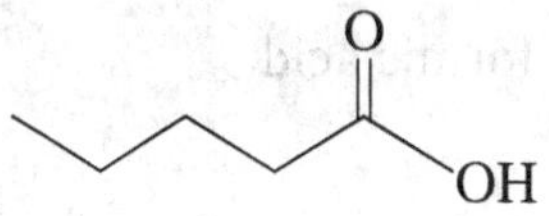

- The name of the carboxylic acid of benzene is benzoic acid. When substituents are bonded to benzoic acid, the ring is numbered from the carboxylic acid on carbon 1 in the direction that gives substituents the smallest possible numbers.

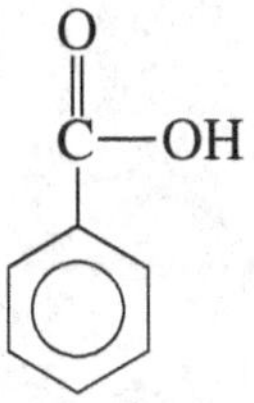

Benzoic acid 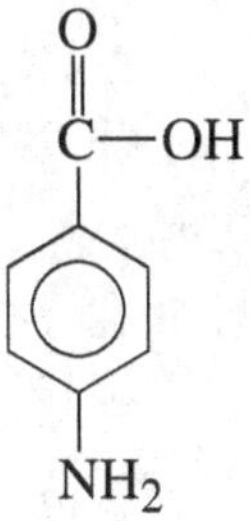   4-Aminobenzoic acid 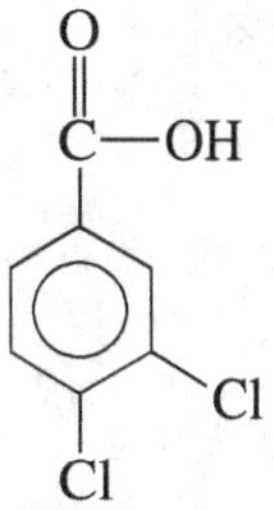   3,4-Dichlorobenzoic acid

- In the presence of a strong acid, a carboxylic acid reacts with an alcohol to produce an ester and water.

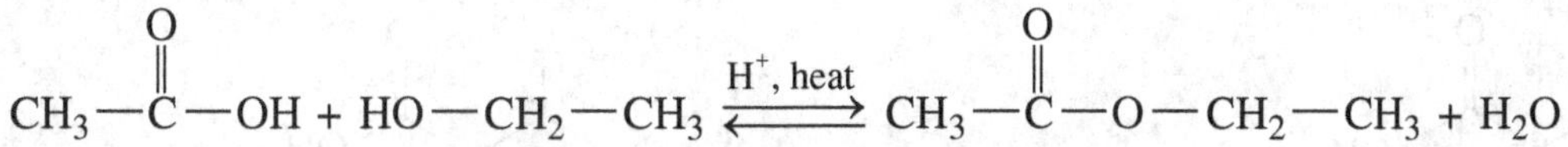

$$CH_3-\overset{\displaystyle O}{\overset{\|}{C}}-OH + HO-CH_2-CH_3 \underset{}{\overset{H^+,\ heat}{\rightleftharpoons}} CH_3-\overset{\displaystyle O}{\overset{\|}{C}}-O-CH_2-CH_3 + H_2O$$

Guide to Naming Carboxylic Acids	
STEP 1	Identify the longest carbon chain and replace the *e* in the alkane name with *oic acid*.
STEP 2	Name and number any substituents by counting the carboxyl carbon as 1.

◆ **Learning Exercise 17.6A**

Write the IUPAC and common name, if any, for each of the following carboxylic acids:

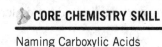

CORE CHEMISTRY SKILL

Naming Carboxylic Acids

a. 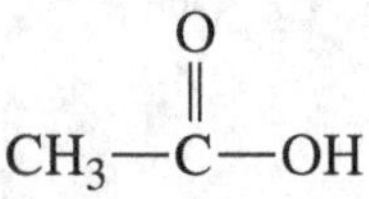 $CH_3-\overset{\displaystyle O}{\overset{\|}{C}}-OH$

b. 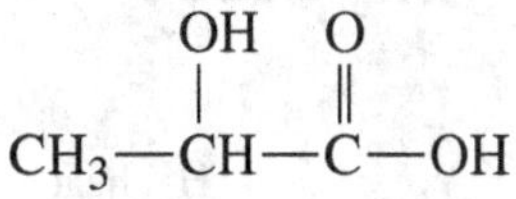 $CH_3-\overset{\displaystyle OH}{\overset{\displaystyle |}{C}H}-\overset{\displaystyle O}{\overset{\|}{C}}-OH$

c.

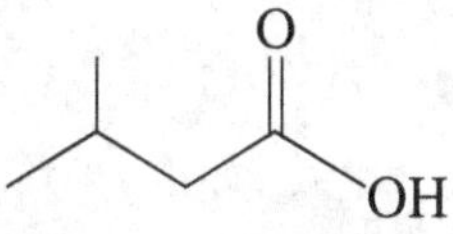

d.

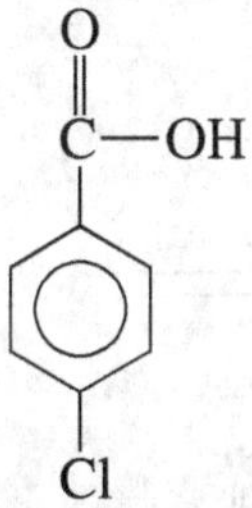

Answers   **a.** ethanoic acid (acetic acid)   **b.** 2-hydroxypropanoic acid
                  c. 3-methylbutanoic acid   **d.** 4-chlorobenzoic acid

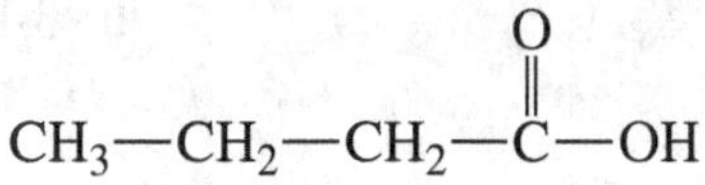

♦ **Learning Exercise 17.6B**

Draw the condensed structural formulas for **a** and **b** and the line-angle structural formulas for **c** and **d**.

a. formic acid

b. 3-hydroxypropanoic acid

c. 2-bromobutanoic acid

d. 3-methylpentanoic acid

Answers

a. 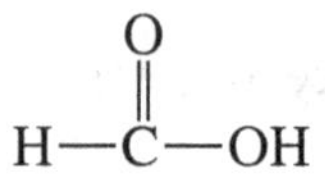 H—C—OH
(with =O above C)

b. 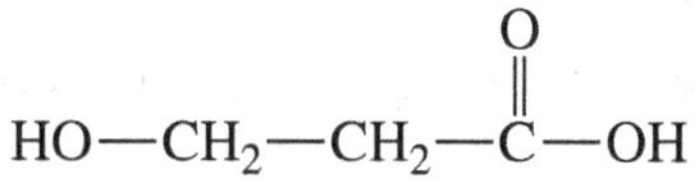 HO—CH$_2$—CH$_2$—C—OH
(with =O above C)

c.

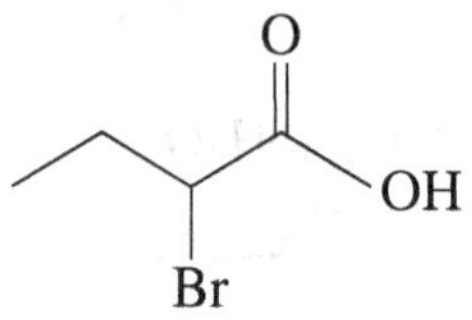

d.

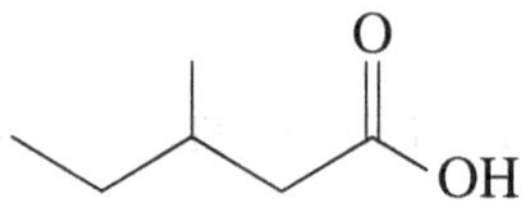

♦ **Learning Exercise 17.6C**

Draw the condensed structural formula for the ester formed when the following carboxylic acids and alcohols react:

⚑ CORE CHEMISTRY SKILL

Forming Esters

a. 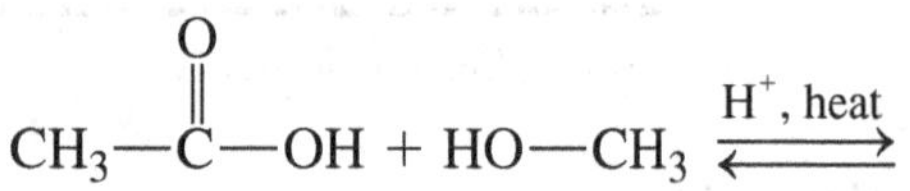 CH$_3$—C—OH + HO—CH$_3$ $\xrightleftharpoons{H^+, \text{heat}}$
(with =O above C)

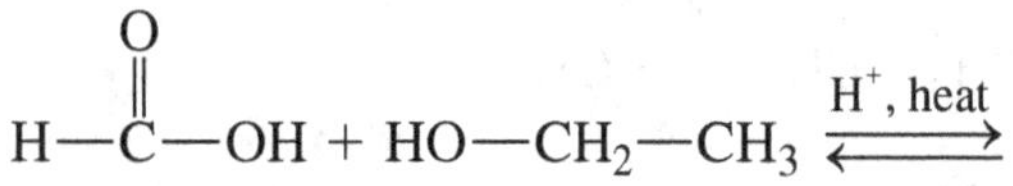

b. H—C—OH + HO—CH$_2$—CH$_3$ $\xrightleftharpoons{H^+, \text{heat}}$
(with =O above C)

c.
OH + HO—CH$_3$ $\xrightleftharpoons{H^+, \text{heat}}$

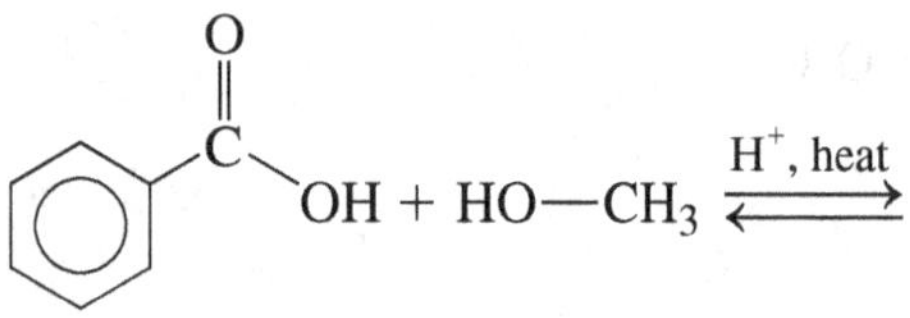

d. propanoic acid and ethanol $\xrightleftharpoons{H^+, \text{heat}}$

Answers

a. 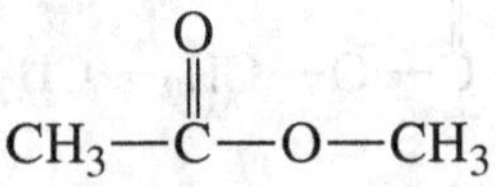 CH$_3$—C(=O)—O—CH$_3$

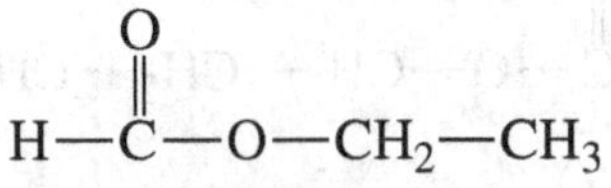

b. H—C(=O)—O—CH$_2$—CH$_3$

c. [benzene ring]—C(=O)—O—CH$_3$

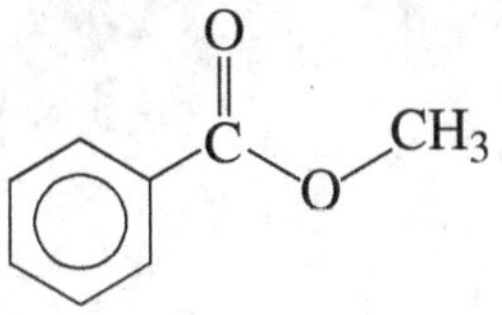

d. CH$_3$—CH$_2$—C(=O)—O—CH$_2$—CH$_3$

The odor of grapes
is due to an ester.

*Credit: Lynn Watson /
Shutterstock*

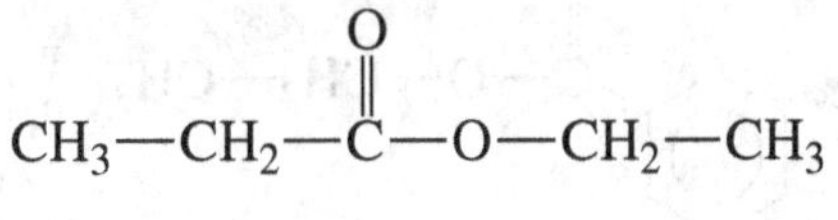

Guide to Naming Esters	
STEP 1	Write the name for the carbon chain from the alcohol as an *alkyl* group.
STEP 2	Change the *ic acid* of the acid name to *ate*.

◆ **Learning Exercise 17.6D**

Write the IUPAC and common name, if any, for each of the following esters:

a. 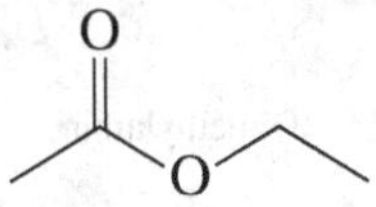 [line-angle ester structure]

b. 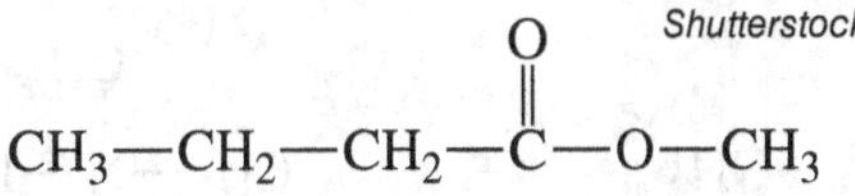 CH$_3$—CH$_2$—CH$_2$—C(=O)—O—CH$_3$

c. CH$_3$—CH$_2$—C(=O)—O—CH$_2$—CH$_2$—CH$_3$

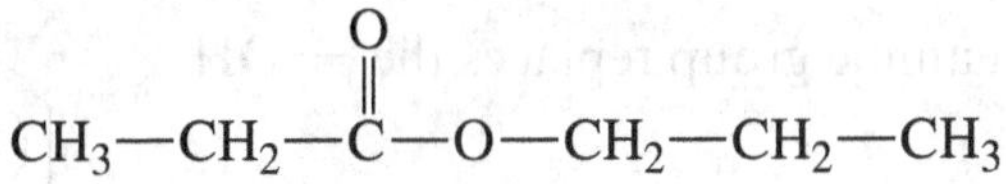

d. 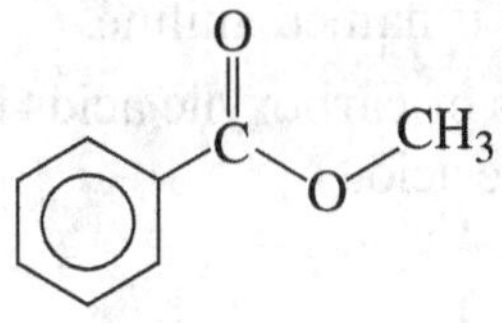 [benzene ring]—C(=O)—O—CH$_3$

Answers   **a.** ethyl ethanoate (ethyl acetate)   **b.** methyl butanoate (methyl butyrate)
c. propyl propanoate (propyl propionate)   **d.** methyl benzoate

◆ **Learning Exercise 17.6E**

Draw the condensed structural formulas for **a**, **b**, and **c** and the line-angle structural formula for **d**.

a. propyl acetate

b. propyl butyrate

c. ethyl benzoate

d. methyl ethanoate

Answers   **a.** CH₃—C(=O)—O—CH₂—CH₂—CH₃   **b.** CH₃—CH₂—CH₂—C(=O)—O—CH₂—CH₂—CH₃

c. 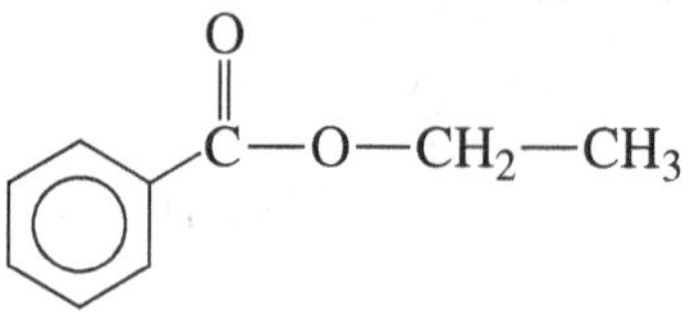 [benzene]—C(=O)—O—CH₂—CH₃   **d.** 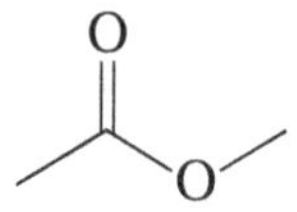 [line-angle ester structure]

17.7   Amines and Amides

Learning Goal: Write the common names for amines and the IUPAC and common names for amides; draw the condensed or line-angle structural formulas when given their names.

Methylamine

- Amines are derivatives of ammonia (NH_3), in which alkyl or aromatic groups replace one or more hydrogen atoms.
- Amines may be named by their common names in which the names of the alkyl groups are listed alphabetically preceding the suffix *amine*.

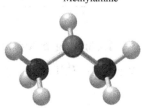

Dimethylamine

CH₃—NH₂   CH₃—N(CH₃)—H   CH₃—N(CH₃)—CH₃

Methylamine   Dimethylamine   Trimethylamine

- The amine of benzene is named aniline.
- Amides are derivatives of carboxylic acids in which an amine group replaces the —OH group in the carboxylic acid.

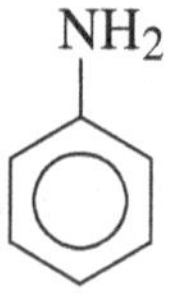

Aniline

CH₃—C(=O)—NH₂   Ethanamide (acetamide)

- Amides are named by replacing the *ic acid* (common) or *oic acid* (IUPAC) ending by *amide*.
- When a carboxylic acid reacts with ammonia or an amine, an amide is produced.

CH₃—C(=O)—OH + NH₃ —Heat→ CH₃—C(=O)—NH₂ + H_2O

CH₃—C(=O)—OH + H₂N—CH₃ —Heat→ CH₃—C(=O)—N(H)—CH₃ + H_2O

Key Terms for Sections 17.4 to 17.7

Match the following key terms with the correct description:

a. carboxylic acid   **b.** amidation   **c.** esterification   **d.** amine
e. alcohol   **f.** ester   **g.** amide

1. ____ a type of organic compound that produces pleasant aromas in flowers and fruits

2. ____ the reaction of a carboxylic acid and an amine

3. ____ an organic compound containing the carboxyl group (—COOH)

4. ____ this compound is formed from a carboxylic acid and an amine

5. ____ an organic compound that contains an amino group

6. ____ a reaction of a carboxylic acid and an alcohol in the presence of an acid catalyst

7. ____ an organic compound that contains a hydroxyl group

Answers   **1.** f   **2.** b   **3.** a   **4.** g   **5.** d   **6.** c   **7.** e

♦ **Learning Exercise 17.7A**

Write the common name for each of the following amines:

a.
$$\underset{\text{H}}{\text{CH}_3 - \overset{\text{H}}{\underset{|}{\text{N}}} - \text{CH}_2 - \text{CH}_3}$$

b. 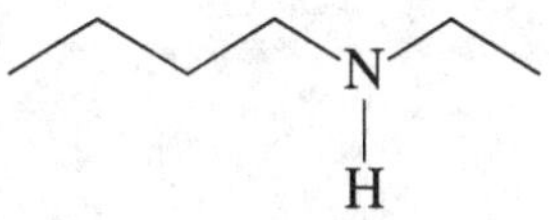 (line-angle structure with N—H and ethyl group)

c. 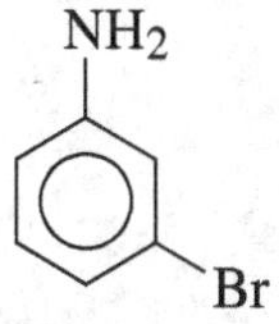 (benzene ring with NH$_2$ and Br substituents)

d.
$$\text{CH}_3 - \text{CH}_2 - \overset{\overset{\text{CH}_3}{|}}{\text{N}} - \text{CH}_3$$

Answers   **a.** ethylmethylamine   **b.** butylethylamine
             c. 3-bromoaniline   **d.** ethyldimethylamine

♦ **Learning Exercise 17.7B**

Draw the condensed structural formulas for **a** and **b** and the line-angle structural formulas for **c** and **d**.

a. isopropylamine   b. butylethylmethylamine

c. 4-bromoaniline   d. triethylamine

Answers

a.
$$\text{CH}_3 - \overset{\overset{\text{NH}_2}{|}}{\text{CH}} - \text{CH}_3$$

b.
$$\text{CH}_3 - \text{CH}_2 - \overset{\overset{\text{CH}_3}{|}}{\text{N}} - \text{CH}_2 - \text{CH}_2 - \text{CH}_2 - \text{CH}_3$$

c. 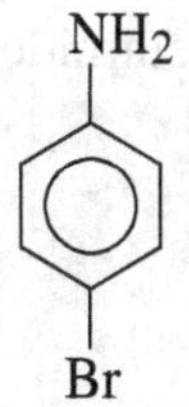 (benzene ring with NH$_2$ and Br substituents, para)

d. 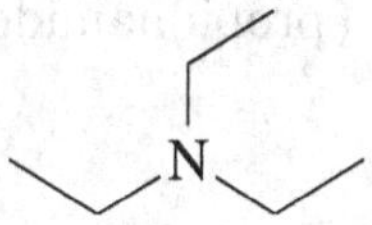 (line-angle structure of triethylamine)

◆ **Learning Exercise 17.7C**

CORE CHEMISTRY SKILL

Draw the condensed structural formula for the amide formed in each of the following reactions:

Forming Amides

a. $CH_3-CH_2-\overset{\overset{O}{\|}}{C}-OH + NH_3 \xrightarrow{\text{Heat}}$

b. 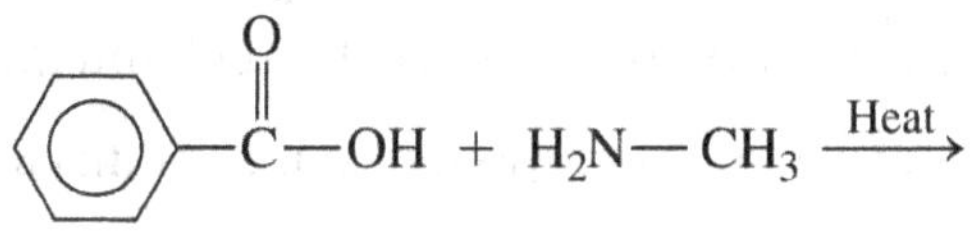 (benzene ring)$-\overset{\overset{O}{\|}}{C}-OH + H_2N-CH_3 \xrightarrow{\text{Heat}}$

c. $CH_3-\overset{\overset{O}{\|}}{C}-OH + H-\overset{\overset{CH_3}{|}}{N}-CH_3 \xrightarrow{\text{Heat}}$

Answers **a.** $CH_3-CH_2-\overset{\overset{O}{\|}}{C}-NH_2$

b. 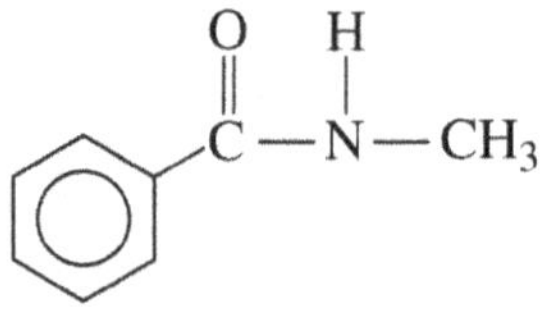 (benzene ring)$-\overset{\overset{O}{\|}}{C}-\overset{\overset{H}{|}}{N}-CH_3$

c. $CH_3-\overset{\overset{O}{\|}}{C}-\overset{\overset{CH_3}{|}}{N}-CH_3$

◆ **Learning Exercise 17.7D**

Give the IUPAC and common name (if any) for each of the following amides:

a. $CH_3-CH_2-\overset{\overset{O}{\|}}{C}-NH_2$ _____

b. 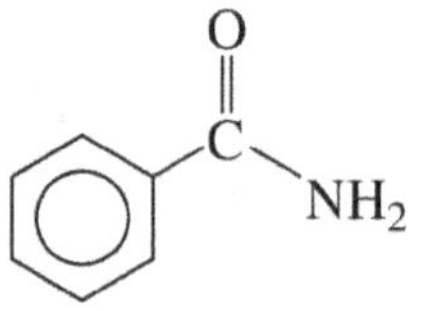 (benzene ring)$-\overset{\overset{O}{\|}}{C}-NH_2$ _____

c. (structure)$-\overset{\overset{O}{\|}}{C}-NH_2$ _____

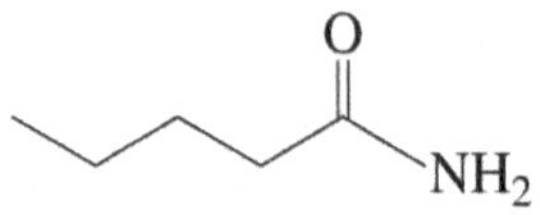

d. $CH_3-\overset{\overset{O}{\|}}{C}-\overset{\overset{H}{|}}{N}-CH_2-CH_3$ _____

Answers **a.** propanamide (propionamide) **b.** benzamide
c. pentanamide **d.** *N*-ethylethanamide (*N*-ethylacetamide)

♦ **Learning Exercise 17.7E**

Draw the condensed structural formula for **a** and the line-angle structural formulas for **b** and **c**.

a. butanamide

b. 3-chloropentanamide

c. benzamide

Answers   **a.** $CH_3—CH_2—CH_2—\overset{\overset{\displaystyle O}{\|}}{C}—NH_2$ **b.**

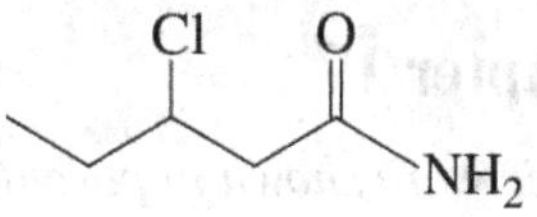

c.

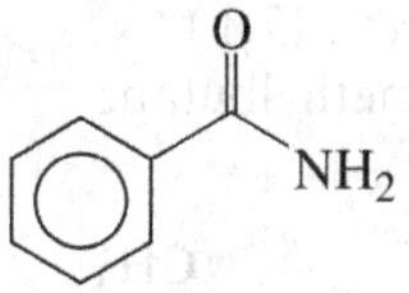

Checklist for Chapter 17

You are ready to take the Practice Test for Chapter 17. Be sure you have accomplished the following learning goals for this chapter. If not, review the section listed at the end of the goal. Then apply your new skills and understanding to the Practice Test.

After studying Chapter 17, I can successfully:

_____ Name and draw condensed or line-angle structural formulas for alkanes with substituents. (17.1)

_____ Identify the structural features of alkenes and alkynes. (17.2)

_____ Name alkenes and alkynes using IUPAC rules and draw their condensed or line-angle structural formulas. (17.2)

_____ Describe the process of forming polymers from alkene monomers. (17.2)

_____ Write the names and draw the condensed structural or line-angle formulas for compounds that contain a benzene ring. (17.3)

_____ Write the IUPAC or common name of an alcohol or phenol and draw the condensed or line-angle structural formula from the name. (17.4)

_____ Write the common name for a simple ether and draw the condensed or line-angle structural formula from the name. (17.4)

_____ Write the IUPAC and common names and draw the condensed or line-angle structural formulas for aldehydes and ketones. (17.5)

_____ Write the IUPAC and common names and draw condensed or line-angle structural formulas for carboxylic acids. (17.6)

_____ Write the IUPAC or common names and draw the condensed or line-angle structural formulas for esters. (17.6)

_____ Draw the condensed or line-angle structural formula for the product of the esterification reaction between an alcohol and a carboxylic acid. (17.6)

_____ Write the IUPAC or common names and draw the condensed or line-angle structural formulas for esters. (17.6)

_____ Write the common names for amines and the IUPAC and common name for amides and draw their condensed or line-angle structural formulas. (17.7)

_____ Draw the condensed or line-angle structural formula for the product of the amidation reaction between a carboxylic acid and ammonia or an amine. (17.7)

Practice Test for Chapter 17

The chapter sections to review are shown in parentheses at the end of each question.

For questions 1 and 2, match the name of the alkane with each structure: (17.1)
A. hexane   **B.** 3-methylbutane   **C.** 2-methylbutane
D. butane   **E.** 2,4-dimethylhexane

$$CH_3$$
$$|$$
1. $CH_3-CH_2-CH-CH_3$

$$CH_3 \qquad CH_3$$
$$| \qquad\quad |$$
2. $CH_3-CH-CH_2-CH-CH_2-CH_3$

For questions 3 through 5, match the name of the alkene or alkyne with the structural formula: (17.2)
A. 1-butene   **B.** 2-butene   **C.** 3-butene
D. 2-butyne   **E.** 1-butyne

3. _____ $CH_3-CH_2-C\equiv CH$

4. _____ $CH_3-CH_2-CH=CH_2$

5. _____ $CH_3-CH=CH-CH_3$

For questions 6 through 9, match the structure of each aromatic compound with the correct name: (17.3)

6. _____ chlorobenzene

7. _____ benzene

8. _____ toluene

9. _____ 1,3-dichlorobenzene

For questions 10 through 13, match the names with one of the following condensed structural formulas: (17.4)
A. 1-propanol   **B.** 3-propanol   **C.** 2-propanol
D. ethyl methyl ether   **E.** diethyl ether

10. CH₃—CH—CH₃
    |
    OH

11. CH₃—CH₂—CH₂—OH

12. CH₃—O—CH₂—CH₃

13. CH₃—CH₂—O—CH₂—CH₃

For questions 14 through 17, match each of the condensed structural formulas with its name: (17.6)

A. CH₃—C(=O)—O—CH₂—CH₃

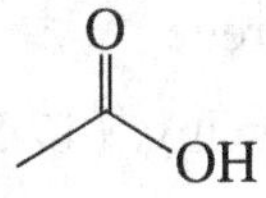

B. CH₃—C(=O)—OH

C. CH₃—CH₂—CH₂—C(=O)—OH

D. CH₃—CH₂—CH₂—C(=O)—O—CH₃

14. _____ butyric acid

15. _____ methyl butanoate

16. _____ ethyl acetate

17. _____ acetic acid

18. Identify the carboxylic acid and alcohol needed to produce (17.6)

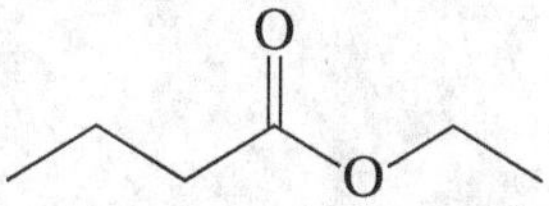

    A. propanoic acid and ethanol   B. acetic acid and 1-pentanol
    C. acetic acid and 1-butanol   D. butanoic acid and ethanol
    E. hexanoic acid and methanol

For questions 19 through 22, match the amines and amides with the following names: (17.7)
A. ethyldimethylamine   B. ethanamide   C. propanamide
D. diethylamine   E. butanamide

19. _____   CH₃—CH₂—N(CH₃)—CH₃

20. _____   CH₃—CH₂—CH₂—C(=O)—NH₂

21. _____   CH₃—CH₂—N(H)—CH₂—CH₃

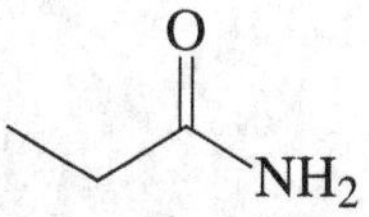

22. _____   (structure: propanamide)

23. Identify the carboxylic acid and ammonia or amine needed to produce the following: (17.7)

$$CH_3—CH_2—CH_2—\overset{\overset{\displaystyle O}{\|}}{C}—NH_2$$

    A. propanoic acid and ammonia   **B.** propanoic acid and methylamine
    C. acetic acid and ammonia   **D.** butanoic acid and methylamine
    E. butanoic acid and ammonia

24. Indicate the monomer needed to produce polyethylene. (17.2)
    A. propene   **B.** chloroethene   **C.** tetrafluoroethene
    D. ethene   **E.** styrene

25. Write the IUPAC name for the following: (17.5)

$$CH_3—CH_2—CH_2—CH_2—\overset{\overset{\displaystyle O}{\|}}{C}—H$$

    A. methylbutanal   **B.** pentanal   **C.** pentaldehyde
    D. hexanal   **E.** pentanone

26. The ester produced from the reaction of 1-butanol and propanoic acid is (17.6)
    A. butyl propanoate   **B.** butyl propanone   **C.** propyl butanoate
    D. propyl butanone   **E.** heptanoate

Answers to the Practice Test

1. C	**2.** E	**3.** E	**4.** A	**5.** B
6. B	**7.** A	**8.** C	**9.** D	**10.** C
11. A	**12.** D	**13.** E	**14.** C	**15.** D
16. A	**17.** B	**18.** D	**19.** A	**20.** E
21. D	**22.** C	**23.** E	**24.** D	**25.** B
26. A				

Selected Answers and Solutions to Text Problems

17.1 **a.** Hexane has a chain of six carbon atoms.
b. Heptane has a chain of seven carbon atoms.
c. Pentane has a chain of five carbon atoms.

17.3 Two structures are isomers if they have the same molecular formula, but different arrangements of atoms.
a. These condensed structural formulas represent the same molecule; the only difference is due to rotation of the structure. Each has a CH_3— group attached to the middle carbon in a three-carbon chain.
b. The molecular formula of both these condensed structural formulas is C_6H_{14}. However, they represent structural isomers because the C atoms are bonded in a different order; they have different arrangements. In the first, there is a CH_3— group attached to carbon 3 of a five-carbon chain, and in the other, there are CH_3— groups attached to carbon 2 and carbon 3 of a four-carbon chain.
c. The molecular formula of both these line-angle formulas is C_8H_{18}. However, they represent structural isomers because the C atoms are bonded in a different order; they have different arrangements. In the first, there is a CH_3— group attached to carbon 4 of a seven-carbon chain, and in the other, there is a CH_3— group attached to carbon 2 of a seven-carbon chain.

17.5 **a.** 2-fluorobutane
b. dimethylpropane
c. 2-chloro-3-methylpentane

17.7 Draw the main chain with the number of carbon atoms indicated by the alkane name. For example, butane has a main chain of four carbon atoms, and hexane has a main chain of six carbon atoms. Attach substituents on the carbon atoms indicated. For example, in 3-methylpentane, a CH_3— group is bonded to carbon 3 of a five-carbon chain.

a.
$$CH_3-\overset{\overset{\displaystyle CH_3}{|}}{CH}-CH_2-CH_3$$

b.
$$CH_3-CH_2-\overset{\overset{\displaystyle Cl}{|}}{\underset{\underset{\displaystyle Cl}{|}}{C}}-CH_2-CH_3$$

c.

d.

17.9 **a.** $CH_4(g) + 2O_2(g) \xrightarrow{\Delta} CO_2(g) + 2H_2O(g) +$ energy
b. $2C_6H_{14}(l) + 19O_2(g) \xrightarrow{\Delta} 12CO_2(g) + 14H_2O(g) +$ energy

17.11 **a.** An alkene has a carbon–carbon double bond.
b. An alkyne has a carbon–carbon triple bond.
c. An alkene has a carbon–carbon double bond.

17.13 a. The two-carbon compound with a double bond is named ethene.
  b. This is a three-carbon alkene with a methyl substituent. The name is 2-methylpropene.
  c. The five-carbon compound with a triple bond between carbon 2 and carbon 3 is named 2-pentyne.

17.15 a. Propene is the three-carbon alkene. $H_2C=CH-CH_3$
  b. 1-Hexyne is the six-carbon compound with a triple bond between carbon 1 and carbon 2.
    $HC\equiv C-CH_2-CH_2-CH_2-CH_3$
  c. 2-Methyl-1-butene has a four-carbon chain with a double bond between carbon 1 and carbon 2 and a methyl group attached to carbon 2.

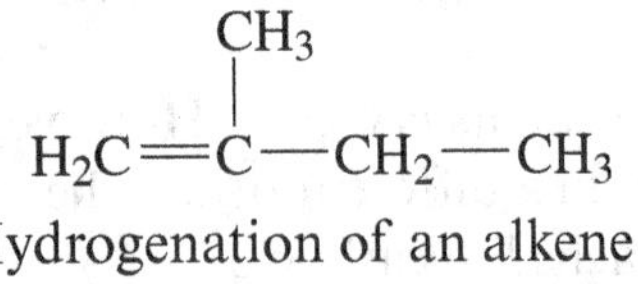

$$\begin{array}{c} CH_3 \\ | \\ H_2C=C-CH_2-CH_3 \end{array}$$

17.17 Hydrogenation of an alkene gives the saturated compound, the alkane.
  a. $CH_3-CH_2-CH_2-CH_2-CH_3$
  b. $CH_3-CH_2-CH_2-CH_3$

  c. ⟍⟋⟍⟋

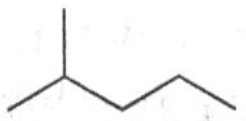

17.19 A polymer is a very large molecule composed of small units (monomers) that are repeated many times.

17.21 $3\ H_2C=CH \longrightarrow$

$$\begin{array}{c} H\quad CH_3\ H\quad CH_3\ H\quad CH_3 \\ |\quad\ \ |\quad\ |\quad\ \ |\quad\ |\quad\ \ | \\ -C-C-C-C-C-C- \\ |\quad\ \ |\quad\ |\quad\ \ |\quad\ |\quad\ \ | \\ H\quad H\quad H\quad H\quad H\quad H \end{array}$$

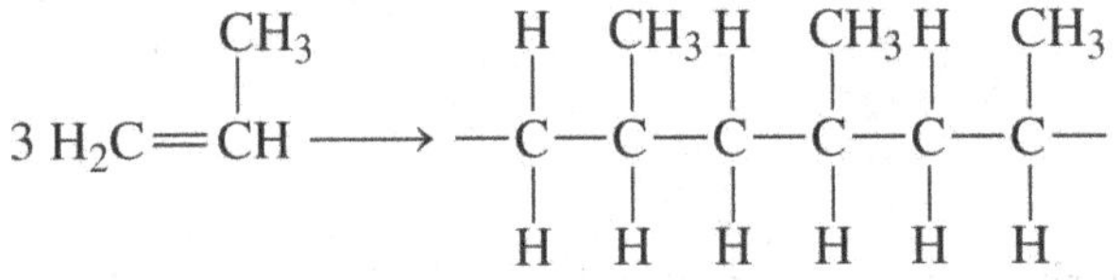

17.23

$$\begin{array}{c} F\quad H\quad F\quad H\quad F\quad H \\ |\quad\ |\quad\ |\quad\ |\quad\ |\quad\ | \\ -C-C-C-C-C-C- \\ |\quad\ |\quad\ |\quad\ |\quad\ |\quad\ | \\ F\quad H\quad F\quad H\quad F\quad H \end{array}$$

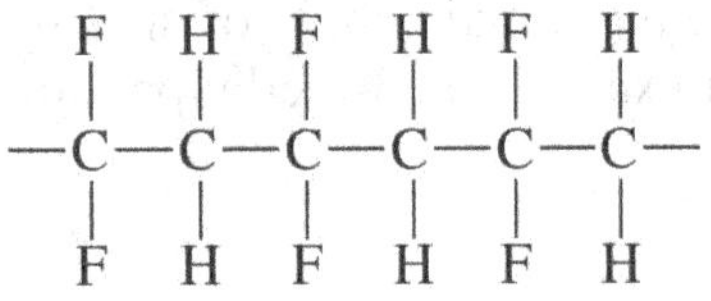

17.25 Aromatic compounds that contain a benzene ring with a single substituent are usually named as benzene derivatives. A benzene ring with a methyl substituent is named toluene. The methyl group is attached to carbon 1, and the ring is numbered to give the lower numbers to other substituents.
  a. 2-chlorotoluene
  b. ethylbenzene
  c. phenol

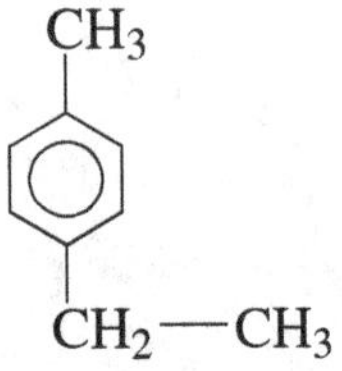

17.27 a. NH_2

  b. Cl ... F

  c. CH_3 ... CH_2-CH_3

17.29 a. This compound has a two-carbon chain. The final *e* from ethane is dropped, and *ol* is added to indicate an alcohol. The IUPAC name is ethanol.

b. This compound has a four-carbon chain with a hydroxyl group attached to carbon 2. The IUPAC name is 2-butanol.

c. This compound is the aromatic alcohol phenol with a chlorine atom attached to carbon 2 of the ring (the —OH group is on carbon 1). The IUPAC name is 2-chlorophenol.

17.31 a. 1-Propanol has a three-carbon chain with a hydroxyl group attached to carbon 1.
CH_3—CH_2—CH_2—OH

b. Methyl alcohol has a hydroxyl group attached to a one-carbon alkane. CH_3—OH

c. 3-Pentanol has a five-carbon chain with a hydroxyl group attached to carbon 3.

$$CH_3-CH_2-\underset{\underset{OH}{|}}{CH}-CH_2-CH_3$$

d. 2-Methyl-2-butanol has a four-carbon chain with a methyl and hydroxyl group attached to carbon 2.

$$CH_3-\underset{\underset{CH_3}{|}}{\overset{\overset{OH}{|}}{C}}-CH_2-CH_3$$

17.33 a. The common name of the ether with a one-carbon alkyl group and a two-carbon alkyl group attached to an oxygen atom is ethyl methyl ether.

b. The common name of the ether with two three-carbon alkyl groups attached to an oxygen atom is dipropyl ether.

c. The common name of the ether with a three-carbon alkyl group and a four-carbon alkyl group attached to an oxygen atom is butyl propyl ether.

17.35 a. The common name of the aldehyde with two carbons is acetaldehyde.

b. The common name of the ketone with a one-carbon alkyl group and a three-carbon alkyl group attached to the carbonyl carbon is methyl propyl ketone.

c. The common name of the aldehyde with one carbon is formaldehyde.

17.37 a. The IUPAC name of the aldehyde with three carbons is propanal.

b. The IUPAC name of the five-carbon ketone with the carbonyl group on carbon 3 and a methyl group on carbon 2 is 2-methyl-3-pentanone.

c. The IUPAC name of the aromatic aldehyde is benzaldehyde.

17.39 a. Acetaldehyde is the common name of the aldehyde with two carbons.

$$CH_3-\overset{\overset{O}{\|}}{C}-H$$

b. 2-Pentanone has a ketone group on carbon 2 of a five-carbon chain.

$$CH_3-\overset{\overset{O}{\|}}{C}-CH_2-CH_2-CH_3$$

c. Butyl methyl ketone has a four-carbon group and a one-carbon group on either side of the carbonyl carbon.

$$CH_3-\overset{\overset{O}{\|}}{C}-CH_2-CH_2-CH_2-CH_3$$

17.41 a. Ethanoic acid (acetic acid) is the carboxylic acid with two carbons.
   b. Propanoic acid (propionic acid) is the three-carbon carboxylic acid.
   c. 3-Methylhexanoic acid is a six-carbon carboxylic acid with a methyl group on carbon 3 of the chain.
   d. 3-Bromobenzoic acid has a carboxylic acid group on the benzene ring and a bromine atom on carbon 3.

17.43 a. Butyric acid is the common name of the four-carbon carboxylic acid.

$$CH_3—CH_2—CH_2—\overset{\overset{\textstyle O}{\|}}{C}—OH$$

   b. Benzoic acid has a carboxylic acid group attached to a benzene ring.

$$\overset{\overset{\textstyle O}{\|}}{C}—OH$$

   c. 2-Chloroethanoic acid is a carboxylic acid that has a two-carbon chain with a chlorine atom on carbon 2.

$$Cl—CH_2—\overset{\overset{\textstyle O}{\|}}{C}—OH$$

   d. 3-Hydroxypropanoic acid is a carboxylic acid that has a three-carbon chain with a hydroxyl group on carbon 3.

$$HO—CH_2—CH_2—\overset{\overset{\textstyle O}{\|}}{C}—OH$$

17.45 a. The ester has a two-carbon part from the alcohol ethanol and a two-carbon part from the carboxylic acid acetic (ethanoic) acid.

$$CH_3—\overset{\overset{\textstyle O}{\|}}{C}—O—CH_2—CH_3$$

   b. The ester has a two-carbon part from the alcohol ethanol and a four-carbon part from the carboxylic acid butyric (butanoic) acid.

$$CH_3—CH_2—CH_2—\overset{\overset{\textstyle O}{\|}}{C}—O—CH_2—CH_3$$

17.47 A carboxylic acid and an alcohol react to give an ester with the elimination of water.

   a.

   b. $CH_3—CH_2—CH_2—CH_2—\overset{\overset{\textstyle O}{\|}}{C}—O—\overset{\overset{\textstyle CH_3}{|}}{CH}—CH_3$

17.49 a. The alcohol part of the ester is from methanol (methyl alcohol), and the carboxylic acid part is from methanol acid (formic acid). The ester is named methyl methanoate (methyl formate).

b. The alcohol part of the ester is from ethanol (ethyl alcohol), and the carboxylic acid part is from ethanoic acid (acetic acid). The ester is named ethyl ethanoate (ethyl acetate).

c. The alcohol part of the ester is from ethanol (ethyl alcohol), and the carboxylic acid part is from propanoic acid (propionic acid). The ester is named ethyl propanoate (ethyl propionate).

17.51 a. The alkyl part of the ester comes from the one-carbon methanol, and the carboxylate part comes from the two-carbon acetic acid.

$$CH_3-\overset{\overset{\displaystyle O}{\|}}{C}-O-CH_3$$

b. The alkyl part of the ester comes from the four-carbon 1-butanol, and the carboxylate part comes from the one-carbon formic acid.

$$H-\overset{\overset{\displaystyle O}{\|}}{C}-O-CH_2-CH_2-CH_2-CH_3$$

c. The alkyl part of the ester comes from the two-carbon ethanol, and the carboxylate part comes from the five-carbon pentanoic acid.

$$CH_3-CH_2-CH_2-CH_2-\overset{\overset{\displaystyle O}{\|}}{C}-O-CH_2-CH_3$$

d. The alkyl part of the ester comes from the three-carbon 1-propanol, and the carboxylate part comes from the three-carbon propanoic acid.

$$CH_3-CH_2-\overset{\overset{\displaystyle O}{\|}}{C}-O-CH_2-CH_2-CH_3$$

17.53 The common name of an amine consists of naming the alkyl groups bonded to the nitrogen atom in alphabetical order followed by *amine*.

a. A one-carbon alkyl group attached to $-NH_2$ is methylamine.

b. A one-carbon and a three-carbon alkyl group attached to the nitrogen atom form methylpropylamine.

c. A one-carbon and two two-carbon alkyl groups attached to the nitrogen atom form diethylmethylamine.

17.55 Carboxylic acids react with amines to form amides with the elimination of water.

a. $CH_3-\overset{\overset{\displaystyle O}{\|}}{C}-NH_2$

b. $CH_3-\overset{\overset{\displaystyle O}{\|}}{C}-\overset{\overset{\displaystyle H}{|}}{N}-CH_2-CH_3$

c. $\langle\!\bigcirc\!\rangle-\overset{\overset{\displaystyle O}{\|}}{C}-\overset{\overset{\displaystyle H}{|}}{N}-CH_2-CH_2-CH_3$

17.57 a. Ethanamide (acetamide) has a two-carbon carbonyl portion bonded to an amino group.
b. 2-Chlorobutanamide (2-chlorobutyramide) is a four-carbon amide with a chlorine atom on carbon 2.
c. Methanamide (formamide) has only the carbonyl carbon bonded to the amino group.

17.59 a. Propionamide is the common name of the amide of propanoic acid, which has three carbon atoms.

$$CH_3-CH_2-\overset{\overset{\displaystyle O}{\|}}{C}-NH_2$$

b. 2-Methylpentanamide is an amide of the five-carbon pentanoic acid with a methyl group attached to carbon 2.

$$CH_3-CH_2-CH_2-\overset{\overset{\displaystyle CH_3}{|}}{CH}-\overset{\overset{\displaystyle O}{\|}}{C}-NH_2$$

c. Methanamide is the simplest of the amides, with only one carbon atom.

$$H-\overset{\overset{\displaystyle O}{\|}}{C}-NH_2$$

17.61

$$-\overset{\overset{\displaystyle Cl}{|}}{\underset{\underset{\displaystyle F}{|}}{C}}-\overset{\overset{\displaystyle F}{|}}{\underset{\underset{\displaystyle F}{|}}{C}}-\overset{\overset{\displaystyle Cl}{|}}{\underset{\underset{\displaystyle F}{|}}{C}}-\overset{\overset{\displaystyle F}{|}}{\underset{\underset{\displaystyle F}{|}}{C}}-\overset{\overset{\displaystyle Cl}{|}}{\underset{\underset{\displaystyle F}{|}}{C}}-\overset{\overset{\displaystyle F}{|}}{\underset{\underset{\displaystyle F}{|}}{C}}-$$

17.63 a. $C_9H_{20}(l) + 14O_2(g) \xrightarrow{\Delta} 9CO_2(g) + 10H_2O(g) + energy$
b. $C_7H_8(l) + 9O_2(g) \xrightarrow{\Delta} 7CO_2(g) + 4H_2O(g) + energy$

17.65

$$-\overset{\overset{\displaystyle F}{|}}{\underset{\underset{\displaystyle F}{|}}{C}}-\overset{\overset{\displaystyle F}{|}}{\underset{\underset{\displaystyle F}{|}}{C}}-\overset{\overset{\displaystyle F}{|}}{\underset{\underset{\displaystyle F}{|}}{C}}-\overset{\overset{\displaystyle F}{|}}{\underset{\underset{\displaystyle F}{|}}{C}}-\overset{\overset{\displaystyle F}{|}}{\underset{\underset{\displaystyle F}{|}}{C}}-\overset{\overset{\displaystyle F}{|}}{\underset{\underset{\displaystyle F}{|}}{C}}-$$

17.67 a. 3,7-dimethyl-6-octenal
b. The *en* signifies that a double bond is present.
c. The *al* signifies that an aldehyde is present.

17.69 a.
$$CH_3-CH_2-\overset{\overset{\displaystyle CH_2-CH_3}{|}}{CH}-CH_2-CH_2-CH_3$$
b. $CH_3-CH=CH-CH_2-CH_3$
c. $CH_3-C\equiv C-CH_2-CH_2-CH_3$

17.71 a. This compound contains a four-carbon chain with two methyl groups attached to carbon 2. The IUPAC name is 2,2-dimethylbutane.
b. This compound has a four-carbon chain with a triple bond between carbon 1 and carbon 2. The IUPAC name is 1-butyne.
c. This compound contains a five-carbon chain with a double bond between carbon 2 and carbon 3. The IUPAC name is 2-pentene.

17.73 a. ketone   **b.** alkene
c. ester   **d.** amine

17.75 a. An alcohol contains a hydroxyl group.
   b. An alkene contains one or more carbon–carbon double bonds.
   c. An aldehyde contains a carbonyl group bonded to a hydrogen.
   d. Alkanes are hydrocarbons that contain only carbon–carbon single bonds.

17.77 a. aromatic, aldehyde   **b.** aromatic, alkene, aldehyde

17.79 Aromatic compounds that contain a benzene ring with a single substituent are usually named as benzene derivatives. A benzene ring with a methyl substituent is named toluene. The methyl group is attached to carbon 1, and the ring is numbered to give the lower numbers to other substituents.
   a. 3-bromotoluene
   b. 1,2-dichlorobenzene
   c. 4-ethyltoluene

17.81 a. CH_2—CH_3

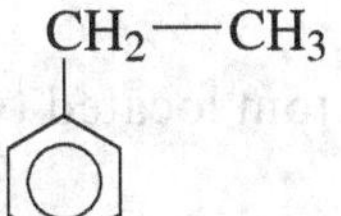

   b. CH_3

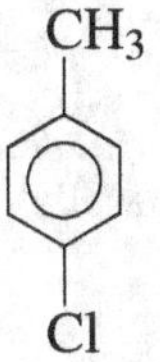

   c. Br

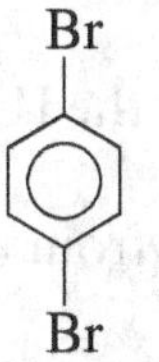

17.83 a. This compound has a four-carbon chain with a hydroxyl group attached to carbon 2. The IUPAC name is 2-butanol.
   b. This compound has a five-carbon chain with a hydroxyl group attached to carbon 2 and a methyl group attached to carbon 3. The IUPAC name is 3-methyl-2-pentanol.
   c. The common name of the ether with a two-carbon alkyl group and a four-carbon alkyl group attached to an oxygen atom is butyl ethyl ether.

17.85 a. OH ... Cl

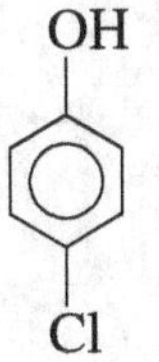

   b. CH_3—CH—CH—CH_2—CH_3 with CH_3 and OH

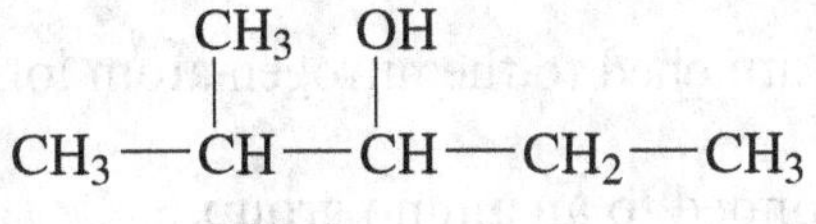

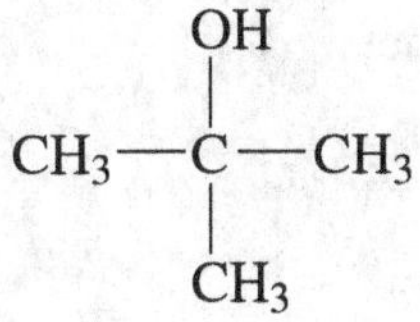

   c. CH_3—C—CH_3 with OH and CH_3

17.87 a. With a chlorine atom on carbon 4 of the aromatic aldehyde benzaldehyde, the IUPAC name is 4-chlorobenzaldehyde.

   b. With a chlorine atom attached to carbon 3 of the three-carbon aldehyde, the IUPAC name is 3-chloropropanal.

   c. With a chlorine atom attached to carbon 2 of the five-carbon ketone, the IUPAC name is 2-chloro-3-pentanone.

17.89 a. 2-Bromobenzaldehyde has a bromine atom attached to carbon 2 of the ring in the aromatic aldehyde.

$$\begin{array}{c} O \\ \| \\ C-H \\ \end{array} \quad Br$$

   b. 3-Chloropropionaldehyde is a three-carbon aldehyde with a chlorine atom located two carbons from the carbonyl group.

$$Cl-CH_2-CH_2-\overset{\overset{\displaystyle O}{\|}}{C}-H$$

   c. Ethyl methyl ketone (butanone) is a four-carbon ketone.

$$CH_3-CH_2-\overset{\overset{\displaystyle O}{\|}}{C}-CH_3$$

17.91 a. With a methyl group attached to carbon 3 of the four-carbon carboxylic acid, the IUPAC name of this compound is 3-methylbutanoic acid.

   b. The alcohol part of the ester is from methanol, and the carboxylic acid part is from the aromatic benzoic acid. The IUPAC name of this ester is methyl benzoate.

   c. The alcohol part of the ester is from ethanol (ethyl alcohol), and the carboxylic acid part is from propanoic acid. The IUPAC name of this ester is ethyl propanoate.

17.93 a. A two-carbon alkyl group is attached to $-NH_2$.
$$CH_3-CH_2-NH_2$$

   b. This is an amide of hexanoic acid, which has six carbon atoms.

$$CH_3-CH_2-CH_2-CH_2-CH_2-\overset{\overset{\displaystyle O}{\|}}{C}-NH_2$$

   c. Three two-carbon alkyl groups are bonded to a nitrogen atom.

$$\begin{array}{c} CH_2-CH_3 \\ | \\ CH_3-CH_2-N-CH_2-CH_3 \end{array}$$

17.95 a. One four-carbon and two one-carbon alkyl groups attached to the nitrogen atom form butyldimethylamine.

   b. Pentanamide has a five-carbon carbonyl portion bonded to an amino group.

   c. A five-carbon alkyl group attached to $-NH_2$ is named pentylamine.

17.97 CH_3—CH_2—CH_2—CH_2—CH_2—OH 1-Pentanol

CH_3—$\overset{\displaystyle OH}{\underset{\displaystyle |}{CH}}$—$CH_2$—$CH_2$—$CH_3$ 2-Pentanol

CH_3—CH_2—$\overset{\displaystyle OH}{\underset{\displaystyle |}{CH}}$—$CH_2$—$CH_3$ 3-Pentanol

HO—CH_2—$\overset{\displaystyle CH_3}{\underset{\displaystyle |}{CH}}$—$CH_2$—$CH_3$ 2-Methyl-1-butanol

HO—CH_2—CH_2—$\overset{\displaystyle CH_3}{\underset{\displaystyle |}{CH}}$—$CH_3$ 3-Methyl-1-butanol

CH_3—$\overset{\displaystyle CH_3}{\underset{\displaystyle \underset{\displaystyle OH}{|}}{\overset{\displaystyle |}{C}}}$—$CH_2$—$CH_3$ 2-Methyl-2-butanol

CH_3—$\overset{\displaystyle OH}{\underset{\displaystyle |}{CH}}$—$\overset{\displaystyle CH_3}{\underset{\displaystyle |}{CH}}$—$CH_3$ 3-Methyl-2-butanol

CH_3—$\overset{\displaystyle CH_3}{\underset{\displaystyle \underset{\displaystyle CH_3}{|}}{\overset{\displaystyle |}{C}}}$—$CH_2$—$OH$ 2,2-Dimethyl-1-propanol

17.99

17.101   **a.** $2CH_3$—CH_2—$C{\equiv}CH + 11O_2 \xrightarrow{\Delta} 8CO_2 + 6H_2O + energy$

       b. CH_3—CH_2—$CH{=}CH$—CH_2—$CH_3 + H_2 \xrightarrow{Pt}$

                           CH_3—CH_2—CH_2—CH_2—CH_2—CH_3

18

Biochemistry

Credit: Pearson Education, Inc.

Rebecca has learned that her extremely high cholesterol level is a result of familial hypercholesterolemia. A lipid specialist at a lipid clinic helped change her diet to include less beef and chicken, more fish, low-fat dairy products, and fiber, and to avoid egg yolks, coconut, and palm oils. She increased her exercise and lost 35 lb.

However, her most recent blood tests at the lipid clinic indicate her cholesterol is lower, but is still in the high range. She was prescribed Pravachol and TriCor that are used for treating high cholesterol.

Pravachol

Draw the structure of cholesterol and explain why it is classified as a lipid.

CHAPTER READINESS*

🌶 Core Chemistry Skills

- Writing Equations for Hydrogenation and Polymerization (17.2)
- Naming Carboxylic Acids (17.6)
- Forming Amides (17.7)

*These Core Chemistry Skills from previous chapters are listed here for your review as you proceed to the new material in this chapter.

LOOKING AHEAD

18.1   Carbohydrates

Learning Goal: Classify a carbohydrate as an aldose or ketose; draw the open-chain and Haworth structures for glucose, galactose, and fructose.

- Photosynthesis is the process of using water, carbon dioxide, and the energy from the Sun to form monosaccharides and oxygen.
- Carbohydrates are classified as monosaccharides (simple sugars), disaccharides (two monosaccharide units), and polysaccharides (many monosaccharide units).

- Monosaccharides are aldehydes (aldoses) or ketones (ketoses) with hydroxyl groups on all other carbon atoms.
- Monosaccharides are classified as *aldo* for an aldehyde or *keto* for a ketone and by the number of carbon atoms as *trioses*, *tetroses*, *pentoses*, or *hexoses*.
- Important monosaccharides are the aldohexoses glucose and galactose and the ketohexose fructose.
- The Haworth structure is a representation of the cyclic, stable form of monosaccharides, which are rings of five or six atoms.
- The Haworth structure forms by a reaction between the —OH group on carbon 5 of hexoses and the carbonyl group on carbon 1 or 2 of the same molecule.
- The formation of a new —OH group on carbon 1 (or 2 in fructose) gives α and β forms of the cyclic monosaccharide.

♦ **Learning Exercise 18.1A**

Identify each of the following monosaccharides as an aldotriose, a ketotriose, an aldotetrose, a ketotetrose, an aldopentose, a ketopentose, an aldohexose, or a ketohexose:

a. 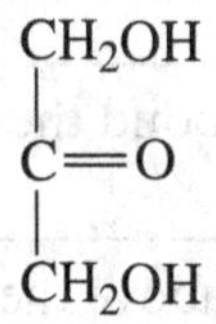
$$CH_2OH$$
$$C=O$$
$$CH_2OH$$

b.
$$H\text{—}C\text{=}O$$
$$H\text{—}C\text{—}OH$$
$$H\text{—}C\text{—}OH$$
$$HO\text{—}C\text{—}H$$
$$CH_2OH$$

c.
$$CH_2OH$$
$$C=O$$
$$HO\text{—}C\text{—}H$$
$$HO\text{—}C\text{—}H$$
$$H\text{—}C\text{—}OH$$
$$CH_2OH$$

d. 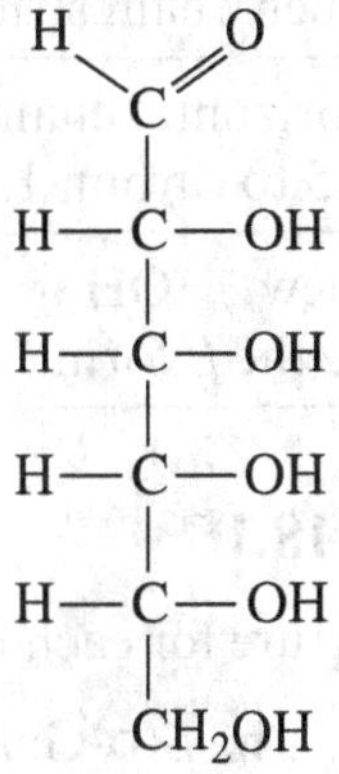
$$H\text{—}C\text{=}O$$
$$H\text{—}C\text{—}OH$$
$$H\text{—}C\text{—}OH$$
$$H\text{—}C\text{—}OH$$
$$H\text{—}C\text{—}OH$$
$$CH_2OH$$

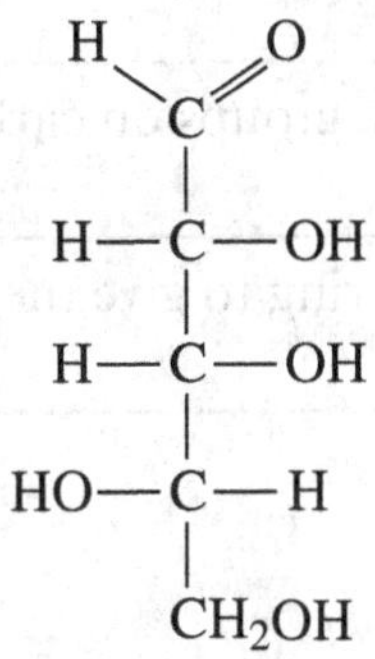

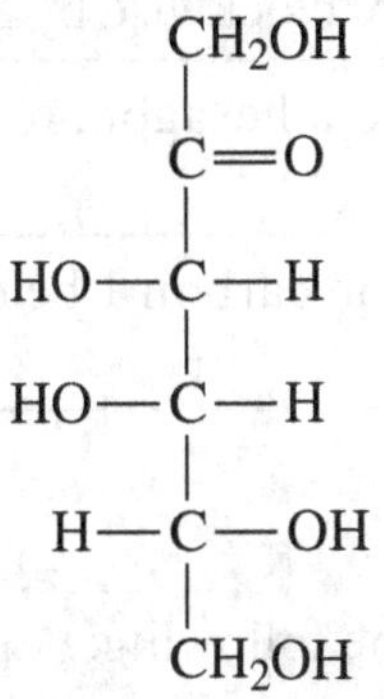

e. 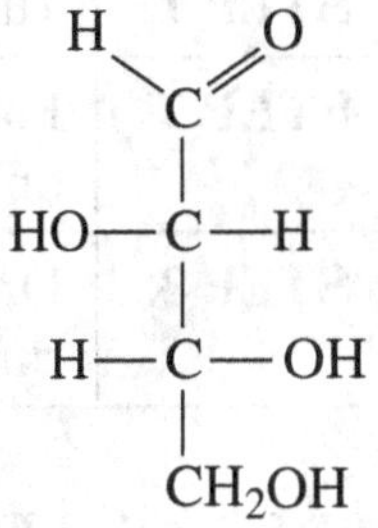
$$H\text{—}C\text{=}O$$
$$HO\text{—}C\text{—}H$$
$$H\text{—}C\text{—}OH$$
$$CH_2OH$$

Dihydroxyacetone Lyxose Tagatose Allose Threose

_____   _____   _____   _____   _____

Answers   **a.** ketotriose   **b.** aldopentose   **c.** ketohexose
   d. aldohexose   **e.** aldotetrose

♦ **Learning Exercise 18.1B**

Draw the condensed structural formula for each of the following monosaccharides:

D-Glucose   D-Galactose   D-Fructose

Answers

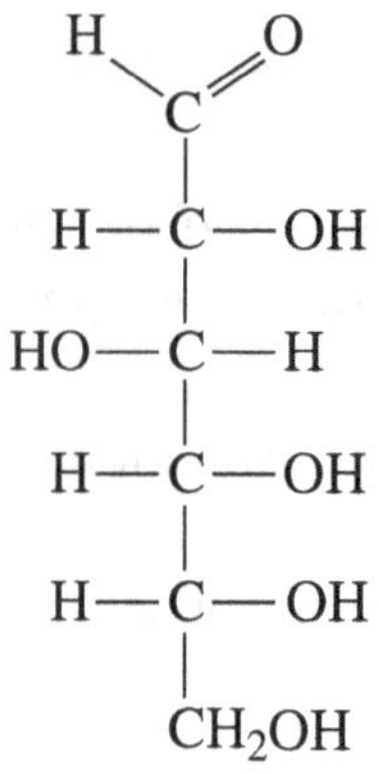

H—C=O H—C=O CH₂OH

$$\text{D-Glucose} \qquad \text{D-Galactose} \qquad \text{D-Fructose}$$

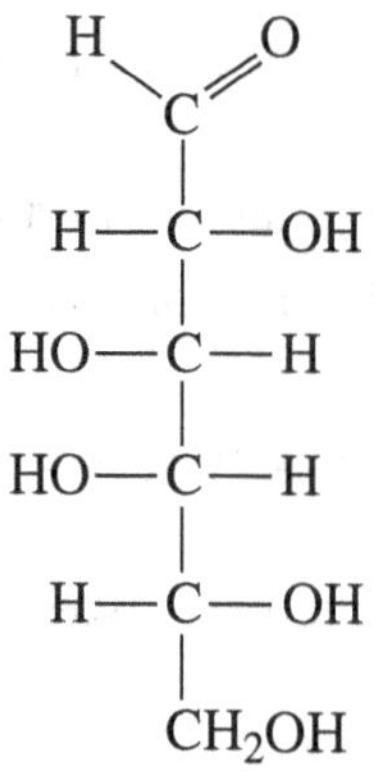

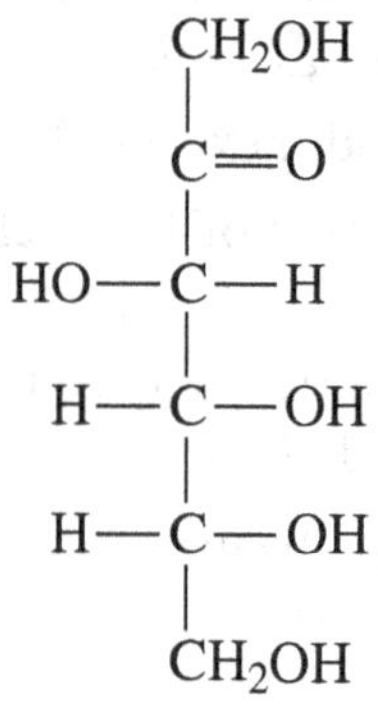

Guide to Drawing Haworth Structures	
STEP 1	Turn the open-chain structure clockwise by 90°.
STEP 2	Fold the horizontal chain into a hexagon, rotate the groups on carbon 5, and bond the O on carbon 5 to carbon 1.
STEP 3	Draw the new —OH group on carbon 1 below the ring to give the α form or above the ring to give the β form.

♦ **Learning Exercise 18.1C**

Draw the Haworth structure for each of the following:

a. α-D-Glucose   **b.** β-D-Galactose   **c.** α-D-Fructose

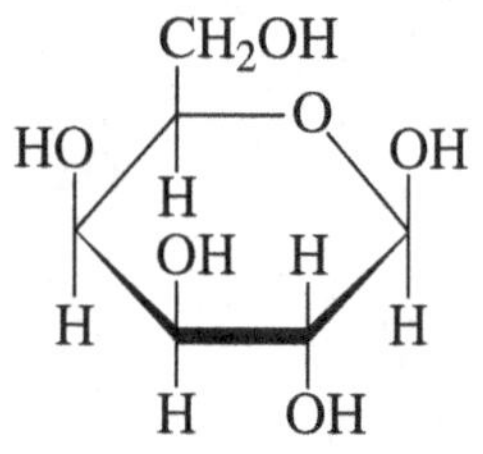

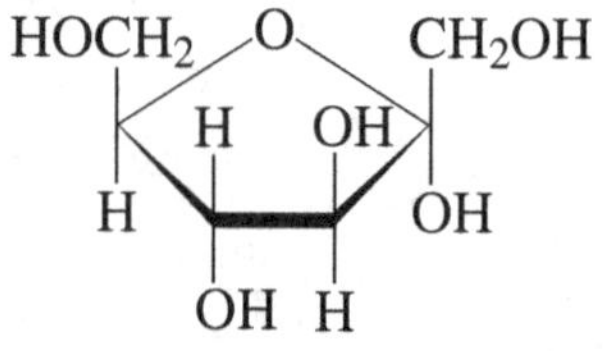

Answers

a.   **b.**   **c.**

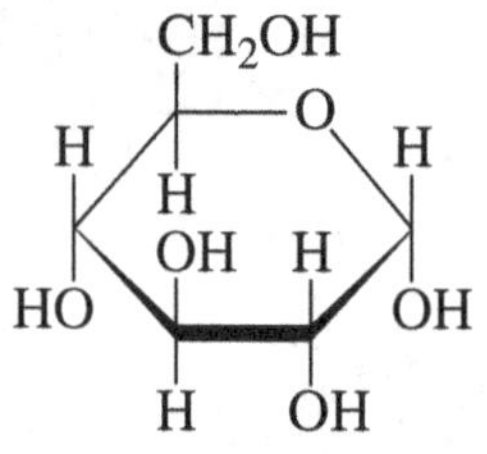

$$\alpha\text{-D-Glucose} \qquad \beta\text{-D-Galactose} \qquad \alpha\text{-D-Fructose}$$

18.2 Disaccharides and Polysaccharides

Learning Goal: Describe the monosaccharide units and linkages in disaccharides and polysaccharides.

◢ **CORE CHEMISTRY SKILL**

Drawing Haworth Structures

- Disaccharides are two monosaccharide units joined together by a glycosidic bond.

  Monosaccharide (1) + monosaccharide (2) ⟶ disaccharide + H_2O

- In the most common disaccharides, maltose, lactose, and sucrose, there is at least one glucose unit.

- In the disaccharide maltose, two glucose units are linked by an $\alpha(1 \rightarrow 4)$ glycosidic bond. The $\alpha(1 \rightarrow 4)$ indicates that the —OH group on carbon 1 of α-D-glucose is bonded to carbon 4 of the other glucose molecule.

  Glucose + glucose ⟶ maltose + H_2O

  Glucose + galactose ⟶ lactose + H_2O

  Glucose + fructose ⟶ sucrose + H_2O

Lactose is a disaccharide found in milk and milk products.

Credit: Pearson Education, Inc.

- Starches consist of amylose, an unbranched polymer of glucose, and amylopectin, which is a branched polymer of glucose. Glycogen, the storage form of glucose in animals, is similar to amylopectin, but has more branching.

- Cellulose is also a polymer of glucose, but in cellulose the glycosidic bonds are β bonds rather than α bonds as in the starches. Humans can digest starches to obtain energy, but not cellulose.

Key Terms for Sections 18.1 and 18.2

Match the following key terms with the correct description:

a. carbohydrate **b.** glucose **c.** disaccharide **d.** Haworth structure **e.** cellulose **f.** polysaccharide

1. _____ a simple or complex sugar composed of a carbon chain with an aldehyde or ketone group and several hydroxyl groups

2. _____ a carbohydrate that contains many monosaccharides linked by glycosidic bonds

3. _____ an unbranched polysaccharide that cannot be digested by humans

4. _____ an aldohexose that is the most prevalent monosaccharide in the diet

5. _____ a carbohydrate that contains two monosaccharides linked by a glycosidic bond

6. _____ a cyclic structure that represents the closed-chain form of a monosaccharide

Answers   **1.** a   **2.** f   **3.** e   **4.** b   **5.** c   **6** d

♦ **Learning Exercise 18.2A**

Indicate the number of monosaccharide units (one, two, or many) in each of the following carbohydrates:

a. sucrose _____   **b.** cellulose _____

c. glucose _____   **d.** amylose _____

e. maltose _____   **f.** fructose _____

Answers   **a.** two   **b.** many   **c.** one   **d.** many   **e.** two   **f.** one

◆ Learning Exercise 18.2B

For the following disaccharides, state (**a**) the monosaccharide units, (**b**) the type of glycosidic bond, and (**c**) the name of the disaccharide:

1.

2.

3.

4.

1. a. _____	b. _____	c. _____
2. a. _____	b. _____	c. _____
3. a. _____	b. _____	c. _____
4. a. _____	b. _____	c. _____

Answers

1. **a.** two glucose units **b.** $\alpha(1 \rightarrow 4)$-glycosidic bond **c.** β-maltose

2. **a.** galactose + glucose **b.** $\beta(1 \rightarrow 4)$-glycosidic bond **c.** α-lactose

3. **a.** fructose + glucose **b.** $\alpha,\beta(1 \rightarrow 2)$-glycosidic bond **c.** sucrose

4. **a.** two glucose units **b.** $\alpha(1 \rightarrow 4)$-glycosidic bond **c.** α-maltose

♦ Learning Exercise 18.2C

List the monosaccharides and describe the glycosidic bonds in each of the following carbohydrates:

	Monosaccharides	Type(s) of Glycosidic Bonds
a. amylose	_____	_____
b. amylopectin	_____	_____
c. glycogen	_____	_____
d. cellulose	_____	_____

Answers

    a. glucose; $\alpha(1 \rightarrow 4)$-glycosidic bonds

    b. glucose; $\alpha(1 \rightarrow 4)$- and $\alpha(1 \rightarrow 6)$-glycosidic bonds

    c. glucose; $\alpha(1 \rightarrow 4)$- and $\alpha(1 \rightarrow 6)$-glycosidic bonds

    d. glucose; $\beta(1 \rightarrow 4)$-glycosidic bonds

The polysaccharide cellulose is the structural material in plants such as cotton.

Credit: Danny E Hooks / Shutterstock

18.3   Lipids

Learning Goal: Draw the condensed or line-angle structural formula for a fatty acid, a triacylglycerol, and the products of hydrogenation or saponification. Identify the steroid nucleus.

- Lipids are nonpolar compounds that are not soluble in water but are soluble in organic solvents.
- Fatty acids are unbranched carboxylic acids that typically contain an even number (12 to 20) of carbon atoms.
- Fatty acids may be saturated, monounsaturated with one double bond, or polyunsaturated with two or more carbon–carbon double bonds. The double bonds in naturally occurring unsaturated fatty acids are almost always cis.
- Saturated fatty acids have higher melting points than unsaturated fatty acids because they pack together more tightly.
- A wax is an ester of a long-chain fatty acid and a long-chain alcohol.
- The triacylglycerols in fats and oils are esters of glycerol and three long-chain fatty acids.
- Fats from animal sources contain more saturated fatty acids and have higher melting points than fats found in most vegetable oils.
- The hydrogenation of unsaturated fatty acids converts double bonds to single bonds.
- In saponification, a fat heated with a strong base produces glycerol and the salts of the fatty acids or soaps.
- Steroids are lipids containing the steroid nucleus, which is a fused structure of four rings. Cholesterol is one of the most important and abundant steroids.

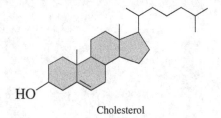

Cholesterol

Key Terms for Section 18.3

Match the following key terms with the correct description:

   a. lipid   **b.** fatty acid   **c.** triacylglycerol   **d.** saponification   **e.** steroid   **f.** cis isomer

1. _____ a type of compound that is not soluble in water, but is soluble in organic solvents

2. _____ the reaction of a triacylglycerol with a strong base producing salts, called soap, and glycerol

3. _____ a fatty acid with carbon groups on the same side of the double bond

4. _____ a lipid composed of a fused four-ring system

5. _____ a long-chain carboxylic acid found in triacylglycerols

6. _____ a lipid consisting of glycerol bonded to three long-chain fatty acids

Answers   **1.** a   **2.** d   **3.** f   **4.** e   **5.** b   **6.** c

♦ **Learning Exercise 18.3A**

⟩ **CORE CHEMISTRY SKILL**

Identifying Fatty Acids

Draw the condensed and line-angle structural formulas for the following fatty acids:

a. linoleic acid (18:2)

b. stearic acid (18:0)

c. oleic acid (18:1)

d. Identify which of the three fatty acids linoleic (L), stearic (S), and/or oleic (O)

    1. is the most saturated   _____   **2.** is the most unsaturated   _____

    3. has the lowest melting point   _____   **4.** has the highest melting point   _____

    5. is/are from plant sources   _____   **6.** is/are from animal sources   _____

Answers

a. $CH_3—(CH_2)_4—CH=CH—CH_2—CH=CH—(CH_2)_7—\overset{\displaystyle O}{\overset{\displaystyle \|}{C}}—OH$

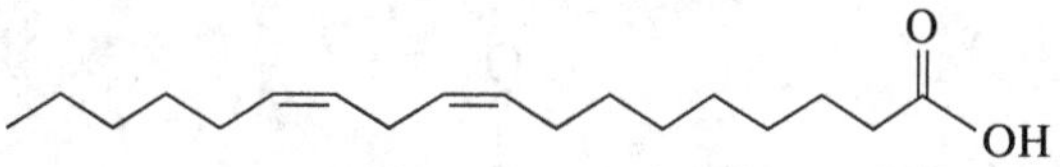

b. $CH_3—(CH_2)_{16}—\overset{\displaystyle O}{\overset{\displaystyle \|}{C}}—OH$

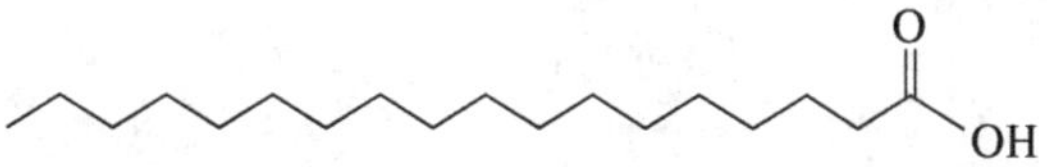

c. $CH_3—(CH_2)_7—CH=CH—(CH_2)_7—\overset{\displaystyle O}{\overset{\displaystyle \|}{C}}—OH$

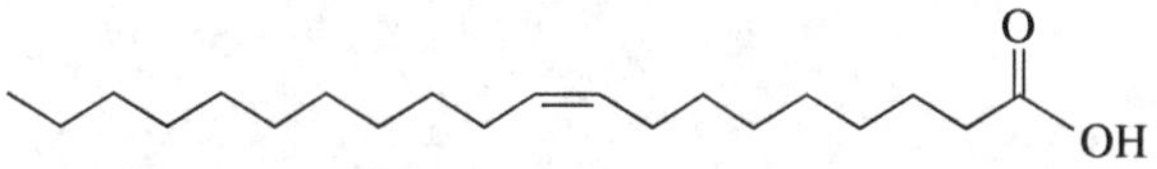

d. 1. S   **2.** L   **3.** L   **4.** S   **5.** L and O   **6.** S

♦ **Learning Exercise 18.3B**

For the following questions, refer to the line-angle structural formula for a fatty acid:

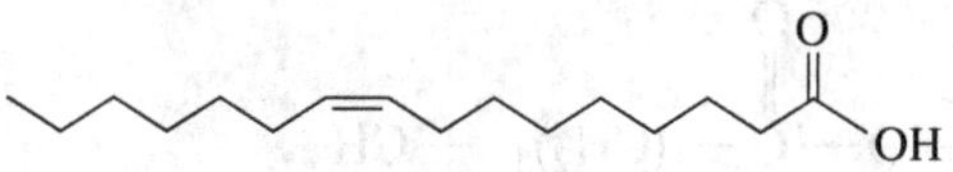

a. How many carbons are in this fatty acid?   _____

b. What is the name of this fatty acid?   _____

c. Is it a saturated or an unsaturated compound? Why?   _____

d. Is the double bond cis or trans?   _____

e. Why is it insoluble in water?   _____

Answers   **a.** 16 carbons   **b.** palmitoleic acid   **c.** unsaturated; double bond
                 d. cis   **e.** It has a long hydrocarbon chain.

Guide to Drawing Triacylglycerols	
STEP 1	Draw the condensed structural formulas for glycerol and the fatty acids.
STEP 2	Form ester bonds between the hydroxyl groups on glycerol and the carboxyl groups on each fatty acid.

♦ Learning Exercise 18.3C

CORE CHEMISTRY SKILL
Drawing Structures for Triacylglycerols

Draw the condensed structural formula for the triacylglycerol formed from glycerol and three molecules of each of the following fatty acids and give the name:

a. palmitic acid, $CH_3—(CH_2)_{14}—\overset{\overset{\displaystyle O}{\|}}{C}—OH$

b. myristic acid, $CH_3—(CH_2)_{12}—\overset{\overset{\displaystyle O}{\|}}{C}—OH$

Answers

a.

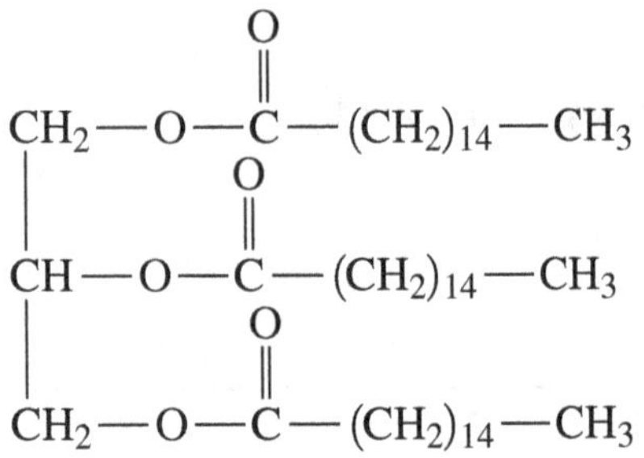

Glyceryl tripalmitate
(tripalmitin)

b.

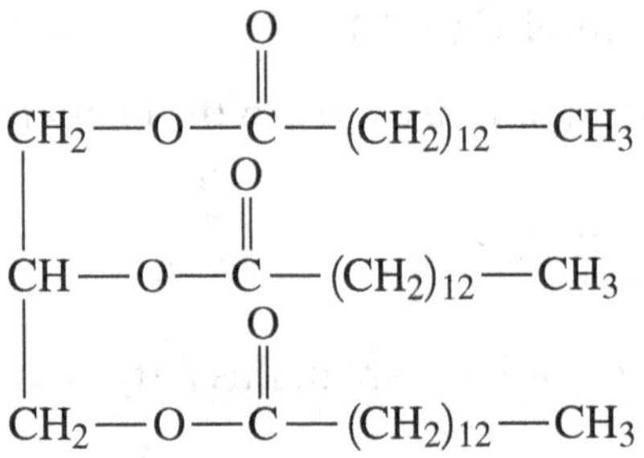

Glyceryl trimyristate
(trimyristin)

♦ Learning Exercise 18.3D

Draw the line-angle structural formulas for the following triacylglycerols:

a. glyceryl tristearate (tristearin)

b. glyceryl trioleate (triolein)

Answers

a.

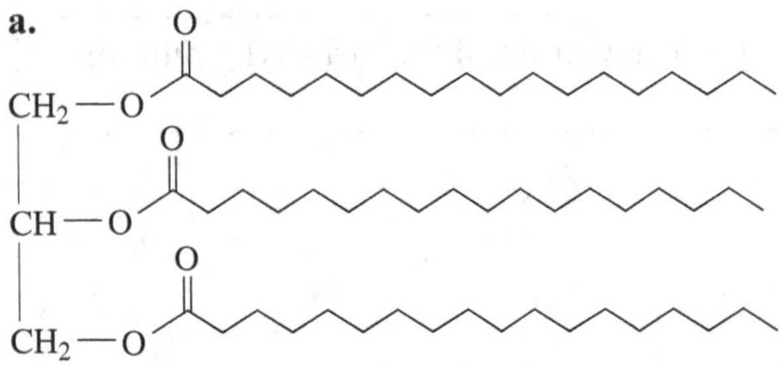

b.

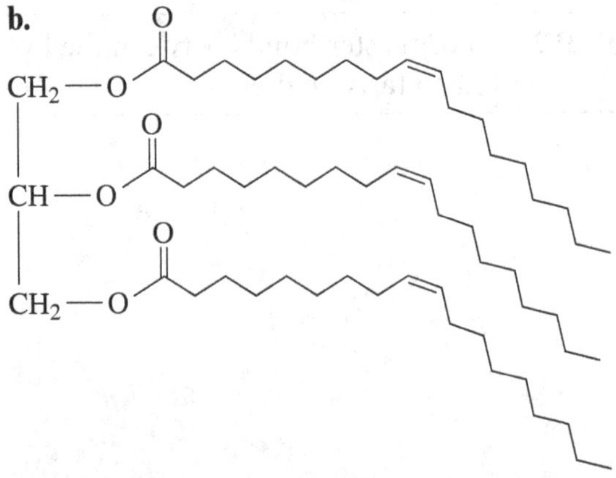

◆ Learning Exercise 18.3E

Draw the condensed structural formulas for the reactants and products for the following reactions of glyceryl tripalmitoleate (tripalmitolein):

a. hydrogenation with a nickel catalyst

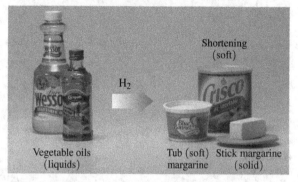
b. saponification with NaOH

Answers

a.

$$CH_2-O-\overset{\overset{\displaystyle O}{\|}}{C}-(CH_2)_7-CH=CH-(CH_2)_5-CH_3$$

$$CH-O-\overset{\overset{\displaystyle O}{\|}}{C}-(CH_2)_7-CH=CH-(CH_2)_5-CH_3 + 3H_2 \xrightarrow{\text{Ni}}$$

$$CH_2-O-\overset{\overset{\displaystyle O}{\|}}{C}-(CH_2)_7-CH=CH-(CH_2)_5-CH_3$$

$$CH_2-O-\overset{\overset{\displaystyle O}{\|}}{C}-(CH_2)_{16}-CH_3$$

$$CH-O-\overset{\overset{\displaystyle O}{\|}}{C}-(CH_2)_{16}-CH_3$$

$$CH_2-O-\overset{\overset{\displaystyle O}{\|}}{C}-(CH_2)_{16}-CH_3$$

b.

$$CH_2-O-\overset{\overset{\displaystyle O}{\|}}{C}-(CH_2)_7-CH=CH-(CH_2)_5-CH_3$$

$$CH-O-\overset{\overset{\displaystyle O}{\|}}{C}-(CH_2)_7-CH=CH-(CH_2)_5-CH_3 + 3NaOH \longrightarrow$$

$$CH_2-O-\overset{\overset{\displaystyle O}{\|}}{C}-(CH_2)_7-CH=CH-(CH_2)_5-CH_3$$

$$CH_2-OH$$

$$CH-OH \quad + \quad 3Na^+ \ ^-O-\overset{\overset{\displaystyle O}{\|}}{C}-(CH_2)_7-CH=CH-(CH_2)_5-CH_3$$

$$CH_2-OH$$

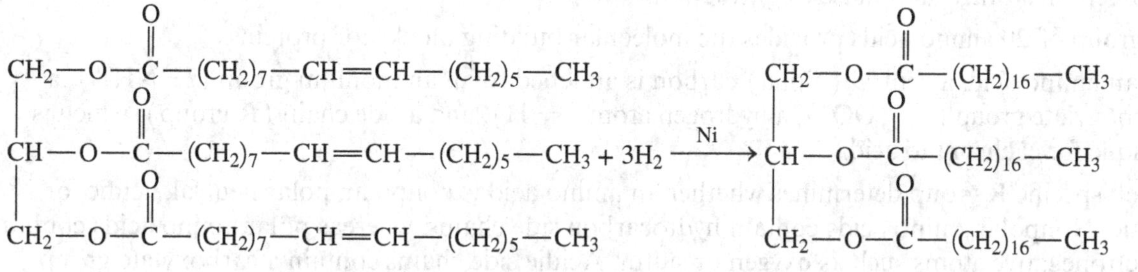

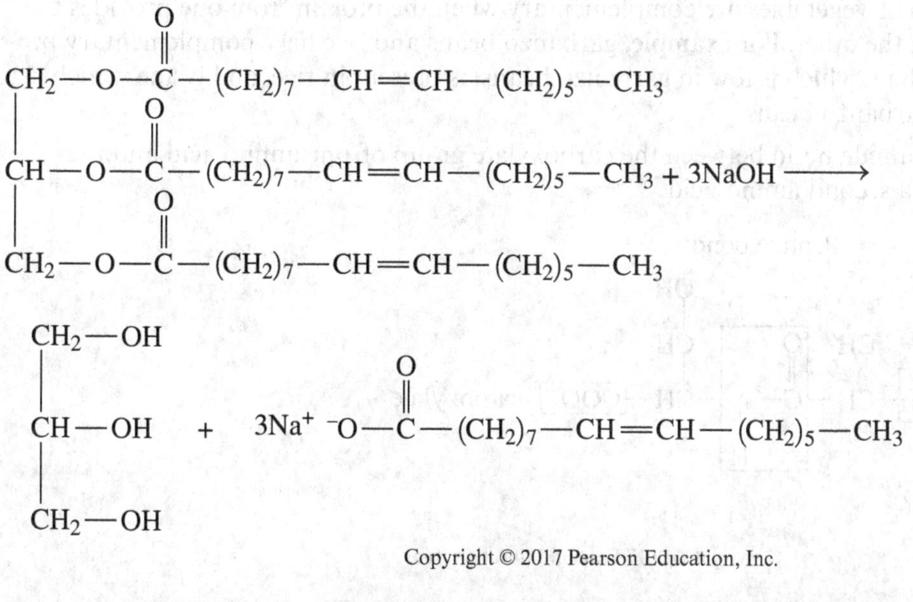

♦ **Learning Exercise 18.3F**

a. Draw the line-angle structural formula for the steroid nucleus.

b. Draw the line-angle structural formula for cholesterol.

Answers

a.

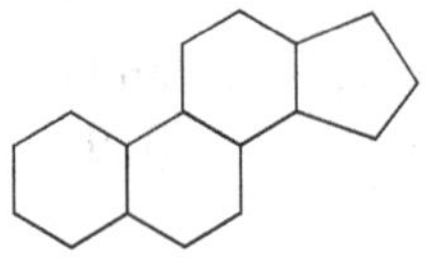

b.

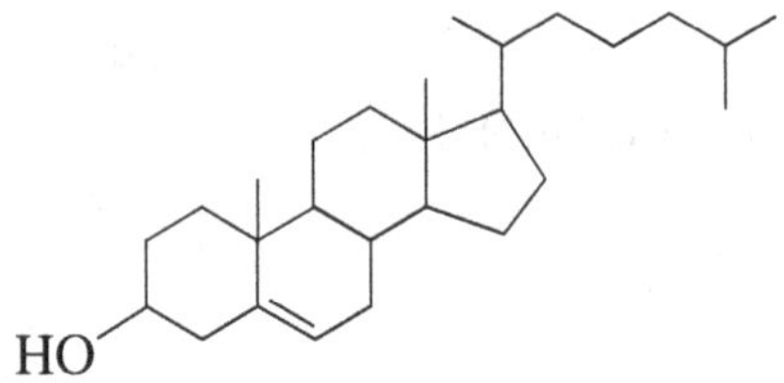

HO

18.4 Amino Acids and Proteins

Learning Goal: Describe protein functions and draw structures for amino acids and peptides.

- Some proteins are enzymes or hormones, whereas others are important in structure, transport, protection, storage, and muscle contraction.
- A group of 20 amino acids provides the molecular building blocks for proteins.
- In an amino acid, a central (alpha) carbon is attached to an ammonium group ($-NH_3^+$), a carboxylate group ($-COO^-$), a hydrogen atom ($-H$), and a side chain (R group), which is unique for each amino acid.
- Each specific R group determines whether an amino acid is nonpolar, polar neutral, acidic, or basic. Nonpolar amino acids contain hydrocarbon side chains, whereas polar amino acids contain electronegative atoms such as oxygen or sulfur. Acidic side chains contain a carboxylate group ($-COO^-$), and basic side chains contain an ammonium group ($-NH_3^+$).
- Certain combinations of vegetables are complementary when the protein from one provides the missing amino acid in the other. For example, garbanzo beans and rice have complementary proteins because tryptophan, which is low in garbanzo beans, is present in rice, and lysine, which is low in rice, is present in garbanzo beans.
- A peptide bond is an amide bond between the carboxylate group of one amino acid and the ammonium group of a second amino acid.

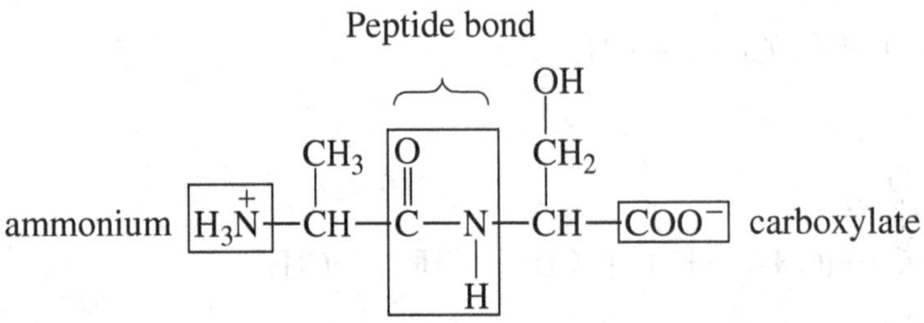

Peptide bond

ammonium $H_3\overset{+}{N}-CH-\underset{\underset{H}{|}}{\overset{\overset{O}{||}}{C}}-N-CH-COO^-$ carboxylate

with CH_3, CH_2, OH side groups

- Short chains of amino acids are called peptides. Long chains of amino acids that are biologically active are called proteins.
- In the name of a peptide, each amino acid beginning from the N-terminus (written on the left) has the *an*, *ate*, or *ine* replaced by *yl*. The last amino acid at the C-terminus (written on the right) of the peptide uses its full name.
- Structures of peptides can be shown using the three-letter and one-letter abbreviations for the component amino acids written from the N-terminus to the C-terminus (left to right).

◆ **Learning Exercise 18.4A**

Using the appropriate R group, complete the condensed structural formula of each of the following amino acids. Indicate whether the amino acid would be nonpolar, polar neutral, acidic, or basic, and give its three- and one-letter abbreviations:

CORE CHEMISTRY SKILL

Drawing the Ionized Form for an Amino Acid

a. 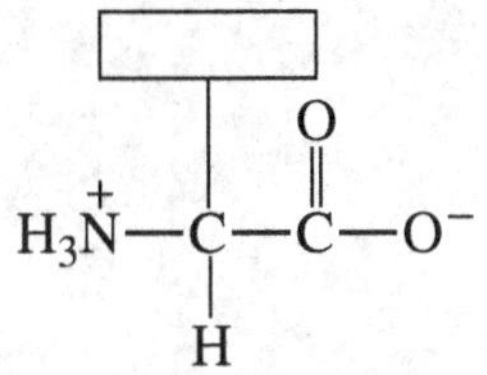 Glycine _____

b. 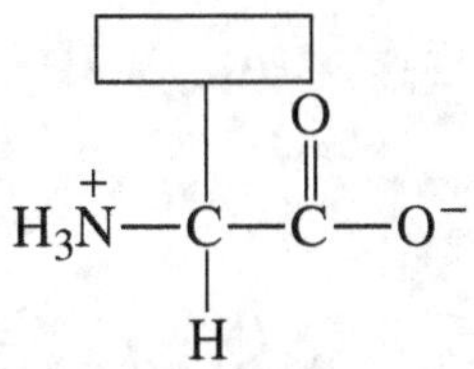 Alanine _____

c. 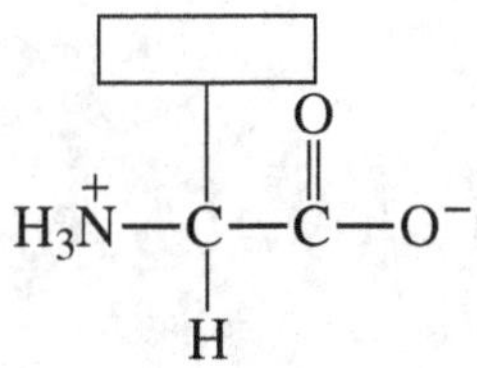 Serine _____

d. 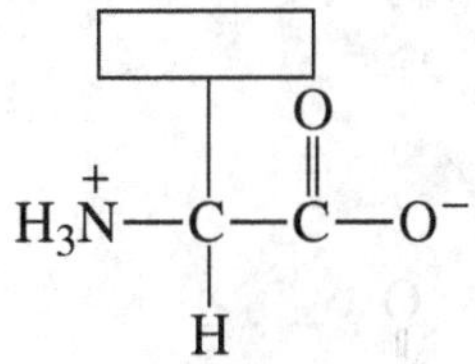 Aspartate _____

Answers

a. 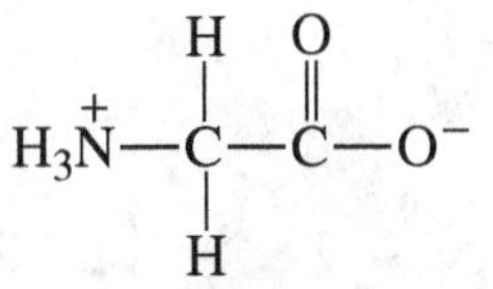 Glycine, nonpolar, Gly, G

b. 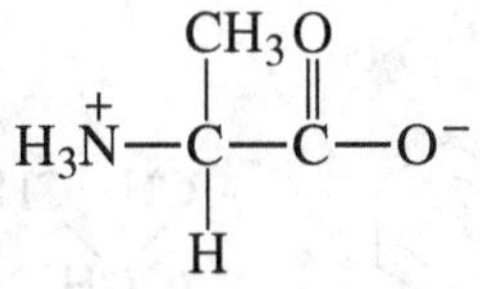 Alanine, nonpolar, Ala, A

c. 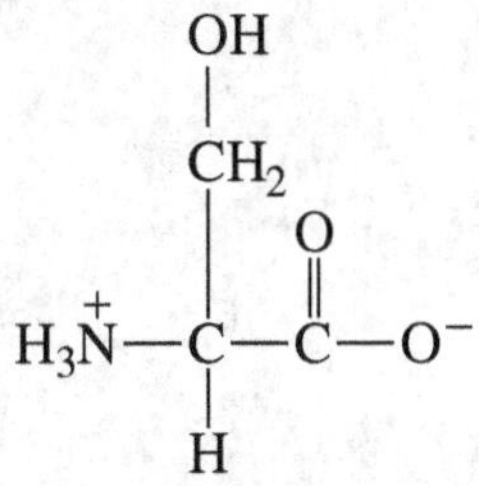 Serine, polar neutral, Ser, S

d. 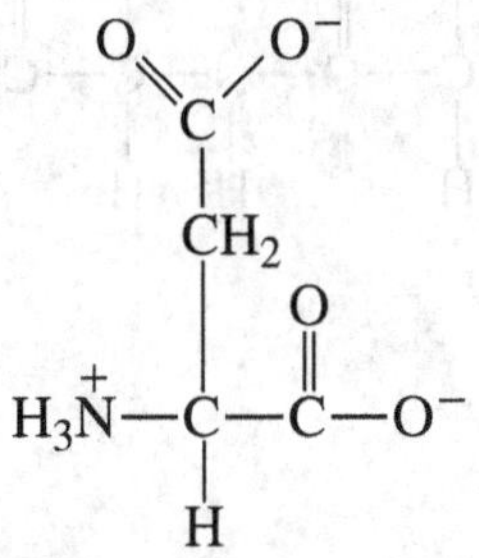 Aspartate, acidic, Asp, D

Guide to Drawing a Peptide	
STEP 1	Draw the structures for the amino acids in the peptide, starting with the N-terminus.
STEP 2	Remove the O atom from the carboxylate group and two H atoms from the ammonium group in the adjacent amino acid. Use peptide bonds to connect the amino acids.

♦ **Learning Exercise 18.4B**

Draw the structure for each of the following peptides:

a. serylglycine

b. CV

c. Met–Phe–Leu

Answers

a.

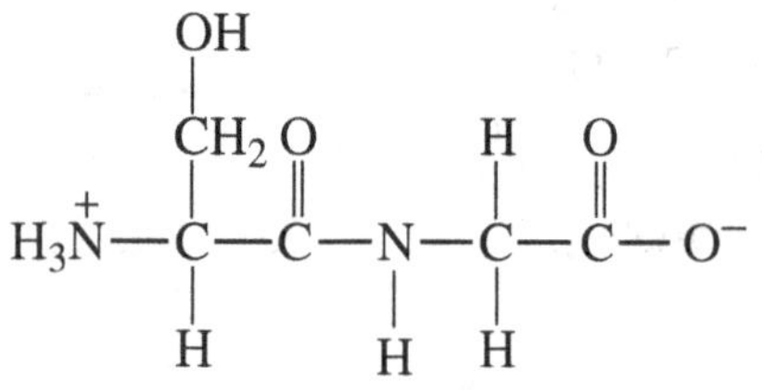

b.

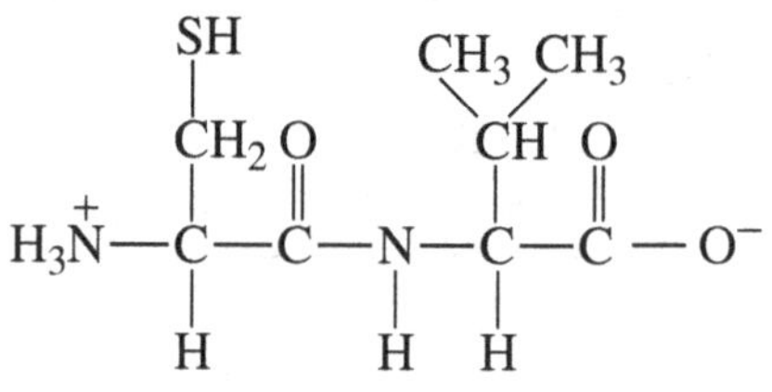

c.

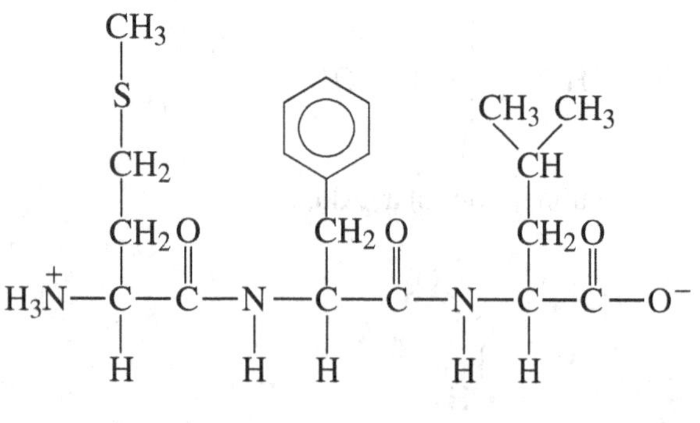

18.5   Protein Structure

Learning Goal: Identify the levels of structure of a protein.

- The primary structure of a protein is the sequence of amino acids connected by peptide bonds.
- In the secondary structure, hydrogen bonds between different sections of the peptide or between different polypeptides produce a characteristic shape such as an α helix or a β-pleated sheet.
- Collagen, which contains a triple helix of peptide chains, makes up as much as one-third of all the protein in the body.
- In the tertiary structure, the peptide chain folds upon itself to form a three-dimensional shape with hydrophobic side groups on the inside and hydrophilic side groups on the outside surface. The tertiary structure is stabilized by interactions between side groups.
- The interactions can be hydrophobic, hydrophilic, salt bridges (ionic bonds), hydrogen bonds, or disulfide bonds.
- In a quaternary structure, two or more subunits combine for biological activity.

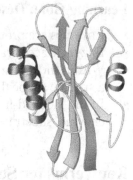

Protein with alpha helices and beta-pleated sheets

Key Terms for Sections 18.4 and 18.5

CORE CHEMISTRY SKILL

Identifying the Primary, Secondary, Tertiary, and Quaternary Structures of Proteins

Match the following key terms with the correct descriptions:

a. secondary structure   **b.** transport protein   **c.** primary structure   **d.** enzyme   **e.** storage protein

1. _____ the sequence of amino acids linked by peptide bonds

2. _____ lipoprotein that carries lipids in blood

3. _____ an α helix

4. _____ amylase that hydrolyzes starch

5. _____ albumin, a protein in egg white

Answers   **1.** c   **2.** b   **3.** a   **4.** d   **5.** e

◆   Learning Exercise 18.5

Indicate whether each of the following interactions is responsible for the tertiary and/or quaternary structures of proteins:

a. _____ a disulfide bond joining distant parts of a single peptide

b. _____ the combination of four protein subunits

c. _____ hydrophilic side groups attracted to water

d. _____ a salt bridge forms between two oppositely charged side chains

Answers   **a.** tertiary   **b.** quaternary   **c.** tertiary and quaternary   **d.** tertiary and quaternary

18.6   Proteins as Enzymes

Learning Goal: Describe the role of an enzyme in an enzyme-catalyzed reaction.

- Enzymes are proteins that act as biological catalysts.
- Enzymes accelerate the rate of biological reactions by lowering the activation energy.
- Within the tertiary structure of the enzyme, there is a small pocket called the active site, which has a specific shape that fits a specific substrate.
- In the induced-fit model, a substrate and a flexible active site adjust their shapes to form an enzyme–substrate complex in which the reaction of the substrate is catalyzed to give product(s).

Key Terms for Section 18.6

Match the following key terms with the correct description:

  a. active site   **b.** substrate   **c.** enzyme–substrate complex   **d.** induced-fit

1. _____ the combination of an enzyme with a substrate

2. _____ the molecule that reacts in the active site in an enzyme-catalyzed reaction

3. _____ a model of enzyme action in which the shape of a substrate and the active site of the enzyme adjust to give an optimal fit

4. _____ the portion of an enzyme that binds to the substrate and catalyzes the reaction

Answers   **1.** c   **2.** b   **3.** d   **4.** a

♦ Learning Exercise 18.6

Indicate whether each of the following is a characteristic of an enzyme: Yes or No.
An enzyme

a. _____ is a biological catalyst

b. _____ is smaller than a substrate

c. _____ does not change the equilibrium of a reaction

d. _____ must be obtained from the diet

e. _____ greatly increases the rate of a cellular reaction

f. _____ is needed for every reaction that takes place in the cell

g. _____ provides an alternate pathway for the biological reaction with a lower activation energy

Answers   **a.** Yes   **b.** No   **c.** Yes   **d.** No   **e.** Yes   **f.** Yes   **g.** Yes

18.7   Nucleic Acids

Learning Goal: Describe the structure of the nucleic acids in DNA and RNA.

- Nucleic acids are composed of nitrogen-containing bases, five-carbon sugars, and phosphate groups.
- Deoxyribonucleic acid (DNA) and ribonucleic acid (RNA) are polymers of nucleotides.
- Each nucleotide consists of a base, a sugar, and a phosphate group.

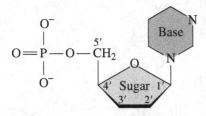

- In DNA, the bases are adenine, thymine, guanine, or cytosine. In RNA, uracil replaces thymine. In DNA, the sugar is deoxyribose; in RNA, the sugar is ribose.
- The two strands of DNA are held together by hydrogen bonds between complementary base pairs: A with T, and G with C.
- During DNA replication, new DNA strands form along each original DNA strand.
- Complementary base pairing ensures the correct pairing of bases to give identical copies of the original DNA.

♦ **Learning Exercise 18.7A**

a. Write the names and abbreviations for the bases in each of the following:

DNA _____

RNA _____

b. Write the name of the sugar in each of the following:

DNA _____

RNA _____

Answers   **a.** DNA: adenine (A), thymine (T), guanine (G), cytosine (C)
              RNA: adenine (A), uracil (U), guanine (G), cytosine (C)
      b. DNA: deoxyribose
              RNA: ribose

♦ **Learning Exercise 18.7B**

Name and write the abbreviation for each of the following bases, and classify each as a purine or a pyrimidine:

a.

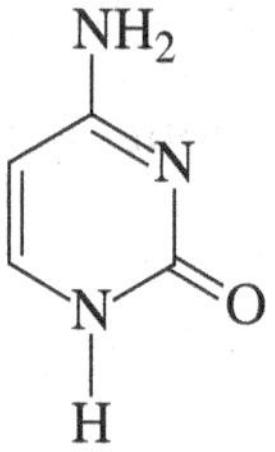

b.

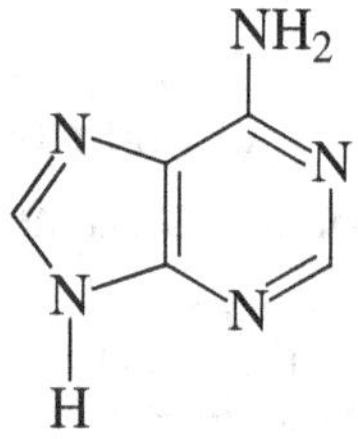

c.

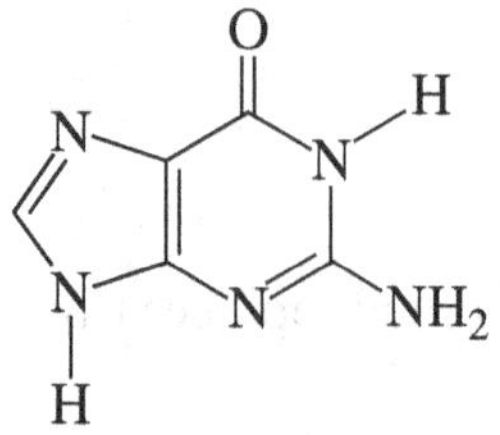

d.

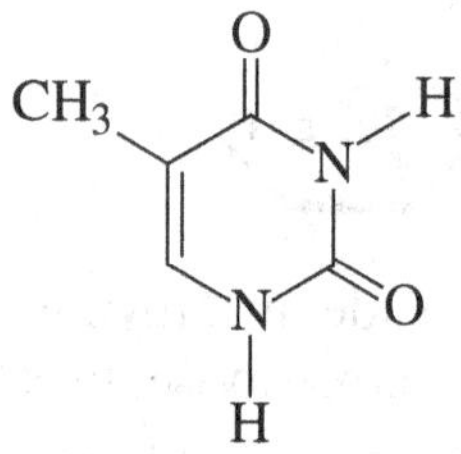

Answers **a.** cytosine (C); pyrimidine **b.** adenine (A); purine
c. guanine (G); purine **d.** thymine (T); pyrimidine

♦ **Learning Exercise 18.7C**

Identify the nucleic acid (DNA or RNA) in which each of the following is found:

a. _____ adenosine monophosphate **b.** _____ guanosine monophosphate

c. _____ dCMP **d.** _____ cytidine monophosphate

e. _____ deoxythymidine monophosphate **f.** _____ UMP

g. _____ dGMP **h.** _____ deoxyadenosine monophosphate

Answers **a.** RNA **b.** RNA **c.** DNA **d.** RNA **e.** DNA **f.** RNA **g.** DNA **h.** DNA

♦ **Learning Exercise 18.7D**

Identify the nucleotides in the following:

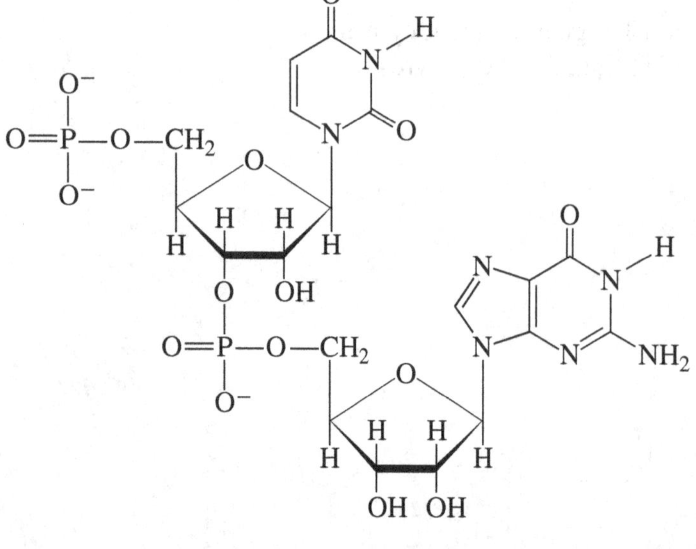

Answer Uridine monophosphate and guanosine monophosphate

♦ Learning Exercise 18.7E

CORE CHEMISTRY SKILL
Writing the Complementary DNA Strand

Write the complementary base sequence for the following segment of a strand of DNA:

a. T A C G A A C C G A G G

b. A A A T T T C C C G G G

c. G C G C T C A A A T G C

Answers

a. A T G C T T G G C T C C

b. T T T A A A G G G C C C

c. C G C G A G T T T A C G

♦ Learning Exercise 18.7F

How does the replication of DNA produce identical copies of the DNA?

Answer In the replication process, the bases on each strand of the separated parent DNA are paired with their complementary bases. Because each complementary base is specific for a base in DNA, the new DNA daughter strands are complements of the original DNA strands and when they pair to form the double helix, they produce an identical copy of the original DNA.

18.8 Protein Synthesis

Learning Goal: Describe the synthesis of protein from mRNA.

- The three types of RNA differ by their function in the cell: ribosomal RNA (rRNA) makes up most of the structure of the ribosomes and is the site of protein synthesis; messenger RNA (mRNA) carries genetic information from the DNA to the ribosomes; and transfer RNA (tRNA) places the correct amino acids in the protein.
- In transcription, the information contained in DNA is transferred to mRNA molecules.
- The bases in the mRNA are complementary to the DNA, except U is paired with A in DNA.
- The genetic code consists of a sequence of three nucleotides (triplets) called codons that code for a specific amino acid in the protein.
- The codon AUG signals the start of transcription, and codons UAG, UGA, and UAA signal the stop.
- Proteins are synthesized at the ribosomes in a translation process in which tRNA molecules bring the appropriate amino acids to the ribosome to which mRNA is bound. Each amino acid is bonded by a peptide bond to a growing peptide chain.
- When the peptide is released, it takes on its secondary and tertiary structures to become a functional protein in the cell.

Key Terms for Sections 18.7 and 18.8

Match the following key terms with the correct description:

a. DNA   **b.** RNA   **c.** rRNA   **d.** mRNA   **e.** double helix   **f.** transcription
g. translation   **h.** hydrogen bonds   **i.** complementary base pair   **j.** daughter DNA

1. _____ the shape of DNA with a sugar–phosphate backbone and base pairs linked in the center

2. _____ the genetic material composed of nucleotides containing adenine, cytosine, guanine, or thymine

3. _____ the nucleic acid that is a single strand of nucleotides containing adenine, cytosine, guanine, or uracil

4. _____ the most abundant type of RNA in a cell

5. _____ the attractions between base pairs that connect the two DNA strands

6. _____ carries information from DNA to the ribosomes for protein synthesis

7. _____ the interpretation of the codons in mRNA as amino acids in a peptide

8. _____ the bases guanine and cytosine in the double helix

9. _____ the new DNA strand that forms during DNA replication

10. _____ the transfer of genetic information from DNA by the formation of mRNA

Answers   **1.** e **2.** a **3.** b **4.** c **5.** h **6.** d **7.** g **8.** i **9.** j **10.** f

♦ **Learning Exercise 18.8A**

Write the corresponding section of an mRNA produced from each of the following sections of DNA:

🦫 **CORE CHEMISTRY SKILL**

Writing the mRNA Segment for a DNA Template

a. C A T T C G G T A

b. G T A C C T A A C G T C C G

c. G C C G T A A G C G A T

d. T A C T A G G C A C C C A A A

Answers   **a.** G U A A G C C A U

b. C A U G G A U U G C A G G C

c. C G G C A U U C G C U A

d. A U G A U C C G U G G G U U U

♦ Learning Exercise 18.8B

CORE CHEMISTRY SKILL
Writing the Amino Acid for an mRNA Codon

Give the three-letter abbreviation for each of the amino acids coded for by the following mRNA codons:

a. UUU ______ **b.** GCG ______
c. AGC ______ **d.** CCA ______
e. GGA ______ **f.** ACA ______
g. AUG ______ **h.** CUC ______
i. CAU ______ **j.** GUU ______

Answers **a.** Phe **b.** Ala **c.** Ser **d.** Pro **e.** Gly
f. Thr **g.** Start/Met **h.** Leu **i.** His **j.** Val

♦ Learning Exercise 18.8C

The following is a segment of the DNA template. Write the corresponding mRNA segment, then write the three-letter abbreviations for the amino acids.

a. DNA template:	CCC	TCA	GGG	CGC	
mRNA:	___	___	___	___	
Amino acids:	___	___	___	___	
b. DNA:	ATA	GCC	TTT	GGC	AAC
mRNA:	___	___	___	___	___
Amino acids:	___	___	___	___	___

Answers **a.** mRNA: GGG AGU CCC GCG
Gly—Ser—Pro—Ala
b. mRNA: UAU CGG AAA CCG UUG
Tyr—Arg—Lys—Pro—Leu

Checklist for Chapter 18

You are ready to take the Practice Test for Chapter 18. Be sure you have accomplished the following learning goals for this chapter. If not, review the section listed at the end of the goal. Then apply your new skills and understanding to the Practice Test.

After studying Chapter 18, I can successfully:

____ Classify carbohydrates as monosaccharides, disaccharides, and polysaccharides. (18.1)

____ Classify a monosaccharide (aldose or ketose), and indicate the number of carbon atoms. (18.1)

____ Draw the open-chain structures for glucose, galactose, and fructose. (18.1)

____ Draw the Haworth structures for monosaccharides. (18.1)

____ Describe the monosaccharide units and linkages in disaccharides. (18.2)

____ Describe the structural features of amylose, amylopectin, glycogen, and cellulose. (18.2)

_____ Identify a fatty acid as saturated or unsaturated. (18.3)

_____ Draw the triacylglycerol produced by the reaction of glycerol and fatty acids. (18.3)

_____ Draw the product from the hydrogenation or saponification of a triacylglycerol. (18.3)

_____ Describe the structure of a steroid and cholesterol. (18.3)

_____ Classify proteins by their functions in the cells. (18.4)

_____ Draw the ionized structure for an amino acid. (18.4)

_____ Classify amino acids by the characteristics of the side chain (R group). (18.4)

_____ Describe a peptide bond and write the order of a peptide using the three- and one-letter abbreviations for amino acids. (18.4)

_____ Distinguish between the primary and secondary structures of a protein. (18.5)

_____ Distinguish between the tertiary and quaternary structures of a protein. (18.5)

_____ Describe the lock-and-key and induced-fit models of enzyme action. (18.6)

_____ Identify the components of nucleic acids DNA and RNA. (18.7)

_____ Describe the nucleotides contained in DNA and RNA. (18.7)

_____ Describe the primary structure of nucleic acids. (18.7)

_____ Describe the structures of DNA and RNA. (18.7)

_____ Explain the process of DNA replication and write a complementary strand for a DNA template. (18.7)

_____ Describe the structures and characteristics of the three types of RNA. (18.8)

_____ Describe the synthesis of mRNA (transcription) and write the mRNA section for a DNA template. (18.8)

_____ Describe the role of translation in protein synthesis and write the amino acid sequence for an mRNA section. (18.8)

Practice Test for Chapter 18

The chapter sections to review are shown in parentheses at the end of each question.

1. The name *carbohydrate* came from the fact that (18.1)
   A. carbohydrates are hydrates of water
   B. carbohydrates contain carbon, hydrogen, and oxygen in a 1:2:1 ratio
   C. carbohydrates contain a great quantity of water
   D. all plants produce carbohydrates
   E. carbon and hydrogen atoms are abundant in carbohydrates

2. What functional group(s) is/are found in the open chains of monosaccharides? (18.1)
   A. hydroxyl groups
   B. aldehyde groups
   C. ketone groups
   D. hydroxyl and carbonyl groups
   E. carbonyl group

3. What is the classification of the following monosaccharide? (18.1)

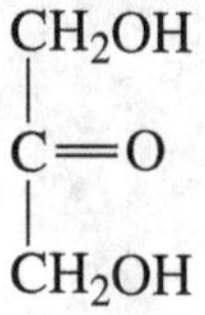
CH2OH
C=O
CH2OH

A. aldotriose **B.** ketotriose **C.** aldotetrose
D. ketotetrose **E.** ketopentose

4. The structure shown below is the Haworth structure of (18.1)

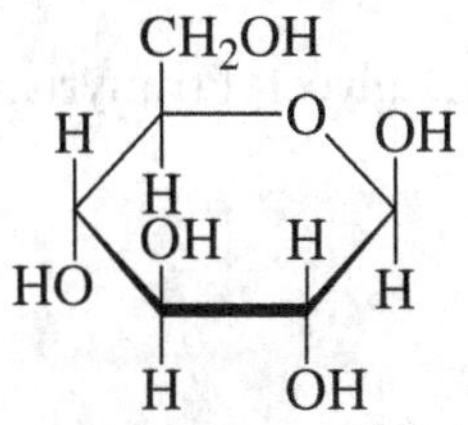
CH2OH
H
O
OH
H
OH
H
HO
H
H
OH

A. fructose **B.** glucose **C.** ribose
D. glyceraldehyde **E.** galactose

For questions 5 through 10, identify the carbohydrate that each statement describes: (18.2)

A. amylose **B.** cellulose **C.** glycogen **D.** lactose **E.** sucrose

5. ________ composed of many glucose units linked by $\alpha(1 \rightarrow 4)$-glycosidic bonds

6. ________ contains glucose and galactose

7. ________ composed of glucose units joined by both $\alpha(1 \rightarrow 4)$- and $\alpha(1 \rightarrow 6)$-glycosidic bonds

8. ________ composed of glucose units joined by $\beta(1 \rightarrow 4)$-glycosidic bonds

9. ________ produced as a storage form of energy in plants

10. ________ used for structural purposes by plants

11. A triacylglycerol is a (18.3)
A. carbohydrate **B.** lipid **C.** protein
D. oxyacid **E.** soap

12. A fatty acid that is unsaturated is usually (18.3)
A. from animal sources and liquid at room temperature
B. from animal sources and solid at room temperature
C. from plant sources and liquid at room temperature
D. from plant sources and solid at room temperature
E. from both plant and animal sources and solid at room temperature

For questions 13 through 15, consider the following compound: (18.3)

$$CH_2-O-\overset{\overset{O}{\|}}{C}-(CH_2)_{16}-CH_3$$

$$CH-O-\overset{\overset{O}{\|}}{C}-(CH_2)_{16}-CH_3$$

$$CH_2-O-\overset{\overset{O}{\|}}{C}-(CH_2)_{16}-CH_3$$

13. The name of this molecule is:
  A. glyceryl tristearate   B. glyceryl trioleate   C. glyceryl trimyristate
  D. glyceryl tripalmitate   E. triglycerol

14. If this molecule is heated with a strong base such as NaOH, the products are
  A. glycerol and fatty acids
  B. glycerol and water
  C. glycerol and soap
  D. an ester and salts of fatty acids
  E. an ester and fatty acids

15. This compound would be
  A. saturated, and a solid at room temperature
  B. saturated, and a liquid at room temperature
  C. unsaturated, and a solid at room temperature
  D. unsaturated, and a liquid at room temperature
  E. polyunsaturated, and a liquid at room temperature

16. Which amino acid is nonpolar? (18.4)
  A. serine   B. aspartate   C. valine   D. cysteine   E. glutamine

17. Which amino acid will form disulfide bonds in a tertiary structure? (18.4)
  A. serine   B. aspartate   C. valine   D. cysteine   E. glutamine

18. Which amino acid has a basic R group? (18.4)
  A. serine   B. aspartate   C. valine   D. cysteine   E. lysine

19. The sequence Tyr–Ala–Gly (18.5)
  A. is a tripeptide
  B. has two peptide bonds
  C. has tyrosine with a free $-NH_3^+$ end
  D. has glycine with the free $-COO^-$ end
  E. all of these

20. What type of interaction is expected between lysine and aspartate? (18.5)
  A. salt bridge (ionic bond)   B. hydrogen bond   C. disulfide bond
  D. hydrophobic interaction   E. hydrophilic attraction

21. What type of bond is used to form the α helix structure of a protein? (18.5)
  A. peptide bond   B. hydrogen bond   C. ionic bond
  D. disulfide bond   E. hydrophobic interaction

22. What type of interaction takes place between two phenylalanine amino acids in a tertiary structure? (18.5)
  A. hydrogen bonds   B. ionic bonds   C. disulfide bond
  D. hydrophobic interaction   E. hydrophilic attraction

For questions 23 through 27, identify the protein structural level(s) that each of the following statements describes: (18.5)

A. primary **B.** secondary **C.** tertiary **D.** quaternary **E.** pentenary

23. _____ peptide bonds

24. _____ a pleated sheet

25. _____ two or more protein subunits

26. _____ an α helix

27. _____ disulfide bonds

28. Enzymes are (18.6)
   A. biological catalysts
   D. lipids
   B. polysaccharides
   E. named with an *ose* ending
   C. insoluble in water

29. The first step in enzyme action is (18.6)
   E = enzyme; S = substrate; P = product
   A. S $\longrightarrow$ P
   B. EPS $\longrightarrow$ E + P
   C. E + S $\longrightarrow$ ES
   D. ES $\longrightarrow$ E + P
   E. EP $\longrightarrow$ ES

30. The final step in enzyme action is (18.6)
   E = enzyme; S = substrate; P = product
   A. S $\longrightarrow$ P
   B. EP $\longrightarrow$ E + P
   C. E + S $\longrightarrow$ ES
   D. ES $\longrightarrow$ E + P
   E. EP $\longrightarrow$ ES

31. A nucleotide contains (18.7)
   A. a base
   C. a phosphate and a sugar
   E. a base, a sugar, and a phosphate
   B. a base and a sugar
   D. a base and a deoxyribose

32. Which of the following are purine bases? (18.7)
   A. adenine and uracil
   D. thymine and adenine
   B. cytosine and thymine
   E. guanine and adenine
   C. guanine and uracil

33. The process of producing DNA in the nucleus is called (18.7)
   A. complementation
   D. transcription
   B. replication
   E. mutation
   C. translation

34. Which occurs in RNA but **not** in DNA? (18.7)
   A. thymine
   D. phosphate
   B. cytosine
   E. uracil
   C. adenine

For questions 35 through 42, select answers from the following nucleic acids: (18.8)

A. DNA **B.** mRNA **C.** tRNA **D.** rRNA

35. _____ a double helix consisting of two chains of nucleotides held together by hydrogen bonds between bases

36. _____ a nucleic acid that uses deoxyribose as the sugar

37. _____ a nucleic acid produced in the nucleus, which migrates to the ribosomes to direct the formation of a protein

38. _____ a nucleic acid that brings the proper amino acid to the ribosome to build the peptide chain

39. _____ a nucleic acid that contains adenine, cytosine, guanine, and thymine

40. _____ a nucleic acid that is a major component of the ribosomes

41. _____ a nucleic acid that contains triplets of nucleotides called codons

42. _____ a nucleic acid that is replicated during cellular division

For questions 43 through 46, select answers from the following: (18.8)

A. A G C C T A
⋮ ⋮ ⋮ ⋮ ⋮ ⋮
T C G G A T

B. AUU GCU C

C. A G T U G U
⋮ ⋮ ⋮ ⋮ ⋮ ⋮
T C A A C A

D. GUA

E. A T G T A T

43. _____ a section of mRNA

44. _____ an impossible section of DNA

45. _____ a codon for valine

46. _____ a section from a DNA molecule

For questions 47 through 50, indicate the correct order of protein synthesis: (18.8)
A. mRNA goes to the ribosomes.
B. Protein is formed and is released.
C. tRNA picks up specific amino acids.
D. mRNA is made from a DNA template.

47. _____ first step

48. _____ second step

49. _____ third step

50. _____ fourth step

Answers to the Practice Test

1. B	**2.** D	**3.** B	**4.** B	**5.** A
6. D	**7.** C	**8.** B	**9.** A	**10.** B
11. B	**12.** C	**13.** A	**14.** C	**15.** A
16. C	**17.** D	**18.** E	**19.** E	**20.** A
21. B	**22.** D	**23.** A	**24.** B	**25.** D
26. B	**27.** C, D	**28.** A	**29.** C	**30.** D
31. E	**32.** E	**33.** B	**34.** E	**35.** A
36. A	**37.** B	**38.** C	**39.** A	**40.** D
41. B	**42.** A	**43.** B	**44.** C	**45.** D
46. A	**47.** D	**48.** A	**49.** C	**50.** B

Selected Answers and Solutions to Text Problems

18.1 Hydroxyl groups are found in all monosaccharides, along with a carbonyl on the first or second carbon of the chain to give an aldehyde or ketone functional group, respectively.

18.3 A ketopentose contains hydroxyl and ketone functional groups and has five carbon atoms.

18.5 In the ring portion of the Haworth structure of D-glucose, there are five carbon atoms and an oxygen atom.

18.7 **a.** This is the α form because the —OH group on carbon 2 is down.
    b. This is the α form because the —OH group on carbon 1 is down.

18.9 **a.** This six-carbon monosaccharide has a carbonyl group on carbon 2; it is a ketohexose.
    b. This five-carbon monosaccharide has a carbonyl group on carbon 1; it is an aldopentose.

18.11 In the open-chain structure of D-galactose, the —OH group on carbon 4 extends to the left. In the open-chain structure of D-glucose, this —OH group goes to the right.

18.13 **a.** When this disaccharide is hydrolyzed, galactose and glucose are produced. The glycosidic bond is a $\beta(1 \rightarrow 4)$ bond since the ether bond is drawn up from carbon 1 of the galactose unit, which is on the left in the drawing, to carbon 4 of the glucose on the right. β-Lactose is the name of this disaccharide since the —OH group on carbon 1 of the glucose unit is drawn up.
    b. When this disaccharide is hydrolyzed, two molecules of glucose are produced. The glycosidic bond is an $\alpha(1 \rightarrow 4)$ bond since the ether bond is drawn down from carbon 1 of the glucose unit on the left to carbon 4 of the glucose on the right. α-Maltose is the name of this disaccharide since the —OH group on the rightmost glucose unit is drawn down.

18.15 **a.** Isomaltose is a disaccharide.
    b. Isomaltose contains two molecules of α-D-glucose.
    c. The glycosidic link in isomaltose is an $\alpha(1 \rightarrow 6)$-glycosidic bond.
    d. The downward position of the —OH group on carbon 1 of the second glucose makes it α-isomaltose.

18.17 **a.** Another name for ordinary table sugar is sucrose.
    b. Lactose is the disaccharide found in milk and milk products.
    c. Maltose is also called malt sugar.
    d. When lactose is hydrolyzed, the products are the monosaccharides galactose and glucose.

18.19 **a.** Cellulose is a polysaccharide that is not digestible by humans.
    b. Amylose and amylopectin (in starch) are the storage forms of carbohydrates in plants.
    c. Amylose is the polysaccharide which contains only $\alpha(1 \rightarrow 4)$-glycosidic bonds.
    d. Glycogen is the most highly branched polysaccharide.

18.21 **a.** Lauric acid contains only carbon–carbon single bonds; it is saturated.
    b. Linolenic acid has three carbon–carbon double bonds; it is polyunsaturated.
    c. Stearic acid contains only carbon–carbon single bonds; it is saturated.

18.23 Glyceryl tricaprylate (tricaprylin) has three caprylic acids (an 8-carbon saturated fatty acid, 8:0) forming ester bonds with glycerol.

$$
\begin{array}{l}
\text{CH}_2-\text{O}-\overset{\displaystyle\overset{\text{O}}{\|}}{\text{C}}-(\text{CH}_2)_6-\text{CH}_3 \\[4pt]
\text{CH}-\text{O}-\overset{\displaystyle\overset{\text{O}}{\|}}{\text{C}}-(\text{CH}_2)_6-\text{CH}_3 \\[4pt]
\text{CH}_2-\text{O}-\overset{\displaystyle\overset{\text{O}}{\|}}{\text{C}}-(\text{CH}_2)_6-\text{CH}_3
\end{array}
$$

18.25 Safflower oil contains fatty acids with two or three carbon–carbon double bonds; olive oil contains a large amount of oleic acid, which has only one (monounsaturated) carbon–carbon double bond.

18.27 Hydrogenation of an unsaturated triacylglycerol adds H_2 to each of the double bonds, producing a saturated triacylglycerol containing only carbon–carbon single bonds.

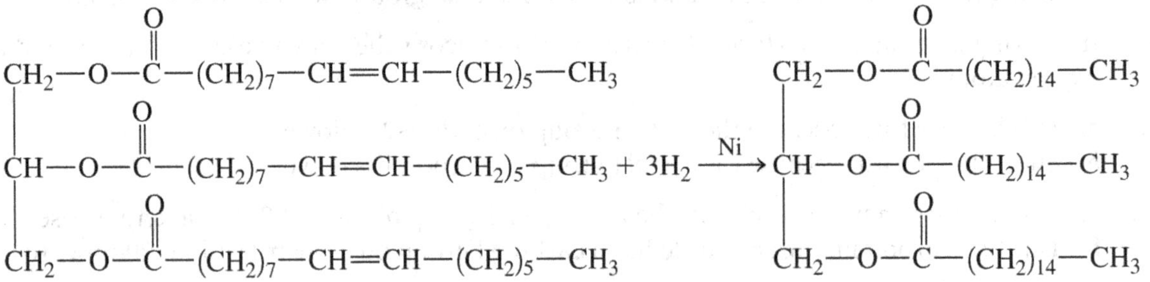

18.29 Saponification of a fat gives glycerol and the salts of the fatty acids.

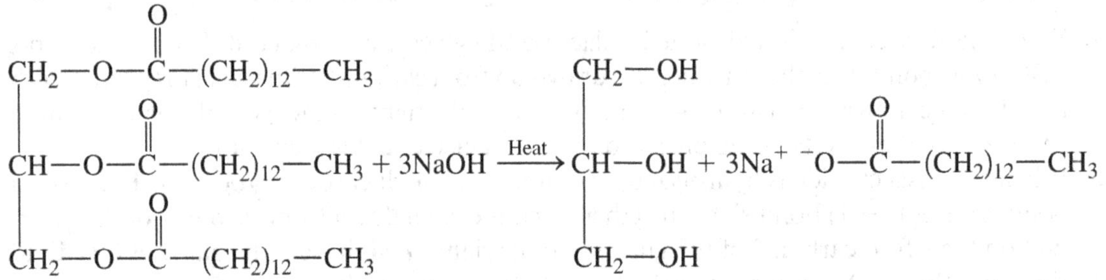

18.31

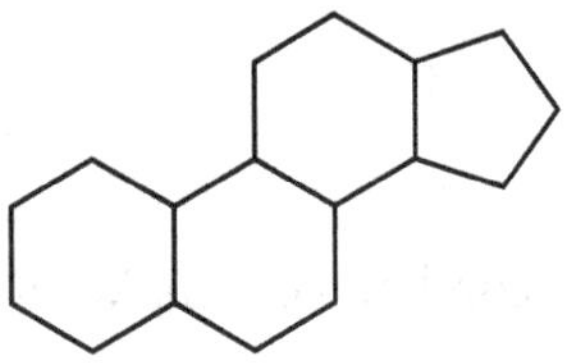

18.33 a.

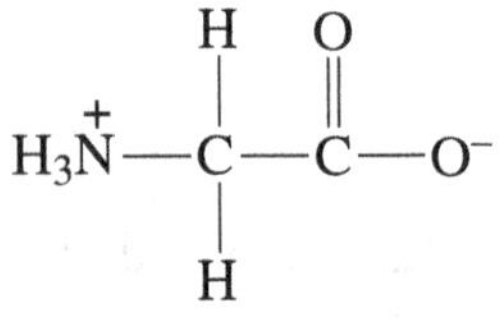

b.

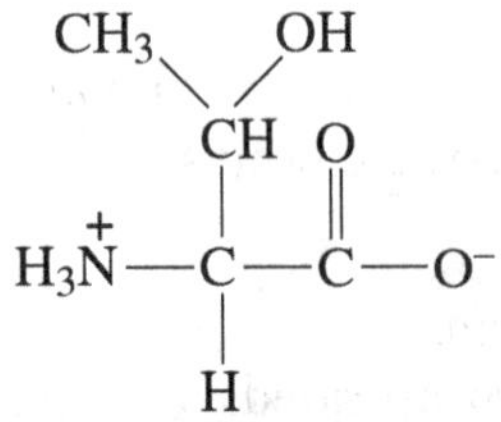

c.

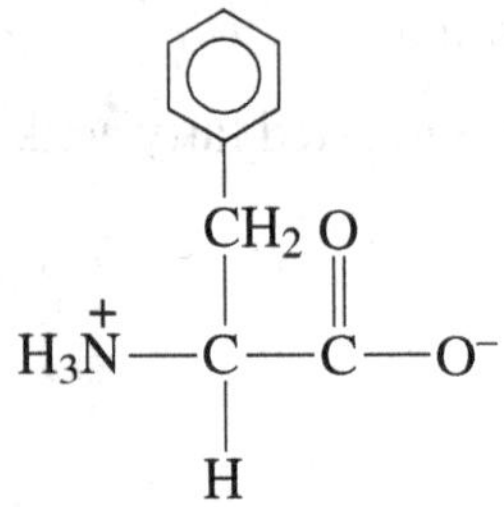

18.35 a. Glycine, which has only a —H as the R group, is hydrophobic (nonpolar).
  b. Threonine has an R group that contains the polar —OH group; threonine is hydrophilic (polar, neutral).
  c. Phenylalanine has an R group with a nonpolar benzene ring; phenylalanine is hydrophobic (nonpolar).

18.37 Amino acids have both three-letter and one-letter abbreviations.
  a. Ala is the three-letter abbreviation for the amino acid alanine.
  b. V is the one-letter abbreviation for the amino acid valine.
  c. Lys is the three-letter abbreviation for the amino acid lysine.
  d. C is the one-letter abbreviation for the amino acid cysteine.

18.39 In a peptide, the amino acids are joined by peptide bonds (amide bonds). At physiological pH (7.4), the first amino acid has a free —NH_3^+ group, and the last one has a free —COO^- group.

a. Ala–Cys, AC

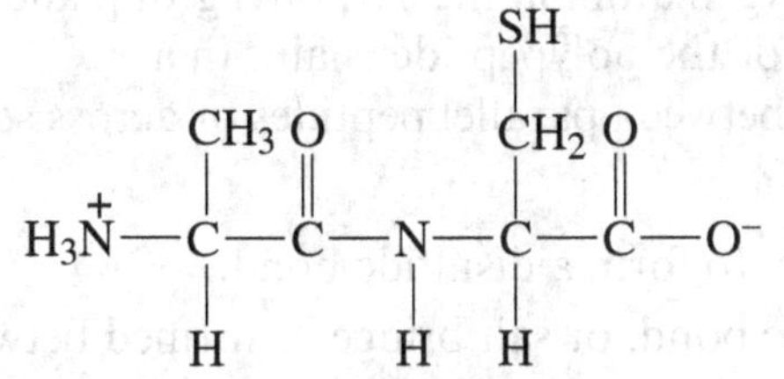

b. Ser–Phe, SF

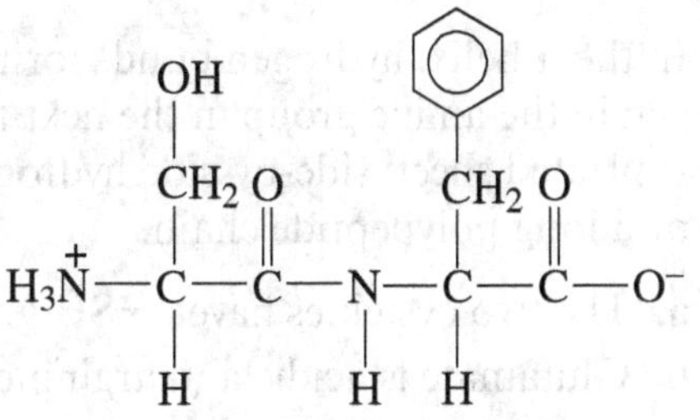

c. Gly–Ala–Val, GAV

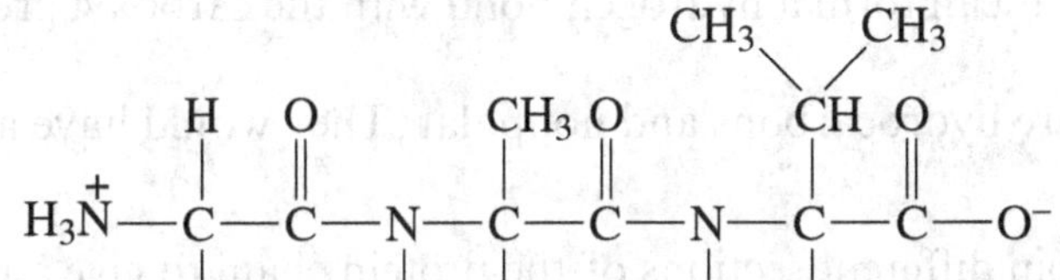

18.41 a. RIY

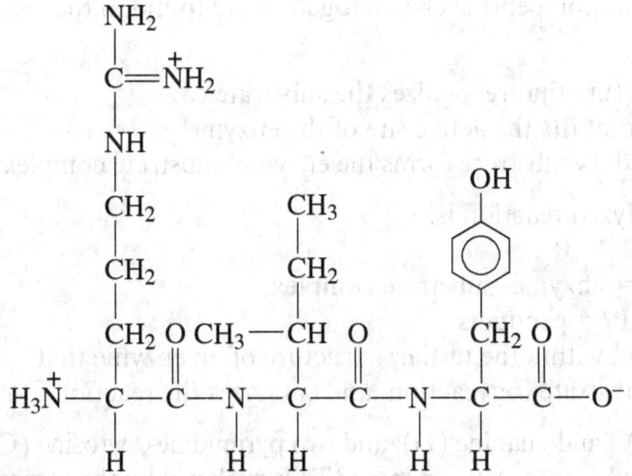

b. $H_3\overset{+}{N}$—C—C—N—C—C—N—C—C—N—C—C—O$^-$

VWIS

18.43 The possible primary structures of a tripeptide of one valine and two serines are Val–Ser–Ser (VSS), Ser–Val–Ser (SVS), and Ser–Ser–Val (SSV).

18.45 In the α helix, hydrogen bonds form between the oxygen atom in the carbonyl group and hydrogen in the amide group in the next turn of the helix of the polypeptide chain. In a β-pleated sheet, side-by-side hydrogen bonds occur between parallel peptides or across sections of a long polypeptide chain.

18.47 a. The two cysteines have —SH groups, which react to form a disulfide bond.
  b. Glutamate is acidic and arginine is basic; an ionic bond, or salt bridge, is formed between the —COO$^-$ in the R group of glutamate and the $=\overset{+}{N}H_2$ in the R group of arginine.
  c. Serine has a polar —OH group that can form a hydrogen bond with the carboxyl group of aspartate.
  d. Two leucines have R groups that are hydrocarbons and nonpolar. They would have a hydrophobic interaction.

18.49 a. Disulfide bonds and ionic bonds join different sections of the protein chain to give a three-dimensional shape. Disulfide bonds and ionic bonds are important in the tertiary and quaternary structures.
  b. Peptide bonds join the amino acid building blocks in the primary structure of a polypeptide.
  c. Hydrogen bonds that hold adjacent polypeptide chains together are found in the secondary structures of β-pleated sheets.

18.51 a. An enzyme (1) has a tertiary structure that recognizes the substrate.
  b. The substrate (3) has a structure that fits the active site of the enzyme.
  c. The combination of an enzyme and its substrate forms the enzyme–substrate complex (2).

18.53 a. The equation for an enzyme–catalyzed reaction is:
  E + S $\rightleftharpoons$ ES $\longrightarrow$ EP $\longrightarrow$ E + P
  E = enzyme, S = substrate, ES = enzyme–substrate complex,
  EP = enzyme–product complex, P = products
  b. The active site is a region or pocket within the tertiary structure of an enzyme that accepts the substrate, aligns the substrate for reaction, and catalyzes the reaction.

18.55 DNA contains two purines, adenine (A) and guanine (G), and two pyrimidines, cytosine (C) and thymine (T). RNA contains the same bases, except thymine (T) is replaced by the pyrimidine uracil (U).
  a. Thymine is present in DNA.
  b. Cytosine is present in both DNA and RNA.

18.57 The two DNA strands are held together by hydrogen bonds between the complementary bases in each strand.

18.59 a. Since T pairs with A, if one strand of DNA has the sequence A A A A A A, the second DNA strand would be T T T T T T.

b. Since C pairs with G, if one strand of DNA has the sequence GGGGGG, the second DNA strand would be CCCCCC.

c. Since T pairs with A and C pairs with G, if one strand of DNA has the sequence AGTCCAGGT, the second DNA strand would be TCAGGTCCA.

d. Since T pairs with A and C pairs with G, if one strand of DNA has the sequence CTGTATACGTT, the second DNA strand would be GACATATGCAA.

18.61 The two DNA strands separate (in a way that is similar to the unzipping of a zipper) to allow each of the bases to pair with its complementary base (A binds with T, C binds with G), which produces two exact copies of the original DNA.

18.63 The strand of mRNA would have the following sequence: GGCUUCCAAGUG.

18.65 a. The codon CUG in mRNA codes for the amino acid leucine (Leu).
b. The codon UCC in mRNA codes for the amino acid serine (Ser).
c. The codon GGU in mRNA codes for the amino acid glycine (Gly).
d. The codon AGG in mRNA codes for the amino acid arginine (Arg).

18.67 a. AAA CAC UUG GUU GUG GAC
b. Lys–His–Leu–Val–Val–Asp, KHLVVD

18.69 $\dfrac{80.\ \text{mg Pravachol}}{1\ \text{day}} \times \dfrac{1\ \text{g Pravachol}}{1000\ \text{mg Pravachol}} \times \dfrac{7\ \text{days}}{1\ \text{week}} = 0.56\ \text{g of Pravachol/week (2 SFs)}$

18.71 a. Melezitose is a trisaccharide.
b. Melezitose contains two glucose molecules and a fructose molecule.

18.73

$$
\begin{array}{l}
\text{CH}_2\!-\!\text{O}\!-\!\overset{\overset{\displaystyle O}{\|}}{\text{C}}\!-\!(\text{CH}_2)_{14}\!-\!\text{CH}_3 \\[2ex]
\text{CH}\!-\!\text{O}\!-\!\overset{\overset{\displaystyle O}{\|}}{\text{C}}\!-\!(\text{CH}_2)_{14}\!-\!\text{CH}_3 \\[2ex]
\text{CH}_2\!-\!\text{O}\!-\!\overset{\overset{\displaystyle O}{\|}}{\text{C}}\!-\!(\text{CH}_2)_{14}\!-\!\text{CH}_3
\end{array}
$$

18.75 a. polyunsaturated fatty acid
b. polyunsaturated fatty acid

18.77 a. Yes, a combination of rice and garbanzo beans provides all the essential amino acids; garbanzo beans contain the lysine missing in rice.
b. No, a combination of lima beans and cornmeal does not provide all the essential amino acids; both are deficient in the amino acid tryptophan.
c. No, a combination of garbanzo beans and lima beans does not provide all the essential amino acids; both are deficient in the amino acid tryptophan.

18.79 a. The polar R groups of asparagine and serine would interact by hydrogen bond.
b. The R groups of aspartate and lysine would form a salt bridge (ionic bond).

18.81 a.

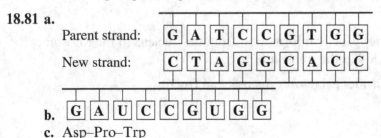

Parent strand: G A T C C G T G G
New strand: C T A G G C A C C

b. G A U C C G U G G
c. Asp–Pro–Trp

18.83 Glucose and galactose differ only at carbon 4: in the open-chain structure of glucose the —OH group on carbon 4 is drawn on the right side, and in the open-chain structure of galactose, it is drawn on the left side.

18.85

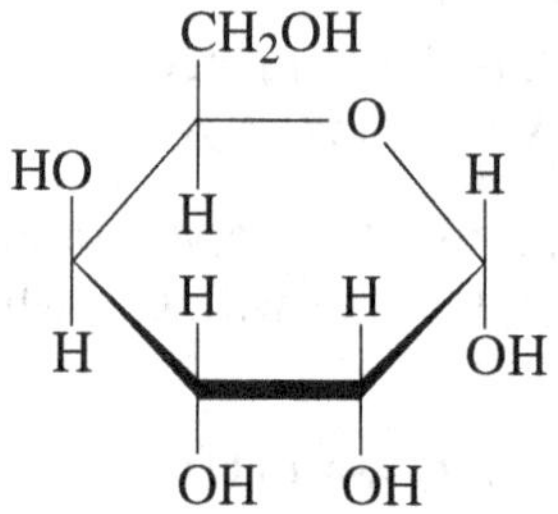

α-D-Gulose β-D-Gulose

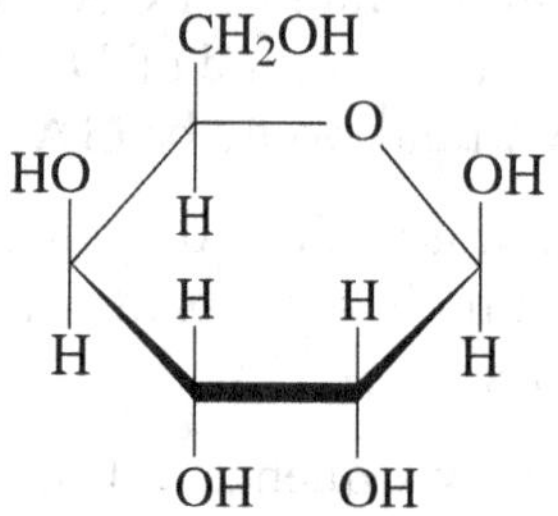

18.87

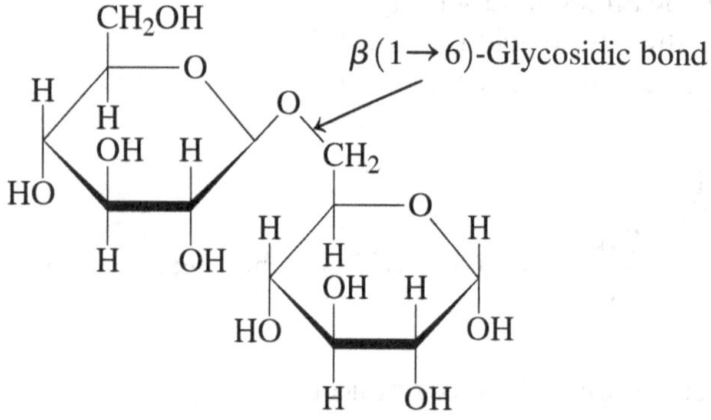

$\beta(1\rightarrow6)$-Glycosidic bond

18.89

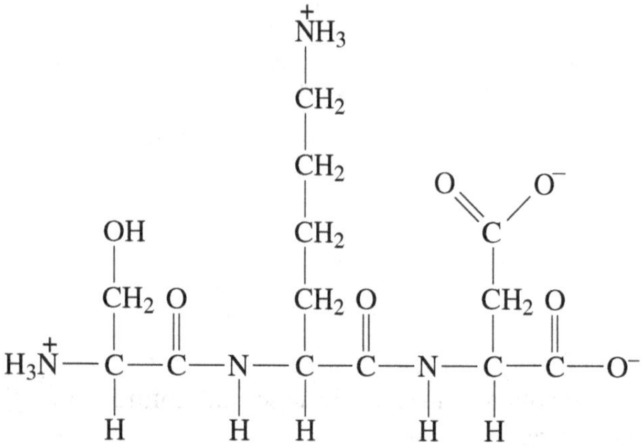

18.91 a. thymine and deoxyribose
    b. adenine and ribose
    c. cytosine and ribose
    d. guanine and deoxyribose

18.93 a. CTGAATCCG
    b. ACGTTTGATCGA
    c. TAGCTAGCTAGC

18.95 a. Messenger RNA (mRNA) carries genetic information from the nucleus to the ribosomes.
    b. Ribosomal RNA (rRNA) is found in the ribosome.

18.97 Raffinose contains the monosaccharides galactose, glucose, and fructose.

18.99 a.

$$CH_2-O-\overset{\overset{\displaystyle O}{\|}}{C}-(CH_2)_7-CH=CH-CH_2-CH=CH-(CH_2)_4-CH_3$$

$$CH-O-\overset{\overset{\displaystyle O}{\|}}{C}-(CH_2)_7-CH=CH-(CH_2)_7-CH_3$$

$$CH_2-O-\overset{\overset{\displaystyle O}{\|}}{C}-(CH_2)_7-CH=CH-CH_2-CH=CH-(CH_2)_4-CH_3$$

$$CH_2-O-\overset{\overset{\displaystyle O}{\|}}{C}-(CH_2)_7-CH=CH-CH_2-CH=CH-(CH_2)_4-CH_3$$

$$CH-O-\overset{\overset{\displaystyle O}{\|}}{C}-(CH_2)_7-CH=CH-CH_2-CH=CH-(CH_2)_4-CH_3$$

$$CH_2-O-\overset{\overset{\displaystyle O}{\|}}{C}-(CH_2)_7-CH=CH-(CH_2)_7-CH_3$$

b.

$$CH_2-O-\overset{\overset{\displaystyle O}{\|}}{C}-(CH_2)_7-CH=CH-CH_2-CH=CH-(CH_2)_4-CH_3$$

$$CH-O-\overset{\overset{\displaystyle O}{\|}}{C}-(CH_2)_7-CH=CH-(CH_2)_7-CH_3 \qquad + 5H_2 \xrightarrow{\ Ni\ }$$

$$CH_2-O-\overset{\overset{\displaystyle O}{\|}}{C}-(CH_2)_7-CH=CH-CH_2-CH=CH-(CH_2)_4-CH_3$$

$$CH_2-O-\overset{\overset{\displaystyle O}{\|}}{C}-(CH_2)_{16}-CH_3$$

$$CH-O-\overset{\overset{\displaystyle O}{\|}}{C}-(CH_2)_{16}-CH_3$$

$$CH_2-O-\overset{\overset{\displaystyle O}{\|}}{C}-(CH_2)_{16}-CH_3$$

18.101 a. The secondary structure of a protein depends on hydrogen bonds to form a helix or a pleated sheet. The tertiary structure is determined by the interactions of R groups such as disulfide bonds and salt bridges and determines the three-dimensional structure of the protein.

 b. Nonessential amino acids can be synthesized by the body, but essential amino acids must be supplied by the diet.

 c. Polar amino acids have hydrophilic R groups, while nonpolar amino acids have hydrophobic R groups.

18.103 Using the genetic code, the codons indicate the following amino acid sequence:

START–Tyr–Gly–Gly–Phe–Leu–STOP

18.105 a. The amino acids in aspartame are aspartate and phenylalanine.

 b. The dipeptide in aspartame would be named aspartylphenylalanine.

 c. The abbreviated form of the name for the dipeptide in aspartame would be named Asp–Phe or DF.

18.107 a. The mRNA sequence would be: CGA AAA GUU UUU.

 b. From the table of mRNA codons, the amino acids would be:
Arg–Lys–Val–Phe, RKVF.

Selected Answers to Combining Ideas from Chapters 17 and 18

CI.41 a. The —OH group in BHT is bonded to a carbon atom in an aromatic ring, which means that BHT has a phenol functional group.

b. BHT is referred to as an "antioxidant" because it, rather than the food, reacts with oxygen in the food container, thus preventing or retarding spoilage of food.

c. $15 \text{ oz cereal} \times \dfrac{1 \text{ lb}}{16 \text{ oz}} \times \dfrac{1 \text{ kg}}{2.205 \text{ lb}} \times \dfrac{50. \text{ mg BHT}}{1 \text{ kg cereal}} = 21 \text{ mg of BHT } (2 \text{ SFs})$

CI.43 a. Adding NaOH will saponify lipids such as glyceryl tristearate (tristearin), forming glycerol and salts of the fatty acids that are soluble in water and would wash down the drain.

b.

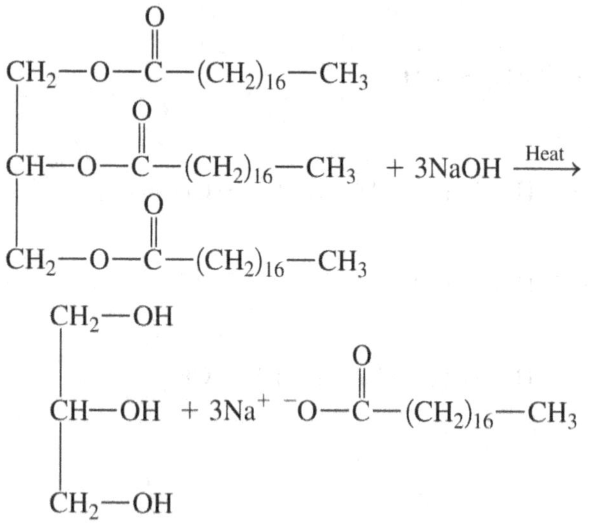

Glycerol   Salts of stearic acid

c. Tristearin $= C_{57}H_{110}O_6$
Molar mass of tristearin $(C_{57}H_{110}O_6)$
$= 57(12.01 \text{ g}) + 110(1.008 \text{ g}) + 6(16.00 \text{ g}) = 891.5 \text{ g/mol}$

$10.0 \text{ g tristearin} \times \dfrac{1 \text{ mol tristearin}}{891.5 \text{ g tristearin}} \times \dfrac{3 \text{ mol NaOH}}{1 \text{ mol tristearin}} \times \dfrac{1000 \text{ mL solution}}{0.500 \text{ mol NaOH}}$

$= 67.3 \text{ mL of NaOH solution } (3 \text{ SFs})$

CI.45 a.

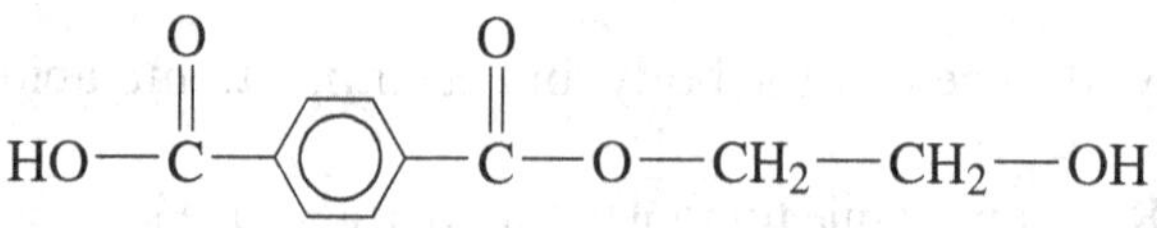

b.

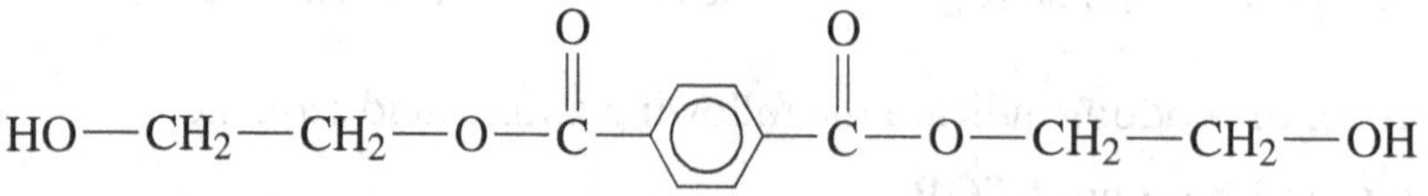

c. $1.5 \times 10^9 \text{ lb PETE} \times \dfrac{1 \text{ kg PETE}}{2.205 \text{ lb PETE}} = 6.8 \times 10^8 \text{ kg of PETE } (2 \text{ SFs})$

d. $6.8 \times 10^8 \text{ kg PETE} \times \dfrac{1000 \text{ g PETE}}{1 \text{ kg PETE}} \times \dfrac{1 \text{ mL PETE}}{1.38 \text{ g PETE}} \times \dfrac{1 \text{ L PETE}}{1000 \text{ mL PETE}}$
$= 4.9 \times 10^8 \text{ L of PETE } (2 \text{ SFs})$

e. $4.9 \times 10^8 \text{ L PETE} \times \dfrac{1 \text{ landfill}}{2.7 \times 10^7 \text{ L PETE}} = 18 \text{ landfills } (2 \text{ SFs})$